Internet of Things

Technology, Communications and Computing

Series Editors

Giancarlo Fortino, Rende (CS), Italy
Antonio Liotta, Edinburgh Napier University, School of Computing
Edinburgh, UK

The series Internet of Things - Technologies, Communications and Computing publishes new developments and advances in the various areas of the different facets of the Internet of Things. The intent is to cover technology (smart devices, wireless sensors, systems), communications (networks and protocols) and computing (theory, middleware and applications) of the Internet of Things, as embedded in the fields of engineering, computer science, life sciences, as well as the methodologies behind them. The series contains monographs, lecture notes and edited volumes in the Internet of Things research and development area, spanning the areas of wireless sensor networks, autonomic networking, network protocol, agent-based computing, artificial intelligence, self organizing systems, multi-sensor data fusion, smart objects, and hybrid intelligent systems.
** Indexing: *Internet of Things* is covered by Scopus and Ei-Compendex **

More information about this series at https://link.springer.com/bookseries/11636

Souvik Pal • Debashis De • Rajkumar Buyya
Editors

Artificial Intelligence-based Internet of Things Systems

Springer

Editors
Souvik Pal
Department of Computer Science
and Engineering
Sister Nivedita University
Kolkata, West Bengal, India

Debashis De
Department of Computer Science
and Engineering
Maulana Abul Kalam Azad University
of Technology, West Bengal
Kolkata, West Bengal, India

Rajkumar Buyya
School of Computing
and Information Systems
The University of Melbourne
Melbourne, VIC, Australia

ISSN 2199-1073 ISSN 2199-1081 (electronic)
Internet of Things
ISBN 978-3-030-87058-4 ISBN 978-3-030-87059-1 (eBook)
https://doi.org/10.1007/978-3-030-87059-1

This Springer imprint is published by the registered company Springer Nature Switzerland AG
The registered company address is: Gewerbestrasse 11, 6330 Cham, Switzerland

Preface

This book aims to bring together leading academic scientists, researchers, and research scholars to exchange and share their experiences and research results on all aspects of Internet of Things (IoT)-enabled artificial intelligence-based technologies. It also provides a premier interdisciplinary platform for researchers, practitioners, and educators to present and discuss the most recent innovations, trends, and concerns as well as practical challenges encountered and solutions adopted in the fields of AI-based IoT. This book aims to attract researchers and practitioners who are working in information technology and computer science. This book is about basics and high-level concepts regarding artificial intelligence paradigm in the context of Internet of Things. This book covers a wide range of AI-enabled IoT technologies. This book aims to explore the insight paradigm of the AI-based IoT technologies which will bring a smooth platform for the scope of industry-academia. The wide-range contents will differentiate this edited book from others. The contents include functional framework and protocols for IoT-based system, intelligent object identification, intelligent sensors, learning and analytics in intelligent IoT-enabled systems, CRISP-DM frame work, RFID technology, wearable sensors, IoT semantics, knowledge extraction, applications of linear regression, classification, vector machines and artificial neural networks for IoT devices, Bayesian learning, decision trees, deep learning frameworks, computational learning theory, multi-agent systems for IoT-based ecosystem, machine learning algorithms, nature-inspired algorithms, computational intelligence for cloud-based Internet of Things, and trustworthy machine learning for IoT-enabled system in IoT related topics. The above topics are likely to be embedded with the AI-enabled IoT technologies for future generation automation.

Chapter 1 explores IoT architecture; analyzes IoT network's technical details; and describes communication enabled technologies. Moreover, this chapter deals with various AI-based technologies integrated into IoT, edge computing, and trust models for IoT appliances. Recent AI-based projects and research challenges concludes this chapter.

Chapter 2 has formulated an overview of the IoT environment which illustrates IoT architecture, gateways, nodes, middleware, OSs, framework, protection, storage and computation, communication or networking technologies of IoT, and interfaces for the efficient utilization of data in an ecosystem. This chapter moreover illustrates the hierarchy of the intelligence of the IoT ecosystem, which describes the process of generation of data, derivation of desired information from those raw data, processing, and manipulation.

Chapter 3 illustrates a detailed view of ML and DL applicability in WSN and IoT. This chapter also describes a complete view of various neural networks (NN) and support vector machine (SVM) types that incorporate frequent, deep neural networks, quarter and ellipsoidal SVMs, and subspace-SVM forms, which are relevant to wireless and IoT appliances. This chapter provides an in-depth summary of various communication issues in IoT that are addressed by neural networks and SVM, and application and motivation for using those techniques. Followed by intrusion detection in IoT with NN and SVM, a case study on outlier detection WSNs data and future research implementations is discussed.

Chapter 4 evaluates the different methods of machine learning that deal with the challenges posed in the handling of IoT data. Big data is generated through the communication of Internet of Things/smart devices, and this data stored at cloud. The taxonomy of machine learning algorithms is described in this chapter, explaining how different techniques are applied to data generated using IoT devices. It will also address the future problems of machine learning for IoT data analytics.

Chapter 5 aims to explore DL frameworks for IoT. The chapter begins with a discussion on the development and architecture of the DL framework. This chapter then discusses about various DL models associated with deep reinforcement learning approaches for IoT. The potential applications, including smart grid management, road traffic management, industrial sector, estimation of crop production, and detection of various plant diseases are discussed. Various design issues and challenges in implementing DL are also discussed. The findings reported in this chapter provide some insights into DL frameworks for IoT that can help network researchers and engineers to contribute further towards the development of next-generation IoT.

Chapter 6 addresses the technique that combines the capability to learn and evolve solutions for large-scale dynamic systems. The chapter deals with the extended classifier system (XCS) which is an amalgamation of reinforcement learning (RL) and genetic algorithms (GA). While RL learns the model-free problem environment, the nature-inspired GA evolves better decision-making rules and improves the existing ones. The motive is to provide intelligent computation for fog-cloud-based IoT systems through XCS. The chapter reveals how the XCS algorithm estimates the optimal number of IoT workload that is to be processed in fog, the remaining of which is transferred to the cloud. The optimal number of workloads estimated by the XCS algorithm balances the energy cost and delay in the fog-cloud based resource allocation (RA) system.

Chapter 7 integrates machine learning and IoT in a portable scale to perform high-accuracy verification system. This model uses a pre-trained convolutional neural network (CNN) on a Raspberry Pi. The CNN will analyze pixels from a signature image taken by the Pi camera to recognize abnormalities and differences and to identify false signature. Other than requiring a secure digital authentication to operate, it also informs the user immediately on the app execution and image being scanned via a cloud-based system. The system is expected to provide on-the-spot signature verification and minimize any logistic issue that stems from faulty signature to an organization.

Chapter 8 illustrates the facilitators of Internet of Things like machine to machine (M2M), radio frequency identification (RFID), and software-defined networking (SDN). Machine to machine (M2M) is a communication system in IoT that endorses the group of devices to communicate with each other. The mobile communication system is optimized by M2M and standardized by 3GPP. The motivation of this chapter is that the communication system facilitated with IoT has performed their actions autonomously without the assistance of a human.

Chapter 9 discusses different types of framework, pros and cons of every framework, architecture, and different criteria to choose the better framework which will be useful for Internet of Things–based applications. Moreover, this chapter discusses architecture, generative models, and deep reinforcement learning for IoT applications.

Chapter 10 presents the active ongoing research in optimizing deep learning models for inference at the edge using connection-pruning, model quantization, and knowledge distillation. This chapter describes the techniques to train/retrain the deep learning models at the resource-constrained edge device using new learning paradigms such as federated learning, weight imprinting, and training smaller models on fewer data.

Chapter 11 presents a survey of techniques that have been introduced to exploit the pros and mitigate the cons of NVMs when used for designing IoT systems. This chapter classifies these techniques along several dimensions to highlight their similarities and differences. Keeping consideration that NVMs are rapidly growing in IoT systems, this chapter will encourage and motivate further researcher and scientists in the field of software technology for NVMs-based IoT.

Chapter 12 describes the digital abstraction of the physical aspects of a city using digital twin to simulate scenarios to understand behaviors of a particular event. This study analyzes the use of artificial intelligence techniques and IoT used in digital twin approaches to analyze cyber security risks in the smart city environment.

Chapter 13 discusses Cognitive Internet of Things (CIoT) which inherited numerous challenges from artificial intelligence, IoT, and cognitive systems. Therefore, the challenges of these fields should be studied to extract the challenges in designing CIoT. In the literature, there is no study on extracting the challenges considering associated technologies to CIoT. In this chapter, the challenges of the associated technologies are summarized. Then, some important challenges in designing CIoT are obtained.

Chapter 14 uses reinforcement learning techniques to find patterns of user dynamics and to determine the incentive prices. Specifically, the authors adapt the state-of-the-art reinforcement learning framework for dock-less BSS rebalancing. Different from existing research, the authors make full use of the benefits of destination incentives. In addition, they further extend the reinforcement learning framework to docked BSSs by adding station capacities to the state space of the reinforcement learning agent. They have examined the performance of our schemes based on real-world datasets. An experiment result reveals that the hybrid incentive scheme outperforms the source-incentive-only scheme.

Chapter 15 discusses vital applications of IoT and Bayesian learning to the monitoring, messaging, and accident analysis on highways. The chapter adopts the case approach in presenting advances in IoT and cloud technologies and builds a concept around a scenario to demonstrate real-life applications and contextual relevance of Bayesian learning models.

Chapter 16 discusses the processes, challenges, and solutions concerning designing an airport smart parking system. IoT parking sensors, Open Automatic License Plate Recognition (OpenALPR) library, and the IBM cloud-based IoT platform are integrated to tackle technical challenges, including the automatic identification of plate numbers, models, and colors of vehicles in parking spaces, in both indoor and outdoor parking environments. The chapter also addresses several issues related to the system, that is, the system architecture design, the selection of sensing technologies, and hardware and software platforms, while taking into account specific characteristics of IoT and AI technologies.

Chapter 17 presents an overview of research on using end-to-end deep learning technologies for computer vision-based autonomous driving systems. It briefly discusses the ethics of autonomous driving; it also describes autonomous driving paradigms and the associated deep learning methodologies. Furthermore, it proposes an IoAT-compatible low-cost, low-latency, high-accuracy, and high-reliability CNN-LSTM based autonomous driving model that incorporates temporal information, transfer learning, and navigational command. It also provides a detailed analysis against existing models. Finally, the chapter draws its conclusions and discusses future research directions to further improve system performance.

In Chap. 18, the Bayesian learning and decision trees are presented in respect of their ability to entrench optimum intelligent prediction in IoT-enabled domain. Succinct elucidation of the potential application of an intelligent IoT-driven system is presented as a possible panacea to address some of the problems in food production cycle especially in post-harvest storage and wastage.

We are sincerely thankful to the Almighty for supporting and standing by us at all times, through thick and thin, and guiding us. Starting from the call for chapters till the finalization of chapters, all the editors have given their contributions amicably, which is a positive sign of significant teamwork. The editors are sincerely thankful to the series editors Prof. Giancarlo Fortino and Prof. Antonio Liotta for providing constructive inputs and allowing an opportunity to edit this important book. We are thankful to reviewers around the world who shared their support and stood firm toward quality chapter submission.

Kolkata, West Bengal, India — Souvik Pal
Kolkata, West Bengal, India — Debasish De
Melbourne, VIC, Australia — Rajkumar Buyya

Key Features

1. Addresses the complete functional framework workflow in AI-enabled IoT ecosystem.
2. Explores basic and high-level concepts, thus serving as a manual for those in the industry while also helping the beginners to understand both basic and advanced aspects in AI-enabled IoT ecosystem related technology.
3. Based on the latest technologies, and covering the major challenges, issues, and advances in AI-based IoT environment.
4. Explores intelligent object identification and object discovery through IoT ecosystem and its implications to the real world.
5. Explains concepts of IoT communication protocols, intelligent sensors, statistics and exploratory data analytics, nature-inspired algorithms, computational intelligence, and machine learning algorithms in IoT environment for betterment of the smarter humanity.
6. Explores intelligent data processing, deep learning frameworks, game theory, and multi-agent systems in IoT-enabled ecosystem.
7. Explores vector machines and artificial neural networks for IoT devices, and big data analytics in IoT-based environment.
8. Explores security and privacy issues and trustworthy machine learning related to data-intensive technologies in AI-based IoT ecosystem.

About the Book

The edited book *Artificial Intelligence-based Internet of Things Systems* is intended to discuss the evolution of future generation technologies through Internet of Things in the scope of artificial intelligence. The main focus of this volume is to bring all the related technologies in a single platform, so that undergraduate and postgraduate students, researchers, academicians, and industry people can easily understand the AI algorithms, machine learning algorithms, and learning analytics in IoT-enabled technologies.

This book uses data and network engineering and intelligent decision support system-by-design principles to design a reliable AI-enabled IoT ecosystem and to implement cyber-physical pervasive infrastructure solutions. This book will take the readers on a journey that begins with understanding the insight paradigm of AI-enabled IoT technologies and how it can be applied in various aspects. This proposed book will help researchers and practitioners to understand the design architecture and AI algorithms through IoT and the state-of-the-art in IoT countermeasures.

It provides a comprehensive discussion on functional framework and knowledge hierarchy for IoT, object identification, intelligent sensors, learning and analytics in intelligent IoT-enabled systems, CRISP-DM frame work, RFID technology, wearable sensors, IoT semantics, knowledge extraction, applications of linear regression, classification, vector machines and artificial neural networks for IoT devices, Bayesian learning, decision trees, deep learning frameworks, computational learning theory, multi-agent systems for IoT-based ecosystem, machine learning algorithms, nature-inspired algorithms, computational intelligence for cloud-based Internet of Things, and trustworthy machine learning for IoT-enabled systems. This book brings together some of the top IoT-enabled AI experts throughout the world who contribute their knowledge regarding different IoT-based technology aspects. This edited book aims to provide the concepts of related technologies and novel findings of the researchers through its chapter organization. The book explores AI-enabled IoT paradigms which will be utilized as a part of betterment of mankind in the future era. Specifically, the far-reaching references of various works and executions will be observed to be significant accumulations for engineers and organizations. The primary audience for the book incorporates specialists, researchers, graduate understudies, designers, experts, and engineers who are occupied with research on Internet of Things, artificial intelligence, machine learning, and applications.

Contents

About the Editors

Souvik Pal is an associate professor in the Department of Computer Science and Engineering at Sister Nivedita University (Techno India Group), Kolkata, India. Prior to that, he was associated with Global Institute of Management and Technology; Brainware University, Kolkata; JIS College of Engineering, Nadia; Elitte College of Engineering, Kolkata; and Nalanda Institute of Technology, Bhubaneswar, India. Dr. Pal received his MTech and PhD degrees in the field of computer science and engineering from KIIT University, Bhubaneswar, India. He has more than a decade of academic experience. He is author or co-editor of more than 15 books from reputed publishers, including Elsevier, Springer, CRC Press, and Wiley, and he holds 3 patents. He is serving as a series editor for Advances in Learning Analytics for Intelligent Cloud-IoT Systems, published by Scrivener-Wiley Publishing (Scopus-indexed); Internet of Things: Data-Centric Intelligent Computing, Informatics, and Communication, published by CRC Press, Taylor & Francis Group, USA; Conference Proceedings Series on Intelligent Systems, Data Engineering, and Optimization, published by CRC Press, Taylor & Francis Group, USA. Dr. Pal has published a number of research papers in Scopus/SCI/SCIE Journals and conferences. He is the organizing chair of RICE 2019, Vietnam; RICE 2020 Vietnam; ICICIT 2019, Tunisia. He has been invited as a keynote speaker at ICICCT 2019, Turkey; ICTIDS 2019, 2021 Malaysia; and ICWSNUCA 2021, India. His professional activities include roles as associate editor and editorial board member for more than 100+ international journals and conferences of high repute and impact. His research area includes cloud computing, big data, internet of things, wireless sensor network, and data analytics. He is a member of many professional organizations, including MIEEE; MCSI; MCSTA/ACM, USA; MIAENG, Hong Kong; MIRED, USA; MACEEE, New Delhi; MIACSIT, Singapore; and MAASCIT, USA.

Debashis De obtained his MTech from the University of Calcutta in 2002 and his PhD (Engineering) from Jadavpur University in 2005. He is a professor and director of the Department of Computer Science and Engineering at Maulana Abul kalam Azad University of Technology, West Bengal (Former West Bengal University of

Technology), India, and adjunct research fellow at the University of Western Australia, Australia. He is a senior member of the IEEE, life member of CSI, and member of the International Union of Radio Science. He was awarded the prestigious Boyscast Fellowship by the Department of Science and Technology, Government of India, to work at the Herriot-Watt University, Scotland, UK. He received the Endeavour Fellowship Award during 2008–2009 by DEST Australia to work at the University of Western Australia. He received the Young Scientist award both in 2005 at New Delhi and in 2011 at Istanbul, Turkey, from the International Union of Radio Science, Head Quarter, Belgium. His research interests include mobile cloud computing and green mobile networks. He has published in more than 260 peer-reviewed international journals and 200 international conference papers, 6 researches monographs, and 10 textbooks. His h index is 27 and i10 index is 100. Total citation is 3350. He is associate editor of the journal *IEEE ACCESS*, and editor of *Hybrid Computational Intelligence* and *Array*.

Rajkumar Buyya is a Redmond Barry Distinguished Professor and director of the Cloud Computing and Distributed Systems (CLOUDS) Laboratory at the University of Melbourne, Australia. He is also serving as the founding CEO of Manjrasoft, a spin-off company of the University, commercializing its innovations in cloud computing. He served as a Future Fellow of the Australian Research Council during 2012–2016. He has authored over 625 publications and 7 textbooks including *Mastering Cloud Computing* published by McGraw Hill, China Machine Press, and Morgan Kaufmann for Indian, Chinese, and international markets, respectively. He also edited several books including *Cloud Computing: Principles and Paradigms* (Wiley Press, USA, Feb 2011). He is one of the highly cited authors in computer science and software engineering worldwide (h-index = 132, g-index = 294, 92,500+ citations). "A Scientometric Analysis of Cloud Computing Literature" by German scientists ranked Dr. Buyya as the World's Top-Cited (#1) Author and the World's Most-Productive (#1) Author in Cloud Computing. Dr. Buyya is recognized as a "Web of Science Highly Cited Researcher" for 4 consecutive years since 2016, a fellow of IEEE, and Scopus Researcher of the Year 2017 with Excellence in Innovative Research Award by Elsevier, and he has recently (2019) received "Lifetime Achievement Awards" from two Indian universities for his outstanding contributions to cloud computing and distributed systems.Software technologies for grid and cloud computing developed under Dr. Buyya's leadership have gained rapid acceptance and are in use at several academic institutions and commercial enterprises in 40 countries around the world. Dr. Buyya has led the establishment and development of key community activities, including serving as foundation chair of the IEEE Technical Committee on Scalable Computing and five IEEE/ACM conferences. These contributions and international research leadership of Dr. Buyya are recognized through the award of "2009 IEEE Medal for Excellence in Scalable Computing" by the IEEE Computer Society TCSC. Manjrasoft's Aneka Cloud technology developed under his leadership has received "2010 Frost & Sullivan New Product Innovation Award." Recently, Dr. Buyya received "Mahatma Gandhi

Award" along with gold medals for his outstanding and extraordinary achievements in the field of information technology and services rendered to promote greater friendship and cooperation between India and the world. He served as the founding editor-in-chief of the *IEEE Transactions on Cloud Computing*. He is currently serving as co-editor-in-chief of the journal *Software: Practice and Experience*, which was established 50 years ago. For further information on Dr. Buyya, please visit his cyberhome: www.buyya.com

Part I
Architecture, Systems, and Services

Artificial Intelligence-based Internet of Things for Industry 5.0

Bhanu Chander, Souvik Pal, Debashis De, and Rajkumar Buyya

1 Introduction

Nowadays, wireless communications, IoT devices, intelligent sensors, industrial IoT, mobile edge computing, and communication protocol are the buzz words in industry-academia. In general, IoT works through implanting short-range moveable transceivers into an eclectic arrangement of devices and everyday objects, enabling novel communication procedures among people and things and things themselves. Therefore, IoT would add a new dimension to information and communication. IoT devices are interconnected devices through a piece of inventive communication machinery such as RFID, Wi-Fi, GSM, Bluetooth, and many more, which can help improve people's living standards [1–4]. The latest survey reports that the number of IoT devices like embedded devices, sensors, game consoles, laptops, and smart devices anticipated to reach more than 60 billion in 2025. In general, IoT expertise's evolution is very similar to the current society, where people and devices practically

B. Chander
Department of Computer Science and Engineering, Pondicherry University, Pondicherry, India

S. Pal (✉)
Department of Computer Science and Engineering, Sister Nivedita University, Kolkata, West Bengal, India

D. De
Department of Computer Science and Engineering, Maulana Abul Kalam Azad University of Technology, West Bengal, Kolkata, West Bengal, India
e-mail: dr.debashis.de@ieee.org

R. Buyya
School of Computing and Information Systems, The University of Melbourne, Melbourne, VIC, Australia
e-mail: rbuyya@unimelb.edu.au

S. Pal et al. (eds.), *Artificial Intelligence-based Internet of Things Systems*, Internet of Things, https://doi.org/10.1007/978-3-030-87059-1_1

integrated into information systems over wireless sensors technology. IoT integration's main intention is information sharing, enabling smart surroundings to identify objects then retrieve information. Embedded devices play an essential part in IoT, which mainly connect with intelligent sensors for information gathering. In detail, embedded devices interact with the physical environment with these sensor nodes [3–5]. Nowadays, the IoT platform provides advanced control and monitoring services for novel appliances to expand their working efficiency.

The *Internet of Things* (IoT) is defined and used by a well-known researcher Kevin Ashton in the early days of 2000. From Kevin Ashton's explanations, the IoT is a system/structure of material things in the real world to link to the Internet through intelligent sensors. Ashton also conceived the *RFID* technology, which heavily applied to transportation tracking services without any human interventions. Now there are different definitions available based on their specific idea in the real-world scenario. For instance, from IEEE, "IoT is a framework of connected devices thru the internet, for new appliances and services enable the interaction in the physical and virtual world in the form of M2M communication" [3]. From Internet Architecture Boards (IAB) definition, "IoT is the networking of smart objects, meaning many devices reasonably communicate in the presence of internet protocol that not directly operated by human beings but exist as components in buildings, vehicles or the environment" [6].

As discussed above, most IoT systems are becoming increasingly dynamic, mixed, and multifaceted; thus, the organization of such an IoT system/model is challenging. IoT System-oriented services need to enhance efficiency and variability to attract more abusers. In recent times, artificial intelligence (AI) reaches tremendous success in numerous domains by employing modifications in computing technologies [5]. Machine learning (ML) is another unique technology and a sub-part in AI applied on IoT for better services. Both AI and ML are recognized as the critical parts for IoT to make intelligent network management and operations. Many kinds of research work produced better results by applying AI and ML in pattern recognition, natural language processing, object detection, and network sharing. Hence, the IoT domain can also benefit from leveraging support from AI and ML. There are huge chances by employing AI- and ML-based models to IoT to make profound analytical and in-depth progress of well-organized smart real-world appliances [6, 7].

Before knowing the technical research trends of IoT, everyone needs to take a look and understand how an IoT works and impacts our everyday life. Every researcher and data scientist tries to import and understand IoT preliminaries according to their visualizations and then requirements. After all, there is no universal definition of IoT and its visualization requirements. Internet of Everything (IoE), Internet of Cloud Things (IoT), and Web of Things (WoT) come from the IoT visualizations and have their respective definitions of working protocols. However, IoT is designed based on integrating various standards and enabling technologies with dissimilar sensing, computational capabilities, connectives, and storage capacities. Here in IoT systems, the integration standards in employed devices present high-rated challenges while authentic connections of everything. The challenges on integration in IoT devices are considered significant IoT issues since those are fundamental to the further development of IoT projects [7–9]. Nowadays, numerous standardization administrations, associations, researchers, and manufacturing industries make an effort on IoT expansions, modernization, and setting things in the right way. However, there is still a lack of a broad context with combined ethics beneath one IoT.

2 Industry 5.0 Paradigm

When it comes to the twenty-first century, most of the domains turn into digitalization. However, we admit that companies struggle to digitalize their business by incorporating AI, IoT, and Industry 4.0 technologies. Apart from the mentioned technologies, the subsequent step of the Industrial Revolution seems in the upcoming days and is named Industry 5.0 [2–6, 10–14]. The term Industry 5.0 was familiarized in early 2015; however, it was called the Fifth Industrial Revolution, which built tremendous influence in different domains, especially day-to-day business, because of the velocity of added industrial, technical enhancement and shifting human process integration [15–18].

The First Industrial revolution or Industry 1.0 started at the end of the eighteenth century; it symbolized industrialized mechanical arrangements consuming coal, human, water, and stream power. The Second Industrial Revolution or Industry 2.0 commenced in the last quarter of the nineteenth century, and it represented mass manufacture through the use of electrical energy [19–22]. The discovery of the telephone, mass production, telegraph, introduction of assembly lines, and mechanization are few features of Industry 2.0. The Third Industrial Revolution or Industry 3.0 started in the early twentieth century. It established computerization and then microelectronic skills into the industrialized field. A higher level of automation is accomplished using robots, information technology, and microprocessors – most of these twentieth-century initiatives are closely related to information and communication technology (ICT). Computer-integrated manufacturing, computer-aided processing planning, computer-aided design, and flexible manufacturing systems are some of the fields taking advantage of the third revolution. In recent times at the start of the twenty-first century, Fourth Industrial Revolution or Industry 4.0 started with the inclusion of cyber-physical systems (CPS), which makes revolutionary changes in manufacturing. Industry 4.0 was predominantly characterized by CPS, cloud computing, big data analytics, augmented reality, IoT, simulation, and intelligent devices. This means it entirely focuses on end-to-end digitalization and incorporating digital industrial ecosystems by seeking completely integrated solutions [20–24]. Besides, it highly focused on IoT objects that connect with the industrial plant.

Industry 5.0 emphasizes collaboration among humans and machinery types, which means the Fifth Industrial Revolution is more captivated by forward-thinking human-machine interfaces through human-machine interaction. Industry 5.0 main intention is to progress Industry 4.0 to an advanced level. For this, it brings the concept of collaborative robots which are also known as cobots. With the successful integration, cobots will fulfil today's need for enterprises that produce personalized products [20–24]. Hence, with improved manufacturing, software tools, the Internet of Everything, and robotics using technical progressions, Industry 5.0 is familiarized in manufacturing and medicine than other allied areas.

It provides chances for a customer to experience mass customization in different groups' collaboration across the world. Technology innovations do not consider the foundation of revolution for the organization, and there is a need for customer goals. To fulfil the customer goals, Industry 5.0 follows some set of principles:

Mass customization – suggest actual price and comfortability of various product or service customization to customers.
Customer-centric – concentrate on customer goals and try to resolve hurdles in business expansion through reengineering
Green computing – also an emphasis on environmental conditions.
Cyber-physical systems – prepare an intelligent system from the human serving the customers by gaining maximum benefits from the human with machine intelligence [16–18, 20–24] (Table 1).

Reasons for Adopting Industry 5.0 in Manufacturing
Industry 5.0 will advise or solve the issues associated with removing human workers from dissimilar manufacturer procedures from the discussions mentioned above. However, there is a need for advanced technologies to boost the Industry 5.0 manufacturer [16–18, 20–24]:

Multiscale modeling and simulation – advances of digital twin with intelligent autonomous schemes arise difficulties in valuation monitoring of manufacturing sites. In this context, visualization tools play a crucial role in constructing the

Table 1 From Industry 1.0 to Industry 5.0

Phase	Period	Description	Identification by	Key point
Industry 1.0	1780	Industrial manufacture based on stream and water machines	Mechanization Water and stream	First mechanical loom
Industry 2.0	1870	Mass production with electrical energy	Electrification Division of labor Mass production	First assembly line
Industry 3.0	1970	Automation with electronic and IT system	Automation Electronics IT systems	The first programmable logic controller
Industry 4.0	2011	The connected device, data analytics, computerized machinery programs to automate the industry production	Globalization Digitalization IoT, robotics, big data, cloud computing	Cyber-physical systems
Industry 5.0	Future	Cooperation among human intelligence with a machine to improve products and services	Personalization Robotics and AI Sustainability	Human-robot co-working Bio-economy

policies for managing and personalizing genuine products and then product outlines.

Miniature sensor data interoperability – usage of sensor nodes highly increased from smart homes to autonomous manufacture cobots and distributed intelligent systems. These intelligent sensor nodes sense and collect real-world raw data, which is an unavoidable asset to the next Industrial Revolution. However, with the progress of energy optimization, fast and effective customization process, selection of a local agent for pre-processing data, and creating high modeled distributed intelligence in IoT, Industry 4.0 is still an open research issue.

Virtual reality with digital twin – with the result of continuous growth in big data and AI-based cobots, it is even more feasible to create more realistic digital twins. It properly allows industry experts to allow reduced wastage in the process flow and system design. Hence, the digital twin with advanced visualization techniques will tremendously increase the throughput of all the sectors.

Real-time trackers – will boost real-time production tracking, facilitating the customers' sales orders with manufacture orders and supplementary material. Virtual training will assist in some cases: when trainee or trainer on different locations but learns a specific job in a virtual/simulated atmosphere. This type of training pointedly decreases the costs than time for both parties.

Intelligent autonomous systems – artificial intelligence models have great deals in autonomously controlling production lines in the manufacturing industry. Up-to-date AI-related ML and DL models effectively make changes in intelligent systems and solutions that assist in decision-making scenarios.

Transfer learning – transfer learning policies guide the schemes mentioned earlier, securely and progressively in Industry 5.0.

Computer vision with DL and RL and GPU-based computation has shown great potential in reproducing primitive vision besides sensory abilities. However, for advanced performances of Industry 5.0, cobots proficiencies must be improved suggestively.

Problems and Limitations in Industry 5.0

Industry 5.0 resolves most of the manufacturing issues associated with removing human workers from different procedures. However, it must incorporate additional forward-thinking skills since humans may add innovative manufacturing skills in the coming days. There are numerous skills in the developing stage, some of them pointed in this section.

1. Before incorporating advanced skills into industrial management, there is a need for how an autonomous system can incorporate ethical principles.
2. There is a need for proper verification and validation of ethical behavior inside the autonomous system model.
3. Implementation operation transparencies and fast and competent manufacturing might have significance in an overproduction phenomenon.

4. The outcome results must be understandable ethical behavior solutions in an autonomous scheme. In particular, industrial experts are facing adapting and implementation issues.
5. Tuning and validation will avert somewhat serious problems among technology, experts, stockholders, society, and businesses.

3 Elements of IoT

As we mentioned in the introduction, understanding IoT building blocks will give some visualization and a better perception of the IoT's actual meaning than functionality [23]. We listed six fundamental elements of IoT, which are noted in Fig. 1.

Identification
In any communication or data transmission network, the term identification plays a considerable role. The precise identification is key to the IoT structure to name and match services with their claim. However, it is tricky to addressing object ID and its corresponding IP address in the IoT system. An ID indicates a particular object or device's name, and an address indicates its present address inside the network territory. Differentiation among object identification then addresses authoritative since identification models are not inimitable; moreover, objects might practice with public IP addresses inside the network. Hence the designed models must overcome the hurdles mentioned above and identify every object inside the network correctly.

Sensing
IoT setup intends to gather information from a particular region/area, organized through sensing devices. Sensing devices/objects collect real-world data from the surrounding atmosphere and send it back to the database or cloud for additional

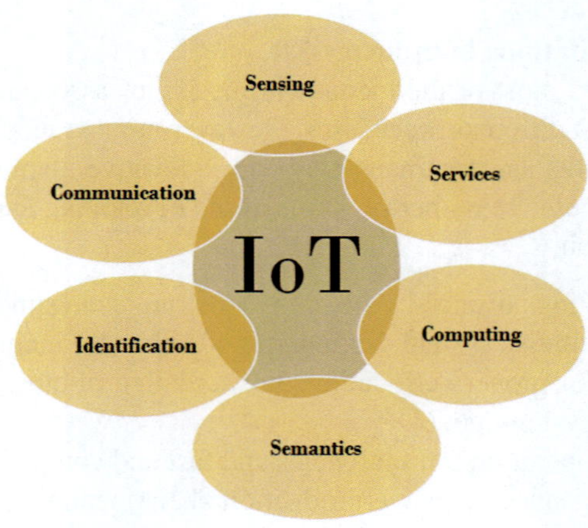

Fig. 1 Internet of Things elements

processing: sensors, wearable devices/sensors, and actuators utilized mainly for sensing purposes. For example, single-board computers (SBCs) like Arduino Yun and Raspberry PI combined with sensors and integral TCP/IP and safety functionalities are naturally used to grasp IoT products. Such devices characteristically attach to a central managing portal to deliver the essential data by clients.

Communication
In general, most IoT objects contain adequate resources; with these limited resources, objects connect with heterogeneous devices/objects in the company of lossy, noisy connotations. Wi-Fi, Bluetooth, NFC, RFID, and IEEE standards some IoT communications; in the next section, a brief description provided a better understanding.

Computing
The computing power of hardware devices is also an essential concern in IoT. The computation components like microprocessors, microcontrollers, and software-oriented appliances represent the brain to a particular appliance. Arduino, Raspberry PI, UDOO, MULLE, and Gadgeteer are hardware platforms designed for IoT appliances. Some other platforms are real-time-software operating systems (RTSOS), for real-time IoT functions; TinyOS, for lightweight operations; and cloud platforms, for too big data processing in real time. Still, some of the computing components have drawbacks, and research community is working on them to perform well.

Services
IoT offers a wide variety of services. Most of them are divided into identity-based services, in which most of the real-time appliances come in this category; information-aggregative services, which accumulate real-world raw sensor data connected with appropriate IoT applications; collaborative-aware services, which use the collected data to data analytics for decision-making; and ubiquitous-based services, aimed to represent collaborative systems to work anytime, anywhere when they are required by clients. Still, the above mentioned services are not reached or achievable to a comfortable stage; many complications besides challenges have to be answered.

Semantics
Semantic operation in IoT performs to useful abstract information smartly from different objects. It is similar to knowledge extraction, like finding resources that improve the model performances. Resource Description Framework (RDF), World Wide Web Consortium (W3C), Efficient XML Interchange (EXI), and Web Ontology Language (OWL) are some of the well-known semantic technologies adopted in IoT systems.

4 IoT Architecture

IoT and its variant inclusion into various domains and organizations will enhance the product or working performances. However, these proposals are severe and complicated to implement when it comes to real life since the number of devices, protocols, and working conditions is entirely dissimilar from one device to another. In other words, the problem of creating a consistent architecture of IoT unavoidably arrives in this phase. Before designing IoT architecture, it is better to understand the factors that affect IoT behavior, making it easier to find reliable IoT solutions. Moreover, it will reduce the various resources spent on IoT design. Before revealing the enigmas and providing an explicit construction of this creativity, it is vital to recognize what this idea means [23–28]. In essence, IoT architecture is the combination of great fundamentals network tools. It is measured as a global network setup collected of several allied devices that rely on communication, networking, sensory, and then information processing types of machinery. See Fig. 2.

4.1 Perception Layer

IoT is a kind of worldwide physical interrelated system in which things can couple and then be measured remotely. The perception layer is considered an initial stage for IoT schemes, and it is like a bridge between real and digital worlds. In some cases it is acknowledged as a sensing layer. Most of the perception layer deals with intelligent wireless devices like intelligent sensors, tags, and actuators. These wireless schemes with tags or sensors are now talented to inevitably sense and then exchange info among different devices. Devices may diverge in procedure and size

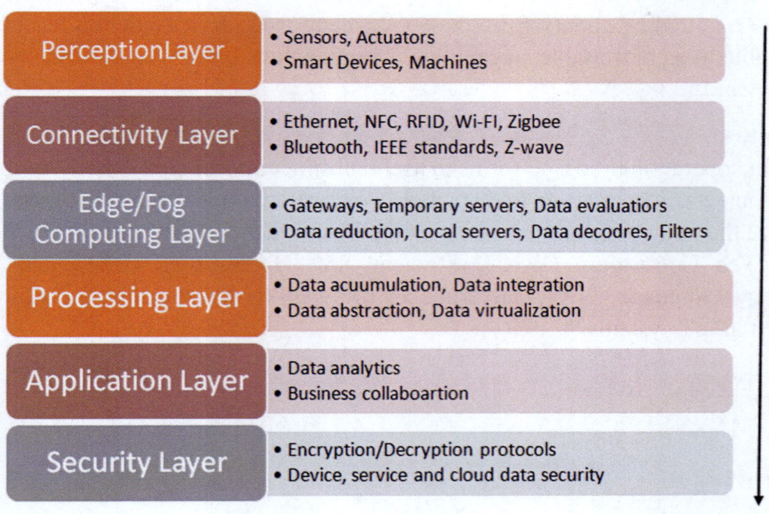

Fig. 2 Internet of Things architecture

from miniature to ad hoc vehicles. Sensors accumulate environmental conditions, transform them into electrical signals, and then forward them to IoT schemes. Actuators transform electrical signals collected from IoT scheme toward physical activities. It must note that IoT architecture does not make any limitations on elements and their deployed locations. It means objects/devices can lace in a small place to corners of the world.

4.2 Connectivity Layer

The connectivity layer is considered the second phase of the IoT scheme since it takes care of complete communications across devices, systems, and then cloud centers that made the perfect IoT scheme. The communication connectivity among physical layers to cloud centers can be achieved thru TCP/UDP or software/hardware modules. Ethernet connects fixed IoT devices; Wi-Fi are widespread wireless connectivity applied on home IoT setups; NFC is data transmission among two devices; Bluetooth is used to transfer small-size data, not applicable for large data files. In some unique scenarios, IoT uses message-oriented protocols depending on the application requirement for data connectivity. Advanced Message Queuing Protocol (AMQP), Constrained Application Protocol (CoAP), Data Distribution Service (DDS), and Message Queue Telemetry Transport (MQTT) are some of them.

4.3 Edge or Fog Computing Layer

Edge/fog computing is vital for IoT systems to satisfy the increasing volume of connected devices and real-world services. The intention of designing edge/fog computing is to store and pre-process the sensed data as fast as it sensed and adjacent to its sources as possible. So it can save time and resources for IoT devices; also, it will decrease the scheme latency time, which can improve performance accuracy. Usually, edge/fog computing takes place on gateways and local servers distributed over the network.

4.4 Processing Layer

The processing layer collects all the data across the IoT schemes. It applies preprocessing models to use abstract information for decision-making or make data available for any further operations. Real-time data is observed with API and used for non-real purposes, and it stands like a hub among event-based and query-based data ingesting. After collecting multidimensional data from various devices and applying data abstraction methods, at that moment, only other connected devices can understand the data.

4.5 *Application Layer*

Data analysis is done through software applications to bound with appropriate answers for main business questions/requirements in the application layer. In IoT, hundreds of IoT requests diverge in intricacy and function, using different expertise stacks than functioning schemes. In present days, various applications are constructed right on top of IoT stages that can suggest software-related advance setups through ready-to-use utensils for data mining, pattern, and forward-thinking analytical skills.

Business Layer
The information collected and pre-processed in IoT schemes can only help problem-solving/decision-making systems achieve excellent results. The business layer is well-defined as a distinct stage, advanced, and challenging to describe in a single application layer for this motive.

4.6 *Security Layer*

In any network-related application, the word security has its place. In IoT, the security layer plays a crucial part, covering all the services mentioned above/layers. It is tough to discuss the security topics of IoT in one single paragraph or a section. There are different security levels in IoT schemes: in *device security*, IoT-related devices need low-resourced authentication services, physical metal shields, and chips that can boost procedures to avoid unauthorized code. *Connection security* is mostly data transfer in IoT done through wireless channels, which is easy for attackers to steal or alter the data. Hence, when the data sent over a device or network, it must be in an encrypted format. In *cloud security*, sensed information kept in the cloud must be encoded to mitigate hazards of revealing delicate info to trespassers. Hence, always pay attention to security protocols to certify that security is high at all stages, from the smallest devices to multifaceted analytical schemes.

5 Enabling Technologies

5.1 *Radiofrequency Identification (RFID)*

RFID communication technology is specially designed for transportation tracking made of tags and readers. RFID is considered an automatic identification mechanism that involuntarily identifies the target tag signal with suitable data. Hence, it was employed extensively in various hazardous and impassive atmospheres. As we mentioned above, the RFID structure completes with tags and readers. The tag consists of address bars attached to objects as a small microchip handled by the antenna. The electromagnetic pitch is applied to send and collect data records from an entity over a tag. The data records stored on a tag can only be read or abstracted by readers only when both tags and readers are placed at a specific angle or range. The reader forwards a signal to read the tag's information, and the antenna on the patch receives it, acknowledging the signal by sending appropriate data. In record, three tags are available in RFID communication: passive tag, which obtains signals from tags working on batteries; active tag, in which tags abstract energy from readers' signals, which means those do not have batteries; and, finally, active reader active tag, which works on both low and high frequencies. RFID tags are professionally applied on real-world appliances since they automatically monitor payments, goods or baggage tracking, inventory management, tracking of products, and product lifecycle supervision and then update the information without any third-party or human interference. RFID technologies can fit into different domains to design and enhance model/systems accessibility and then efficiency. However, there are some drawbacks for implementation of RFID because most IoT WSNs appliances are built in harsh environments, where the signals are disturbed and intercepted and there is a chance for the entire device to collapse.

5.2 *Power-Line Communication (PLC)*

In PLC, data records are forwarded through the attached cables. It means a sender modulates the data records into the transfer medium; when it reaches, receivers demodulate the data records and then read them. By doing this, data transfer with power cables, where one can both power it up and then at a similar time control/retrieve data from it in a half-duplex style. Hence, PLC attracts a communications model in intelligent meters (AMI), HEMS, BEMS, and solar panel-intensive care schemes that understand smart society. There are low-speed and then high-speed kinds of PLC, each of which uses a different communication procedure.

5.3 *Electronic Product Code (EPC)*

EPC was utilized to recognize RFID tags; it is in string type 96-bit long and placed on tag/patch. Out of these 96-bits, 8 bits represent header which aimed to identify the version of the protocol, 28 bits refer to the unique address of the system that manages the data on tags, 24 bits hold the type of product to be recognized, 36 bits mention the serial number of the tag. Finally, the last 2 bits are being hold the by the organization that created the tag.

5.4 Wireless Sensor Network

Wireless sensor networks collect small tiny sensor nodes employed to gather sensed data from surrounding atmospheres. Computer networks, micro-electro-devices, and wireless technology combinations made the formation of WSNs. In the past, wired sensor networks/nodes used for communication, which places very local amenities, overcome WSN technology developed and produce possible results with various appliances. It is a known fact that WSNs drive IoT systems and enhance performance precision. Due to the node's resource constraints, deployment topology, connection, detection of neighbors, and transmission paths are the essential tasks in WSN formation. WSN is a vital element of IoT as it combines mixed sensor data, systems, and appliances. Researchers designed various inclusion techniques for IoT, the Internet, and WSNs, but still face many challenges that need optimal solutions and research under study.

5.5 Near-Field Communication

NFC technology is applied for data transmission and small communication setup when two objects are near to each other. It is similar to radio communication but works by touching or two objects closer to the exact location. The communication range of the NFC depends on the scale of the object's antenna. Hence, NFC technology is mostly not recommended to isolated locations, and it also not safer due to its limitations easily vulnerable to attackers.

5.6 Actuator

Actuators apply to specialized appliances, and they work significantly when the objects are in motion. It creates various motions like rotary, spherical, linear, and oscillator; then, it creates power from using them into kinetic energy. Actuators consider three types: electrical-based, employed on motors; hydraulic-based, hydraulic fluids; and pneumatically based, which use compressed air.

5.7 Machine to Machine (M2M)

M2M communication is similar to LAN and WAN networks; devices gather data from various sources and sent it back to other devices within the network. In M2M, stored data records are monitored and automatically take some assigned tasks depending on the applications. Moreover, the performance of M2M depends on software-controlled communications among machines and devices.

5.8 *ZigBee*

The main intention of ZigBee's innovation is to expand the application regions of WSN and IoT. It is a special kind of flexible wireless networking technology, better performance for short communicated appliances like intelligent home automation, healthcare, and industrial appliances. ZigBee is designed with MAC and IEEE protocols; besides, it has four-layer architecture, namely, physical, MAC, network, and application layer.

5.9 *Wireless Fidelity (Wi-Fi)*

Wi-Fi is a famous wireless network ability and an excellent fit for data-intensive IoT-based solutions. It has high wireless access for a small area with an intelligent transportation system. It has collective versions, and some of them are as follows: IEEE 802.11a delivers a data rate of 54 Mbps, and IEEE 802.11a data rates up to 2.4 GHz.

5.10 *IEEE 802.15.4*

IEEE 802.15.4 (low-rate wireless personal area networks – LRWPANs) act as a sublayer for the MAC layer. It provides effectual communication for low-power consumption data rate, high security, and low cost and supports a vast number of sensor nodes at a time. Based on these specifications, IEEE 802.15.4 is considered a basis for various communication technologies like ZigBee, Z-Wave, Bluetooth, etc. However, it does not provide QoS; also, this is a fascinating topic to research.

5.11 *Z-Wave*

Z-Wave communication technology initially designed smart home automation appliances like door switches to a central controller. The working procedure of Z-Wave is quite similar to ZigBee, both employed with mesh topology and low wireless standards to improve the low-resource devices. Z-Wave functions in the 868 MHz frequency band, while ZigBee functions in 2.4 GHz. Besides, Z-Wave left the software-side encoding, but ZigBee practices the 128bit AES on the hardware side.

5.12 *Bluetooth LE*

Bluetooth or IEEE 802.15.1 stands for the information exchange among fixed and mobile devices over a short distance using industrial, scientific, and medical (ISM) bands. It heavily applied to smart home, smart city, healthcare, security, military appliances, fitness, and industries. Bluetooth SIG, Bluetooth BLE, Bluetooth 4.0, and Bluetooth 5.0 are the latest versions of collecting and aggregating sensed data from IoT-based sensor nodes. Bluetooth technology was very much suitable for short-range monitoring appliances.

6 Artificial Intelligence (AI) in the Internet of Things (IoT)

The operative functions of the Internet are insistently from the "Internet of Computers (IoC)" to the "Internet of Things (IoT)." There is a need to deliberate the importance of AI techniques to allow intelligent Internet communications. In present days, wireless sensor networks are becoming hot research topics because of their reality applications, incredible remote monitoring of events in fields like healthcare, weather report, seawater levels, event predictions, etc. Besides, intelligent sensors were employed heavily in electronic-based home appliances, smart cities, and gadgets to mobiles [22–25, 29–36].

The idea of IoT is "the pervasive presence around us of a variety of things or objects – such as Radio-Frequency Identification (RFID) tags, sensors, actuators, mobile phones, Etc. Through unique addressing schemes, they can interact with each other and cooperate with their neighbors" [36]. Hence, the changes in IoT protocols and services will surely have a good impact throghout the world. AI approaches also help IoT build robots whose situatedness evolves roles that avoid persistent human command [37–40].

Figure 3 deals with the IoT data flow diagram. Data initially comes from the IoT-enabled devices and IoT appliances, and then through IoT gateway, it goes to a cloud-based server. Here data has been analyzed via different analytic tools and learning and training methods. Then recommendation systems come into the picture for optimal actions; actuators are there for transferring the flow toward IoT appliances for further processing. However, we can say that innovative IoT standards are vital for shuffling from today's sensor networks into systems of intelligent sensors permitted with actuation types of machinery. These kinds of upcoming schemes will involve the "Internet of Intelligent Things (IoIT)." These are the successive evolutions of networking to create the experienced ubiquitous, living, intelligent Internet connections. It seems necessary to give familiar objects the capability to understand their backgrounds and make conclusions freely [36–45]. At present, decisions or conclusions no need to forward to central decision-making nodes. Through great intellect of sensors and giving them the skill to turn by affording to the incentive professed by sensors, empowering the IoIT to reply improved time-critical conditions, since the conclusions complete in a noncentralized manner.

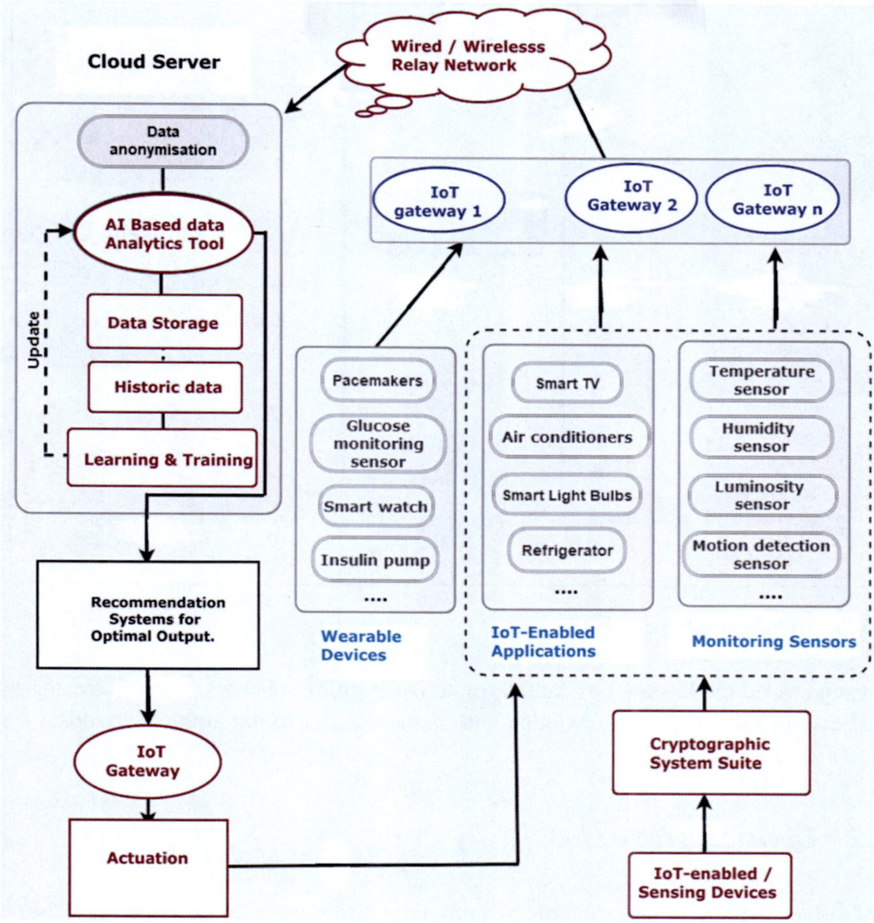

Fig. 3 IoT Data workflow diagram

The Internet of Things helps us collect data, which is one of the most valuable resources today due to its role as a catalyst. When paired with another catalyst, artificial intelligence, as shown in Fig. 4, vast volumes of unstructured data can be easily shifted through, resulting in industry insights and well-informed decisions.

6.1 *Artificial Intelligence for Intelligent Sensing*

AI for intelligent sensing talks about the ML models to recognize valuable patterns or forecasts from the information collected by the intelligent sensors. For example, active sensor learning dynamically increases class volumes identified by the model. As the data is composed in a real-time environment, a predefined approach for data acquisition must follow episodic retraining or careful querying [32–36].

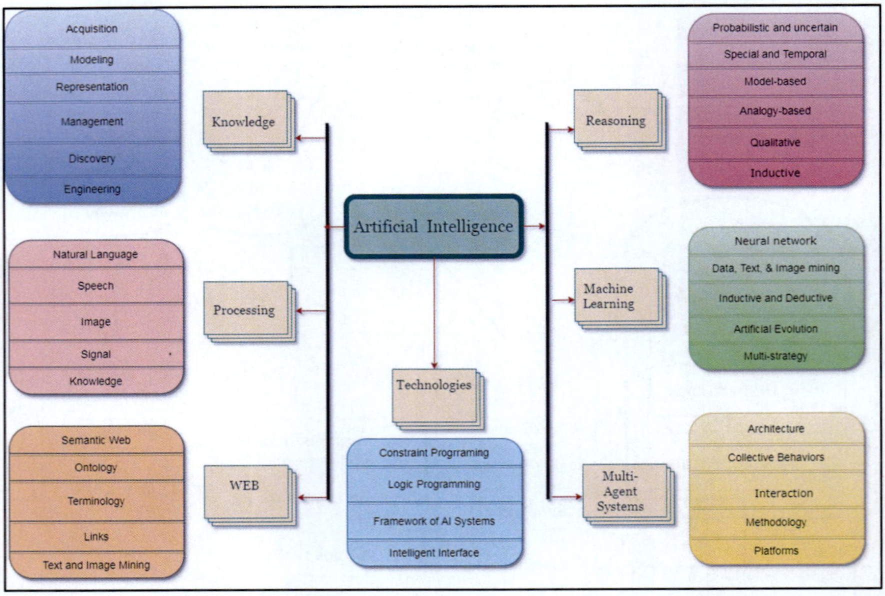

Fig. 4 Artificial intelligence classification

Unsupervised models are best suited for active learning schemes, aiming to categorize events with no prior knowledge with numerous dissimilar ambient sounds.

6.2 *Decision Tree in IoT*

DT solves classification problems by applying sorting techniques based on respective features. In DT, numerous procedures are used to find the most attractive feature that best splits the training examples; some are information gain and Gini index. The complete process of a DT is as follows: first, pre- and post-pruning applied to reduce the tree size; second, searched space among the objects adjusted; third, optimized search model employed to eliminate the redundant features; and fourth, a structure of resultant tree transformed into an appropriate data structure like set of rules. DT effectively applied IoT-based real-time applications such as pattern recognition, decision-making, environment monitoring, detection of security parameters, healthcare management, etc.

6.3 *Random Forest in IoT*

Random forest (RF) comes under the category of supervised learning models. RF consists of numerous trees built randomly and then skilled to make a vote for a good class. At last, the most voted class is elected as the final classification result. In general, RF utilizes decision trees to build subset rules for voting, so the result classification is the average of the DT results. Besides, RF computational accuracy overcomes the feature selection where it requires the fewest input parameters, but it cannot be applicable in real-world applications. RF models were highly applicable to IoT devices in various domains. For instance, RF models skilled with features obtained from the network traffic correctly recognize IoT device categories because RF precisely holds real-world implications for correctly categorizing unauthorized IoT devices.

6.4 Clustering

K-means – the main goal of k-means is to cluster the unlabeled data features into K number of clusters or sets; here, data points fitting to the identical cluster must have some likenesses. Typically, k-means is a fast and highly scalable ML technique. In some cases, employed MapReduce to analyze the several minor datasets then offers a cluster approach for a high dimensional of small data based on the K-means procedure [39–41]. Researchers designed the K-means cluster and then categorize travel pattern consistencies.

Density-based spatial clustering of applications with noise (DBSCAN) clusters the unlabeled datasets based on the data point density (data point with the highest count of close neighbors) values. It is a widely used clustering system with several real-world requests like anomaly exposure in temperature data, traffic control, emotions recognition, and X-ray crystallography.

6.5 One-Class Support Vector Machine (OC-SVM)

OC-SVM comes under a semi-supervised technique, and it is an extension for SVMs. It generates a boundary line among the trained data when the new data after some operations lie outside the boundary line commented as an outlier or anomaly. Because of its nature of work, OC-SVMs are useful in anomaly detection in WSN, network intrusion detection, and IoT-based machine performance evaluations.

6.6 Ensemble Learning Models in IoT

Ensemble learning (EL) combines various basic classification practices and produces a collective, effectual output. Research work on EL experiments shows that learning models vary by precise application. So, the research community starts combining various dissimilar classifiers to expand precision. Moreover, EL models use numerous learning techniques which lessen the variance, robust in contradiction of overfitting. EL has been effectively used for online intrusion and anomaly detection in IoT-based environmental datasets and evaluating real-time datasets for accurate IoT devices' decision-making.

6.7 Neural Networks

Neural networks (NNs) with the condensed representations are the quickest models to process the new data instances. From the innovations, NN has diverse NNs with a distinct structure and appliances. The feed-forward neural network (FFNN), also called multilayer perceptron, is considered the most common type of neural network in functional appliances. In FFNN, every layer's activity is determined through the nonlinear function or active function. An FFNN with a minimum of two hidden layers can estimate a random mapping from a finite input space to a finite output space with sufficient hidden units. Nevertheless, the issue is detecting the optimal weights for an FFNN comes under the NP-complete problem. The model has various learning approaches, like adaptive moment estimation, stochastic gradient descent, adaptive slope, Nesterov's accelerated gradient, adaptive delta, and RMSProp. FFNN in IoT applies as a solution for energy efficiency, decision-making, feature selection, energy management, reducing computation complexities, etc.

6.8 Support Vector Machine (SVM) in UIoT

SVMs perform classification by forming the splitting hyperplane among two distinct classes by calculating the distance's data attributes. SVMs are chosen for large datasets with many feature attributes but contain tiny sample points. The main advantage of SVMs can perform in real-time intrusion exposure and then inform the training patterns energetically. SVM variants like QS-SVM, CE-SVM, and SVDD are widely used in numerous security applications like an outlier and intrusion detection; moreover, they are effective in memory storage with less time complexity.

6.9 Internet of Intelligent Things (IoIT) for Social Networks

Social media plays a crucial role in a current digital world, where millions of people regularly participate, connect, and express their ideas, views, and suggestions. With this connection and sharing of ideas, many people can answer complex issues more efficiently than single individuals. Nowadays, intelligent sensors are automatically categorizing the actions of crowded people in real time. IoT turns out to be an enabled model for other networking forms than computation like IoIT and robotics as a service in another side of the networking world. These novel models efficiently add cleverness to the things linked to the Internet or consider things as robots, cobots, services, and users. Employing principles studied in social networking to the IoT might bring tremendous changes as well as advantages.

Usually, humans and robots, or a mixture of them, form web communities – however, such groups are shaped by intelligent avatars in the virtual world of the IoT. Continuous research links other biological creatures, automated to be proficient in intelligent processing, into social networks. Co-location object relationship, social object relationship, and ownership object relationship are examples of SIoT.

6.10 Principal Component Analysis

Principal component analysis (PCA) orthogonally plans data facts onto an L-dimensional linear subspace, termed the principal subspace. PCA deals with high-dimensional datasets based on reiterative expectation expansion practice and data compression; data visualization comes under PCA applications. Hence, PCA is considered the most crucial pre-processing procedure in ML.

Canonical correlation analysis (CCA) variant of PCA deals with two or more variables. Here the main goal is to recognize a consistent pair of extremely cross-correlated linear subspaces. Hence, inside one of the subspaces, there is a relationship among each factor along with a solitary component from the other subspace.

6.11 Bagging

The bagging objective is to enhance the precision and stability of ML-based techniques and then diminish the overfitting. In this method, training datasets are engendered by arbitrarily selecting data points from the unique training set with substitutes. So, on each originally produced training set, an ML practice is trained. In ML, there are numerous approaches like DT, RF, and neural networks, for which the bagging method advances the outcomes.

6.12 Artificial Intelligence in Analytical Skills (IoT)

Various business organizations have hired analytical skills for quite a few decades; nowadays, numerous organizations attract to planning their AI abilities. From the past few decades, organizations/companies synthesize their skills for efficient utilization of data and their statistical analytics and quantitative procedures to progress decision-making. However, currently, those companies are mainly engrossed in discovering and operating AI to strengthen each other. AI is not statistical, like ML and

DL, which quickly increases supremacy besides demand [29–34]. Analytics-oriented clusters inside the administrations may want to concentrate their care mainly on these machineries or obtain novel skills in nonstatistical portions. The innovation analytics has transformed into various versions, some of them mentioned below.

Analytics 1.0 – it is an age of artisanal expressive talent and the initiation of scrutiny and writing utensils. In this stage, it conquered the commercial analytics for years, and the price stayed mainly determined by the objective of interior decision provision rather than progressive analytical abilities. Analytics 2.0 – this stage big data analytics stands like Hadoop, and then information-based innovations such as Google and Facebook led to data experts' advent. The main intention is to shift from "internally designed decision support" to "data products" designed for the data and then analytics for use by clients. Analytics 3.0 – large-scale corporations make data, then analytics-based productions, and then analytical events with numerous ML models. Analytics 4.0 – AI and cognitive-based models are heavily applied in analytical sophistication by various organizations. It adopts various model accuracy levels and applies with AI models and superior use of self-rule in the performance of approaches as automated ML. Some reasons for adopting AI into analytics are a mixture of skills and internal partnerships needs AI [34–42]. For instance, computer knowledge requires understanding the embedded learning data models. Another reason is that accurate data analytics with immense data processing and cutting-edge statistical models are required. ML is the central part of many approaches to AI and analytical techniques. The usage of ML in analytical procedures is started several years back and may be more aware of predictive analytics. ML uses supervised learning where both the creation and results from values are known.

AI is steadfast on the rise and will play a considerable role in analytics 4.0 because of its potential in transforming business models; hence, the influence of analytics 4.0 will possibly be greater and also higher unsettling than preceding automation evolutions. Moreover, organizations that shift to analytics 4.0 more fast-track than those that do not apply any AI model. The procedure toward understanding AI achievement starts with the primitive consideration of AI, how AI will influence creativity, the new abilities, and what a workable act policy should be applied. Businesses that control their present analytical abilities can have a much quicker and more active start with AI.

6.13 Deep Learning for Analytics (IoT)

Due to the development of various networks and miniature technologies, IoT-based devices collect massive sensed data from surrounding environments where they deploy. Moreover, depending on the applications, these IoT devices/objects will result in fast and real-time data streams. Here, deploying analytical models on such substantial data streams for finding original information, forecasting forthcoming structures, and then taking control of results is vital. It makes IoT applications a well-intentioned standard for business and a quality-of-life enlightening skill.

From the past few years, most of the IoT appliances are designed with dissimilar research fields such as military, smart city, healthcare, and agriculture. The success of these applications is because of intelligent learning mechanisms for prediction or data analytical outlines. DL has been aggressively employed in many IoT appliances in present times with consideration of various ML tactics. The combination of DL-IoT is considered a top strategic technology in future applications. The main reason for implementing DL in traditional ML is that it quickly addresses the emergent analytical services needed in real-time IoT appliances [40–44]. Besides its variant's derivatives big data, the expansion of IoT needs stakeholders to identify their meaning, building blocks, abilities, and challenges. There is a strong collaborative relationship between IoT and big data: IoT is a significant information producer for big data. Similarly, it is a significant mark for extensive data analytical skills to expand the methods and then services of IoT applications.

To better understand IoT-based data analytics requirements, it needed to determine the features of IoT data and how they dissimilar from those of big standard data [40–44]. Some of them are mentioned below:

1. Large-Scale Streaming Data: IoT deployed with vast numbers of devices placed in distributed manner collects enormous data from IoT applications, leading to a high volume of continual streaming data.
2. Heterogeneity: IoT is a heterogeneous connected network, so numerous IoT data acquisition device assembles dissimilar result in data heterogeneousness.
3. Time and Space Relationship: At present, a maximum of IoT appliances real-world based, here sensor devices involved to a definite position, then have a position and timestamp for every single data substance.
4. High-Noise Data: Due to dynamic environment changes, miniature error bits, and noisy data produced in IoT requests, before applying them to any decision-making systems, it needs to eliminate them; otherwise, it will affect the outcome results.

While obtaining confidential information from big data is a talented technique to improve our lives' excellence, it is not a simple, straightforward job. There is a need to go outside the outdated inference learning models' abilities, innovative skills, practices, and infrastructures to deal with such composite and thought-provoking tasks. Fortunately, due to the contemporary developments in ML and DL variants, it is easy for big data analytics and information abstraction appropriate for IoT appliances. IoT appliances like fire detection and vehicular identification need fast and continuous streaming data for quick movements to accomplish their targets.

Numerous researchers have projected methods and outlines for fast real-time streaming data analytics that influence cloud setups' abilities than its services. As mentioned earlier, appliances need fast analytics in slighter-scale platforms such as fog/edge computing for the IoT. For example, healthcare-related applications must make quick decisions on a particular time instance; otherwise, it may cause a patient's loss. These kinds of decisions should be maintained thru quick analytics with multivariate datasets. Hence, accurate identification must be performed in IoT quickly and in real time to prevent fatal mistakes.

6.14 Edge Computing in IoT

In general, the connected objects in IoT generate vast amounts of data, collecting and processing that much of data at one of the appropriate/suitable objects to turn data into useful information. Hence, nowadays, the entire IoT setups apply significant data operations; big data support IoT applications since the tribunals of gigantic sensing then stimulate information maintained in IoT. Also, IoT collects unstructured, multivariate data that needs additional analysis to be abstract the valuable information [38–44] because of the heterogeneous connections. With the rapid development of various technologies, IoT becomes the next technology revolution; however, it will confuse ample data storage, processing, and systematic analytical skills. IoT employs real-time applications to work with continuous streaming, disturbing the data storage dimensions in numerous establishments. Hence there is a need for additional data centers for handling collected data from IoT appliances. One probable answer is to transfer the information to the cloud via leveraging the application platform as a service. Nowadays, cloud computing is one of the well-established technologies, and it offers computing facilities or data storing on the Internet.

IT companies like Google Cloud, Amazon Web Services, and IBM Cloud analytics present cloud services. Cloud computing offers various advantages like proficiency, capability, and flexibility to store and then use sensed data information. However, data from vast quantities of objects spanning a vast geographical region must be stored, managed, and analyzed proficiently in IoT-related appliances. However, when cloud computing is employed in IoT, new encounters will come into action. Fog or edge computing is talented enough to outspread cloud computing faster than it assists in overcoming mentioned issues. In brief, as a replacement for performing entire computational processes at the center of the cloud, fog/edge computing offers computing and then storage facilities to devices at the edge of the system. The node/object with fog computing capability of any network can efficiently perform the data storage, computation, and heterogeneous network connectivity. These devices/objects are employed at any place of the network and assemble the IoT things with connected applications [38–45]. Usually, different kinds of data are collected from IoT objects and transferred to suitable object/place for additional analysis based on the application necessities. Here, the priority-based information that is required to be forwarded instantly can be managed on fog/edge computing nodes, which are nearer to the IoT campaigns that create significant evidence. The low-priority data records can then be forwarded to some collective nodes/objects for additional processing and then scrutiny. Besides the advantages of fog/edge computing, it has limitations and tribunals while integrating IoT with fog/edge computing. Establishing fog/edge computation and assigning adequate resources to IoT things is the most significant task. In IoT, every time, a minor number of services are demanded by IoT devices; hence each fog/edge service node contains inadequate communication, computation, and storage capabilities. In this context, every fog/edge computing node should be optimally accomplished and composed for IoT devices to deliver demanded service resourcefully. How to adjust the allocated resources of a fog/edge node is also a tricky task. It means focusing on the resource managing between the fog and edge nodes is the hot research topic in IoT fog/edge computing. Hence, when applying fog/edge computing nodes to a particular service, there is a need to verify the different requirements like energy consumption, node cost, and service availability. Also, safety and confidentiality problems in fog/edge computing structure are also vital problems.

6.15 Federated Learning

Federated learning (FL) [46–48] is a machine learning methodology in which an algorithm trains across numerous decentralized edge devices or servers that keep local data samples without exchanging them.

As shown in Fig. 5, users use local data to train local models to update the global model at the base station. The global model aggregates and sends to the local models for training. These processes carry out again and again until the global model converges.

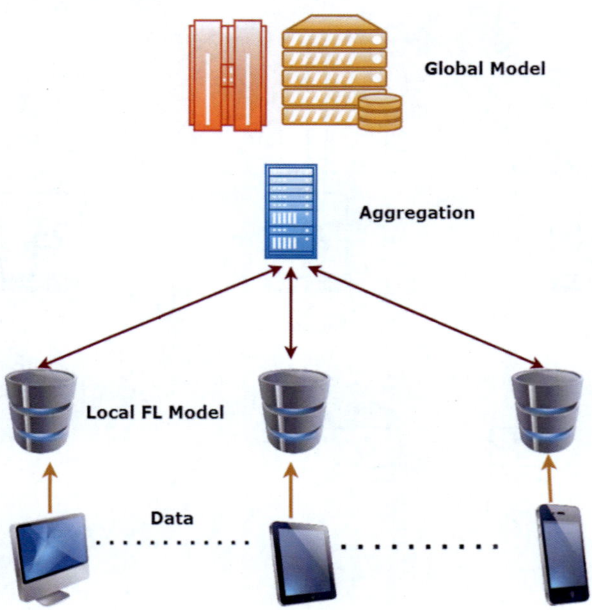

Fig. 5 Fundamental FL architecture

Federated learning allows devices to learn from a shared model collectively. With proxy data, the shared model first trains on a server. The model is then downloaded and improved by each device utilizing data – federated data. The device uses locally available data to train the model. The model's modifications compile into an update, which delivers to the cloud. The device retains the training data and individual updates [48, 49]. The model compresses via random rotations and quantization to ensure faster downloads of these updates. When the devices communicate their models to the server, the models integrate to form a single model. It is repeated for multiple cycles until the model is of good quality.

The following is the technique for federated learning as shown in Fig. 6:

1. A training model will send to the devices.
2. The devices are programmed to learn from local data.
3. The devices give the server encrypted updates on the parameters.
4. The devices are grouped by the server. The server aggregates the updates it receives from each set of devices to conduct a single update to the current model for each group.
5. The new updated model is delivered to the devices for on-device testing (again, the notion of decentralization is at play here), and a fresh round of training follows.

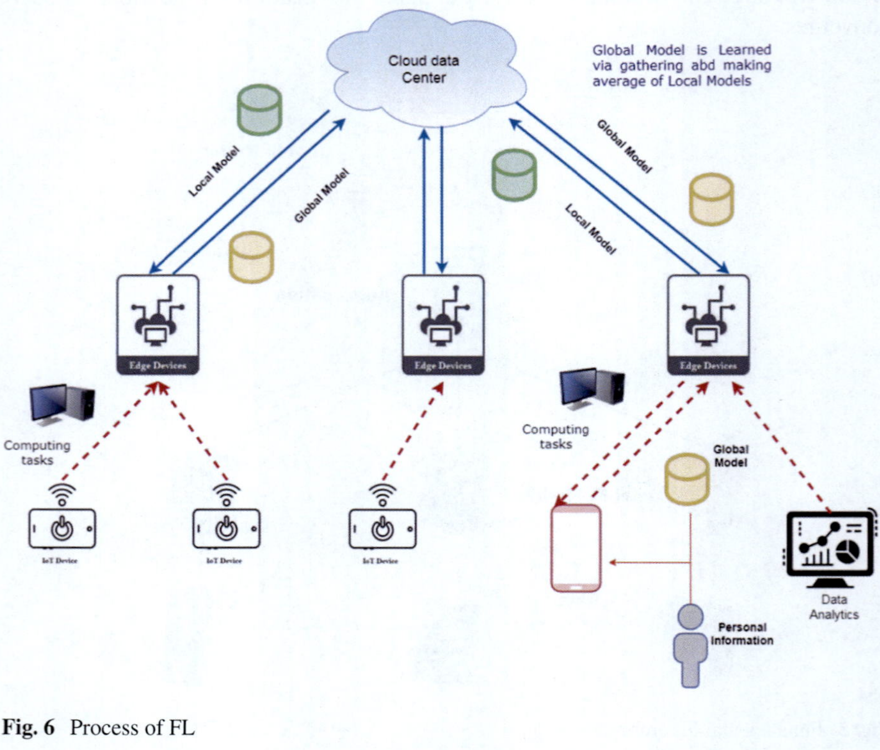

Fig. 6 Process of FL

Advantages of Federated Learning

Here we are discussing some of the FL benefits [47]. Firstly, FL allows smartphones to learn a shared prediction model cooperatively while keeping the training data on the device rather than uploading and storing it on a central server. Secondly, it brings model training to the edge, to devices like smartphones, tablets, IoT, and even "organizations" like hospitals that must adhere to tight privacy regulations. It is a significant security benefit to keep personal data local. Thirdly, it allows for real-time prediction because prediction takes place on the device itself. FL eliminates the time lag caused by sending raw data to a central server and then shipping the findings back to the device. Fourthly, because the model stores on the device, the prediction process can continue even if there is no Internet connection.

7 AI-Based Trustworthiness in IoT Systems

From the inclusion of IoT, many of us thought IoT turns human living more comfortable and stress-free. However, some researchers stated that IoT means "Internet of Garbage" because it consists of malware, copyrights, spam, etc. However, it builds with improved communications, better address, strict moderation, and effective community administrations. After collecting the information from the garbage-liked network, finding appropriate value is the most significant task. It is a known fact that IoT is rapidly growing and makes novel demands. From the above discussion, big data analytics, real-time monitoring with streamed data, and another critical discussion are efficient communication capabilities and ensuring security requirements in such a large-scale network. The deployed software applications with appropriate network connections should also be secure.

Clients and operative workers of smart IoT objects will be highly susceptible since their data is accessible on a network. IoT devices and services have three key issues – data confidentiality, privacy, and trust. The IoT object/device must authorize with an entity or person before starting data sharing and access to service. The model of securing IoT systems and their related components is called cybersecurity. Cybersecurity protocols have the most important in dealing with miniature devices, where IoT-based cybersecurity systems mostly avoid attackers stealing sensitive data. There are countless cybersecurity approaches like cryptographic protocols, firewalls, antivirus, intrusion detection systems, and scanners, which secure socket layers. ML, DL, blockchain, and quantum-resistant crypto-techniques profoundly apply to IoT schemes for better security. Also, some issues have happened recently, like small IoT wearable devices collect user's data, which are transformed to device providers since it connects with their respective databases [43–45]. Then these device providers sell the collected user data to other business companies without the user's permission. Business companies make continuous notifications based on the information and advertisements via social networks to that particular user. How to avert these kinds of data ethics in IoT-based schemes is also the most significant challenge apart from security requirements.

8 AI Tools for IoT

Sl. no.	AI tools	Features
1	SensiML Analysis Toolkit [50]	1. Build small algorithms that run on IoT endpoints rather than in the cloud 2. Acquire datasets that are reliable, traceable, and version-managed systems 3. To quickly generate autonomous working computer code, this tool can be used for advanced AutoML code generation 4. Option to select our desired interface and AI knowledge level, keep complete control over our algorithm, and design edge tuning models
2	Vertica Analytics Platform [51]	1. Analysis of communication and network 2. Analysis of embedded systems and customer behavior 3. Analysis of IoT systems and scalable, SQL-compliant time series analysis
3	DewSim [52]	1. SCE resource management – computing capabilities of intelligent devices fluctuate due to owner contact computing capabilities 2. Computing and networking practices use many resources on intelligent devices 3. Data is uploaded/downloaded from/to nodes via WLAN
4	iFogSim [53]	1. Network communication can be done 2. Mobility and edge processing can be simulated 3. Some parameters like energy efficiency, network protocols, and heterogeneity cannot be exhibited
5	IoTSim [53]	1. In IoTSim, IoT devices are modeled, and performance analysis realized 2. But, edge devices, energy efficiency, mobility, communication protocols cannot be modeled
6	IoTSim-Edge [54]	1. It allows researchers to model mobile IoT devices 2. It allows researchers to model a variety of IoT protocols 3. It is in favor of a high-energy-consumption profile 4. It allows for the abstraction of graph modeling

9 Applications of the Internet of Things

From the introduction, Internet of Things (IoT) applications convey incredible value into our daily life. With the new-fangled innovations in wireless networks, intelligent sensors, and revolutionary computing capabilities every day, a new IoT-enabled product proclaims. IoT applications project to train billions of everyday things/objects with connectivity as well as intelligence. This section attempts to overview and discuss numerous domains like intelligent homes, structural health monitoring, environment, logistics, agriculture, health, lifestyle, and industry domains with IoT applications.

9.1 Agriculture

As the world's population increases, the demand for a food source tremendously raised. Developed governments and research institutions are helping agriculturalists to use cutting-edge methods to raise food production. Smart farming is one of the fastest-growing fields in IoT. Here farmers are using expressive visions from the data to yield a healthier return on investment. Smart irrigation determines the quantity of moisture in the soil, releases the water with irrigation pipes for controlling water-usage, and regulate traditional peats with the help of IoT sensors. In greenhouse control, the weather-related information of a greenhouse could monitor and control to harvest the most delicate situations for growing plants. The stored sensible facts from various sensors in a centralized server where they analyze then improve different control strategies.

9.2 Augmented Reality

Augmented reality (AR) enhances how persons require, realize, and display information without disturbing the real world. Mobile augmented reality (MAR) with superimposing virtual elements over fundamental substances on the screen gives added value and enhances the interface with reality. It can increase efficiency and manufacture services by allowing staff to see the most relevant sensor data in the control panel like the view option. US-based DAQRI designed a helmet that can protect workers from falling objects and assist them in avoiding mistakes. Besides, the DAQRI device is also proficient in diagnostics besides sensing risks with thermal vision. Caterpillar, the heavy machinery company, uses AR technology to look at the machine and also instantly see a visual overlay that states when several mechanisms need to replace filter operations and how much fuel is needed. User booklets and technical papers are infamously tedious, so Bosch company incorporated AR to create overlay text, videos, and augmented 3D simulations over a piece of equipment.

9.3 Virtual Reality

With the continuous growth of virtualization, the number of connected devices with the Internet is increasing significantly. Since virtual reality (VR) shows great potential to revolutionize the market, compared with traditional video systems, VR has ultrahigh definition with apparent, dynamic changes that possess significant challenges for realizing such potential. Smart cities are highly involved with virtual reality technologies, and China has previously established VR-based smart cities with virtual and real-world guidelines for emergency department fire monitoring.

Simultaneously, Japan implemented the Tokyo virtual lab, which simulates traffic situations by integrating street and traffic information; moreover, it also assists the vehicle driver in critical situations.

9.4 Mixed Reality

Mixed reality (MR) combines VR and AR services to build physical, virtual objects that exist and intermingle in real time. Recent surveys concluded that companies' investment in MR would reach more than 4.4 billion dollars by the end of 2020. MR-based Microsoft HoloLens and wearable holographic computers are used in the education and training phases. With 3D modeling with MR, professionals can effortlessly shape their projects up in a shared virtual atmosphere. In healthcare, MR has numerous training and education applications like surgeries taught remotely by professionals as they do them in real time.

9.5 Smart Locks

IoT in smart home security has empowered operators to do away with traditional locks and make interest in smart locks. Since smart locks do not require any physical key to open, an alternative operator can open the doors with biometric info like iris scans, fingerprints, and face mappings.

9.6 Smart Factories

Smart factories involve enterprise asset management – IoT-based power-driven asset management increases operational efficiency, optimizes resources, and better controls the sales lifecycle, compliance procedures, and receptive bright atmospheres. WebNMS is an example of an IoT smart factory platform that affords energy managing to improve businesses' energy ingesting.

9.7 Intelligent Road Toll and Traffic Monitoring

With the accumulated data from the implanted sensors, cameras, traffic regulators, and IoT devices, we can effortlessly automate the timings of road traffic lights on busy roads and highways. IoT devices enable collecting road toll when a car enters into its zone and automatically lift the barrier after successful toll collection like fasting.

9.8 Smart Intelligent Grid

In electric power generation, IoT is used resourcefully to monitor the power generation of various power plants. Moreover, IoT-based schemes are effectively applied to observe substations, towers, electricity consumption, and dispatch lines. IoT devices also assist clients with intelligent meters by measuring different parameters and networks. High processing-capable IoT devices can enhance the intelligent grid performance in processing, disaster recovery, reliability, and warnings.

9.9 Intelligent Robotics

The Internet of Robotic Things (IoRT) has numerous applications and can analyze and optimize machine performances in real time from the data facts collected from intelligent sensors. Service and humanoid robots use logistics delivery, rescue, agriculture, security, health and defense, and entertainment. However, the recent pandemic crisis has shown that much more advances are needed in IoRT technologies.

9.10 Waste Management

Most metropolitan cities face waste management issues, one of the most inefficient actions in a city. The techniques used in waste management are not identical; IoT strategies can support municipal waste hoarders in monitoring their trucks' schedules, the volume of waste dumps, and the course's overall proficiency.

9.11 Near-Field Communication (NFC) Payment

Nowadays, every retail payment is made over NFC, since in NFC-based payments, the client/customer can use his/her NFC-enabled intelligent devices to make contactless payments. It reduces the time required to make the payment and increases the security and indemnity of payment.

9.12 AI-Enabled Internet of Underwater Things

IoUT (Internet of Underwater Things) with intelligent sensors and autonomous underwater vehicles are relevant to detect underwater treasure and enemy submarines. IoUT also assists in the detection of minerals, corals, reefs, and metals. In general, finding underwater resources requires sensors with video capturing devices that are fulfilled by IoT schemes.

9.13 Intelligent UAV

Because of the UAVs' high-range agilities and autonomy, they can offer a wide range of amenities to IoTs. UAV-based IoT schemes efficiently apply for crowded surveillances with face recognition and mobile edge computing with inadequate energy power and dimensions. Moreover, in auto spacing, UAVs with theory-based game platforms are highly applied for locating terrestrial stations. UAVs with AR/VR/MR technologies allow isolated operators to navigate in explicit scenes of interest. UAVs are also helpful in optimal clustering of IoT devices and reduces transmission power.

9.14 IoT-Based Forensic Applications

There are countless models implanted for IoT security and privacy with available resources; however, it is still an open research issue. Nowadays, a little focus shifted toward digital forensics in IoT. Since IoT security is still developing, there are high probabilities of breaches in IoT. Active digital forensics procedures must be established in equivalent with security explanations to track attacks and find reliable digital evidence to expose perpetrators. Inspecting the VitalPatch will disclose associated forensics objects of individuals like ECG trends, heart rate, activity monitoring, port scans, timeline logs, etc.

9.15 Intelligent Healthcare Systems Using IoT Systems

Wearable IoT devices allow constant monitoring of physiological constraints, which assist in ongoing health than fitness monitoring. Moodable is a mood-enhancing device to monitor and improve our mood in a day. In detail, moodable is a head-mounted wearable that sends low-intensity current to the brain, elevating our mood. Ingestible sensors – miniature-sized sensors – monitor the medicine inside our body and advise us if it notices any anomalies, which helps diagnostic patients with early warnings. Moreover, it is applicable in reducing emergency room wait time, enhancing drug management, tracking patients, and ensuring critical hardware staff availability.

9.16 Intelligent Disaster Management

Intelligent disaster management helps in minimizing potential damage from upcoming disasters. Besides, it confirms instant and suitable recommendations to the victims for fast regaining. Nowadays, the IoT skill has reached its advanced level and has probability to be very beneficial in disaster conditions. IoT system with satellite communication and geographic information arrangements helps in risk minimization. It suggests prevention, makes early warning, and utilizes social media to avoid awareness creation, relief and response measures, and missing person search.

9.17 Music

Up to date, most of the IoT applications are designed for environmental monitoring, industrial manufacturing, energy optimization, intelligent home automation, intelligent healthcare, and transportation. But in present day, it is gradually valuable for the music technology industry also. Google's Universal Orchestra and MIT's patchwork are some notable examples of IoT-based music innovations. SoundWire and JackTrip are some of the remote performances designed by well-known multinational companies, enabling instrumentalists in different locations to accomplish as if they were in a similar room. In rhythmic vibration, actuators collect data and then start to tremble with a rhythm and then intensity relative to that of the music playing, and this aids wearer to sense the rhythm of the music. With auto-tune instruments, instrumentalists can play tools over the allied device. The device is implanted with sensors that sense the traces on the screen of the devices. It then auto-initiates the similar movement on essential musical tools, letting instrumentalists play tools remotely over the device.

Different applications have been described in Fig. 7.

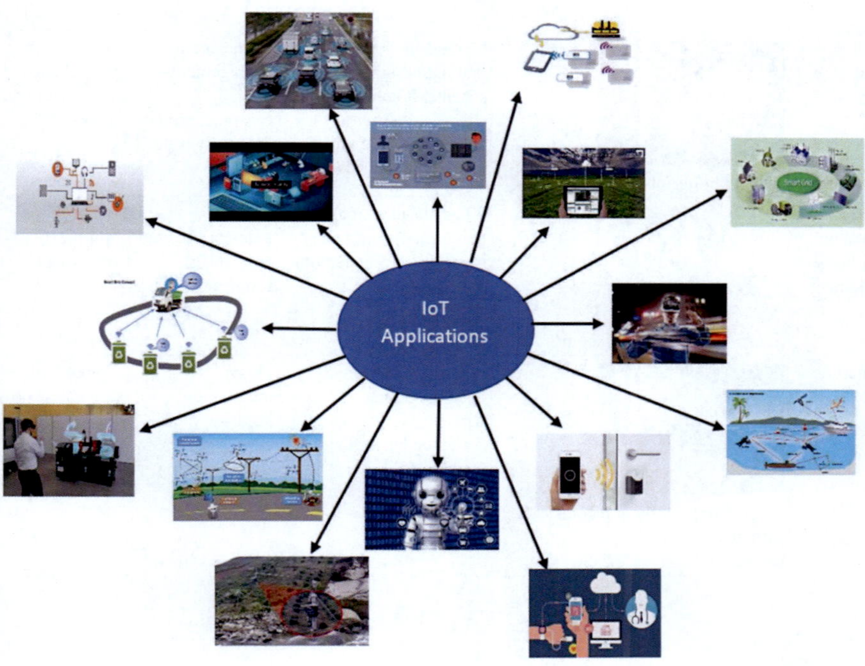

Fig. 7 Internet of Things application domain

10 Consumer Electronic Products for IoT

Product name	Product picture	Features	Applications	Limitations
Bit Defender Box	 *Source*: http://iotlineup.com/	Provides high-security solutions for scan and efficiently blocks incoming threats. It protects all our IoT devices, even when we go out! Acts as an intelligent wireless router	Smart home safety, smart automation network safety	Battery optimization Bugs, overreliance on technology
SmartMat intelligent yoga mat	 Source: http://iotlineup.com/category/iot_health_and_fitness	SmartMat notices when we are out of position and then gives us immediate advice on precisely our position	Health and fitness, exercise equipment, smart healthcare	Compatibility and complexity, cost, security
Lockitron Bolt – smart lock	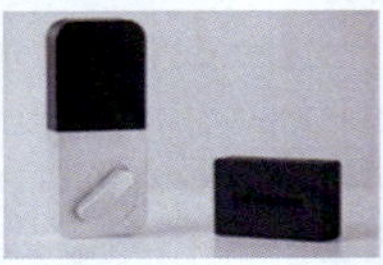 *Source*: http://iotlineup.com/category/iot_consumer_security_cameras	Provides home security with intelligent lock options when far away from our home environment	Home security and safety, home smart healthcare, consumer smart locks	Cost, integration, lack of connectivity standards
Philips Hue Hue Go	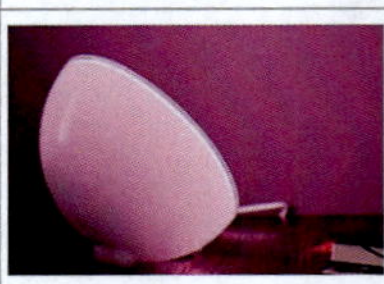 *Source*: http://iotlineup.com/	It comes under an intelligent lighting scheme. It continuously changes the way we experience light with intelligent controls	Home energy management, home computerization, indoor lighting	Data breach, overreliance on technology, security
Airfy iBeacon for home automation	 *Source*: http://iotlineup.com/	It permits us to make our home smart using one or more Wi-Fi routers with optimal WLAN connections	Home automation, home appliances, robotics	Battery optimization Bugs, cost
Smart door locks	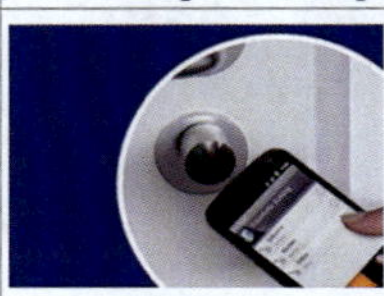 *Source*: http://iotlineup.com/category/iot_smart_locks	It allows a console to open the door, wireless-based custom access codes for explicit members	Home automation, security, smart city	Compatibility and longevity

Product name	Product picture	Features	Applications	Limitations
Smart Bluetooth Trackers	 *Source*: http://iotlineup.com/category/iot_smart_locks	Smart Bluetooth-based devices, with the help of short-range indications, digitally tie necessary items to our smartphones. We will get an immediate alert (chirp, beep, bleat, or make noise) if you start to leave somewhat behind, and a moveable app will direct you back to the unstable object	Health and fitness, smart city, consumer smart locks	Lack of standards for authentication, data breach, connectivity issues
Smart bike tracker	 *Source:* http://iotlineup.com/category/iot_smart_locks	These devices track a bike's place and then send alarms if the bike leaves a nominated zone. In addition, smart locks also allow multiple riders to share a single bike	Home security and safety, home computerization, consumer smart locks	Need efficient results on ride analytics and crash alerts
Amazon Dash Button	 *Source*: http://iotlineup.com/IoT_home_applinaces	It is one of the finest inventions of IoT, which makes life simple and relaxed. It helps us make our orders quickly and correctly without missing and reorder from a high brand	Home security and safety, health and fitness	Overreliance on technology, security, cost
Ring Alarm Smoke and CO Listener	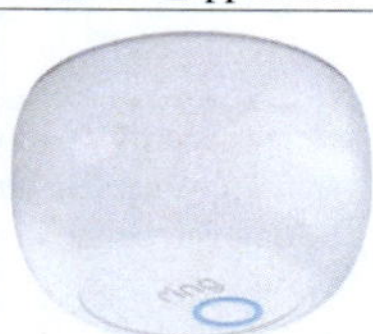 *Source*: http://iotlineup.com/	It is a device that allows manufacturing companies to accomplish their carbon monoxide indicators with more comfort. It offers a warning when our smoke indicator alarms	Home security and safety, consumer appliances, home automation, consumer smart locks	Overreliance on technology, connectivity
WeMo Insight Smart Plug	 *Source*: http://iotlineup.com/IoT_home_applinaces	It is an IoT-based invention that aids us in regulating our lights and applications by revolving them on or off. It also creates guidelines, timetables, and energy consumed by our devices and helps us protect our home through providing the required information	Home security and safety, home automation	Data breach, compatibility, and complexity, cost, security

Product name	Product picture	Features	Applications	Limitations
YI Security Home Camera Baby Monitor	 *Source*: http://iotlineup.com/category/iot_home_security_and_safety	This device allows us to access our camera with a PC and alerts based on gesture or sensitivity. Moreover, increases the night vision and provides the storage option on the cloud	Home security and safety, home automation	Battery optimization, cost, connectivity
Foobot Air Quality Monitor	 *Source*: http://iotlineup.com/IoT_home_applinaces	It is an IoT-based air quality monitor device designed primarily for pollution sensitivity and calculates the humidity and temperatures It continuously monitors made changes to improve the air quality and is easily implanted into homes or workplaces	Home security and safety, home automation	Overreliance on technology, unstructured data
Google Home Voice Controller	 *Source*: http://iotlineup.com/IoT_home_applinaces	It is a voice-based intelligent IoT device that allows us to control alarms, media (TV and speaker) volumes, and light	Home security and safety, garden equipment, home automation	Cost, complexity
Kuri Mobile Robot	 *Source*: http://iotlineup.com/IoT_home_applinaces	Kuri is known as the first home robot explicitly programmed for entertaining purposes. In addition, it intermingles with specific abusers to capture daily movements	Home security and safety, garden equipment, home automation	Overreliance on technology

Product name	Product picture	Features	Applications	Limitations
Logitech Harmony Universal Remote	 *Source*: http://iotlineup.com/IoT_home_applinaces	This IoT-based smart device allows us to remotely control every media device, lighting device, and other intelligent devices. It contains nearly eight innovative remotes capabilities, which efficiently reduce complexities inside the house	Home automation, consumer smart locks	Overreliance on technology, cost
Particle Photon Wi-Fi with Headers	 *Source*: http://iotlineup.com/IoT_home_safety	It provides resources for an abuser to design accessible and build connection projects like easy plugins	Home security and safety, home automation	Overreliance on technology, cost, complexity
Logitech Pop – smart button controller	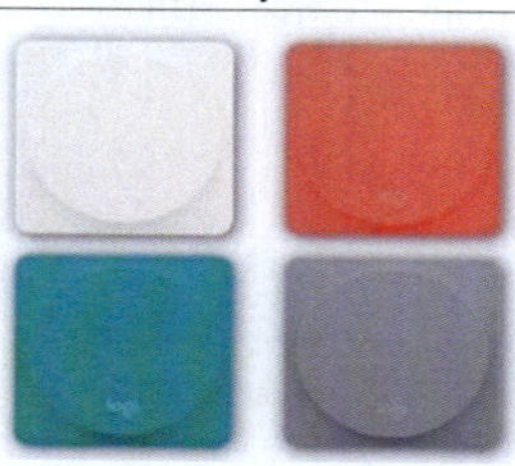 *Source*: http://iotlineup.com/IoT_home_applinaces	POP allows users to control bright lighting, music, and speakers inside the house with push buttons	Home automation, home appliances, home energy management, remote controls, indoor light	Dara breach, security, and privacy

Product name	Product picture	Features	Applications	Limitations
Ninja Sphere	 *Source*: http://iotlineup.com/IoT_home_applinaces	It is a miniature-type IoT device that produces intelligent home connections without any wired arrangements	Home automation, home appliances Other smart household appliances	Overreliance on technology, security
Hydrawise	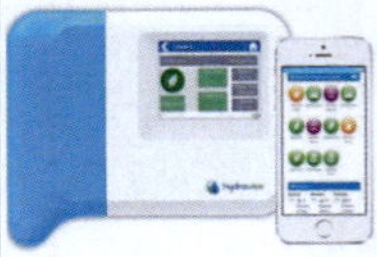 *Source*: http://iotlineup.com/	It allows users to arrange schedulers for irrigation control and water saving with a connected mobile device	Home appliances, home automation Garden equipment	Connectivity, security, and privacy
Singlecue Gesture Controller	 *Source*: http://iotlineup.com/	It is used to manage TV, media, and smart home devices via touch-free motions	Home automation, home appliances Remote controls	Overreliance on technology
Skydrop	 *Source*: http://iotlineup.com/IoT_home_applinaces	Skydrop is a fascinating, intelligent IoT device that automatically adjusts to local weather and controls sprinkler watering with perfect lawn season	Home automation, home appliances Garden equipment	Overreliance on technology, data breach
Mr. Coffee smart coffee maker	 *Source*: http://iotlineup.com/IoT_home_applinaces	It is an intelligent kitchen applicant who takes control from anywhere, makes schedules, monitors, and modifies the coffee-making procedure	Home appliances, home automation Kitchen appliances	Overreliance on technology, connectivity, complexity
Edyn Garden sensor	 *Source*: http://iotlineup.com/IoT_home_applinaces	Edyn virtually connects peoples with the garden with intelligent sensors, and it tracks humidity, temperature, moisture, and lighting and then sends notices to our mobile	Home automation, home appliances Indoor plants, garden equipment	Connectivity, cost
Prodigio – connected espresso machine	 *Source*: http://iotlineup.com/IoT_kitchen_appliances	It takes guidelines from the mobile phone to brew rapidly, list brews, receive preservation warnings, and then track the capsule stock	Home appliances, home automation Kitchen appliances	Cost, complex connections

Product name	Product picture	Features	Applications	Limitations
Anova – precision cooker	 *Source*: http://iotlineup.com/IoT_home_applinaces	It cooks our food based on a schedule timer and stops; no overcooking entertained	Home appliances, home automation Kitchen appliances	Overreliance on technology, cost
Withings – blood pressure monitor	 *Source*: http://iotlineup.com/	It is an IoT-based wireless blood and heart rate checker anytime from anywhere	Health and fitness Healthcare	Data breach, security, and privacy, cost
Mimo – smart baby monitoring	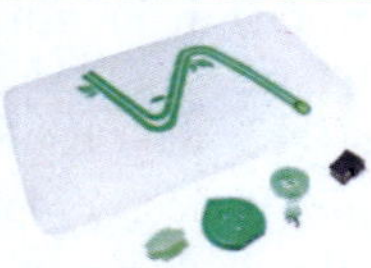 *Source*: http://iotlineup.com/	Mimo is an IoT-based baby monitor that uses a wise washable crib sheet, and parents can follow baby activity and movements from their connected, intelligent mobiles and tabs	Health and fitness Baby monitors	Cost

11 Open Research Challenges for AI-Based IoT Systems

Challenge 1: How Computing Power Is Handled in AI-Based Industry 5.0?
Technical as well as industrial companies are suffering from computing power challenges. IoT devices collect a vast volume of data to build AI models to examine these massive data with ML and DL; there is a need for consistent power consumption. It is a big trouble for product manufacturing and start-up companies. In most cases, the quantity of power need for a learning algorithm makes the developer away. However, ML and DL are outstanding AI components with high precision but efficient but require increasing cores than GPUs. The researcher has implemented numerous ideas and schemes to progress ML and DL models in a wide variety of appliances. Besides, cloud computing with AI and parallel system processing with AI is also used for efficient power ingesting.

Challenge 2: How AI Overcomes the Technical Challenges in IoT?
Up-to-date many research-oriented models are planned for IoT. However, it still has many practical tasks like a heterogeneous network – IoT consists of numerous connections to contact/communicate with different networks. Presently, there are proficient, recognized platforms to hide the network arrays, but they suffer from complexity and power consumption. In SOA (Service Oriented Architecture), IoT faces enormous challenges from both performances and then cost restrictions.

Depending on the applications, the number of connected devices increases, and scalability, data pre-processing, service provisioning, and networking are complex. In particular, the transmission of sensed data across heterogeneous networks also causes recurrent delays and communication problems. There is a need for high automated standards that easily allow data collected from different devices and transform proficiently inside IoT systems. More importantly, connected device collaboration among different entities needs proper addressing; identification and optimization are still an open research issue. An IoT is slightly affected by newly implanted objects; hence it needs suitable integrated mechanisms like unified data structure. AI-based approaches employ meaningful IoT data features, but they suffered from complexity and power ingestings.

Challenge 3: How Will Industry 5.0 Handle Dissimilar Implementation Strategies?
Based on the recent progression in various network technologies, AI can transform any industry and business field to digitalization. However, one main issue of AI is the lack of ideas for implementation. A strategic approach is needed to succeed in a business environment like detecting progressive areas, objects with benefits, continuous failures, etc. To get the knowledge of over-mentioned issues, company managers, supervisors, and technical teams must have a broad knowledge of AI skills, advances, and limitations and keep an eye on present issues faced by AI. If companies follow these updates of AI-related information working styles, organizations effortlessly know the zones that AI can enhance.

Challenge 4: What Are the Challenges IoT Data Analytics Faced in Implementation, and How They Overcome?
In the above section, we discussed IoT analytics with ML and DL approaches. In some cases, the implementation of IoT analytics faces difficulties while handling time series-based data structures. IoT-based intelligent sensor nodes continuously gather massive static data samples for a long time, making it tough to find reasons for extrapolative analysis. Moreover, sighting proper storage space and then rapidly analyzing the stored data is difficult. It is a known fact that data is sensitive, with proper and exact approaches, and can provide helpful information of any company's product. Hence, data scientists are required to be very expert in data analysis, and DBA-oriented skills.

Challenge 5: How AI Solves the System Complexity and Security Challenges of Wireless Networks and IoT?
In any network, complexity plays a huge role, with the implementation of AI in communication architectures driving more upsurges of the intricacy of schemes. In industrial network case-specific ML and DL approaches employed to achieve the solitary goal ignore the remained objectivities. Moreover, with adequate resources, IoT devices transmit the data to higher levels without computing pre-processing operations. So, the implanted AI approaches in WSNs and IoTs must optimize solitary objectives while superintending other limits like storage, link, processing, and

latency. The employed AI approach in one layer could also benefit or help in optimization in another layer.

The main intention of utilizing AI for IoT is to examine network flow, finding security breaches like intrusion and anomaly detection before decision-making. AI helps to generate high-quality datasets which contain information of attack categories and then outlines. So, AI approaches based on first-rate datasets to secure IoT are highly infeasible. It should be renowned that obtaining a dataset for IoT safety training is more complicated, tough other domains.

Data Privacy concerns the proper handling of sensitive data, which is expected to be exposed in the era of AI-capable systems. There are numerous ways for information leakage; in some cases, AI itself leaks data samples while performing different tasks at a time. So, maintaining privacy in AI is a risky task from both algorithmic and human perspectives. Encoding, decoding, blockchain, quantum models, and shuffling algorithms will undoubtedly handle privacy issues in IoT and WSNs.

Challenge 6: How Will AI Produce High-Speed Intelligent Communication in UIoT?

UIoT is a complex heterogeneous network that includes dynamic AUVs, underwater magnetic inductions, and acoustic networks. The volatile atmosphere of the ocean is an unforeseen issue for the UIoT system: the topology and localization accuracy underwater affected by the actions of tides, wind power, and temperature. By employing AI-based approaches in UIoT, every network element of the UIoT is optimized and coordinated for communication and then makes the best use of deployed network.

Challenge 7: How AI-Based Industry 5.0 Changes in Product Management?

Radiofrequency identification (RFID) tags employ many industrial sectors for supply chain, product tracking, and delivery management. RFID tags are implanted on delivery products, and readers placed on the entire way to monitor. IoT pieces of machinery can afford improved flexibility in reader's locations while permitting continuous interoperability among RFID-based appliances used by dissimilar performers. These IoT appliances are primarily applied to retail companies to display product availability and then accurate stock records. More importantly, an amalgamation of sensors and biosensor with RFID technology may permit control production methods and final product value in the food industry.

Challenge 8: Man Power

No matter how big a company it is, it matters how well a particular company employs AI in its zones where development is needed. AI is developing technology, so there is a need for specialists capable of handling and implementing the AI models. Hence, the companies make an additional budget for the training and hiring of a specialist in AI.

Challenge 9: Training AI

If the installed AI system collects sufficient information, it is ready for the training phase. It is dissimilar for every AI-based approach like model type, data structure, outcomes, and decision-making. In AI, there is no perfect model that is suitable in all cases, and we must try every model; based on the outcome result, we determine which works better among all. Hence, if we push machines to learn more, we must first analyze more from the natural environment by implanting intelligent sensors with new features. Also, make reruns on proposed models repeatedly with every new choice, so AI models will learn automatically to improve performance.

Challenge 10: Connected Devices
Almost every field in the world transformed into digitalization with connected devices. However, simple IoT devices like intelligent sensors communicate with other sources with Bluetooth/ZigBee, and it still stands as a challenge to connect them with the Internet. Since connecting every device to the Internet is not an easy task, which needs reinstallation, replacement with new equipment, and advanced hardware machines. AI-based MQTT is a simple example for connecting devices with the Internet, not a simple procedure.

Challenge 11: Efficient Sensing
The urban environment consists of numerous factor combinations; for efficient monitoring of those factors, there is a need for heterogeneous multiple sensing models. In detail, they need a generalized framework for sensed data collection and display them in spatial and temporal characteristics from fixed and mobile sensing infrastructure and constant, arbitrary sampling. For instance, urban noise and air pollution zone detection need constant noise and air quality data samples from fixed intelligent sensor nodes. Compressive wireless sensing (CWS) exploits synchronic messages to diminish the broadcast power of each and convey noisily. Quality of air measurement forecasted data samples to a central node for data aggregation.

12 Conclusions

This chapter introduced a journey that started with understanding the vision pattern of AI-enabled IoT skills and then how it can be helpful in many parts. It also supports researchers and experts that recognize the design construction and AI algorithms through IoT and state-of-the-art IoT countermeasures. It offers a complete discussion on a functional framework, then knowledge hierarchy for IoT, object identification, intelligent sensors, learning, and analytics in intelligent IoT-enabled systems. This chapter explores AI-enabled IoT paradigms that will be utilized to better humankind in the future era. Specifically, the far-reaching references of numerous works and then implementations will be observed to be essential accumulations for engineers and administrations.

References

1. Lee, S. K., Bae, M., & Kim, H. (2017). Future of IoT networks: A survey. *Applied Sciences, 2017*(7), 1072. https://doi.org/10.3390/app7101072
2. Garcia, C. G., Nunez-Valdez, E. R., Garcia-Diaz, V., PelayoGBustelo, B. C., & Lovelle, J. M. C. (2018). A review of artificial intelligence in the internet of things. *International Journal of Interactive Multimedia and Artificial Intelligence, 5*(4).
3. Al-Fuqaha, A., Guizani, M., Mohammadi, M., Aledhari, M., & Ayyash, M. (2015). Internet of things: A survey on enabling technologies, protocols, and applications. *Ieee Communication Surveys & Tutorials, 17*(4).
4. Lin, J., Yu, W., Zhang, N., Yang, X., Zhang, H., & Zhao, W. (2017). A survey on internet of things: Architecture, enabling technologies, security and privacy, and applications. *Ieee Internet of Things Journal, 4*(5).
5. Samie, F., Bauer, L., & Henkel, J. (2019). From cloud down to things: An overview of machine learning in internet of things. *Ieee Internet of Things Journal, 6*(3, June).
6. Altexsoft. (2020). *IoT architecture: The pathway from physical signals to business decisions*. https://www.altexsoft.com/blog/iot-architecture-layers-components/
7. Chander, G. B., & Kumaravelan, G. (2018). Introduction to wireless sensor networks. *Soft Computing in Wireless Sensor Networks, 1*.
8. Chander, B., & Kumaravelan, G. (2020). Internet of things: Foundation. In *Principles of Internet of Things (IoT) ecosystem: Insight paradigm* (pp. 3–33). Springer.
9. Gopalakrishnan, K. (2020). Security vulnerabilities and issues of traditional wireless sensor networks in IoT. In *Principles of Internet of Things (IoT) ecosystem: Insight paradigm* (pp. 519–549). Springer.
10. Lin, Y. C. (2017). Development of Advanced Manufacturing Cloud of Things (AMCoT) - A intelligence manufacturing platform. *IEEE Robotics and Automation Letters, 2*(1), 1809–1816.
11. Chen, C. C. (2018). A novel automated construction scheme for efficiently developing cloud manufacturing services. *IEEE Robotics & Automation Letters, 3*(3), 1378–1385. https://doi.org/10.1109/LRA.2018.2799420
12. Gupta, H., Dastjerdi, A. V., Ghosh, S. K., & Buyya, R. (2015). iFogSim: A toolkit for modeling and simulation of resource management techniques in the Internet of Things, Edge and Fog computing environments. *Journal Software: Practice and Experience, 47*(9), 1275–1296.
13. Zhou, B., & Buyya, R. (2018). Augmentation techni9u es for mobile cloud computing: A taxonomy, survey, and future directions. *ACM Computing Surveys (CSUR)., 51*(1), 1–38.
14. Ranjan, R., Rana, O., Nepal, S., Yousif, M., James, P., & Wen, Z. (2018). *The next grand challenges: Integrating the Internet of Things and data science*. IEEE Cloud Computing.
15. Gubbi, J., Buyya, R., Marusic, S., & Palaniswami, M. (2013). Internet of Things (IoT): A vision, architectural elements, and future directions. *Future Generation Computer Systems, 29*(7), 1645–1660.
16. Ungureanu, A. V. (2020). The Transition from Industry 4.0 to Industry 5.0. The 4Cs of the Global Economic Change. In C. Nastase (Ed.), *Lumen proceedings: Vol. 13* (pp. 70–81). 16th Economic International Conference NCOE 4.0 2020.
17. Adi, E., Anwar, A., Baig, Z., & Zeadally, S. (2020). Machine learning and data analytics for the IoT. *Neural Computing and Applications, 32*, 16205–16233.
18. Sharma, I., Garg, I., & Kiran, D. (2020). Industry 5.0 and smart cities: A futuristic approach. *European Journal of Molecular & Clinical Medicine, 07*(08), 2515–8260.
19. Aslam, F., Wang, A., Li, M., & Rehman, K. U. (2020). Innovation in the era of IoT and industry 5.0: Absolute Innovation Management (AIM) Framework. *Information, 11*, 124. https://doi.org/10.3390/info11020124
20. Nahavandi, S. (2020). Industry 5.0—A human-centric solution. *Sustainability 2019, 11*, 4371. https://doi.org/10.3390/su11164371

21. Qiu, T., Zhao, Z., Zhang, T., Chen, C., & Chen, C. L. P. (2019). Underwater internet of things in smart ocean: System architecture and open issues. *IEEE Transactions On Industrial Informatics*, 1551–3203. (c) IEEE.
22. Özdemir, V., & Hekim, N. (2018). Birth of industry 5.0: Making sense of big data with artificial intelligence, 'The Internet of Things' and next-generation technology policy. *OMICS A Journal of Integrative Biology, 22*(1) Mary Ann Liebert, Inc. https://doi.org/10.1089/omi.2017.0194
23. Chander, B., & Kumaravelan, G. (2021). Cyber security with AI—Part I. In *The "Essence" of network security: An end-to-end panorama* (pp. 147–171). Springer.
24. Skobelev, P. O., & Borovik, S. Y. (2017). On the way from Industry 4.0 to Industry 5.0: From digital manufacturing to digital society. *International Science Journal, 2*(6), 307e311.
25. Pflanzner, T., & Kertesz, A. (2018). A taxonomy and survey of IoT cloud applications. *EAI Endorsed Transactions on Internet of Things, 3*(12).
26. Lee, I., & Lee, K. (2015). The Internet of Things (IoT): Applications, investments, and challenges for enterprises. *Elsevier, Business Horizons, 58*, 431–440.
27. Uviase, O., & Kotonya, G. (2018). IoT architectural framework: Connection and integration framework for IoT systems. In D. Pianini & G. Salvaneschi (Eds.), *First workshop on Architectures, Languages, and Paradigms for IoT EPTCS 264*, (pp. 1–17). https://doi.org/10.4204/EPTCS.264.1.
28. Tiwary, A., Mahato, M., & Chidar, A. (2018). Internet of Things (IoT): Research, Architectures and Applications. *International Journal on Future Revolution in Computer Science & Communication Engineering, 4*(3), 23–27, ISSN: 2454-4248.
29. Patel, K. K., & Patel, S. M. (2016). Internet of things-IOT: Definition, characteristics, architecture, enabling technologies, application & future challenges. *International Journal of Engineering Science and Computing, 6*(5).
30. Kibria, M. G., Nguyen, K., Villardi, G. P., Zhao, O., Ishizu, K., & Kojima, F. (2018). Big data analytics, machine learning and artificial intelligence in next-generation wireless networks. *IEEE Access, 6*, 32328–32338. https://doi.org/10.1109/ACCESS.2018.2837692
31. Mohammadi, M., AI-Fuqaha, A., & SamehSorour, M. G. (2018). Deep learning for IoT big data and streaming analytics: A survey. *IEEE Communications Surveys & Tutorials, 20*(4), 2923–2960.
32. Meruje, M., GwaniSamaila, M., Virginia, N. L., Franqueira, M. M. F., & MoraisInâcio, P. R. (2018). A tutorial introduction to IoT design and prototyping with examples, Internet of Things A to Z: Technologies and applications. In Q. F. Hassan (Ed.), *The Institute of Electrical and Electronics Engineers, Inc* (1st ed.). Wiley & Sons, Inc.
33. Siemens, G. (2013). Learning analytics: The emergence of a discipline. *American Behavioral Scientist, 57*(10), 1380–1400. c 2013 PAGE Publications.
34. Davenport, T. H. (2018). From analytics to artificial Intelligence. *Journal of Business Analytics, 1*, 2,73–2,80. https://doi.org/10.1080/2573234X.2018.1543535
35. Ghosh, A., Chakraborty, D., & Law, A. (2018). Artificial intelligence in Internet of Things. *ET Research Journals*, 1–11.
36. Lv, Z., Han, Y., Singh, A. K., Manogaran, G., & Haibin. (2020). Trustworthiness in industrial IoT systems based on artificial intelligence. *IEEE*, 1551–3203.
37. Davenport, T. H. (2018a). *The AI advantage*. MIT Press.
38. Davenport, T. H., & Harris, J. G. (2017). *Competing on analytics*. Harvard Business Review Press (revised and updated).
39. Davenport, T. H., & Kirby, J. (2016). *Only humans need to apply: Winners and losers in the age of smart machines*. Harper Business.
40. Davenport, T. H., & Mahidhar, V. (2018). *What's your cognitive strategy?* MIT Sloan Management Review, (Summer). Retrieved from https://sloanreview.mit.edu/article/whats-your-cognitive-strategy/
41. Jha, S., & Seshia, S. A. (2017). A Theory of Formal Synthesis via Inductive Learning. *Acta Informatica, 54*(7), 693–726.

42. Hassan, Q. F., Khan, A. R., & Madani, S. A. (2017). *Internet of things: Challenges, advances, and applications*. CRC Press.
43. Fortino, G., & Trunfio, P. (2014). *Internet of things based on smart objects: Technology, middleware, and applications*. Springer.
44. Yang, L. T., Di Martino, B., & Zhang, Q. (2017). Internet of everything. *Mobile Information Systems, 8*, 201.
45. Chander, B. (2020). Clustering and Bayesian networks. In *Handbook of research on big data clustering and machine learning* (pp. 50–73). IGI Global.
46. Yang, Q., Yang, L., Chen, T., & Tong, Y. (2019). Federated Machine Learning: Concept and Applications. *ACM Transactions on Intelligent Systems and Technology, 10*: 2, Article 12 (February 2019), 19 pages. https://doi.org/10.1145/3298981
47. Li, T., Sahu, A. K., Talwalkar, A., & Smith, V. (2020). Federated learning: Challenges, methods, and future directions. *in IEEE Signal Processing Magazine, 37*(3), 50–60. https://doi.org/10.1109/MSP.2020.2975749
48. Wu, Q., He, K., & Chen, X. (2020). Personalized federated learning for intelligent IoT applications: A cloud-edge based framework. *IEEE Open Journal of the Computer Society, 1*, 35–44. https://doi.org/10.1109/OJCS.2020.2993259
49. Pang, J., Huang, Y., Xie, Z., Han, Q., & Cai, Z. (2021). Realizing the heterogeneity: A self-organized federated learning framework for IoT. *IEEE Internet of Things Journal, 8*(5), 3088–3098. https://doi.org/10.1109/JIOT.2020.3007662
50. https://sensiml.com/
51. https://www.vertica.com/
52. Hirsch, M., Mateos, C., Rodriguez, J. M., & Zunino, A. (2020). DewSim: A trace-driven toolkit for simulating mobile device clusters in Dew computing environments. *Softw: Pract Exper., 50*, 688–718. https://doi.org/10.1002/spe.2696
53. Zeng, X., Garg, S. K., Strazdins, P., Jayaraman, P. P., Georgakopoulos, D., & Ranjan, R. (2017). Iotsim: A simulator for analysing iot applications. *Journal of Systems Architecture, 72*, 93–107.
54. Jha, D. N., Alwasel, K., Alshoshan, A., Huang, X., Naha, R. K., Battula, S. K., Garg, S., Puthal, D., James, P., Zomaya, A. Y., Dustdar, S., & Ranjan, R. (2019). *IoTSim-Edge: A simulation framework for modeling the behaviour of IoT and edge computing environments*. Software: Practice and Experience.

IoT Ecosystem: Functioning Framework, Hierarchy of Knowledge, and Intelligence

Mobasshir Mahbub

1 Introduction

Via technical advancement, people around the world want to interact more through the Internet. IoT is now one of the most impressive applications as technology develops. This technology efficiently connects trillions of devices like sensors, actuators, gateways, and controllers and provides countless applications for every domain. IoT is a technology that connects the real world via the Internet to the virtual world. In all trades around the globe, IoT has a wide variety spanning from electronics, pharmacy, banking, fuel, electricity, and agriculture. With the progress of IoT, several new problems such as broad accessibility, availability, protection, and scalability emerge as well. The interconnection of items now seems simple to achieve this purpose, which causes challenges a few years ago. Development and obstacles are becoming increasingly relevant as also. Every day, IoT introduces a new feature to its description. It takes time for IoT to stabilize its boundary. IoT paradigms may be commonly categorized as business, manufacturing, and network applications. Any market technologies include household-device control, cars, smart wearables, integrated medical treatment, intelligent buildings, etc. Smart manufacturing and distribution systems, smart market and supply chain, factory automation, etc. are the major industrial technologies. Smart community, ecosystem security, smart climate, smart grid, etc. are the infrastructure-based technologies [1].

IoT environment or ecosystem is such a network that efficiently integrates all IoT equipment. It includes system integration, apps, controlling features, gateway, middleware, and server. All those components are connected to the IPv6 (6LoWPAN, NFC), Bluetooth low-energy (BLE), etc. via the protocol and interfaces of communications, including ZigBee, low-energy Wi-Fi, and MQTT. A large number of

M. Mahbub (✉)
Department of Electrical and Electronic Engineering, Ahsanullah University of Science and Technology, Dhaka, Bangladesh

S. Pal et al. (eds.), *Artificial Intelligence-based Internet of Things Systems*,
Internet of Things, https://doi.org/10.1007/978-3-030-87059-1_2

physical instruments are connected to a single system through the IoT ecosystem. These interconnected devices are rapidly growing. According to numerous ICT estimates, by 2020 the number is projected to increase to 50–100 billion. Various connectivity protocols connect the sensors and actuators with the gateway. The sensors also known as the preceptors collect all the data as environmental parameters and provide the gateway with all the detail. The actuator works on the environment and is directed by the gateway. Hundreds or even thousands of sensors and actuators can be managed through the gateway. The data flow between devices is often controlled. The controller manages several gateways and handles rich data processing such as classification, data measurement, and data conversion. The device then contacts middleware. The middling program carries out the task as to transfer all of your data to the storage server, evaluating your data, producing graphs and notes, maintaining protection and safety, monitoring and handling the whole network, and assisting the processing of all data at the middleware cloud. All implementations require middleware resources and research information, such as smart cities and intelligent houses. The application API requires such facilities to be used. The program provides the customer with a complete IoT view [2].

This work aims to define IoT as an ecosystem by describing the diverse work on IoT and all elements such as apps, OS, middleware, IoT communication, and network gateways. This analysis merged efforts in various sectors and provided an overview of relations between all sectors. This work, after acknowledging advances in IoT, discusses the elements that render IoT infrastructure and enable it to be effective. Several works regarding IoT were previously done. The work includes a comprehensive description of all elements that are integrated to create the IoT environment and the knowledge hierarchy of IoT which includes data generation, purification, processing, storing, and analytics, relative to other IoT-related articles. This will assist new analysts and enthusiasts to obtain a simple understanding of how the various modules work altogether with features without diving through RFCs and standards. It offers a description of different innovations, integrated as an IoT trend, thereby revolutionizing the environment of real-life implementations such as smart cities and towns, e-healthcare, etc.

2 Related Works and Motivation

IoT was examined from different perspectives in this study. IoT's developers demonstrate how layers can be decoupled and the architecture can be tailored to structures, networks, protocols, and technologies. The design also allows IoT structures to calculate and work more modularly with abstraction. A comprehensive overview of various IoT equipment, operating systems, networks, and networking interfaces is performed which allows the best option to suit the need. The significant unanswered problems in the IoT have been answered after all the latest researches and experiments have been reviewed. The priorities of privacy concerns have been explained for IoT applications. For real-time data utilization, maintaining efficiency, quality, and usability, there has been a discussion of the value and need of data analytics, the fog-edge-cloud computing in IoT. At last, the research shows how certain components are selected and how they all work together to provide an efficient IoT ecosystem.

The work of this paper varies from other works in terms of applications, gateways, software, middleware, networks, and networking technologies because it gives an understanding of IoT as a full ecosystem. The IoT framework is generated together by all these components. A recent study [3] highlights IoT developments and diverse IoT implementations. The article [4] addresses various conception dimensions and problems of the operating system. There is an IoT development promoting survey that also discusses IoT testing questions [5]. The paper [6] introduces an IoT design, protocols, and software literature review. Documents address the report on information technologies and protocols [7, 8]. Recent studies describe a detailed analysis of IoT middleware and application components [9]. However, no study or review or survey illustrated a detailed overview of the IoT ecosystem including almost all technical components of IoT with knowledge hierarchy. The work performed on this analysis is described below in comparison to current literature on the IoT study:

- This work first discussed the layered IoT architecture for a better understanding of the IoT ecosystem.
- This work also reveals the applications, middleware, gateways, and different communication solutions accessible at different layers of an IoT ecosystem.
- The work aims to research the role of privacy mechanisms in IoT and explains the relation between IoT and emerging technology, such as data analytics (depending on the machine learning as a term of IoT knowledge hierarchy), cloud storage, fog, and edge computing.

The chapter is elaborated as follows: Section 3 introduces layered IoT architecture. In Sect. 4, the document describes the taxonomy of IoT. In Sect. 5, the core concepts for the IoT ecosystem are discussed in depth in which IoT devices, gateways, computation, and operating systems, communication, middleware, data storage, and IoT platforms are clarified. Section 6 simply explained the knowledge hierarchy of an IoT ecosystem. In Sect. 7, the paradigms of intelligent IoT ecosystem are illustrated including fog-edge-cloud-centric IoT for intelligent storage and computing and incorporation of machine learning algorithms for IoT data analytics and security of IoT ecosystem. Section 8 contains the applications of IoT ecosystems. The work concluded with Sect. 9.

3 IoT Architecture

The architecture is a contour that stipulates essential physical elements, their functionality, and the values underlined. Various researchers have various IoT application-based architectures. IoT's primary or basic design is the architecture with three layers. It consists essentially of perception or sensing, network, and application layer. The layer of sensing is primarily based on sensors and actuators. The layer of the network is responsible for data transfer and delivery. The application layer provides the customer with a particular program. There are two further levels in the five-layer design to provide the IoT design a more abstraction. The layers are sensing or perception, processing, middleware, transport, and application.

In fog-based architecture, there are four levels of control, tracking, preprocessing, storage, and privacy between the physical and transport layers. The monitoring or tracking layer tracks and reviews the sensor details. The layer of preprocessing analyzes the sensed details. All of the analyzed data is then handled by the storage layer. Data confidentiality and protection are ensured by the security layer.

This segment illustrated the four-layer SoA-based IoT architecture [10] in which the applicable allowable technologies are introduced.

3.1 Sensing or Perception Layer

The key role of the perception is to recognize and trace artifacts. The following technology may be applied to accomplish this purpose.

RFID RFID is commonly used to recognize and monitor artifacts without touch as noncontact communication equipment. It facilitates the exchange of data via short-distance radio signals. The RFID program contains RFID identifiers, antennas, and RFID readers. The RFID tag could be an antenna-fixed microchip. Each RFID tag has its identification number and is inserted in an item. An RFID reader will identify an item and get the details by asking for the correct signals from the attached RFID tag. An antenna module is utilized between the RFID tag or card and an RFID reader for signal transmission. RFID has the following advantages compared with other technologies (fast search, robustness, reusability, broad volume, contactless reading, safety, small size, low cost, etc.). Due to all of these advantages, RFID may be useful for recognizing, monitoring, and sharing knowledge in the IoT awareness layer.

Wireless Sensor Networks (WSN) The function of WSN in IoT may be quite significant. WSN can observe and track the system state and transfer state data through multiple hops to the control center or sink nodes. WSN should then be used as the next link between the physical and cyberspace realms. WSN has many benefits, in contrast to other systems, including scalability, automatic reconfiguration, durability, low costs, lower consumption of energy, etc. All of this allows it to be aligned for diverse criteria in multiple fields. WSN and RFID both may be utilized in IoT to acquire the details, while RFID is mainly used for the recognition of objects, whereas WSN is primarily employed to perceive real-world surrounding physical parameters.

Others The one-dimensional encryption barcode holds the details in many black points and white intervals. These lines and spacing are structured with various widths, linearly or one-dimensionally, and unique encoding rules. An infra-beam scanning system will interpret the details found in the barcode. A double-dimensional code tracks the details by the use of black and white interface pixels, of which the binary "1" is a black pixel and the binary "0" is a white pixel. The white and black pixels will hold a notable amount of details using different encoding procedures. Two-dimensional coding provides strong information quality, good durability, excellent robustness, etc., relative to barcode. Therefore, an application between an RFID device and a sensor network is the RFID sensor network (RSN). In the RSN, the RFID device may be used to define and track the objects' positions. In an RSN, tiny RFID-based sensing tools and RFID readers are used to produce data and control network functions as a sink node.

3.2 *Network Layer*

The layer of networking is used for routing determinations and facilitates data transfer across heterogeneous interconnected networks. Some protocols are provided below which will enable safe and efficient IoT communication.

IEEE 802.15.4 This is a physical layer protocol and is also designated as a MAC protocol WPAN. The goal of this protocol is to concentrate on low-cost LRWPANs, which include low-cost, low-energy, low-rate connectivity and low-cost connections in the individual market. The Open System Interconnection (OSI) concept is the basis for the IEEE 802.15.4 protocol stack, where each layer is confined only to portions of the communication functions, and low layers will support the higher layers. The bands 868 or 915 MHz and 2.4 GHz can be assisted by the IEEE 802.15.4, and the peak rate of communication of those bands can be 20, 40, and 250 kbps, respectively. IEEE 802.15.4 is a base for other innovations, such as ZigBee, WirelessHART, and several other WLAN innovations and protocols.

6LoWPAN The LoWPAN (Low-Power Wireless Personal Area Network) is coordinated by a vast variety of inexpensive, wireless communications-connected devices. LoWPAN provides a vast range of advantages (small packet sizes, reduced strength, low latency, etc.) when opposed to other forms of networks. The upgrade was the mixture of IPv6 and LoWPAN by way of the 6LoWPAN protocol. Through IEEE 802.15.3 networks, the 6LoWPAN framework is capable of transmitting IPv6 packets. 6LoWPAN is adaptable to the IoT, which integrates a vast range of low-cost technologies, due to its low prices and small energy usage. 6LoWPAN provides many benefits such as fast accessibility and traditional architecture continuity, low energy usage, self-organization, and so on.

ZigBee ZigBee is a networked wireless system intended to be connected for the short term with low energy usage. Five layers have been used in the ZigBee protocol: physical, MAC, transmission, network, and application layer. Low energy usage, low expense, low data speeds, low latency, stability, and protection are the benefits of ZigBee networks. The ZigBee network will accommodate several topologies, such as the star, mesh, and tree topologies.

Z-Wave This is a short-range, low-cost, low-energy, and high-efficiency wireless networking technology. Z-Wave's key purpose is to ensure stable communication between a driver and one or more terminal computers, whereas Z-Wave is ideal for a low-bandwidth network. Notice that only the 232 nodes (slaves) in a Z-Wave network can be used and the controls of all the nodes (slaves) and routing functions are available. The dynamic routing system is assisted by the Z-Wave network, and every slave stores a memory list of routes maintained by a controller. While both ZigBee and Z-Wave promote short-range, low-cost wireless contact and low energy usage, certain variations remain. The major distinction between ZigBee and Z-Wave is the variation in physical layer operational frequency. In ZigBee, the physical layer's frequency spectrum is typically 2.4 GHz, while in Z-Wave the operational spectrum is less than 1 GHz. ZigBee can accommodate 65,000 terminals, and the Z-Wave infrastructure can accommodate only 232 terminals. Z-Wave is an easier to deploy network relative to the ZigBee architecture.

Message Queue Telemetry Transport (MQTT) MQTT protocol is utilized to gather calculated data from wireless or remote sensing gadgets and relay it through servers by utilizing publish-subscribe methodology. MQTT is a lightweight and simple protocol that facilitates a lower-bandwidth network. This protocol can be deployed on many platforms to link items of IoT to the network or Internet, thereby making MQTT an essential function in IoT as a communications protocol among sensors or actuators and servers.

Constrained-Application Protocol (CoAP) It is a messaging mechanism based on REST (REpresentational State Transfer) architecture. As the majority of IoT devices are resource-constrained (i.e., limited computation and storage capacity), issues regarding HTTP protocol, CoAP introduced certain modifications on HTTP functions to satisfy the IoT specifications. Generally speaking, CoAP is an application layer protocol of the 6LoWPAN framework aimed at enabling resources-restricted devices to achieve RESTful interactions. CoAP facilitates group-based communication and push notification, but not eligible to support broadcasting. Many of the essential features supported by the CoAP are resource monitoring, resource identification, association with HTTP, and protection.

Extensible Messaging and Presence Protocol (XMPP) XMPP is an instant messaging system focused on XML streaming protocols. XMPP has XML protocol features that allow XMPP to be used for multiparty chatting, speech, video sharing, and telepresence, with great scalability, addressing, and protection capability. The following three functions are used in XMPP, client, server, and gateway; among these three two members (client and server) enable bidirectional contact. The server can accomplish the connection management and routing features, the gateway facilitates secure connectivity between the heterogeneous networks, and the users can bind to the network server through a TCP/IP interface and relay context-aware XML streaming interface. It (XMPP) will also be utilized in IoT to enable object-to-object-aware text messaging correspondence in XML.

Data Distribution Service (DDS) DDS is a publish-and-subscribe-based protocol that supports device-to-device communication with a high level of performance. DDS is a data-centered protocol that supports a multicast network to provide a high QoS and robustness and has been created by Object Management Group. DDS is ideal for real-time limited-resource IoT and device-to-device communication with the broker-less publish-and-subscribe technique. DDS may also reach a strong degree of scalability.

Advanced Message Queuing Protocol (AMQP) AMQP is an open-source message queuing protocol used to include message resources (queuing, routing, safe and confident), in the device. AMQP concentrates on message-aware applications and might be viewed as a protocol designated for middleware. Using AMQP, even when the customers and middlewares are developed using different programming architectures, stable communication can be secured. AMQP also provides various forms of interfaces for message sharing, such as saving, sending, publishing, and uploading, message delivery, queuing, context-aware, and point-to-point routing.

Others Certain protocols can also perform essential roles in IoT, in addition to transmitting protocols, collaboration protocols, and messaging protocols. The name resolution can be supported for IoT applications by the multicast DNS (mDNS). Clients may use DNS Service Discovery (DNS-SD) to discover requested resources using mDNS over a specific network. The low-power and lossy network routing mechanism is an independent routing protocol, which can be used in resource-aware frameworks to define routes via low-power and lossy channels. While IoT may implement these protocols, enhanced IoT protocols are needed for the promotion of IoT production with more stability, reliability, and interoperability capabilities.

3.3 Service Layer

This layer is designated within the application, and the network layer offers secure and protected support for artifacts or programs, as mentioned above. The following habilitating technologies shall be used in the application layer to ensure the reliability of service delivery: interfacing technologies, process management, middleware and inventory management, and sharing technology.

Interface For effective and secure sharing of data between communicating devices and applications, the system technologies must be built in the service layer. Moreover, interconnected systems, including user attachment, system disconnection, user connectivity, and computer service, can be easily handled by the interface. An Inter-Face Profile (IFP) can be used as an interface model to help IoT applications and can be used to allow connections between services delivered by specific devices and applications. UPnP should be added to obtain a secure IFP. A variety of attempts have been made in the creation of the IoT interface. To achieve successful interaction among apps and services, SOCRADES Integrating Architecture (SIA) can, for example, be used. As the SoA-IoT method is being created, the service provisioning mechanism provides experiences with applications and services. While IoT has established a range of interface technologies, it still poses a significant challenge in IoT research to introduce more reliable, stable, and scalable low-cost interface technology.

Service Management Service management can identify devices and systems easily and schedule secure and effective facilities for the fulfillment of demands. In order to accomplish a particular purpose, the program can be perceived as a behavior, like information gathering, sharing, and storage. For IoT, only a single provider is required to meet those requirements, and several systems must be combined with certain services. The services may therefore be categorized as primary services and secondary services in terms of IoT. The primary services will show key functionalities on applications or devices, often called simple services. In comparison, the auxiliary feature may be accomplished by secondary operations as a supplementary basis. SoA is used to incorporate infrastructure to protect specifics of the deployment of the systems and to render such infrastructure compliant with heterogeneous applications and devices. This allows for the continuity and uniformity of services. OSGi, for instance, was built by a dynamic SoA architecture as a modular platform for resourcing. The service configuration structures should be built first, and the functionality and communication features of devices should be eliminated to incorporate a SoA-based infrastructure. Finally, it is important to include a growing collection of services. Service delivered by a system or application in SoA-based services is regarded as a common task that can be utilized in specific appliances and applications reliably and effectively without modification. This makes for quicker and more effective fulfillment of specifications in SoA-based IoT.

Middleware It is a program or utility that can ensure the integration of technology and applications interposed within IoT. The middleware disguises the specifics of different technologies and provides standard interfaces that enable developers to concentrate on application creation without contemplating device-infrastructure compatibility. This allows for the sharing of knowledge and services with the usage of middleware, equipment, and applications with various interfaces. The benefits of middleware are as follows: (i) middleware can be used to assist different applications; (ii) middleware may be used in multiple operating systems and services; (iii) centralized code assists communications across heterogeneous networks, equipment, and applications; (iv) basic protocols may be enabled by middleware; (v) basic interfaces may be given through middleware. Such characteristics allow middleware ideal for IoT since many heterogeneous systems and networks have been incorporated, frequently modified, or upgraded. A variety of research activities have been introduced into middleware that can be categorized into five groups; these are (i) message-aware middleware, (ii) location-aware middleware, (iii) semantic web-dependent middleware, (iv) transmission middleware, and (v) pervasive middleware. The efficient sharing of knowledge across the different channels and communication protocols (e.g., AMQP, DDS, MQTT, and the XMPP) can especially be accomplished by message-based middleware. Semantic web-based interconnection between the different sensor networks can be given. SoA-based middleware, computing-aware middleware, etc. are examples. Location-based middleware support and tracking combine system positions and other details with automated utility services. Transmission middleware can provide efficient communications between heterogeneous equipment and applications. Typical instances of this layer are RFID-aware middleware (Fosstrak and so on), sensor network-aware middleware (TinyREST [11]), and SCADA. Pervasive (ubiquitous) middleware is developed for the processing and computing world in general and enables applications on various services.

Resource Management Different heterogeneous network infrastructures are built to distribute data of all IoT applications (intelligent mobility, smart grid, etc.). To cut costs, some applications may share network resources to maximize their usage. In this situation, it is a problem in IoT to make sure that the information needed by specific users is provided on time. To effectively organize several networks with the same frequency to optimize the utilization of the network assets, current resource-sharing systems concentrate mainstream on spectrum-sharing. The distribution of the spectrum is separated into three different dimensions: time, volume, and frequency. Almost all of the existing frameworks have been built for machine-to-machine connectivity; IoT is based on networks where "things" not only describes machines but also the actions of humans and objects. Therefore it's a big challenge for potential growth to build an effective resource sharing system through heterogeneous networks suited to the IoT ecosystem.

3.4 Application Layer

This layer is at the peak of this design and provides total system-based assistance for end users. The application layer is not recognized as a subclass of middleware as compared to the conventional three-layer design but instead orchestrates the middleware layer. This layer offers integrated interfacing across heterogeneously distributed networks and frameworks via regular web interface protocols and platform structure for software applications. Smart housing, intelligent transportation, smart business, smart treatment, etc. are instances of such applications.

4 Taxonomy of IoT

The taxonomy of IoT is a mechanism that defines the impact and affiliation of tools or sectors by grouping them. This taxonomy will describe the most common layers in the IoT setting. In the diagram focused on the elements and design, taxonomy for study into IoT was suggested. Effective IoT devices (sensors and actuators) may be categorized as lower-end, intermediary, and higher-end devices at the perception layer. Data is gathered by the sensors and necessary actions are performed by actuators. Different sensors are used for position, orientation, video and audio detection, flux detection, environmental sensing, chemical recognition, etc. Sensors and actuators are lower-end equipment with simple memory, OS, and battery limitation. The layer of pre-processing includes programs for filtering and capturing data with minimal storage space, a compact processing system, and other protection measures before submitting it to the middleware. The pre-processors are designed with microcontrollers and microprocessors which have scalable OS, higher RAM, and higher processing capacity. The different operating systems and their computing specifications are categorized as lower-end and higher-end based on various resources. Diverse accessible networking systems, such as NFC, low-power technology, WSN, etc., have been classified for IoT, beginning with restricted to unrestricted functions. Gadgets are interconnected with each other via a complex protocol across the wireless network. RFID, Bluetooth, NFC, ZigBee, and Wi-Fi which are known as short-range communication technologies are only briefed on this work. A vital aspect of middleware is that the programmer uses several dynamic interrelationship structures as an abstraction layer, which illustrates the features of the framework. Middleware may be split into service-aware, actor-based, and cloud-based services. The integrated middleware components are IoT platforms, storage, and privacy mechanisms.

The key functions of the middleware layer include storage (cloud) and analytics which manage data and require complex computations to improve scalability, robustness, usability, and functionality. The core factors for IoT's progress and at the same time a big obstacle for IoT services are protection and privacy issues. To maintain consistency, the entire IoT architecture is configured to function with middleware. Applications can be classified in terms of client-level applications, web or cloud applications, and APIs that permit the transition of information. Such IoT technologies have proven useful in intelligent infrastructure, intelligent safety, smart roads, smart agriculture, intelligent towns, intelligent grids, etc. Thus, developers have studied various algorithms and deployed them to enhance the IoT ecosystem.

5 Core Elements

This section highlights the core components of an IoT ecosystem. The whole ecosystem is divided into five parts which are (i) devices, (ii) gateways, (iii) computation, (iv) communication, and (v) middleware.

5.1 IoT Devices

IoT devices are the IoT system's basic building block. All the levels of IoT architecture conceive all these devices. IoT devices can be categorized as proprietary or open-source. The freely accessible open-source IoT devices are one of them. All of the devices related to open-source services can be reviewed, updated, and repeated as needed, while proprietary IoT devices are required to be licensed, and no information on such devices is publicly accessible. Because of poor bandwidth, storage, and computing resources and limited power, IoT devices have restricted functionality. IoT systems utilize different memory types, based on the device's restricted capacity, expense, and service. Dynamic RAM (DRAM), static RAM (SRAM), synchronous DRAM, DDR RAM, and embedded multimedia card (eMMC) are the significant kinds of RAM (random access memory).

Unless power is given, SRAM keeps the data in static form. The pace of processing is relatively high and costly. This sort of memory is integrated into microcontrollers and is often found in a microprocessor as an L1/L2 cache. A kind of dynamic memory is SDRAM. It holds the binary data as electrical charges. This RAM is coordinated with the microprocessor and microcontroller clock directly. This sort of memory is used in applications where high processing speed is required as more instructions are to be processed per second. DDR is an advanced SDRAM, almost having twice the capacity of SDRAM. In contrast with SDRAM, the data of a DDR RAM is processed in both cycles of the clock. The cost of DDR is high and is used in devices where clock synchronization with a higher data transmission rate is needed. eMMC is a special kind of internal storage that operates in combination with embedded controllers integrated with the eMMC cards. It is usually used in mobile phones, tabs, etc.

The IoT ecosystem provides a variety of utilities, protocols, and network architectures to handle multimillions of IoT data sharing tools. IoT will embrace heterogeneous architectures with a broad range of services available. IoT devices are classified into three classes: Class 0, Class 1, and Class 2 [12].

Class 0 or low-end IoT devices or gadgets are such kinds of gadgets having minimal resources such as memory, electricity, processing, etc. Most of these gadgets are accessible in the first layer or stage of the IoT ecosystem or environment. They are capable of perception or sensing and required actuating. They utilize lightweight protocols for correspondence. They consist of very simple computing mechanisms. The memory of RAM is between 1 and 50 kilobytes, and the memory of the flash drive is between 10 and 50 kilobytes. These systems are very susceptible to attacks, and in these lower-end devices, privacy is the main concern. They are closely oriented with the nearby environmental parameters such as temperature, moisture, strain, and so on.

Class 1 or mid-end IoT gadgets are more resourceful than low-end gadgets. They deliver more flexibility than lower-level IoT gadgets, but they are not adequate to meet highly complex demands with computational power. Such microcontrollers are important. These gadgets are mounted on low-end IoT gadgets to control and improvise low-end IoT system functionality. These gadgets can process images, filter data, etc. Different networking technologies may be mounted on these devices [13]. The performance of such gadgets is greater than that of low-end machines with clock frequencies between 100 MHz and 1.5 GHz. The size of RAM ranges between 100 kilobytes and 10 megabytes, and the memory of flash drive varies from 10 kilobytes to 100 megabytes. Such gadgets may be protected partly through data encryption because of extra features. Arduino and Netduino are some of the mid-end IoT gadgets. These gadgets are available either at the first or second layer of the IoT ecosystem. Such gadgets are having peripheral interfaces such as USB or micro-USB, which may be known to be the simple gateways of the IoT ecosystem.

Single-board computers with a high-end CPU, RAM, and flash memory are tier 2 or high-end IoT machines. These gadgets are compliant with standard OS such as LINUX and UNIX. Such gadgets help the advancement of technology such as AI, machine learning, analysis of natural languages, etc. Such gadgets also have a production interactive user interface. Up to 64-bit architectures are available in these sorts of gadgets. Such systems have everything on-board and almost every networking protocol is enabled. These gadgets provide a wide range of communication interfaces including HDMI, Wi-Fi, USB, LAN, Bluetooth, etc. Such gadgets are capable of multimedia and graphical processing and analysis of data. These highly sophisticated gadgets are deployed in the IoT ecosystem as specific gateways and controllers. Owing to the vast amount of resources, they have comparatively fewer privacy issues.

5.2 *IoT Gateways*

IoT gateway is an interface for various sensor networks and higher-end IoT applications or the storage (cloud) server. The key function is the handling of the variability attributable to different types of data gathered from sensors and the transfer of data to the higher-level controller or server. The data needed to be processed and analyzed which are obtained by the gateway. High-level data analysis is performed based on the requirements, categories of commands, and data. The gateway or gateways serve as a bridge across various layers of IoT in various networks. To boost efficiency and form a stable layer in the IoT framework, the gateway and its end point nodes are required to be operated efficiently. As regards, data transmission gateways, data centers, or the cloud database serves as a proxy for lower-end IoT gadgets such as sensors. Gateways can work as high-end equipment, but where a large IoT network is present mostly perform as a mid-end gadget while a limited number of low-end gadgets are in the IoT ecosystem. Gates are designed to be resilient in terms of hardware to environmental factors. They will also solve weaknesses and seek to resolve the networking gaps between lower-end systems and orchestrators or controllers. There is also a minimal OS with restricted memory, processing ability, and power usage features and services. The gateway then carries out smaller operations, such as filtering content, translating data from various formats to a single format, and managing the power failure. The gateway will provide a limited power buffer to store the system state in nonvolatile memory and lead the system to hibernation. And the gateway will restart from the same state when the power is returned. Gateways must have the capability to solve problems and, if an issue is not resolved, be able to relay user-vision knowledge to the IoT network. Gateways should be able to reboot as well. Gateways provide a high-performance control framework in the broad IoT ecosystem. It allows the system to manage several gateways. Therefore, a gateway should support the following sorts of communication: (i) lower-end IoT gadgets to gateway or gateways, (ii) gateway or gateways to IoT platforms, (iii) gateway or gateways to orchestrator or controller, and (iv) gateway-to-gateway communication.

Gateways typically establish networking via GPS, WLAN, Ethernet protocols, and Bluetooth. Moreover, local Internet on the gateway and controller level is accessible in a broad IoT ecosystem or network. Finally, the data obtained on this limited intranet level is transmitted through the Internet to the high-end gadgets or computers, networks, or cloud. Throughout certain gateways, the tracking of devices such as auto-detection, insertion, and elimination of IoT devices is enabled. It often tracks data and integrates minimal data such that accurate knowledge may be distributed to the ecosystem. This ensures that the operating system, the servers, and other IoT gadgets collect sufficient data to enhance their real-time efficiency. Throughout the restricted IoT network, the CoAP protocol-dependent correspondence between IoT devices may be facilitated. IoT-relevant gateways operate in three diverse modes: (i) passive, (ii) semiautomated, and (iii) automated [14].

Passive The new gadgets must be inserted or removed by manual approaches. The customer permits the gateway to connect and remove hardware. Network activity can be controlled and configured. Every node or sensor can be reached by the gate. Moreover, it accesses any network without altering the protocols for communication. Hence, these are not versatile and not reconfigurable according to the requirement of an IoT ecosystem.

Semiautomated In such an operational mode of gateways, there is an interconnection between the included gadgets and the communication gateways. It increases the efficiency of the network in terms of real-time performance, the stability of data processing, and management. This is based on an interface with a plug-in configuration to demonstrate that devices can be linked with a network requirement. Due to the external interface, it is versatile than passive gateways.

Automated The gadgets in this mode can reconfigure themselves. No authorization is required. IoT gateways may connect or delete gadgets themselves. These gateways may be linked to different networking protocols and interfaces such as ZigBee, Bluetooth, CoAP, and Wi-Fi. These kinds of gateways can interact with multi-variety networks easily. They also facilitate the tracking of real-time sensors so that systems can be modified and reviewed with ease by attaching or removing equipment.

5.3 Computation and OS

OS is a combination of operations or programs serving as an interface between the consumer and the applications. Various programs are built and operate on the so-called OS. OS is mounted on IoT devices to operate and control the programs on the IoT ecosystem. OS may therefore be specified as a part, leading hardware to be a complete IoT gadget. Thus, OS is responsible for handling and controlling power usage, guiding operations, programming IoT gadgets for different attributes, and providing connectivity with different machines. In introducing a robust and secure IoT environment, OS plays a key function.

Owing to the heterogeneous existence of IoT systems, OS should be modified to suit each sub-IoT system's requirements. The computing, storage, energy consumption, and memory capacity of IoT gadgets are significant characteristics. In the future, all these properties exist with all IoT gadgets. OS is better adapted when it embraces the control and communication functionality that shifts when IoT gadgets adjust their properties [15]. Additionally, basic programming tools are expected that can facilitate applications' portability across diverse platforms and that can be easily recreated and maintained. IoT OS are categorized as:

(a) High-level OS – Most of which are based on Linux. These operating systems operate on high-end or mid-end IoT machines; they are high-power, storage-, and capacity-based single-board computers. For instance, RPi (Raspberry).
(b) Low-level OS – That may be based on Linux OS or non-Linux. These operating systems run on low- and mid-end IoT gadgets, which have minimal computational resources, energy requirements, and processing capacity, for example, a tiny board machine, such as Arduino.

To build an effective, secure, scalable, and interoperable code, OS plays an important role in IoT. The IoT devices need both high-level and low-level OS. OS is defined in the following terms: architecture, operating model, kernel, scheduling, management of storage, protocols for networking, stability, energy consumption, and multimedia support. For developing an efficient operating system, the following aspects must be taken into consideration.

Architecture The operating system architecture comprises the kernel and stipulates software facilities. Architecture may be graded according to the following parameters:

(a) Monolithic – Designed for multitier application ecosystems. These can perform dynamic computations of higher order and hence provide a great processing speed. The kernel space performs all operations.
(b) Microkernel – This design presents only main features such as scheduling, process-to-process communication, and kernel space synchronization. Many other functions of the operating system operate via multiple threads. The kernel system and user system both perform processing tasks here. Due to the acces-

sible plug-in, they provide high processing versatility and allow the operating system to be installed upon the base environment.

(c) VM-based architecture – This kind of device is a virtualized device that operates with the existing framework. This design is slower because of its virtualization, but it provides a higher degree of portability, versatility, and extension.
(d) Modular architecture – It requires components to be inserted and removed dynamically in the kernel during runtime. The configuration of each module is different. The modules may then be linked in and out as needed.
(e) Layered architecture – Multitier architecture is built for a specific purpose, and implementation of such architecture is not much flexible. But running and maintaining this form of the operating system is simple.

Development Model This specifies the requirements for the simulation of an application or system. Multi-threading and event-driven, efficient managers and memory architecture are influences that affect. It depends on the IoT system in which the operating system will be utilized and leads the OS to be versatile and extendable, and the requirement of the system is achieved. The operating system is needed to be configured in such a way that it can be updated and revamped. A software or program development kit in the operating system deals with library modeling, OS design improvisation, memory layout, and configuration of control.

Scheduling
The preparation approach is proportional directly to the capacity of the operating system. Priority and non-priority algorithms are kinds of scheduling mechanisms collaborated by preemptive architecture and non-preemptive architecture. Preemptive executes the highest-priority function to interrupt all activities, whereas non-preemptive begins a new job after the existing functionality has been completed.

Memory Management This applies to the consolidation and decommissioning of information. The memory is classified according to the need for versatility. Static memory is a kind of memory that is fixed, and another one is called dynamic memory which is required for dynamic frameworks. A separate processing degree dependent on processing speed and efficiency is a significant consideration. Efficiency and memory are critical concerns when an operating system is built. Furthermore, another essential OS memory is cache memory, which is relevant for operations at the OS level. This memory contains OS-required details and metadata.

Interfaces and Communication Protocols For IoT, the operating systems are designed to ensure system interfacing with devices. Various hardware interfaces and different networking protocols allow interfacing. The form of IoT unit, deployment site, environment needs, speed, and data transmission protection now depend on which helps to finalize the OS protocols. Various protocols for connectivity include CoAP, MQTT, ZigBee, WLAN, Bluetooth, and other interfaces such as micro-USB, USB, etc., which not only help to interact with interfaces but also help to establish a link with other devices that are contained in the IoT ecosystem or Internet. The heterogeneous complexity of the large IoT environment is often taken into account.

Power Management This is another major concern of the IoT ecosystem. IoT gadgets are constrained machines. Such gadgets are deployed in areas commencing from the surrounding ecosystem to the servers (Internet). Power orchestration has to be reliable for energy efficiency, recycling, and prolonging battery life, and physical assaults also should be mitigated. This will also be able to conserve electricity and push the control in a different direction if a failure occurs. To maintain a stable state of operation, OS is required to be powerful enough to withstand power loss.

The "brain" and computing capabilities of the IoT are defined by processing, communication, and computation systems (such as microprocessors, SOCs, microcontrollers, and FPGAs) and software applications. Various IoT hardware platforms, including Arduino, FriendlyARM, UDOO, Intel Galileo, Gadgeteer, Raspberry Pi, CubieBoard, BeagleBone, WiSense, Z1, T-Mote Sky, and Mulle, have been developed for the service of IoT applications.

Moreover, several software frameworks provide IoT features. Operating systems are important to such platforms because they work during a device's initialization period. There are many real-time operating systems (RTOS) that are strong candidates for designing IoT software focused on RTOS. Contiki RTOS, for example, has been commonly used for IoT scenarios. Contiki has a Cooja emulator, which simulates IoT frameworks for researchers and developers [16]. Lightweight OSs for IoT environments are also supported by TinyOS [17], LiteOS [18], and RIOT OS [19]. Google is part of the Open Auto Alliance (OAA) where several members of the automotive industry are preparing to include additional technologies to the Android applications to enhance the features of the Internet of Vehicles (IoV).

5.4 IoT Communication

Heterogeneous artifacts are linked to different intelligence resources via IoT communication technologies. In the case of noisy and lossy contact connections, usually, IoT nodes will work utilizing low power. For example, WLAN, Bluetooth, Z-Wave, IEEE 802.15.4, and LTE Advanced are the sources for IoT networking protocols. Some unique networking systems are also being employed, such as RFID, NFC, and UWB. A basic M2M system can be utilized with RFID. The RFID tag is a simple chip attached to the item. RFID readers transmit the request signal to the tag and receive a mirrored signal from the tag which is transmitted to a nearby database. The database is connected to a processing center to recognize items based on the mirrored signals within proximity of 10 cm to 200 m. RFID tags are classified as active, passive, and semi-active or passive tags. Active tags are battery-powered and the passive tags are not equipped with battery power. Semi-active or passive tags use on-board power during the time of need.

The NFC protocol operates at the 13.56 MHz frequency and allows the data exchange rate up to 424 kbps. The range of NFC is up to 10 cm between passive tags and active readers or between two active readers during the communication. To facilitate shorter coverage and larger bandwidth, UWB connectivity is recently evolved which can ensure communication among sensors. Wi-Fi, which uses radio waves to exchange data between objects within 100 meters radius, is another networking technology. Wi-Fi permits shrewd gadgets to impart and trade data without utilizing a router in ad hoc architectures. Bluetooth represents a communication innovation that is utilized to trade information between gadgets over short separations utilizing short-length waves to limit power utilization. As of late, the special interest group (SIG) working for Bluetooth technology created Bluetooth 4.1 that ensures Bluetooth Low Energy for faster connectivity and IP availability to help IoT. The specifications of IEEE 802.15.4 include both physical layer and access control mechanisms for low-performance wireless networks that achieve stability and scalability.

LTE (Long-Term Evolution) is a cellular communication framework for rapid data exchange between cell phones dependent on GSM/UMTS network innovations. It supports quick voyaging gadgets and gives broadcasting and multicasting facilities. LTE-A (LTE Advanced) is introduced as an improved rendition of LTE including data transfer capacity expansion (up to 100 MHz), uplink and downlink spatial multiplexing, broadened coverage, lower latencies, and higher throughput [20].

5.5 *IoT Middleware*

IoT middleware is an IoT ecosystem feature that provides the consumer with raw data in a web application format. The data generated on IoT gadgets must be configured in such a way that it suits all forms of IoT applications. The IoT middleware features address the above challenges. Nevertheless, it has minimal interpretation and data incorporation capabilities. It poses a great challenge to connect, capture, and view inter-device data and merge the data gathered from many devices into a scalable framework. The IoT middleware must be built to be scalable, adaptable, modular, safe, and open-source. This form of middleware will enable researchers and specialists to customize and compile improvised applications and also to incorporate new IoT gadgets into the ecosystem. The usage, functionality, and versatile design of IoT middleware may be commonly classified into three kinds [21].

Service-Oriented Middleware End users and developers may use this form of middleware to connect and adjust IoT devices as utilities in the IoT ecosystem. It provides resources such as control of entry, storage, and event management system. Security models are costly and most of them have minimal protection, privacy, and trust service. User data processing is constrained in such a program. Protection strategies are not meant to help restricted resources in the SOA architecture.

Cloud-Oriented Middleware This middleware makes it simple to capture and process information. But in terms of different types of IoT gadget-based systems, it limits the productivity of the whole ecosystem. The cloud-based protection model is described in a cloud-based architecture. Therefore the cloud structure determines anonymity and protection and cannot be controlled by consumers. Sensitive data are the key problems in such kind of system and are not structured to meet resource constraints.

Actor-Oriented Middleware This form of middleware is open-source and has a plug-and-play architecture. IoT devices can be introduced as a plug-in to the IoT ecosystem and can easily be disabled without disrupting the IoT environment when an IoT device is not required. The safety model here can be configured through the plug-and-play system by users. The actor-oriented middleware is designed to assist restricted resource services.

IoT Platforms One of the vital components of IoT middleware is IoT platforms. IoT platforms are sophisticated applications developed primarily to control the whole IoT network. It's the foundation of the IoT. IoT architecture connects gadgets, gateways, cloud networks, servers, and applications. The interface allows all the gadgets to be recognized. They supply all the IoT devices with a software development kit. In the form of IoT administration, architecture, and deployment, these programs allow the visualization of IoT products. They handle privacy as well. This is where the IoT ecosystem's rules are mentioned for evaluation. Such platforms describe the IoT ecosystem's purpose. It's regarded as a bridge connecting the gadgets of an IoT ecosystem to the cloud. The whole environment management is controlled here concerning complexity, cost, market, and data flow. IoT systems require the connectivity of cross-device and the automation of linked equipment. IoT platform will serve as middleware in IoT architecture, to render an IoT (ecosystem) environment robust, stable, secure, and scalable. The IoT middleware framework may also be viewed as a bridge from software to hardware. The main activities of IoT platforms include the detection, configuration recognition, and modification of sensors and heterogeneous inputs for specific applications. It promotes the integrated use of artificial intelligence among completely different instruments for communication and for taking very complex measures. With the form of layers, the platform might be identified. It is focused on the IoT platform's characteristics that provide a foundation for the IoT ecosystem. To define, connect, compute, and react, the IoT platforms are responsible for an IoT ecosystem.

Storage One of the middleware functions is to preserve the data in multiple ways. The amount of IoT data traffic is high and therefore requires enhanced measurement and evaluation techniques to accommodate big data. Techniques of data mining such as machine learning and decision-making algorithms are required to be deployed, which provide valuable knowledge to improve raw IoT data. The diverse specifications may be managed and fulfilled through cloud technologies. Cloud facilitates heterogeneous programming mechanisms, such as disk and dynamic computation. However, several blockers of such techniques also have to be identified. Transition of edge devices' (e.g., sensors, client gadgets like phones) data to the fog or cloud servers can generate issues regarding network efficiency (in terms of latency, capacity, congestion, trustworthiness, quality, etc.), Internet prices, server protection, storage costs, secure data transfer and secrecy, cloud interoperability, and cloud-gadget security and stability.

Big Data – The interconnection of sensors and a wide number of artifacts helps in the collection of bulk data. IoT gadget-generated data is growing quickly. This large amount of data is referred to as "big data" reflecting the scale of the data collected. It is a mechanism in which data is checked and analyzed, and eventually the trends and relationships are created. Big data's four fundamentals include safety, data protection issues, broad data processing, and everyday life impacts of data. Such kinds of data include numerical data, documents, video, audio, statistical data, etc. These data can be classified as unstructured, structured, or semi-structured. The best way to process such enormous data is to adopt "big data analytics." Big data methodology is capable of processing multiple details and providing them to the analytical platforms. This detail is then passed to the data center. Massive data infrastructure is prone to multiple protection challenges, such as protected transfers, safe data processing, safe retrieval, and computing. Big data research is split into three specific factors: velocity, variety, and volume. The size of the data stream is defined by the volume. Big data versatility includes varieties of data audio, picture, text, etc. Velocity reflects the processing speed of data. There are a vast variety of big data and IoT apps that have a significant effect on our everyday lives, such as intelligent watches, which constantly monitor a person's well-being by gathering data and sending the alert to the doctor whether an individual becomes ill or has any issues. Large data research evaluates a vast amount of data to allow choices that are valid and effective. To boost decision-making, IoT and big data mix are used. There is a great deal of data gathered from the "linked gadgets." Big data engineering that handles data utilizing query, report, training and test data set, and analytical software is therefore important. The data must be processed intelligently and efficiently to execute theoretical and logical operations. A single processor and minimal storage cannot execute a big data operation. The massive data generated by the IoT ecosystems sometimes cannot be supported by data analytics platforms such as Apache Hadoop. Performance often includes the analysis of real-time data in IoT. IoT requires a broad, universal computational interface for all IoT applications. The total IoT environment will not be an overhead of such an empirical machine. An IoT big data approach is to monitor just the appropriate and required data.

Cloud Computing – Cloud infrastructure is an emerging platform for addressing the on-demand virtual network which is capable of configuring tools including network gateways, servers, software, and services. Cloud storage provides efficient and expense controlled distributed utilization and control of services. Cloud storage and computing resources allow IoT to handle this huge data with archiving and processing capabilities. OpenIoT, ThingWorx, Google Cloud, GENI, and Amazon are significant cloud services used by IoT. Cloud storage handles large data that enables data to be stored and useful knowledge to be retrieved. It is a daunting job to implement cloud infrastructure in IoT.

6 IoT Knowledge Hierarchy

A primary purpose of communicating and gathering data from inter-connected gadgets (e.g., sensors) is to increase knowledge of the condition and to allow gadgets, computers, or humans to be more conscious of the climate. Knowing the scenario or context may enable wise choices and adapt to the complexities of their environments by delivering resources and applications. Data obtained by various sensors and instruments typically are multimodal (temperature, flux, voice, video, etc.) and diversified in nature (data quality may be time-dependent on multiple devices). The complexity, uncertainty, and omnipresence are major difficulties in the collection, application, and analysis of real-world data. The amount of data on the web and the cloud has also grown at an impressive rate: around 2500 trillion kilobytes of data are produced per day, and 90% of today's data is reported to have been created in the last 2 years (IBM, 2012). Tactile data (including the sensors) identified with various occasions and events can be examined and transformed into significant information to provide better comprehension about the real world and to make more value-added services, for instance, information from meters might be utilized to all the more likely anticipate and balance power utilization in intelligent grids; analyzing traffic, climate, and congestion tangible information records can ensure better city and traffic management; checking and handling sensory gadgets appended to patients can ensure enhanced healthcare. These processes of transformation of data can be wisely elaborated using the "knowledge hierarchy" concept.

The lower tier implies that IoT gadgets and equipment generate a significant amount of data. To improve interoperability, the layer above allows organized and machine-readable information from different sorts of raw data. However, the information and services often demanded by people and high-level applications are not able to ensure high-level abstractions and expectations which offer the underlying data meanings and insights to the individual and the system. The high-level abstractions and assumptions will then be turned into actionable insights with sector and context information to take maximum advantage of the IoT ecosystem's ability and build end-to-end solutions.

Cloud systems and "big data" applications will include the technology and resources required to process, manage, and interpret the IoT data. However, we do need effective tools and technologies that can organize, register, exchange, and interpret IoT data in diverse applications and promote the transition into actionable information and intelligence. As several IoT equipment and infrastructure (e.g., battery-powered computers, small processing nodes, and memories) are widely dispersed, heterogeneous, and infrastructure restricted, requirements for developing IoT ecosystems and applications are very different from those commonly used the Internet and network framework (in particular, for interoperability, scalability, efficiency). The latest architecture and design initiatives for the future Internet and web are reflected by this. Figure 1 illustrates the knowledge hierarchy of an IoT ecosystem.

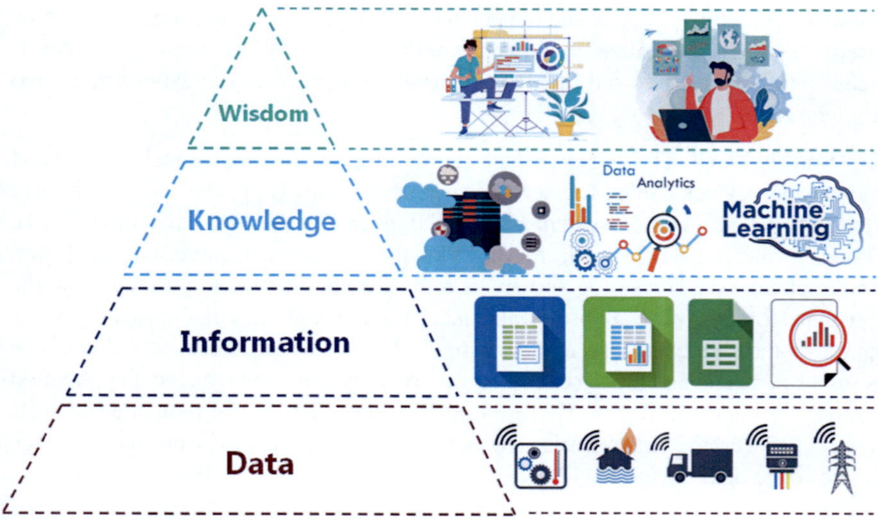

Fig. 1 Knowledge hierarchy of an IoT ecosystem

7 Paradigms of Intelligent IoT

7.1 Generalized Fog-Edge-Cloud-Enabled IoT

The cloud-IoT framework generally consists of five phases: data capturing, pre-processing, analysis, storage, and retrieval. IoT sensors or tracking systems are used to capture data. Data sharing among devices typically takes place via Bluetooth, NFC, Wi-Fi, etc., as the primary communication between gadgets. Depending on the program, the wearable systems interpret data periodically and send them to the respective processing units. Sensing frequency bands are set, and energy-aware methods are not used while a load of data is less than typical. The second step of pre-processing is needed to filter sensor details. The details would need to be structured to match client requirements. Pre-processing includes the detection of outliers and data noise, the exclusion of pieces of knowledge of low value, etc. Every task requires significant time and computing resources because the detection of outliers or interest points is a task having high complexity. The analytical processes are the most critical among phases in the cloud framework. In general, a standard IoT application is intended to be smart. The analytical process, to integrate the information into the underlying IoT framework, which is appropriately named as an IoT knowledge layer, will use the data to learn effectively. Specific learning algorithms play a key function in data processing. Using the IoT framework, the evaluated data will be sent to the application layer, where it is expected to provide the customer with the required activity. The data storage framework is expected to include several servers in cloud-aware architecture. A resource manager performs as a broker between the servers and the end gadgets. Resource management maintains track of resource scheduling, supply, and dependence. Multiple software instances or virtual machines (VMs) are utilized to make the servers. The VM has access to cloud services and also preserves settings for device setup, memory, processing power, and storage space.

Fog-Enabled IoT The fog-dependent computing applications consist of several networking devices named fog servers or nodes or hubs [22]. These fog nodes can be seen in the cloud-based system as a small duplicate of the virtual machines (VM). Therefore the fog will execute the tasks of measurement, storage, and analytics. Unlike cloud VMs, these fog servers or hubs are not positioned much near to the core data storage. These are positioned near the edge machines that ultimately allow data from the core data storage to be computed and analyzed. Because VMs operate as server instances in cloud computing, the fog hubs are known as local application compatible micro-instances. These micro-instances incorporate requisite memory, power of processing, and resources to measure the data collected from specific edge devices or related devices.

In the fog computing framework, where computation is done near the edge gadgets, the fog devices are not often rendered operational. They're activated as per request to conduct service. That is, the fog servers can be switched off when a load of data is smaller, thereby saving resources. This is not the case where the VMs are eligible for requests from edge users. This is important to ensure the required functionalities are secure and accessible. With a fog-based distributed ecosystem, centralized control can be eliminated. Thus, instead of a core cloud server for performing all the processing and storage functions, fog nodes deployed near the edge gadgets conduct the computational functions.

The data generated by the edge gadgets are grouped according to the computational requirements in this distributed framework. Fog servers should be deployed in appropriate numbers and may be clustered to ensure the required bandwidth. Several fog nodes are meant to execute application-aware processes in every cluster, and others are designated to conduct the management of databases, complex computations, and interchange of information with other fog clusters. Every cluster or group consists of one fog server serving as the head of the cluster or group. The head of the cluster serves as the orchestrator and scheduler of the resources. The head of the cluster distributes the computational tasks among its subordinates. Such computational parallelization results in an enhanced rate of response. The head of the cluster turns off the nodes that aren't allocated with tasks, for saving resources. If any node of the cluster has stopped its activity, its responsibilities will be spread to other nodes to sustain the application's regular services. If the head of the cluster stops operation due to the failure of the system, then another inactive node from the cluster might be given the charge. It reflects that the fog-based computational system is secure.

Edge-Centric IoT As IoT technologies are quickly expanding, by 2025 it is expected to have 77.44 billion IoT gadgets. Various IoT frameworks are being developed from various viewpoints by different organizations for the enormous amount of IoT applications, and the edge computing paradigm has been recognized as significant support for IoT ecosystems [22]. But a recognized and efficient edge-centric architecture for the IoT ecosystem does not exist.

There are four main components of the edge-computing-aware IoT architecture: cloud, IoT end gadgets, edge servers, and clients. The architectural design takes into account both the necessary tools and the specific characteristics of each group. Users utilize smart IoT apps to lead their livelihood comfortably, although they connect more IoT end gadgets via integrated edge-cloud interfaces rather than communicating directly with end IoT gadgets. The end IoT gadgets are rooted profoundly in the real universe. They experience the physical environment and behave to regulate the physical world, but in practice, their capabilities are not much advanced. The cloud has virtually infinite space but is generally located far from the end terminals. Therefore, the cloud-dependent IoT ecosystem is not much efficient, especially during the need for real-time processing. As the edge server is a key component of the whole system, it will integrate and support the other parties for optimum performance in the cloud and IoT applications.

IoT users send queries to get access to IoT data or services to monitor IoT gadgets inside the edge-aware IoT architecture. These queries will ultimately arrive at the edge server from a web server, edge gadgets like sensors, smartphones, etc. They are then managed into the edge node, which then transfers them to end IoT gadgets or manages the queries in exchange for end IoT gadgets at the edge tier. The edge tier communicates with end IoT gadgets that not only connects them with consumers and the web but is also capable to preserve data gathered and transmitted from end IoT gadgets and offload heavy computing tasks, such as massive data and robust IoT ecosystem protection algorithms. Additionally, many current services of end IoT gadgets can be moved from the core cloud to the edge nodes or servers, which can be configured as per client needs. The edge nodes may operate independently without the intervene of the cloud server, or the edge nodes may work in tandem with the core cloud, in terms of the interaction between the core cloud server and edge node. The edge nodes are capable enough to satisfy IoT application requirements according to the first framework. It will, for instance, provide storage and processing resources to satisfy all IoT gadgets' demands. In the second framework, the edge provides cloud supports for controlling the edge tier or for addressing IoT application requirements.

A sophisticated engineered configuration leads the edge-aware IoT ecosystem. The edge tier is a highly optimized zone for the IoT ecosystem dedicated for purposes, such as meeting many of the real-time needs and discharge of heavy computing tasks for end users. First, the edge layer has more capability than IoT gadgets, which allows a range of computational protection mechanisms, such as homomorphous encryption and access control dependent on attributes, to be implemented on the edge tier. Second, the edge layer is directly connected to the end-level IoT gadgets. It can fulfill the criteria of protection for real-time security frameworks. Third, the layer at the edge gathers and stores data from several IoT terminals. Therefore, the edge is best at making security decisions relative to end users, meaning that an optimal protection judgment relies on both the performance of the algorithm and the accessibility of adequate details. For instance, the edge tier can more effectively perform the detection of intrusion with such additional data. Many security mechanisms are transformed into routing policies with the rise of software-defined networks (SDN) and network functions virtualization (NFV), but these technologies might conflict with one another. If the whole network is linked at the edge, such problems may be overcome through the edge tier. Fourth, despite the resource limitations, the expense of service, and the vast range of end gadgets, installing and maintaining firewalls on each IoT end gadget are typically not feasible. The usage of firewalls on the outside of the network makes more efficient filtering and blocking of incoming threats. Fifth, the edge layer can track and provide a simultaneous secure connection to these gadgets because of the mobility of such terminal gadgets. The fairly secure relationship between the edge tier and the end gadgets often helps to create a deep trust among the gadgets. This alleviates the problems of building trust within these gadgets. Moreover, the edge is normally associated with the cloud through a high-bandwidth connection. The edge tier is capable of requesting extended security from the cloud whenever appropriate. The cloud will, for example, provide position and activity control for the edge, and the cloud will establish effective edge-aware security mechanisms.

7.2 Machine Learning-Enabled IoT Intelligence

Taxonomy of Machine Learning Algorithms Machine learning (ML) is a field of computer engineering and a kind of AI that allows devices the opportunity to perform operations without specific programming. Machine learning appeared from the recognition of patterns and the theory of computational learning. Principles of machine learning and the widely utilized ML algorithms for intelligent and efficient data processing are addressed here.

An algorithm for learning takes a variety of instances as an input called a training collection. There are three major learning categories: supervised learning, unsupervised learning, and reinforcement learning. Informally, in the case of supervised learning, a training data set consists of specimens of input vectors and their respective target vectors, also called labels. In terms of unsupervised learning algorithms, the training package (set) needs no labels. Reinforcement learning algorithms deal with the queries of discovering the correct action or series of actions that might be done to optimize reward within a specific scenario. This work focuses on unsupervised and supervised learning algorithms as such learning algorithms are commonly used in IoT intelligent data analytics. Supervised learning algorithms are designed to perceive how to determine the correct output in terms of a specific input vector. Applications under which the target labels comprise of a confined number of distinct groups are considered as classification. Cases under which the target labels consist of one or multiple successive variables are considered as regression.

It is quite difficult to identify the exact goal of unsupervised learning. One of the key goals is to recognize responsive clusters of related samples beneath the input data, called clustering. To move the input value into a latter variable space, the goal can also be the creation of a suitable internal representation through the preprocessing of the initial input variables. The output of a corresponding machine learning algorithm is substantially enhanced by this preprocessing and commonly known as feature extraction.

To make the required decisions to interpret the intelligent data, it is important to define which activity needs to be performed from systemic exploration, locate odd data points, predict values, forecast categories, or extract features. Clustering algorithms might be able to yield the most suitable tools to identify the structure of unlabeled data. K-means algorithm is the popular and widely used algorithm for clustering which can accommodate a huge amount of data assuring support for broad categories of data. Reference [23] suggested the K-means algorithm to handle intelligent data in smart cities and intelligent house. DBSCAN is another unlabeled data clustering method that can be used for sophisticated clustering of citizen-behavior.

Two significant ML algorithms can be used to determine unusual data points and incongruities in intelligent data. The support vector machine (SVM) and principal component analysis (PCA)-based incongruity detection techniques are effective detection methods, allowing the training of incongruities and vexatious data with higher efficiency. Single-class SVM was used by [24] to track and identify incongruities in human activities. The SVM discussed is another common classification algorithm that can handle massive data and identify their different forms. SVM is typically used with the most intelligent data analytics algorithms since it can accommodate a large volume of data and varieties of data. For instance, SVM can be used to categorize traffic info.

The support vector regression (SVR) and linear regression-based techniques are the two frequently used algorithms for forecasting values and classifying sequenced data. The models used in these algorithms are intended to process and train high-velocity data. For instance, the research [25] performed real-time prediction by applying linear regression. The regression tree-based classification mentioned is a rapid training algorithm, which was implemented for the classification of citizen behavior in a smart city.

Neural networks are adequate learning models for feature approximation problems to predict classifications of data. In comparison, a multigrade neural network should offer an appropriate solution as intelligent data will have to be reliable and needs lengthy training time. For example, the feedforward neural network (FFNN) [26] has been used to minimize the potential energy needs, forecast the generation of the data, and remove their redundancy.

The two most widely used algorithms for extracting data characteristics are PCA and canonical-correlation analysis (CCA).

To make the right decisions, the selected algorithms should be applied and optimized.

Machine Learning-Aided IoT Security In recent years, there has been growing curiosity in the research and study of machine learning (ML). For the enhancement of several domains, ML is highly utilized and also deploying ML-aided techniques for IoT protection. ML is a viable solution for shielding IoT systems from cyber threats, as opposed to more conventional approaches, by defending them from cyberattacks or assaults. The followings are the approaches ML provides for coping with such security risks.

DoS Attack – One of the severe security issues of IoT gadgets is DoS assaults (attack). One solution to avoid these attacks is utilizing a protocol focused on multilayer-based perceptron (MLP), which secures IoT networks against threats of DoS. The authors of [27] suggested an expansion of particle swarm and algorithm of backpropagation to lead an MLP that will assist to improve the protection of wireless networks. ML strategies assist to improve the precision of deductions and stable IoT devices resistant to DoS attacks.

Eavesdropping – Attackers can eavesdrop on messages at the time of transmission. ML strategies like Q-learning-dependent offloading technique or Bayesian nonparametric methods may be used to secure the IoT ecosystem from attacks of this type. ML techniques like Dyna-Q and Q-learning might be used to secure gadgets from eavesdropping. Reinforcement learning-aided techniques for the evaluation of these schemes can also be deployed.

Spoofing – Spoofing attacks can be prevented utilizing Dyna-Q, Q-learning, SVM, DNN model, distributed Frank-Wolfe (dFW), and incremental aggregated gradient (IAG) techniques. These methods not only improve the efficiency of identification and diagnosis but also minimized the false alarm and average error rate.

Privacy Leakage – Personal details such as health records, location, or photographs are gathered that jeopardize the user's privacy. Scientific Calculations for Privacy Preservation (PPSC) might be used to prevent issues of privacy. Another technique for improving the trust for IoT applications is an integrity detection algorithm for the commodity (CIDA) focused on the Chinese Remainder Theory (CRT).

Digital Fingerprinting – Digital fingerprinting is an emerging approach for secure IoT ecosystems to allow clients to avail of adequate trust in IoT applications. Fingerprints nowadays are widely deployed to unlock mobile phones, allow payments, unlock home doors and the car, etc. Digital fingerprinting is arising as a dominant form of biometric detection because of its low expense, precision, high security, and acceptability [28]. Besides the beneficiary features of digital or computerized fingerprinting, the effective usage of this methodology in IoT, including fingerprint recognition, enhancement of image, matching of features, etc., are certain problems yet to be solved. Different machine learning algorithms are designed to offer nontraditional approaches to solve these problems. Some of them are discussed below.

Support vector machine is a training algorithm for linear and nonlinear classifications, text categorization, PCA, regression, and speaker identification. It has maximized the disparity between the training patterns and decision boundaries. Details of SVM-based automated fingerprinting method were addressed by the authors. The work has also contrasted it with other conventional fingerprint detection models. A feature vector is constructed using fingerprint pixel values and is used for SVM training. Different fingerprint patterns are analyzed, which are then matched based on identified patterns.

Artificial neural network is a widely deployed machine learning algorithm. It has a variety of advantages such as fault tolerance, integrated learning, and widespread use. A system for automated recognition of fingerprints based on ANN was introduced in [27]. The computerized values of some characteristics such as the ridge, minutiae, and bifurcation in the fingerprint are employed to the neural network-aided backpropagation algorithm for training purposes. The fingerprint is verified based on the previously captured values preserved in a database.

The basic requirement of an IoT ecosystem is to secure the entire system and gadgets that will engage in communication with the network. ML has to train algorithms to find out security breaches in IoT gadgets or to discover unauthorized activities in the IoT ecosystem to avoid data loss or other difficulties. ML is therefore an efficient technique for solving the problems in securing IoT gadgets. To accelerate the enhancement of the IoT ecosystem, additional contributions are needed.

8 Applications of IoT Ecosystems

IoT technologies are expanding rapidly and reaching the majority of established industries. Specific applications of IoT ecosystems are addressed in this segment.

Smart Cities The usage of modern computing and networking technologies to boost the overall efficiency of people's lives in intelligent cities is extensive. It requires smart houses, intelligent traffic control, intelligent disaster prevention, intelligent services, etc. Cities are being made smarter and governments are being encouraged by various incentives.

Smart Home Home automation is one of the evolving services of IoT technology. In most of the markets meanwhile, different intelligent home apps are developed on the hypothesis of the IoT. They provide basic thermostat controls and many more sophisticated control technologies including intelligent metering, maintenance, and home business services involving automatic heating and lighting. It is estimated that the amount of data produced on a standard IoT network such as the intelligent home would be enormous.

Smart Grid IoT is known as the basis for knowledge in intelligent grid networks. IoT integrates networking technologies into all sorts of grid components such as transformers, switches, circuit breakers, meters, relays, smart electronic gadgets, voltage regulators, capacitor banks, and much more. Such IoT gadgets gather the data necessary for automation. Smart grids powered by IoT deliver many advantages, including reduced capital spending, increased renewable power, reduced maintenance expenditure, and increased consumer loyalty.

Smart Healthcare The consequences of chronic illnesses and improving wellness are mitigated by intelligent healthcare. This can be achieved by the integration of sensors in patients to better track clinical problems and offer guidance to the provider of healthcare. Several researches assessed the overall appropriateness of wearable IoT gadgets in systems of healthcare. Present statuses of patients become much simpler to reach with sensors. A perfect instance of intelligent healthcare is the prompt intervention whereby suggestions should be provided to the treatment facilities once a patient's blood pressure crosses the predefined level.

Mechanized Agriculture Mechanized agriculture includes soil moisture analysis, microclimate management, dry-area selective irrigation, and temperature regulation. The use of these advanced characteristics in agriculture can contribute to high returns and save ranchers from financial losses. Temperature and moisture controls in various processing of grains or vegetables may help to avoid fungal and other microbial pollutants. Climate management will also help improve the production and quality of vegetables and crops. Much like crop surveillance, IoT ecosystems are capable to track farm animals' movements and safety through fastening animal sensors.

Industry 4.0 The Industrial IoT (IIoT), also named Industry 4.0, is a manufacturing IoT program. To enhance information and collaboration, IIoT combines several innovative automation and communication technologies, such as M2M, machine learning, and data analytics. For example, IIoT networks will link and forward data of employees to the administrative offices from the production plant. Decision-makers or workers may therefore build a complete and detailed description of their production cycle by utilizing the IIoT network, thus enhancing their capacity for decision-making that is better known.

9 Conclusion

The concept of the IoT makes its way rapidly into everyday life and seeks to improve the standard of life by integrating many intelligent gadgets, applications, and uses. The IoT would enable everything around us to be automated. This chapter provides an overview of the basis of this concept including the prime enablers of IoT ecosystem, protocols, applications, intelligent technologies such as cloud computing and machine learning (AI) that ensures an actual knowledge-based IoT ecosystem, and the ongoing researches and developments that address various characteristics of IoT. However, several issues and problems related to IoT systems architecture and delivery have been highlighted. Moreover, there were discussions of the interplay between IoT, data analytics, and cloud-fog-edge computing. In turn, that should provide a good basis of insights on the IoT technologies and standards for researchers, developers, and enthusiasts who wish to realize the architecture and role of the various IoT protocols and components.

References

1. Moore, S. J., Nugent, C. D., Zhang, S., et al. (2020). IoT reliability: A review leading to 5 key research directions. *CCF Trans. Pervasive Comp. Interact.* https://doi.org/10.1007/s42486-020-00037-z
2. Khanna, A., & Kaur, S. (2020). Internet of things (IoT), applications and challenges: A comprehensive review. *Wireless Personal Communication, 114*, 1687–1762. https://doi.org/10.1007/s11277-020-07446-4
3. Banda, G., Bommakanti, C. K., & Mohan, H. (2016). One IoT: An IoT protocol and framework for OEMs to make IoT-enabled devices forward compatible. *J Reliable Intell Environ, 2*, 131–144. https://doi.org/10.1007/s40860-016-0027-5
4. Javed, F., Afzal, M. K., Sharif, M., & Kim, B. (2018). Internet of things (IoT) operating systems support, networking technologies, applications, and challenges: A comparative review. *IEEE Communications Surveys & Tutorials, 20*(3), 2062–2100. https://doi.org/10.1109/COMST.2018.2817685
5. Noura, M., Atiquzzaman, M., & Gaedke, M. (2019). Interoperability in internet of things: Taxonomies and open challenges. *Mobile Networks and Applications, 24*, 796–809. https://doi.org/10.1007/s11036-018-1089-9
6. Farooq, M. S., Riaz, S., Abid, A., Umer, T., & Zikria, Y. B. (2020). Role of IoT Technology in Agriculture. *A Systematic Literature Review. Electronics, 9*(2). https://doi.org/10.3390/electronics9020319
7. Shah, S. H., & Yaqoob, I. (2016). A survey: Internet of things (IOT) technologies, applications and challenges. In *2016 IEEE smart energy grid engineering (SEGE), Oshawa, ON* (pp. 381–385). https://doi.org/10.1109/SEGE.2016.7589556
8. Sethi, P., & Sarangi, S. R. (2017). Internet of things: Architectures, protocols, and applications. *Journal of Electrical and Computer Engineering, 2017*, 9324035. https://doi.org/10.1155/2017/9324035
9. Ngu, A. H., Gutierrez, M., Metsis, V., Nepal, S., & Sheng, Q. Z. (2017). IoT middleware: A survey on issues and enabling technologies. *IEEE Internet of Things Journal, 4*(1), 1–20. https://doi.org/10.1109/JIOT.2016.2615180
10. Niknejad, N., Hussin, A. R. C., & Amiri, I. S. (2019). Literature review of service-oriented architecture (SOA) adoption researches and the related significant factors. In *The impact of service oriented architecture adoption on organizations. Springer briefs in electrical and computer engineering*. Springer. https://doi.org/10.1007/978-3-030-12100-6_2
11. Phung, C. V., Dizdarevic, J., & Jukan, A. (2020). An experimental study of network coded REST HTTP in dynamic IoT systems. In *ICC 2020–2020 IEEE international conference on communications (ICC), Dublin, Ireland, 2020* (pp. 1–6). https://doi.org/10.1109/ICC40277.2020.9149026
12. Kaur, N., & Aulakh, I. K. (2018). Clean technology: An eagle-eye review on the emerging development trends by application of IOT devices. In *2018 IEEE international conference on smart energy grid engineering (SEGE), Oshawa, ON* (pp. 313–320). https://doi.org/10.1109/SEGE.2018.8499518
13. Triantafyllou, A., et al. (2018). Network protocols, schemes, and mechanisms for internet of things (IoT): Features, open challenges, and trends. *Wireless Communications and Mobile Computing, 2018*, 5349894. https://doi.org/10.1155/2018/5349894

14. Kang, B., et al. (2018). An experimental study of a reliable IoT gateway. *ICT Express, 4*(3), 130–133. https://doi.org/10.1016/j.icte.2017.04.002
15. Zikria, Y. B., Kim, S. W., Hahm, O., Afzal, M. K., & Aalsalem, M. Y. (2019). Internet of things (IoT) operating systems management: Opportunities, challenges, and solution. *Sensors, 19*(8). https://doi.org/10.3390/s19081793
16. Laaouafy, M., Lakrami, F., Labouidya, O., Elkamoun, N., & Iqdour, R. (2019). Comparative study of localization methods in WSN using Cooja simulator. In *2019 7th Mediterranean congress of telecommunications (CMT), Fès, Morocco* (pp. 1–5). https://doi.org/10.1109/CMT.2019.8931399
17. Amjad, M., et al. (2016). TinyOS-new trends, comparative views, and supported sensing applications: A review. *IEEE Sensors Journal, 16*(9), 2865–2889. https://doi.org/10.1109/JSEN.2016.2519924
18. Takahashi, M., et al. (2009). Demo abstract: Design and implementation of a web service for liteos-based sensor networks. In *2009 international conference on information processing in sensor networks, San Francisco, CA, 2009* (pp. 407–408).
19. Baccelli, E., et al. (2018). RIOT: An open source operating system for low-end embedded devices in the IoT. *IEEE Internet of Things Journal, 5*(6), 4428–4440. https://doi.org/10.1109/JIOT.2018.2815038
20. Souri, A., et al. (2019). A systematic review of IoT communication strategies for an efficient smart environment. *Emerging Telecommunications Technologies.* https://doi.org/10.1002/ett.3736
21. Razzaque, M. A., Milojevic-Jevric, M., Palade, A., & Clarke, S. (2016). Middleware for internet of things: A survey. *IEEE Internet of Things Journal, 3*(1), 70–95. https://doi.org/10.1109/JIOT.2015.2498900
22. Omoniwa, B., Hussain, R., Javed, M. A., Bouk, S. H., & Malik, S. A. (2019). Fog/edge computing-based IoT (FECIoT): Architecture, applications, and research issues. *IEEE Internet of Things Journal, 6*(3), 4118–4149. https://doi.org/10.1109/JIOT.2018.2875544
23. Habibzadeh, H., et al. (2019). Smart City system design: A comprehensive study of the application and data planes. *ACM Computing Surveys, 52*(2). https://doi.org/10.1145/3309545
24. Bengalur, M. D. (2013). Human activity recognition using body pose features and support vector machine. In *2013 international conference on advances in computing, communications and informatics (ICACCI), Mysore* (pp. 1970–1975). https://doi.org/10.1109/ICACCI.2013.6637484
25. Nastiti, M. D., Abdurohman, M., & Putrada, A. G. (2019). Smart shopping prediction on smart shopping with linear regression method. In *2019 7th international conference on information and communication technology (ICoICT), Kuala Lumpur, Malaysia* (pp. 1–6). https://doi.org/10.1109/ICoICT.2019.8835271
26. Handley, C. M., et al. (2010). Potential energy surfaces fitted by artificial neural networks. *The Journal of Physical Chemistry A, 114*(10), 3371–3383. https://doi.org/10.1021/jp9105585
27. Zajmi, L., et al. (2018). Concepts, methods, and performances of particle swarm optimization, backpropagation, and neural networks. *Applied Computational Intelligence and Soft Computing, 2018*, 9547212. https://doi.org/10.1155/2018/9547212
28. Li, Y., Wang, D., & Tang, L. (2020). Robust and secure image fingerprinting learned by neural network. *IEEE Transactions on Circuits and Systems for Video Technology, 30*(2), 362–375. https://doi.org/10.1109/TCSVT.2019.2890966

Artificial Neural Networks and Support Vector Machine for IoT

Bhanu Chander

1 Introduction

Wireless communication technology is one of the major revolutions of this century, and it is continuously expanding. In the past, wired connections produce small connections with limited benefits; however, it is now possible to develop massive networks with enormous benefits with wireless technology. Wireless technology is the result of recent improvements like microelectromechanical devices, protocols, computer networks, and network frequencies. The smartphone-centric set of connections from the olden days is steadily changing into an IoT environment. IoT connects integrated, heterogeneous wireless-enabled devices like smart mobiles, computers, special wearable sensors, connected vehicles, virtual reality devices, agricultural machines, etc. These transformations improve the expansion of wireless traffic controllers and show how to materialize modern and untried wireless service use cases, which significantly fluctuate from standard services [1–3]. The high data rate and traffic analysis play a huge role in the past decade's wireless revolution; in preset days, adaptive and instantaneous appliances of IoTs work efficiently with reliable, low-latency communications. In the coming days, wireless sensor nodes will continuously gather vast quantities of data in a real-time method with the numerous sensing plus wearable accessories that supervise the physical atmosphere.

Such enormous short-packet broadcasting resolves direct extensive travel over the wireless uplink, which is usually fewer overcrowded than the downlink. Naturally, wireless technology was extensively applied to cloud computing, real-time video streaming in multimedia services, virtual reality, useful healthcare reports, etc. This eventually generates a dissimilar networking atmosphere whose original claims moreover their various quality of service (QoS) along with

B. Chander (✉)
Department of Computer Science and Engineering, Pondicherry University, Pondicherry, India

S. Pal et al. (eds.), *Artificial Intelligence-based Internet of Things Systems*, Internet of Things, https://doi.org/10.1007/978-3-030-87059-1_3

consistency necessities consent a deep-seated revolution in how wireless sets of connections are replicated, intended, optimized, as well as explored [1–6].

The need to manage and handle this new and continuous progression of wireless technology services and devices has shown the way to extensive research areas to explore the optimal wireless, mobile networks within the circumstance of the talented next-generation wireless-based 5G and 6G setups. To date, there are significant factors of 5G networks like millimeter-wave and device-to-device communications, miniature cell employment techniques designed and implemented. But mixing them within a tuneful wireless scheme that can light up the IoT contests, requirements and implanting intelligent roles through the margin/boundary and the center of the network is tricky [3–5].

These intellectual models and roles should be talented to intelligently operate the wireless scheme resources plus the engendered data, to enhance the scheme actions in real time and the QoS desires of good wireless as well as IoT services. These kinds of computing and intellect can be potentially comprehended by incorporating machine learning- and deep learning-based techniques across wireless transportation and abuser devices.

2 Preliminaries

2.1 History of Neural Networks

Neural networks' innovation started in the early days of 1943 with a single neuron by two well-known researchers McCulloch and Pitts. The designed model describes the neuron as a linear threshold with multiple inputs but produces a single output either 0 (inactive) or 1 (active) if the cell exceeds the specified threshold value. After some years, Hebb states the existing rules to support the links among neurons; in the year 1958, Rosenblatt designed a matured neural network with a perception unit based on McCulloch and Pitt's model, and the designed model produces output ranges from 1 or − 1 depending upon the weighted linear arrangement of inputs [6–10].

Rosenblatt, Widrom, and Hoff discover different perception-based artificial neural networks in the early 1960s; however, it takes rapid encouragement between 1970 and 1980 among various contemporary scientists and researchers. Moreover, these innovations' primary reason is the difficulties the world faces during their period; it stated that perception was incompetent to represent simple functions that are linearly undividable. The era of ANNs started when Hopfield stated new limitations into the neuron field; he introduces major concepts like nonlinearity between total inputs received through a single neuron and the output it generated; another concept was the feedback possibilities coupling of outputs with the appropriate inputs [4–8]. In present days, mostly ANN is used for solving real-life problems than in improvising accurate representations of the neurons; ANNs are extensively applied in chemistry, physics, security, finance, geology, medicine, and engineering.

Initially, ANNs are very complicated nonlinear computational tools which are proficient of modeling hardcore complex function. Notably, the relation among input neurons and corresponding output neurons can be visualized with appropriately selected ANN architecture. Within these structures, neural networks also maintain a continuous check on the curse of dimensionality issues. Also, neural networks will learn by data records, which means the model is automatically trained from representative data points. However, the designers or abusers must have some basic knowledge of how to select and prepare data and choose a suitable neural network and what basis detailed data successfully applied for neural networks to produce effective results.

2.2 Role of ANNs in Wireless Sensor Networks

From the invention, machine learning (ML)-based models are undeniably one of the central mechanisms for wireless networks in finding optimal solutions. However, in the circumstances of WSNs and IoTs, ML algorithms will allow every wireless device to dynamically and wisely observe its surroundings through learning, predicting the progression. With the development of a mixture of ecological features such as abuser context, traffic patterns, content requests, dynamic wireless channels adoptions, etc. and based on these proactive procedures that make the best use of predefined objectives [1–4, 8–13, 60–64]. ML allows the network communications to become skilled at wireless networking background and take adaptive system optimization procedures. As a result, ML is likely to show many parts in the coming production of wireless networks 5G and 6G setups.

There are some extreme reasons for applying machine learning models to WSNs and IoTs; those are:

1. It is a known fact that applying ML to wireless technology is to develop the device or nodes intelligence and improve predictive data analytics to boost up the situational consciousness besides overall network actions. In perspective, ML endow with the particular WSN and make the capability to parse through an enormous volume of data accumulated from several grounds. Sensor interpretations to buzz with surveillance imageries toward building an overall outfitted plan to optimize the enormous number of devices contained by the network.
2. ML will also provide a primary driver of intellectual data-motivated techniques for system optimization. This can report a wide range of glitches equivocating from cell connotation and then radio access expertise collection for power management, frequency sharing, intellectual beamforming, and efficient spectrum management. In the past, traditional optimization models were applied in offline and semi-offline approaches, which produced limited benefits. In contrast, ML showed that resource management devices would be talented in an entirely online style by learning the wireless system's real-time conditions.
3. ML models discover the best features by online learning, which continuously improves their performances, allowing cleverer and self-motivated decision-

making. Such kind of self-learning is essential for IoT, 5G, and 6G services, which require real-time, low-latency operations in drone and industrial device controls. Accurately designed ML and DL techniques will produce effective automated and intelligent optimization solutions for a wide-ranging range of glitches within the background of sensor plus IoT resources.
4. An ML algorithm effectively enhances wireless communication's physical layer appliances like levels of receiver and transmitter, coding, and modulation within the network circumstances. This will improve delivering lower bit error rates along with efficient robustness in frequency channels.
5. Finally, the rapid development of well-designed client-based central wireless services like virtual reality, robotic, and drones where the incorporation among clients with associated network roles is roughly nominal tremendously encourages the prerequisite for WSNs to be adapted to the human user behavior.
6. Moreover, intellectual as well as predictive data analytics improve the wireless network capacity to smartly practice with the tremendous amount of data records, collected from heterogeneous devices, to explore and foretell the situation of the wireless abusers and then the wireless grid's ecological circumstances, as a result empowering data-motivated network-wide equipped decisions.
7. The wireless system's capacity to energetically study the wireless surroundings moreover wisely directs the wireless system and improves its properties affording to data learned regarding the wireless atmosphere to end-users situations.

Based on the above rules, it is visible that ML is the first mechanism that stands proficient to gain knowledge and imitate humankind's actions; this increases or assists wireless schemes to adapt various real-life functions toward creating a right immersive situation for its abusers as well as maximize the overall quality of services. ML kind algorithms designed for wireless technology will recommend to an inexpressibly high standard set of novel system tasks along with wireless services. In the future, 5G and 6G networks will incorporate significant theories and techniques from ML, DL, and nature-inspired approaches. The significance of ML-driven wireless setups has already been provoked through many current wireless networking examples like mobile edge computing, fog computing, and context-aware networking. Nevertheless, a higher percentage of ML techniques is applied for a variety of operations like customer performance analysis along with estimations to determine which contents to assign computing resources [18–23]. On the other hand, regardless of their consequence, these workings have a contracted spotlight and shed light on the confronts moreover prospects connected with ML's utilization for designing an intellectual wireless set of connections.

2.3 ML Models in IoT

Machine learning prepares computer models to learn or educate on their own without providing any explicit code. ML-based approaches examine historical data to execute the desired operations or assignments. For this, ML needs two kinds of data records. The first is called training data, to train the learning algorithm offline or online; another one is test data – this is for evaluating the model's performance on an unseen dataset. From the year 1959 of initiation, ML has been effectively utilized and made tremendous changes in finance, social media, healthcare, education, and smart devices. It is also employed in the manufacture of chatbots plus automated smart devices – software for facial identification, natural language processing, etc. In recent times, ML models have also originated as a well-accepted appliance in the IoT systems [1–6, 14–18].

Internet of Things (IoT) is a connection of mixed devices that are capable of storing and sharing data, and the word Internet of Things (IoT) is first coming from the famous researcher Kevin Ashton in 1999. From the day of introduction, IoT boosted growth rate in diverse fields like healthcare, agriculture, smart-wearable devices, pollution monitoring, and whatnot. An optimal IoT system works with three phases: first sensors, actuators, and smart devices employed to determine physical parameters which then transformed into electrical signals for transmission; second acknowledged signals uploaded on the system with wireless communication arrangements; and third uploaded data – data processed in the course of specific methods for preferred précised results [2–8, 16–24, 60–64]. Sensible devices respond with alterations in their physical environment, which plays the primary role of storing and sharing data records and outfits with ML approaches for more rapid and unconstrained processing. ML methods are used globally to boost the processing of the data gathered by these devices. ML techniques are faster, consistent, and more protected.

The main intention to bring or insert ML techniques in IoT is to diminish errors and simultaneously enlarge the processing speed. Bayesian networks are used for reducing error and noise when data are transmitted from device to device, neural networks to extract the high-level features, then random forest, and clustering for classification of data records and recovering lost data from IoT-based devices.

2.4 *Artificial Neural Network Preliminaries and Types of ANN*

Artificial neural networks (ANNs) are motivated by the formation of human brain functional features of biological neural networks and discover high rated features from the complex raw data. ANNs are highly suitable for WSNs to investigate, forecast network and abuser actions to produce the decision-making information, and solve diverse wireless networking problems like spectrum management, cache information analysis, adopting frequency modulations and computational resource allocations, etc. In recent days, smart IoT devices and mobile appliances have mostly increased users' concentration with mobile Schemes [16–26]. A well-designed and skilled ANN can contemplate as a professional in dealing human-related information. Thus, exhausting ANNs to extort data from the abuser's surroundings can offer a wireless system with the capability to forecast the abuser's potential activities and, therefore, to plan the most acceptable approach to progress the resulting QoS and consistency which is still a tricky task.

Modular Neural Networks (MNN) A MNN is a mix of more than a few autonomous ANNs and conciliators. In MNN, every ANN exploits to solve a solitary subtask of the complete mission an MNN desires to execute. Here, conciliators utilized for practice the output of every autonomous ANN and then produce the productivity of an MNN.

Recurrent Neural Networks (RNNs) RNNs make possible links between neurons in one layer toward neurons in preceding layers. Depending on the various models, activation functions, and theories for the neuron, RNNs are defined in multiple architectures. Stochastic neural networks, long short-term memories (LSTMs), bidirectional neural networks (BNNs), echo state networks (ESNs), fully recurrent neural network (FRNN), gated recurrent units (GRUs), neural Turing machines (NTMs), etc. are some of them.

Generative Adversarial Networks (GANs) Generally, GANs are built with twofold neural nets, one neural system utilized to acquire a record from a hidden space to specific data distribution, whereas an additional neural grid employed to distinguish among the right data distribution and then the dandling plotted via a neural system.

Deep Neural Networks (DNNs) All the artificial neural networks based on multiple hidden layers are acknowledged as DNNs.

Spiking Neural Networks The spiking neural nets contain spike-type neurons that imitate or copy the genetic neural systems.

Feedforward Neural Networks (FFNN) In FFNN, every neuron contains incoming links from the previous layer and, in the same way, makes outgoing links with the successive layer. Convolution neural networks (CNNs), radial basis functions (RBF), and auto-encoders are some of the advanced architectures that come under FFNN.

Physical Neural Networks (PNNs) In PNNs, the primary function of neural activation is design with a fully adaptable, electrically resistant substance. Here both RNN and SNN are utilized for different suitable learning jobs. For example, RNNs are effectual in dealing with time-reliant data records, whereas SNNs successful in trade with invariable data. It should be renowned that utmost of the records composed of wireless nets is time-dependent and continuous data. However, RNNs or SNNs can witness only a partial chronological data size; they may not be talented to resolve every wireless transmission trouble. To resolve composite wireless malfunctions, one can utilize DNNs with a high memory volume for data analytics and then disconnect the composite crisis that desires to be learned into an arrangement of numerous more straightforward difficulties, hence constructing the learning procedure valuable.

2.5 *Modifications of ANN*

Neural Networks Artificial neural network (ANN) or else neural network (NN) is a supervised-based knowledge model that follows the human brain's structure to derive multipart, nonlinear decision limitations for effective classification models.

A simple ANN model contains one input, several hidden layers, as well as one output layer. Here, the input layer corresponds to the input data variables. The hidden layer corresponds to process essentials called neurons that develop their inputs using activations (linear, sigmoid, and hyperbolic tangent functions) deciphers the input signals to output signals. Links excessively link the elements among every layer with numeric weights that are well-read by the model. The output layer shows the predictions for the given inputs based on the interconnection weights described during the hidden layer. Neural networks gain more and more attention from researchers because they can estimate any attention when tuned sound. The ANNs train to recognize how ecological dimensions and the setup condition change the performance knowledge on different channels. Based on this data record, it is feasible to energetically decide on the control channels, which is likely to yield the most excellent concert for mobile abusers.

Deep Learning The neural network which holds numerous hidden layers with the same complexity as a single hidden layer network is called deep neural networks (DNNs), and the procedure of learning acknowledged as deep learning (DL). DNNs hold the latent to restore the progression of manually obtaining important structures that depends on previous domain data with unsupervised or else semi-supervised feature knowledge approaches. CNN, RNN, and DBN are some of the models in DL that have shown success in numerous domains.

DL use cases in wireless technology as follows: RNN has been practical to WSNs; the authors planned a distributed learning form by sharing the neural network into different layers and then positioning them on numerous sensor nodes. This will effectively reduce the number of data records for training. Software-defined networks (SDN) and unsupervised optimization techniques, flexible network connections, error recognition, and many more are dramatically included DL as a good part.

Radial Basis Function (RBF) RBF has tremendous influence in ANNs but is not expressively utilized in modern appliances. Since the mentioned systems/nets are not powerfully built, they logically utilized two layers. The top layer is fabricated in an unsupervised mode, and the subsequent layer proficient with supervised models. RBF, a completely different working procedure from feedforward systems, argues that the former advances its power by increasing the range of the feature's tempo moderately than intensity. RBF models are closely related to the Cover's theorem on the separability of prototypes, and here pattern categorization issues are mostly related to linearly independent when transmitting toward vast amounts with a non-linear alteration. The former layer consists of activation with the comparison of the input data to the archetype. These initiations are then pooled with trained weights of the subsequent layer to generate an ultimate forecast. This advance is very alike to the nearest-neighbor classifier; apart from the second layer's influences, it offers a supplementary administration level. RBF networks are efficient in building simulation-based kernel techniques like support vector machines, especially problems like classification and clustering models. RBF networks' unlimited possibilities remained new since this style has been largely beyond the enlarged concentration on vanilla feedforward approaches.

Restricted Boltzmann Machines Restricted Boltzmann machines (RBMs) construct networks for indicating data info in an unsupervised mode with limited power. These networks closely related or follow probabilistic graphic models, slightly origin from Hopfield networks for storing recalls/memories. Stochastic modifications of these nets were unsystematic to Boltzmann types of machinery, in which hidden layers exhibited multiplicative characteristics of the data. RBMs are effectively utilized for unsupervised modeling and reduction of dimensionality; however, they are also applied for supervised modeling. RBMs are the first approaches used to deep learning and afterward adopted to other models since they have a historical impact in inspiring several training models in-depth.

Recurrent Neural Networks RNNs are specially intended for successive data records like text, time series, plus biological categorizations. For instance, the input is in the form of x1 . . . xn, where xt is a d-dimensional data point acknowledged at the time-stamp t. Here the vector xt encloses the d standards at the tth mark of a multivariate time series. The vector xt contains a one-hot encoded term at the time-stamp. Here a course of distance equivalent to the lexicon scope, besides the ingredient for the appropriate term has a cost of 1, further apparatuses 0. A central point around categorizations is that consecutive words are reliant on each other. So, it is obliging to obtain an input xt only afterward the former inputs take previously established and renewed toward a hidden state. Thus, RNN permits the input xt to interrelate straightforwardly via the hidden state formed from the previously recreated inputs at preceding time-stamps. Consider an example; the question is to solve or try to categorize a sentence's viewpoint as positive or else negative; the productivity only happens at the last time-stamp. The hidden state-run at time t is specified by the input vector's role at time t along with the hidden vector at the time (t − 1): ht. = f(ht − 1, xt). A separate function yt = g(ht) is pragmatic to expose the output prospects since the hidden states, here both roles f(·) and g(·), are similar at every time-stamp.

An essential property of RNN was that they are tuned well and complete means for the appropriate data and resources they can execute or learn any model. The quantity of data plus the scope of the hidden circumstances mandatory for longer categorizations enhances in a move that is not sensible. Additionally, there are real-world questions in sighting the most favorable limitations since the vanishing and the explosion of gradient troubles. So, dedicated modifications of the RNNs have been featured, such as the employment of LSTM.

Convolution Neural Networks CNNs are proficiently applied for image processing and pattern recognition tasks. The fundamental aspects of CNN are attained from Hubel and Wiesel's work of understanding the cat's visual cortex, where explicit parts of the visual field appeared to stimulate finicky neurons. In CNN architecture, every layer of the setup is three-dimensional and has a three-dimensional degree along a depth equivalent to the sum of features. The perception of depth in a CNN single layer is dissimilar from the perception of deepness in terms of the numeral layers. For example, in the input layer, structures match the color frequencies as red, green, and blue, and in the hidden frequencies, these structures symbolize to hidden feature maps that translate a variety of outlines in image. For layers, a convolution procedure is definite as a filter to plot the stimulations from layer to layer as it moves deeper. A convolution process utilizes a three-dimensional filter of weightiness with a similar deepness as the existing layer though a slighter spatial range. The dot product among every weight in the filter with any first rate of a spatial section describes the hidden state's value in the next layer. The filter plus the spatial sections in a layer is achieved at every probable place to define the next layer. The CNN associations are very meager since any initiation in a meticulous layer is only a small spatial section in the preceding layer.

However, the weights in previous layers are still functional since they discover several forms in the imageries that can be valuable for almost each kind of cataloging appliance. Besides, the feature initiations in the last layer can still be utilized for unsupervised appliances. Consider an instance, by building multidimensional demonstrations on uninformed image datasets via forwarding through the CNN, and try to extort high features. These operations will produce fantastic results in image recovery due to the semantic nature of the features learned through the setup.

Generative Adversarial Networks GAN is archetypal for data-generation, generating a generative model on the dataset by employing adversarial among two players like generator along with discriminator. The generator takes Gaussian noise-based data as input and then harvests the corresponding output. They are coming to discriminator, which characteristically works like classifier-based logistic regression to discriminate actual samples from the dataset and then the generated sample. Here the purpose or the role of the generator is to build data samples as realistic; it means trying to bamboozle or fool the discriminator; on the other end, the discriminator recognizes the forged or false samples regardless of how good the generator attempts to jester it. These models could help create sensible fantasy samples by using a dataset, which is utilized regularly in the image area.

Consider a sample, the model trained with the royal house's images; model builds realistic-looking royal homes that are indeed a part of the base data. So the model was effectively employed to generate artistic, creative samples. These techniques are also applied to a definite type of context, which might be labels, text captions, or images with missing facts.

Echo State Networks Echo state network (ESN) is an explanation of RNN models, which measure professionally with the measure of sequential elements but not with dimension of the input. This kind of network helps a small quantity of real-esteemed time successions over a long-time vision. ESN approaches are also acknowledged as liquid state-run machines; apart from that, it next utilizes spiking neurons with binary outputs, while ESNs use old-fashioned activations such as the sigmoid and the tanh.

ESN applies arbitrary weights in hidden-to-hidden layer and the input-to-hidden layer, even though the size of hidden shapes is nearly superior to the size of input states. It is a reminder that the output layer's preparation combines the faults at different output nodes, even though the weights at other output nodes are fixed communal.

Long Short-Term Memory RNNs work proficiently with time-series data but suffered from vanishing gradients. One best solution for this is for neural networks that utilize only multiplicative updates to be better only at learning over the short sequences; moreover, it is intrinsically capable of high-quality rapid-term memory although underprivileged long-term memory. There is a need to modify the recurrence calculation for the hidden vector by utilizing the LSTM's long-term memory to solve the mentioned issues. Here, the LSTM was developed to hold a well-defined switch on the data records transcribed toward long-term memory. These vectors are mostly abstractly utilized, for example, Boolean gates designed for determining whether to affix to a cell state or not recall a cell state or else whether to tolerate leak within a hidden state against a cell state.

In layer-wise neural network background, it is vital to effort through constant operations to ensure the dissimilarity requisite aimed at gradient modernization. The vector encloses with the recently planned substances of the cell status, even though the input plus forget gates normalizes how it is allowable to transform the initial cell state.

Gated Recurrent Units (GRUs) The GRUs slightly looks like the implication model of LSTMs, where it does not utilize any clear cell conditions. LSTM straightforwardly controls information changes in hidden states with its forget plus output gates. Besides, a GRU employs a solitary retune gate to accomplish a similar goal. Here, the simple difference between RU and LSTM was partially resetting the hidden states of the network.

Deep Reinforcement Learning (DRL) The DRL is a reward-driven procedure, where a scheme studies to interrelate with a composite surrounding to attain the best conclusions. It is like a gateway to producing accurate intellectual agents like gaming models, automated cars, devices cooperating with different atmospheric conditions, etc. DL has made a tremendous contribution in various fields, which makes humans live their lives comfortably. Some of them are like DL that has a huge role in video games' background, including raw pixels in videos as feedback. Atari 2600 is a famous video console that comes from DRL; the Atari's deep learner's input demonstrates pixels from the game's current condition. Here, the RL technique forecasts the acts constructed on the demonstration and inputs into the Atari.

Naturally, the system makes many errors, those replicated in the computer-generated rewards through the console. If learners achieve the best things from the faults, it makes superior conclusions. Which precisely in what way do humans absorb to perform in video games? The appearance of the latest model on the Atari plinth was exposed to exceed human-level concerts for an excessive quantity of games. Video games are the best example for viewing the excellencies of DRL algorithms since they observe as extremely edited versions of choices one must formulate in numerous decision-centric backgrounds. DeepMind has trained with DL-based AlphaGo, for the video game Go via using reward outcomes in the changes of games drained from human as well as computer self-play. Generally, Go is a challenging game that needs tremendous humanoid perception and an incredible ability to build a gaming model. One more innovation that got attention in DRL is self-driving cars utilizing different sensors' reaction to make conclusions. These cars now continuously build less faults during driving than do human beings, which will learn and make appropriate actions. Now DRL chases for creating automated robotics; training a robot to walk can be implied as an RL mission; in the DRL, we only incentivize the automaton to acquire from point A to point B as proficiently as achievable by its accessible members along with motors. It concluded as reward-guided, trial-and-error appliances trained to rotate, move slowly, and then finally walk.

In other words, DRL is appropriate for easy-to-estimate tasks but tough to indicate their functionalities. For instance, it was simple to guesstimate the player's act after the game, but it is tough to estimate the player's behavior at every game movement. DRL affords a simplified path for problematic activities by defining the reward and then allowing it study reward-maximizing performances. The complication of these performances or actions is robotically inborn from the surrounding atmosphere.

3 Applications of ANN in WSNs and IoT

Recent innovations confirmed that ANNs have started to draw attention from various real-life appliances in wireless communications and IoT schemes. Moreover, smart device appliances' expansion has drastically amplified wireless systems' independence and abuser's interaction with a wireless-based IoT system. Additionally, the advance of the IoT inspires employment of ANNs to advance how wireless data is practiced, composed, and used for a variety of sense and self-determination reasons. Given wireless communication fields, ANN is applied on wireless nodes or IoT devices based on the two primary purposes: Initially ANN is best for forecast, inference, and intellectual data analytical abilities. This is very needed in WSNs since nodes from the deployment continuously accumulate real-world data from abusers and environments and connected network devices. Moreover, ANNs are also employed to study or estimate the wireless user's mobility prototypes and content requirements. In IoT-based smart cities, inside a smart-city environment, deployed sensors observe the huge amount of data used with the wireless network to enhance its resource practice, realize its system process, observe letdowns, or just distribute smart-services smart transportation and goods tracking. Additionally, ANNs can quickly discover sensitive features from huge sensor data cross-correlations, resulting in more truthful estimated wireless network circumstances and a proficient distribution of the existing resources (see Table 1). The second reason for applying ANNs to WSNs is to enable automated network operations at the edge of base station and IoT devices. These kinds of automated edge computing and data offloading are best suited for resource-constrained WSNs. An

Table 1 Advantages and challenges of ANN models

Model	Input data/ sensed records	Advantages	Challenges	Drawbacks
Recurrent neural networks (RNNs)	Time-series/ time-dependent data	Efficiency in processing time-related correlation data like wireless traffic and channel allocation	Limited power and computation for training ANN	Difficulty in training due to the loose connections among dissimilar neurons
		Proficiently catches self-motivated activities and ability use	Errors in training data limited computational resources	
		Information from sequential data	Real-time training for ANNs	
Stacked neural networks	Continuous data	Ability to perform multiple learning tasks	Data cleaning/ content sorting	Individual training model needed for every SNN training task is complicated
		Effectiveness in performing continuous data	Partial sorting of ANNs for recording every form of insides	
		Large memory offered for data gathering		
Deep neural networks	High-dimensional data/large datasets	Better learning capability in comparison to shallow ANNs	Channel selection Load balancing	Massive data required/ Computationally intensive to train
		Discover low-dimensional data for huge datasets	Mobility prediction, Real-time training of ANNs	

ANN-based reinforcement learning (RL) approaches can be employed to gain knowledge of the abusers or client information like locations, the flow of data rate, and the learned information that discovers the UAV's pathway. As a result, ANN-based RL approaches can efficiently discover complex associations among wireless abusers plus their networking situations to explain demanding problems optimization and resource management.

3.1 UAVs for Wireless Networking Communications with ANN

Possible connection or coverage from the sky to earth-based wireless devices or abusers is an emerging technique in WSNs. In comparison with earth or ground-based transportations, a wireless skill thru low-altitude UAVs is easy to employ. In addition, a resilient reconfiguration is possible to a great practice, and enhanced transmission frequencies suitable to the existence of short-range and line-of-sight links. There are few challenges compared to benefits while using highly developed, energy-constrained UAVs like optimal employment, energy efficiency, security, limitations of UAV-to-UAV communications, and path selection. Thus, it is requiring investigating the most favorable employment of UAVs for coverage expansion plus capacity progression. At present, high parts of UAVs are utilized for data collection, transportation of telemetric, and goods. Therefore, there is a high prerequisite to progress intellectual and automated controller procedures to augment UAVs' airborne paths.

UAVs are broadly utilized in urban areas to collect user's behaviors and collect information associated with the abusers, plus the UAVs inside any distance, anytime or else anyplace, which helps an idyllic setting for the accomplishment of ANN procedures. In UAVs, ANN has two most significant use cases: the first is the ANN-based RL model for automated scheme decisions, path choosing, user's dynamic environment, and resource allocation decisions. In the second, UAVs can plot the ground surroundings and the wireless atmosphere itself to gather information, take gain from ANN-centric methods to triumph the composed sensed information, and execute data scrutiny to forecast the ground abusers' activities. By analyzing behavioral examples of abusers or devices, power management-based UAVs can efficiently regulate their respective ideal positions and plan an ideal flying route to service ground operators. In the meantime, ANNs allow more highly developed UAV appliances in atmosphere documentation.

ANNs can efficiently contract with time-dependent records that formulate them a realistic option for UAV-based wireless communication appliances. However, UAVs face various issues while deploying and designing with inadequate flight time to accumulate data records and authority to perform computation operations. UAV models do not hold enough data to train ANN models, and UAVs are considerably affected by weather conditions and movement. Thus, the composed data contains contaminated data records, faults that might distress the accurateness of the outcome of the ANNs.

Future Works It is a known fact that ANN is an essential tool for tackling critical tasks in UAV-based communication systems. In detail, ANN-centric models and tools efficiently work with various proper UAV appliances like dealing with time-dependent data and user behavior analysis, etc. This permits UAVs to enhance their position based on the changing aspects of the set of connections. Based on this information and predictions, the UAVs can uncover their most favorable route and decide which region to serve at any given time.

3.2 Wireless Virtual Reality Over Wireless Networks with ANN

In recent days, wireless-based technology-based industries are heavily attracted by virtual reality, which is very important in the appliance in 5G and next-generation networks. As a result of various networks, developments inspire industries to investigate wireless VR with ANNs in upcoming wireless systems. If a VR stratagem is performed through a wireless linkage, customers or clients forward the tracking records like the client's localization and dynamic movements to the BS. Then, BS analyzing the collected data to build the 30-degree pictures is forwarding them to abusers and clients. For this reason, both uplinks and downlinks transmissions are considered equal. So while designing proper and efficient VR over wireless networks, challenges like high data rate, tracking accuracy, effective image compressions, user experience modeling must be considered.

The inclusion of ANN-based ML techniques resolves the numerous issues related to wireless virtual reality. Compared to other appliances, VR appliances require clients surrounding and environmental behavior. ANNs are valuable at classifying and predicting the abusers' activities. Based on the estimates of the abusers' atmosphere, procedures, and activities, the BSs can progress the making. As we mentioned in the above sections, ANNs can be intelligent into progress self-organizing models to control and then accomplish wireless VR net, hence discussion troubles like self-motivated resource managing. But it is challenging to apply ANNs to VR: in wireless networks, information or data records collected from abusers might contain many errors; in detail, BS might prerequisite to benefit inaccurate data to train the ANN, then guess the accurateness of ANN considered unnatural. Moreover, VR holds 360-degree images that may be large, so BS must use a significant number of computing properties to practice VR imageries. Therefore, efficiently allotting the computing properties for handling VR imageries and then exercising ANNs is a crucial dispute.

Future Works The above section clarifies that ANNs tools are promising equipment to address various challenges in wireless-based VR appliances. Wireless spectrum allocation can proficiently handle ANN approaches, and then, the system can decrease data volume of all broadcasted VR audiovisual. RNNs can expect and sense the VR abuser's progression like eye and head movement and their connections with the atmosphere. Typically, user and VR scheme interrelate with time-dependent data records, so RNN-based approaches are good choices. Presumption of the abuser's association will straightforwardly shape the VR images sent to the abusers at every time slot, consequently employing RNNs that are calm to training for the calculation of the users' measurements.

3.3 *Coexistence of Manifold Radio Access Machineries with ANN*

The imagined 5G and 6G network services are the multiple choices for every user throughout the network to handle increased deices and data traffic. This kind of augmentation can be achieved over innovative PHY/MAC technologies and professional spectrum organization. The critical advantage of 5G and 6G was integrating multiple dissimilar radiofrequency techniques like cognitive radio schemes and LTE-U setups. So, incorporating different radiofrequencies will improve the exploitation of the existing radio resources and the scheme's abilities. It similarly assures a reliable service practice for particular operators regardless of the assisted radio technologies; furthermore, it assists the system administration. Spectrum supervision is also considered an additional critical section of multi-radio access technology-based networks, unlike initial age group cellular systems that function entirely on the sub-6 GHz (microwave) certified tape. Multi-RAT founded systems predictable to broadcast over the usual sub-6 GHz tape, the unrestricted range, and then the 60 GHz mm-wave frequency tape. Hence, a multi-mode BS working on certified, unlicensed, plus mm-wave frequency ranges can develop special uniqueness and accessibility of the frequency ranges, thus affording healthy and consistent communication associations for the operators.

Neural Systems for Spectrum Administration and Multi-RAT Multi-radio technologies resolved various issues with the inclusion of Artificial neural networks (ANNs). ANNs can consent to the smart employment of particular RATs where a BS could train when to broadcast on individual kind of rate of recurrence collection built on the fundamental system circumstances. Furthermore, ANNs can offer multi-mode BSs with the capability to discover the suitable source managing process over particular RATs or spectrum tapes in an online mode. Most mobile data traffic demonstrates statistically changeable and episodic demand designs, particularly for appliances like file transference, video flooding, and browse styles.

To handle dynamic changes in traffic consequences, ANN deals with RL-based ML models through appropriately retraining the weights at the learning mechanism. With suitable network proposals, ANNs can permit the operator to advance their network's concerts by tumbling the possibilities of congestion and degree of fairness to the other equivalent expertise in the setup.

Future Works Appliances of ANNs to LTE-U schemes can be effortlessly completed with a multi-mode set of connections, where the BSs broadcast on the certified, the unlicensed, plus the mm-Wave range. RNNs can improve mobility in highly dynamic mobile wireless locations via learning the flexibility prototypes of abusers. Antenna design modeling and learning are the most prominent futures in ANN with radio access technology. Since various learning structures of the system state, and the best possible position on obtainability of LoS bonds, can be effectively handled with DNN models. Besides, LSTMs are best suitable for learning long time series; this will permit BSs to forecast the link structure for the mm-Wave system.

3.4 Internet of Things with ANN

IoT is a heterogeneous network with numerous machine-kind devices like sensors, wearable devices, vehicles, and other objects connected with the Internet. Most of these devices connect in wireless connectivity to operate various activities in an automated manner. IoT-based devices collect real-world information to provide appropriate user services in the domain like smart home, smart city, h-healthcare, transportation, etc. However, IoT devices' suitable employment faces many challenges like security, invaluable data transmission, data analytics, connectivity, and computation capacity. The up-to-date central base communication models are not enough to handle such massive connectivity, so there is a requirement for good and intelligent computing replicas for IoT devices connectivity. Moreover, how to adjust or allocate computational and energy resources to all available IoT devices is an additional test.

Neural Networks for the Internet of Things From the past decade, ANN effectively addresses several critical disputes in WSNs and IoT. ANN makes the IoT devices intelligently extort the key patterns and associations from the data collected from different devices, improving the data compression and data retrieval capacity. Moreover, ANN approaches improvise the quality of human life by reducing the interface between humans and IoT machines; by doing this, ANNs effectively predict or guesstimate abuser's actions, which will help IoT devices. By employing various ANNs tools in IoT schemes, we can limit both energy and computational IoT devices to collect dissimilar data types, which may be in different data structures and several various errors. So, in the context of IoT, when data are used on ANN models training, one should believe how to categorize the data then compact with the faults and missing values in the data. In IoT schemes, ANNs take advantage of dissimilar forms of data aimed at forecast along with the automatic controllers. In some cases, for a given job, the data records composed from the IoT strategies might not be correlated to the job; here ANNs pick relevant data for the mission or job. While every IoT device's computational resources because of their manufacture or design or industrial background, these IoT devices with different resources are mapped with different neurons. For instance, an IoT stratagem that has different computing properties will plot to various neurons. There are numerous ways to plot the IoT networks for optimal connections designed based on detached functions and diffuse power.

Future Works ANNs are a useful tool to handle IoT problems like intelligent data analytics and execute smart actions. Additionally, ANNs effectively utilized data compression and the recovery process to forward the bulk of information to the end-user devices. For data compression, ANN-based CNNs are employed to extract the valuable features which are very useful for data firmness and data retrieval besides RNNs appropriate aimed at obtaining the associations from time-supported sequence data. DNNs are accurate for human-device integrations because they have numerous hidden layers to hoard more data records associated with an abuser and evaluate with other ANNs.

4 Support Vector Machine for WSNs and IoTs

SVM is a well-designed, modern, and famous classification approach designed by Boser, Guyon, and Vapnik in 1992. It is extensively applied in big data analytics and gene expression, where it deals with different data structures. SVMs effectively deal with a broad type of kernel approach; usually, a kernel is a model depending on the dot products in high-dimensional feature space. SVM gains two benefits: The first benefit is the power to produce nonlinear decision limitations with the linear classifier models. A second benefit, exploiting kernel functions allows the abuser to affect a classifier to data records with no apparent fixed-dimensional vector space depiction.

When preparing an SVM, the practitioner desires to create several conclusions, like how to process data proficiently, which kernel model is appropriate, and how to fix the SVM model parameters. If any of these conclusions are unsuited, then the results will be diverse.

Authors of [23] designed an SVM-based approach for simple communication in WSNs. The simulation results of the designed model improve the efficiency between the communications among sensor nodes with short message exchange between neighbor nodes. In [24], the authors improve the link quality among nodes by employing an SVM-based decision tree model. Here based on the received signal strength and link quality indicators, SVM determines the communication superiority. In [25], an SVM-based sliding window with PCA is implemented to detect WSNs and IoT outliers. Here authors utilized a modified RBF kernel with least-squares SVM for anomaly detection in nonstationary time series. Authors of [26] extend the SVM to SVDD to better categorize outliers from sensed data and reduce the computational complexity compared to other SVM-based models. Authors designed two-order base sequential minimal optimization (SMO) to lessen the training period's complication in testing a fast-decision-making approach employed

[27]. explains the detailed study of various event or outlier detection approaches based on SVM-based variants and found that quarter sphere model has low complexity and better performances. In [28], a novel DBSCAN model is designed to detect outliers in WSNs accurately; moreover, the authors employed modified SVM for accuracy boost in low-density regions [29]. An essential advantage of SVM classifying with kernel functions utilized for decision-making with incorrect or error included sensor data. Authors of [30] proposed an error prediction model in dynamic situations with an SVM-based cuckoo search approach. In this model, the cuckoo search is implemented to avoid local minima. Routing plays a significant role in data communications, so authors of [31] projected an SVM-based routing protocol for optimal energy consumption. Experimental marks display that the archetypal achieved a high lifetime compared to other existing models. In [32], the authors improve the congestion control in WSNs; in [33], the authors designed a multi-agent-based data collection model to efficiently utilize data traffic models.

4.1 Support Vector Machine in Radio Access Frequency

Radiofrequency-based energy or power harvesting is quite a simple process that transforms the wireless RF energy toward collectible DC power. This will be a cutting-edge technology like deployed sensor nodes; Internet-connected devices will profit from this expertise to influence their respective operations or activities and awaken from sleep mode. However, harvesting wireless devices and sensors facing various limitations like device configurations, locations, and the number of abusers are connected. Hence, based on these issues, there is a need for an advanced technique that assists in resolving device issues for the maximum input power. An appropriate ML approach must be vigorous to noise, precise, and computing resourceful in categorization approach. The decision, random forest trees, and linear regression tree models performed well in real-time energy management in the radio band. Metaheuristic optimization practices like swarm intelligence models and genetic algorithms are employed to discover the best possible ways for energy-harvesting models. One more major confrontation for RF harvesting is enhancing the active rectifier's modification effects to enlarge the harvested power.

4.2 SVM for Energy Harvesting

Power management is one of the vital constraints for sensor nodes and IoT strategies. The life of the network or the device depends on how well it utilizes its energy resources, most of the WSNs nodes require a long range of active status (from months to years), but these nodes lose half of the energy due to unwanted events. To enhance the network lifetime, numerous authors present various energy-efficient protocols like mobile chargers, routing paths, sleep schedules, etc. Still, the energy obligations are not crammed since of high computational resources and inaccessible sensor nodes. Energy gathering provides continuous energy for sensor nodes from nearby entities like mechanical, wind, thermal, and radiofrequency energy. Some ML-based copies have been utilized to follow successful energy gathering models for WSNs.

In [34], the authors developed solar irradiance prediction models with ML-centric approaches. Authors perform several actions on historical sensed data and then find out the relationship coefficients. Authors of [35] designed photovoltaic cell-based moralities for solar-powered WSNs. The proposed model is verified with central and scattered WSNs structures, and results are quite comforting for distributed design. In [36] authors move to learning base solar energy prediction (Q-SEP) to calculate the past sensor node observations within a time. In [37] authors come with location-based range-free SVM (RFSVM); moreover, a transmit matrix announced toward display the association among hops along with distances, which helps find unidentified nodes in WSNs. Authors of [38] extend [37] the work by adding the polar coordinate system, which is separated into polar grids moreover labeled with SVM. In [39], the authors proposed a fast-SVM-based localization model for localization issues. The fast-SVM split the feature space obsessed by SVM clusters created on the highest comparisons. Since the split clusters simple to concern, the SVM then progresses the show. Fast-SVM similarly conquers coverage-hole trouble and then boundary issues.

5 Smart Transportation with SVM, ANN, and DL

With the improvement of dissimilar fields on the IoT, the appliances of allied strategies have got a tremendous rise in their utilization in every part of a modern city. For continuous data generation from IoT devices, ML techniques are superbly useful to improve the cleverness and proficiencies of an appliance. The outlook of smart transport has engrossed numerous scholars, besides it advanced within the cooperation of ML, DL, and IoT practices. Every entity that allied in the IoT scheme is acknowledged as a thing, and these things consist of sensors, embedded systems, and actuators. Things require being in touch with different entities through short-range communications like Bluetooth, radio, Zigbee, LoRa, GSM, WiMAX, 4G, 5G, LTE VoLTE, etc. Smart transportation is one of the hot research issues because it deals with most everyday human problems and the massive footprint in a contemporary smart city.

5.1 Artificial Neural Networks

ANN developed the structure of the human neural system with intelligent learning models. In ANN, an input layer deals with variable input quantity injected into the system; here, the individual neuron epitomizes a variable besides the output layer. Then each neuron represents a label, and also, one or more hidden layers include among input and output layers. If there is no looping in the neuron's construction, the system was acknowledged as feedforward neural network (FF-NN), also known as a simple form of an ANN. In [40], the authors designed the BN-SARIMA model to predict short-range traffic in huge urban-based transportation and relate with other existing ANN-based models.

The prophecies of [40] complete for 5, 10, and then 15 min, and the applied FF-NN specified a minor benefit at the 10- and then 15-min prophecy over the additional NN process; however, it looks to be fewer active than BN-SARIMA method. In addition to the work of [40], authors of [41] use multilayer FF-NN, where it contains some hidden layer and the input and output layers. Outcome results showed that the FF-NN model has the lowest error metrics after the RF model. In [42], a modified FF-NN employ to predict travel time variations and various traffic data records like travel history, number of rounds, and speed accumulated from roadside sensors. The unsupervised learning approach applied to traffic outlines and then employed ANN for training on clustered data. FF-NN trained with backpropagation models, which minimizes prediction errors. The proposed model's analysis is completed with the root-mean-square error (RMSE), and then the results are quite comfortable. The model FF-NN with backpropagation, also utilized in [43], is also employed to detect and predict road coincidences in real-time. Moreover, the K-NN model and simulation outcomes illustrate that the projected model improved proficient precision. Authors [44] fabricated a novel technology that detects potholes; here, accelerometer data is composed of automaton phones, feeding an FF-NN that turns in the automaton phones in real-time. The simulation outcome indicates 93.58 percentage detection accuracy.

5.2 Deep Learning

Deep learning (DL) is considered a subpart of ML; it enfolds powerful models for the progression of a massive volume of unstructured data. DL efficiently works with big data and also computing procedures such as natural language processing, pattern recognition, space analysis, etc. To acquire the best results, DL needs powerful CPUs and GPUs. Here the term deep discusses to the amount of hidden layers that constitute the neural network. The model features are robotically predictable in deep structure, and there is no need for feature scheming and abstraction before applying any technique. Furthermore, a wide variety of network constructions presented with the improvement of DL.

In [45], the authors employed DL-based CNN applied to healthcare image pattern recognition. An essential innate of CNN is that various neurons practice on similar filters. CNN layers are tailed by nonlinear layers, adapting the entire undesirable standards to zeros. Then, size drop is functional with subsampling layers. Authors of [46] build a blind-spot exposure classification for smart automobiles with the help of fully connected networks and equate with CNN. In [47], DBNs assist digital map conception through programmed highway parts recognition like traffic decorations, crossroads, etc. The model takes input from the GPS; first, outliers are determined to make some primary estimations on the street parts. Moreover, the authors included RBM; for every single class, the DBNs abstract a set of structures that correctly define the input statistics. Then DBNs finally diverge input statistics to different classes. The joint research outcomes with a recall of 0.89, then the accuracy of 0.88. In [48], writers designed a DL-based deep recurrent attention model (DRAM) for successive sorting of numerous substances from an image input. Experimental results show that model triumphs over CNN on several digits indebtedness when experienced on Google Street View and house imageries. Also, DRAM has fewer computation than CNN models. In [49], the authors modeled a stacked auto-encoder (SAE) to forecast the traffic movement based on big data. Logistic regression is also included on SAE and fine-tuned with backpropagation in the end. Moreover, the designed model has been tested with the Caltrans Performance Measurement System (PeMS) dataset; then evolution is done with MRSE, MAE, and MRE. The proposed model is compared with other existing AE-based models, and the results are promising. In [50], the authors projected an inception neural network for traffic accident hotpots plus their automatic recognition and classifications. In [52] a systematic comparison of nonlinear autoregressive exogenous model (NAEM), BN-SARIMA, and FF-NN on short-term-traffic forecasts on massive urban stream of traffic is made. The valuation is executed with the MAE, MAPE, and MRSE directories and results better with NAEM. Moreover, in [41], the authors use CNN on smart cameras on car parks for free parking spaces. Parking lot imageries are also processed in [51], with a mixture of SVM along with MRF process. Finally, the PKLot dataset is shaped in [53], which is afterward established by consuming an SVM algorithm.

6 Anomaly Detection Analysis and Prediction

Sensor nodes or IoT devices installed in the real world to compute the valuable uncooked facts will make appropriate decisions by evolution on these data experts. However, due to some events, data may contain errors or miss values. ML and DL models are improved to detect anomalies, errors, or outliers. Support vector machine (SVM) is a widespread model for anomaly or outlier exposure. One-class SVM could categorize the positive data instances from contaminated data instances in feature space. SVM classifiers' models in [54–59] are highly modeled for estimating the outlier behaviors on the real-life data collected from smart environments. In [55], support vector description and regularization techniques are designed to identify outliers for old-age people living home alone. In [56] multi-class SVMs implied for the representation of human actions. In [57], a feature reduction plus the PCA model with fuzzy logic applied for outlier behavior detection checks the inputs and detects the assurance limits. In the subsequent phase, rules are fired, rendering the assurance limits of inputs. Authors of [58] fabricated an SVM based on nonlinear kernel regression for detection and diminished false-positive rates. Authors of [59] designed a CNNA-based RNN with SVM to improve the detect anomalous behavior. They train the CNN archetypal over the fresh signs with motion class plus prepare forecast with SVM. They too malformed the fresh info stream to spectrogram and apply the RNN. They informed mutual representations in their projected system adequate to notice anomalous performance.

7 Challenges and Future Directions

Massive Scale of IoT Data The amount of data collected by IoT-based devices turns to significant disputes because of time and structure complexities. Numerous input samples containing a huge count of choice features and high-degree classification precision requirements produce an enormously complex DL model with huge source requirements. To conquer these disputes, the proposed models must build on the distributed structure through parallel processing.

Automated NN Models In general, IoT systems are applied on real-time situations where data changes dynamically so the future models of ANN and SVM-based models must be automated techniques to avoid the unnecessary implications.

Data Pre-processing Basically, IoT devices collect data records from different devices or resources which may hold noise, errors, or missing values, and this makes data pre-processing turn to be a tricky task.

High Velocity In real-life appliances, the rate or high-speed production is another challenge since real-life IoT appliances must work with fast processing and data analytical tricks on such data. To deal with these kinds of problems, deep learning models must follow online learning capabilities.

Online Mode The decision-making with real-time data is very important for analyzing the upcoming conquests. Thus, the designed models must work in online mode with limited resources.

Adaptableness It is important that designed model must be adaptable mode so that it can change operation according to particular situational data records.

Heterogeneity Optimal DL models can deal with various device-based source data, but marinating the inconsistencies among those data points is another level challenge.

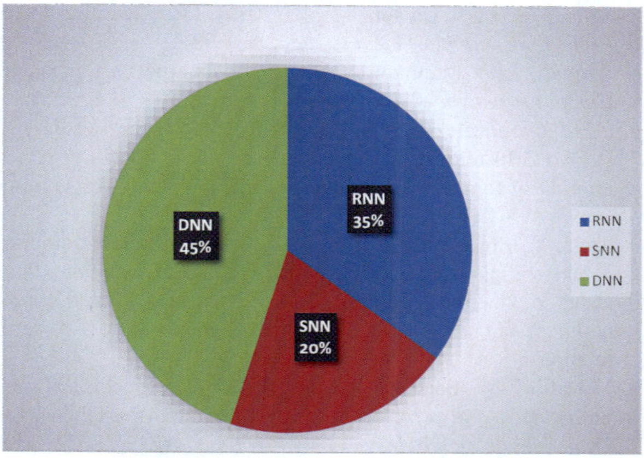

Fig. 1 Percentage-wise neural network models applied in IoT

Deep Learning for IoT Devices The known fact that DL models' inclusion into resource-constrained IoT devices is the central dispute in most IoT systems implementations (Fig. 1).

8 Conclusion

The word IoT is turned to central model that connects entrenched devices, objects, and sensor nodes to the Internet. The technology enhanced the usage of mentioned devices incorporated into our daily life not like ever before. Most of IoT appliances employed in real world which collect huge amount of data and extracting significant features cannot be done effectively by traditional procedures. However, by applying SVM and ANN models, a vast quantity of data is analyzed effectively; moreover, the performance of IoT schemes is also boosted. So, in this chapter we described importance of role of SVM and ANN model in IoT policies, how SVM and ANN provide security in data transmission, and energy-harvesting models for IoT based on SVM and ANN. In addition, applications, future challenges, and research problems are discussed.

References

1. Chen, M., Challita, U., Saad, W., Yin, C., & Debbahk, M. (2019). *Artificial neural networks-based machine learning for wireless networks: A tutorial*. IEEE Communications Surveys and Tutorials.
2. Mahmut Taha Yazici ID, Shadi Basurra ID and Mohamed Medhat Gaber (2018). Edge machine learning: Enabling smart internet of things applications, big data and cognitive computing.
3. Eid, A., Mghabghab, S., Costantine, J., Awad, M., & Tawk, Y. (2019). Support vector Machines for Scheduled Harvesting of Wi-Fi signals. *IEEE Antennas and Wireless Propagation Letters.*

4. Saleem, T. J., & Chishti, M. A. (2019). Deep learning for internet of things data analysis. *Procedia Computer Science, 163*, 381–390.
5. Asghari, P., & Rahmani, A. M. (2019). Internet of things applications: A systematic review. *Computer Networks, 148*, 241–261.
6. Cristian González García, Edward Rolando Núñez-Valdez, Vicente García-Díaz, B. Cristina Pelayo GBustelo, Juan Manuel Cueva Lovelle (2018). A review of artificial intelligence in the internet of things, International Journal of Interactive Multimedia and Artificial Intelligence, Vol. 5, N 4.
7. Zeng, Y., Zhang, R., & Lim, T. J. (2016). Wireless communications with unmanned aerial vehicles: Opportunities and challenges. *IEEE Communications Magazine, 54*(5), 36–42.
8. Yang, Y., Chen, M., Guo, C., Feng, C., & Saad, W. (2019). *Power efficient visible light communication (VLC) with unmanned aerial vehicles (UAVs)*. IEEE Communications Letters, to appear.
9. Mozaffari, M., Saad, W., Bennis, M., & Debbah, M. (2017). Wireless communication using unmanned aerial vehicles (UAVs): Optimal transport theory for hover time optimization. *IEEE Transactions on Wireless Communications, 16*(12), 8052–8066.
10. Liu, C. H., Chen, Z., Tang, J., Xu, J., & Piao, C. (2018). Energy-efficient UAV control for effective and fair communication coverage: A deep reinforcement learning approach. *IEEE Journal on Selected Areas in Communications, 36*(9), 2059–2070.
11. Zhang, H., Cao, C., Xu, L., & Gulliver, T. A. (2018). A UAV detection algorithm based on an artificial neural network. *IEEE Access, 6*, 24720–24728.
12. Cui, J., Liu, Y., & Nallanathan, A. (2018). *Multi-agent reinforcement learning based resource allocation for UAV networks*. arXiv preprint arXiv:1810.10408.
13. Chen, M., Mozaffari, M., Saad, W., Yin, C., Debbah, M., & Hong, C. S. (2017). Caching in the sky: Proactive deployment of cache-enabled unmanned aerial vehicles for optimized quality-of-experience. *IEEE Journal on Selected Areas in Communications, 35*(5), 1046–1061.
14. Chen, M., Saad, W., & Yin, C. (2019). Liquid state machine learning for resource and cache management in LTE-U unmanned aerial vehicle (UAV) networks. *IEEE Transactions on Wireless Communications, 18*(3), 1504–1517.
15. Liu, X., Chen, M., & Yin, C. (2019). *Optimized trajectory design in UAV based cellular networks for 3D users: A double Q-learning approach*. arXiv preprint arXiv:1902.06610.
16. Chen, M., Saad, W., & Yin, C. (2017). Echo state networks for self-organizing resource allocation in LTE-U with uplink-downlink decoupling. *IEEE Transactions on Wireless Communications, 16*(1), 3–16.
17. Challita, U., Dawy, Z., Turkiyyah, G., & Naoum-Sawaya, J. (2016). A chance constrained approach for LTE cellular network planning under uncertainty. *Computer Communications, 73*, 34–45.
18. Chen, M., Saad, W., Yin, C., & Debbah, M. (2019). Data correlation-aware resource management in wireless virtual reality (VR): An echo state transfer learning approach. *IEEE Transactions on Communications, 67*(6), 4267–4280.
19. Du, L., Du, Y., Li, Y., Su, J., Kuan, Y., Liu, C., & Chang, M. F. (2018). A reconfigurable streaming deep convolutional neural network accelerator for internet of things. *IEEE Transactions on Circuits and Systems I: Regular Papers, 65*(1), 198–208.
20. Yu, T., Wang, X., & Shami, A. (2019). UAV-enabled spatial data sampling in large-scale IoT systems using denoising autoencoder neural network. *IEEE Internet of Things Journal, 6*(2), 1856–1865.
21. Zhang, P., Kang, X., Wu, D., & Wang, R. (2018). *High-accuracy entity state prediction method based on deep belief network towards iot search*. IEEE Wireless Communications Letters, to appear.
22. Liang, J., Yu, X., & Li, H. (2018). Collaborative energy-efficient moving in internet of things: Genetic fuzzy tree vs. neural networks. *IEEE Internet of Things Journal*.

23. Kim, W., Stankovi, M. S., Johansson, K. H., & Kim, H. J. (2015). A distributed support vector machine learning over wireless sensor networks. *IEEE Transactions on Cybernetics, 45*(11), 2599–2611.
24. Shu, J., Liu, S., Liu, L., Zhan, L., & Hu, G. (2017). Research on link quality estimation mechanism for wireless sensor networks based on support vector machine. *Chinese Journal of Electronics, 26*(2), 377–384.
25. Gil, P., Martins, H., & Januario, F. (2018). Outliers detection methods in wireless sensor networks. *Artificial Intelligence Review*, 1–26.
26. Feng, Z., Fu, J., Du, D., Li, F., & Sun, S. (2017). A new approach of anomaly detection in wireless sensor networks using support vector data description. *International Journal of Distributed Sensor Networks, 13*(1), 1–14.
27. Shahid, N., Naqvi, I. H., & Qaisar, S. B. (2015). One-class support vector machines: Analysis of outlier detection for wireless sensor networks in harsh environments. *Artificial Intelligence Review, 43*(4), 515–563.
28. Emadi, H. S., & Mazinani, S. M. (2018). A novel anomaly detection algorithm using DBSCAN and SVM in wireless sensor networks. *Wireless Personal Communications, 98*(2), 2025–2035.
29. Zidi, S., Moulahi, T., & Alaya, B. (2018). Fault detection in wireless sensor networks through SVM classifier. *IEEE Sensors Journal, 18*(1), 340–347.
30. Jiang, M., Luo, J., Jiang, D., Xiong, J., Song, H., & Shen, J. (2016). A cuckoo search-support vector machine model for predicting dynamic measurement errors of sensors. *IEEE Access, 4*, 5030–5037.
31. Khan, F., S. Memon, S. H. Jokhio (2016). Support vector machine based energy aware routing in wireless sensor networks. In: *Robotics and Artificial Intelligence (ICRAI), 2016 2nd International Conference on, IEEE* (pp. 1–4).
32. Gholipour, M., Haghighat, A. T., & Meybodi, M. R. (2017). Hop-by-hop congestion avoidance in wireless sensor networks based on genetic support vector machine. *Neurocomputing, 223*, 63–76.
33. Moon, S.-H., Park, S., & Han, S.-j. (2017). Energy efficient data collection in sink-centric wireless sensor networks: A cluster-ring approach. *Computer Communications, 101*, 12–25.
34. Sharma, A., & Kakkar, A. (2018). Forecasting daily global solar irradiance generation using machine learning. *Renewable and Sustainable Energy Reviews, 82*, 2254–2269.
35. Tan, W. M., Sullivan, P., Watson, H., Slota-Newson, J., & Jarvis, S. A. (2017). An indoor test methodology for solar-powered wireless sensor networks. *ACM Transactions on Embedded Computing Systems (TECS), 16*(3), 82.1–82.25.
36. Kosunalp, S. (2016). A new energy prediction algorithm for energy-harvesting wireless sensor networks with Q-learning. *IEEE Access, 4*, 5755–5763.
37. Tang, T., Liu, H., Song, H., & Peng, B. (2016). Support vector machine-based range-free localization algorithm in wireless sensor network. In *International conference on machine learning and intelligent communications* (pp. 150–158). Springer.
38. Wang, Z., Zhang, H., Lu, T., Sun, Y., & Liu, X. (2018). A new range-free localisation in wireless sensor networks using support vector machine. *International Journal of Electronics, 105*(2), 244–261.
39. Zhu, F., & Wei, J. (2017). Localization algorithm for large scale wireless sensor networks based on fast-SVM. *Wireless Personal Communications, 95*(3), 1859–1875.
40. Zantalis, F., Koulouras, G., Karabetsos, S., & Kandris, D. (2019). A review of machine learning and IoT in smart transportation. *Future Internet, 11*, 9.
41. Fusco, G., Colombaroni, C., Comelli, L., Isaenko, N (2015). Short-term traffic predictions on large urban traffic networks: Applications of network-based machine learning models and dynamic traffic assignment models. In *Proceedings of the 2015 IEEE International Conference on Models and Technologies for Intelligent Transportation Systems (MT-ITS), Budapest, Hungary*, 3–5 June 2015; pp. 93–101.
42. Hou, Y., Edara, P., & Sun, C. (2015). Traffic flow forecasting for urban work zones. *IEEE Transactions on Intelligent Transportation Systems, 16*, 1761–1770.

43. Yu, J., Chang, G. L., Ho, H., Liu, Y (2008). Variation based online travel time prediction using clustered neural networks. In *Proceedings of the 11th International IEEE Conference on Intelligent Transportation Systems, Beijing, China*, 12–15, pp. 85–90.
44. Ozbayoglu, M., Kucukayan, G., Dogdu, E. (2016). A real-time autonomous highway accident detection model based on big data processing and computational intelligence. In *Proceedings of the 2016 IEEE International Conference on Big Data (Big Data), Washington, DC, USA*, 5–8, pp. 1807–1813.
45. Kulkarni, A., Mhalgi, N., Gurnani, S., & Giri, N. (2014). Pothole detection system using machine learning on android. *International Journal of Emerging Technology and Advanced Engineering, 4*, 360–364.
46. Amato, G., Carrara, F., Falchi, F., Gennaro, C., Meghini, C., & Vairo, C. (2017). Deep learning for decentralized parking lot occupancy detection. *Expert Systems with Applications, 72*, 327–334.
47. Kwon, D., Park, S., Baek, S., Malaiya, R. K., Yoon, G., Ryu, J. T. (2018). A study on development of the blind spot detection system for the IoT-based smart connected car. In *Proceedings of the 2018 IEEE International Conference on Consumer Electronics (ICCE), Las Vegas, NV, USA*, 12–14 January pp. 1–4.
48. Munoz-Organero, M., Ruiz-Blaquez, R., & Sánchez-Fernández, L. (2018). Automatic detection of traffic lights, street crossings and urban roundabouts combining outlier detection and deep learning classification techniques based on GPS traces while driving. *Computers, Environment and Urban Systems, 68*, 1–8.
49. Ba, J.; Mnih, V.; Kavukcuoglu, K (2014). *Multiple object recognition with visual attention*. arXiv 2014, arXiv:1412.7755.
50. Lv, Y.; Duan, Y.; Kang, W.; Li, Z.; Wang, F.Y (2014). Traffic flow prediction with big data: A deep learning approach. IEEE Transactions on Intelligent Transportation Systems, 16, pp. 865–873.
51. Ryder, B., Wortmann, F. (2017). Autonomously detecting and classifying traffic accident hotspots. In *Proceedings of the 2017 ACM International Joint Conference on Pervasive and Ubiquitous Computing and 2017 ACM International Symposium on Wearable Computers, Maui, HI, USA*, 11–15 September, pp. 365–370.
52. Wu, Q., Huang, C., Wang, S. Y., Chiu, W. C., Chen, T (2007). Robust Parking Space Detection Considering Inter-Space Correlation. In *Proceedings of the 2007 IEEE International Conference on Multimedia and Expo, Beijing, China*, 2–5 July.
53. Almeida, P. R. D., Oliveira, L. S., Britto, A. S., Silva, E. J., Koerich, A. L., & Lot, P. K. (2015). A robust dataset for parking lot classification. *Expert System with Applications, 2015*(42), 4937–4949.
54. Fahim, M., & Sillitti, A. (2019). *Anomaly detection, analysis and prediction techniques in IoT environment: A systematic literature review*. IEEE Access.
55. Palaniappan, A., Bhargavi, R., & Vaidehi, V. (2012). Abnormal human activity recognition using SVM based approach. In *Proceedings of International Conference on Recent Trends Information Technology (ICRTIT)*, (pp. 97–102).
56. Shin, J. H., Lee, B., & Park, K. S. (2011). Detection of abnormal living patterns for elderly living alone using support vector data description. *IEEE Transactions on Information Technology in Biomedicine, 15*(3), 438–448.
57. S. M. Mahmoud, A. Lotfi, and C. Langensiepen (2012). User activities outlier detection system using principal component analysis and fuzzy rule-based system, In *Proceedings of 5th international conference of pervasive technology related to assistive environment*. (p. 26).
58. Yin, J., Yang, Q., & Pan, J. J. (2008). Sensor-based abnormal human-activity detection. *IEEE Transaction Knowledge Data Engineering, 20*(8), 1082–1090.
59. Han, N., Gao, S., Li, J., Zhang, X., & Guo, J. (2018). Anomaly detection in health data based on deep learning. In *Proceedings of International Conference on Network Infrastructure and Digital Content (IC-NIDC)* (pp. 188–192).

60. Chander, B., & Kumaravelan. (2018). A analysis of machine learning in wireless sensor network. *International Journal of Engineering & Technology, 7*(4.6), 185–192.
61. Chander, B., & Kumaravelan. (2019). One class SVMs outlier detection for wireless sensor networks in harsh environments: Analysis. *International Journal of Recent Technology and Engineering (IJRTE)*. ISSN: 2277-3878, *7*(4).
62. Chander, B., & Kumaravelan, G. (2020). Security vulnerabilities and issues of traditional wireless sensors networks in IoT. In *Principles of internet of things (IoT) ecosystem: Insight paradigm* (pp. 519–549). Springer. (ISBN-13: 978-3030335953).
63. Chander, B., & Kumaravelan, G. (2020). *Wearable sensor network for patient health monitoring: Challenges, applications, future directions, and acoustic sensor challenges*. Healthcare Paradigms in the Internet of Things Ecosystem, Elsevier. (Paperback ISBN: 9780128196649).
64. Chander, B. (2019). *Blockchain technology Integration in IoT and Applications*. Secure System Model for Replicated DRTDBS, Security and Privacy Issues in Sensor Networks and IoT, ISBN13: 9781799803737, DOI: https://doi.org/10.4018/978-1-7998-0373-7, IGI global Publications.

The Role of Machine Learning Techniques in Internet of Things-Based Cloud Applications

Shashvi Mishra and Amit Kumar Tyagi

1 Introduction

In recent years, new technologies and dramatic changes to Internet protocols and computing systems have made it easier than ever before to communicate between different devices. Around 25–50 billion devices are expected to be connected to the Internet by 2020 [1], according to various estimates. As IoT will be among the most important sources of new data, data science will make a major contribution to making IoT applications more intelligent. Data science is the fusion of numerous scientific fields which use data mining, machine learning, and other techniques to identify trends and new ideas from data.

1.1 Machine Learning

In the recent decade, machine learning has witnessed great recognition across the globe in multiple fields and has proved to be a rising innovative trend. In order to witness maximum value from big data, both idea and technology must go hand in hand. Much attention should be given on how to exactly pair up the algorithms and tools/processes to create a ML model purely based on iterative learning. Here are listed a few key machine learning algorithms:

- Neural networks
- Random forests
- Decision trees

S. Mishra · A. K. Tyagi (✉)
School of Computer Science and Engineering, Vellore Institute of Technology, Chennai, Tamilnadu, India

S. Pal et al. (eds.), *Artificial Intelligence-based Internet of Things Systems*, Internet of Things, https://doi.org/10.1007/978-3-030-87059-1_4

- SEO
- Discovery of sequence and associations
- SOM (self-organizing maps)
- Nearest neighbour
- SVM (support vector machines)
- Multivariate adaptive regression (linear, logistic, multiple, etc.)
- Boosting and bagging gradients
- Analysis of principal components

Here below are the tools or processes listed with which the machine learning algorithms are to be paired to generate efficient results:

- Data exploration → visualization of model predictions
- Complete data quality and management
- Efficient and easy model deployment to easily generate repeated and reliable outputs
- Formation of graphical user interface for developing process flows and building models
- Comparison of multiple ML models and choosing the best fitting model
- Identification of the best performers with the help of automated ensemble model evaluation
- Automation of data to decision process

1.1.1 Importance of Machine Learning in Present Business Scenario

Machine learning has growing importance in most of the industries dealing with enormous amounts of data. Business can earn a competitive edge and can prove to be more efficient by obtaining the hidden patterns and insights from this quality data [2]. Huge chunks of data with higher level of complexity can be analysed through the machine learning model with the help of affordable and easy computational processing in addition to the cost-effective data storage options so that the models provide higher accuracy and efficiency. Delivery of personalized services and differentiated products in accordance with the varying needs of the customers can be achieved in organizations through ML [3]. In addition to this, companies can be exposed to ample of opportunities that can prove to be profitable in long run through machine learning. Here are a few things to keep in mind, if one really wishes to develop effective machine learning systems in order to augment one's business:

- Superior data preparation capabilities
- Basic and advanced algorithm's knowledge
- Scalability
- Automation and iterations
- Ensemble modelling knowledge

1.1.2 Applications of Machine Learning

Every industry that deals with huge chunks of data has recognized the value of machine learning technology. By leveraging all the hidden trends and insights of the data, it is possible for companies to work efficiently to control costs and obtain a competitive edge in the market [4]. Below are the ways how machine learning is exceling in certain fields:

(a) Financial Services: Machine learning technology in this department may help companies under financial sector to analyse the hidden insights of financial data and identify the occurrences of financial frauds. Machine learning can also be useful in determining opportunities for trades and investments. Cyber surveillance, one of the technologies under machine learning, has sky rocketed in the financial grounds as it helps in identification of the individuals or institutions which are in the range of hitting financial risk, so that necessary preventive actions could be taken in order to prevent fraud.
(b) Marketing and Sales: Departments of utmost importance to be analysed in this sector are purchase history of customers, and where machine learning comes into existence is after the analysis, to produce customized and personalized product recommendations for the next purchase. The capturing technique, proper analysis and implementation of customer data to create a personalized shopping experience is the new level of sales and marketing.
(c) Government: Major areas of interest are utilities and public safety. They are exposed to multiple data sources for mining, in order to observe the useful trends and insights, like analysis of sensor data for identifying ways to minimize cost and enhance efficiency. Similarly, machine learning can be used to detect frauds and thefts and reduce their occurrences.
(d) Healthcare: ML can come up with wearable sensors and devices which intend to use the data to access health of a person in real time and has thus has turned out to be the rapidly growing trend in healthcare. Wearable sensors provide real-time information regarding the patient, some of them being overall health condition, blood pressure, heartbeat, pulse rate and many other more vital parameters. Proper analysis of such information can help medical specialists to extract repeated trends from the patient's past and come up with ailments, cures and preventive measures in the future. Machine learning technology also empowers medical experts and helps them to determine insights from the medical data and produce efficient outcomes and diagnosis techniques and improved treatments.
(e) Transportation: Proper route analysis can be done using machine learning algorithm that is predictions regarding the potential problems on a particular route based on the travel history and pattern of travelling through routes and thus throwing advisory outcomes regarding which route to choose for convenience. Transportation firms and delivery organizations use this analysing machine learning technique to create a smarter city and provide their customers with optimum decisions for travel.

(f) Oil and Gas: Indeed, the neediest industry for machine learning. ML has vast and expanding application in this industry, starting from analysis of underground minerals to extraction of new energy sources and to streaming oil distribution.

1.2 *Internet of Things (IoTs) Technology and Infrastructure*

To understand more about an advanced technology, we need to learn first its basics, its subparts and their internal components (including their working structure). Generally, Internet of Things work based on wireless sensor networks (WSNs), used in almost all applications in today's era. On a larger scale, many IoTs make a huge network in connection with the Internet, which can be considered a combination of physical space and cyber space, called cyber-physical systems (CPSs). Today in this post-COVID-19 scenario, wireless sensor network and cyber-physical system use have been increased. Each term is discussed below.

Wireless Sensor Networks (WSNs) in the Post-COVID-19 Era

Digital tools are being used by countries all over the world to tackle this global crisis. In one way or another, these emerging developments strongly depend on the availability of wireless communication systems. Wireless networking systems, including virus spread control, health automation and interactive education and conferencing, are helping to fight this pandemic:

(a) IoT platforms: During this global emergency, emerging technology applications are various, including tactile robots to assist hospital medical doctors and nurses, crowd-monitoring aircraft, artificial intelligence and deep learning to recognize healthcare patterns, supply chain automation of the Internet of Things (IoT) and virtual learning to continue education. IoT is everywhere, considering the rise of smart watches that allow individuals to monitor their health, track their sleeping habits and measure their heart rate. IoT is everywhere to smart sensors that go beyond human control in industrial maintenance operations.
(b) IoT infrastructure: In order to cope with the effects of COVID-19 and future public health emergencies, our digital infrastructure needs strengthening. Better incorporation of artificial intelligence into the public health response could be a priority; support for preventive programs could be used to evaluate big data relating to citizen movement, disease transmission trends and health monitoring [5].
(c) IoT architecture: There is no single, widely accepted consensus on IoT architecture. Different architectures have been proposed by different investigators:

  - Three-layer architecture: The three-layer architecture describes the primary concept of the Internet of Things, but it is not appropriate for IoT research because research also focuses on finer aspects of the Internet of Things. It was introduced in the early stages of research in this field. It has three layers, namely, the layers of perception, network and application.

- Five-layer architecture: In addition, the processing and business layers are included in the five-layer architecture. The five layers are understanding, transport, production, implementation and business layers. The function of the perception and application layers is the same as the architecture of the three layers.

(d) IoT data processing and intelligence: In order to make sense of the vast amount of data our IoT sensors collect, we need to process it. In other words, the purpose of data processing is to transform raw data into something useful. "Wikipedia explains data processing as" the compilation and manipulation of information objects to generate meaningful information.

Cyber-Physical Systems (CPSs) in the Post-COVID-19 Era

Computational, networking and physical process integrations are cyber-physical systems (CPSs). Embedded computers and networks track and track physical processes with feedback loops in which physical processes influence computations and vice versa. Machine learning and the attendant biosensors, big data and digital health approaches can be conceptualized within the overarching cyber-physical system (CPS) paradigm. CPS brings greater proximity to the virtual and physical worlds:

(a) IoT data management: Management of IoT data helps organizations understand how the performance of their products can be affected by environmental conditions and user behaviour. It is also possible to use IoT sensors to measure product performance metrics. Data collected by these sensors can be used to enhance future product versions.
(b) IoT testing: A form of testing to check IoT devices is IoT testing. The need to deliver better and faster services is increasing today. Accessing, creating, using and sharing data from any device is a huge demand. The goal is to provide greater insight and control over different IoT devices that are interconnected.
(c) IoT data analytics: Simply put, the analysis of enormous data volumes generated by connected devices is IoT data analytics. Organizations can derive a number of advantages from it, i.e. optimize activities, automatically control processes, engage more clients and empower employees. In retail, healthcare, telematics, manufacturing and smart cities, the combination of IoT and data analytics has already proved to be beneficial.

1.3 Cloud Technology and Infrastructure

Cloud infrastructure refers to the hardware and software components required, such as servers, storage, network and virtualization software, to meet the computing requirements of a cloud computing model. Cloud infrastructure refers to the backend components in a cloud computing architecture – the hardware elements found within most enterprise data centres. These include multi-socket, multicore servers, permanent storage and local area network equipment, such as switches and routers, but on a much larger scale:

(a) Edge computing: Edge computing is transforming the way data from millions of devices around the world is handled, processed and delivered. Edge-computing systems continue to drive the explosive growth of Internet-connected devices – the IoT – along with new applications that require real-time computing power. Gartner defines edge computing as "a component of a distributed computing topology in which the processing of information is located close to the edge where information is produced or consumed by things and people".
(b) Fog computing: Fog computing extends the cloud computing concept to the edge of the network, making it ideal for real-time interactions required for the Internet of Things (IoTs) and other applications. Fog computing is the idea of a network fabric that extends from the outer edges of where data is generated to where it will ultimately be stored, whether in the cloud or in the data centre of a customer.
(c) Resource pooling: Resource pooling is an IT term used to describe a situation in cloud computing environments in which providers serve provisional and scalable services to multiple customers, customers or tenants. Without any changes being apparent to the client or end user, these services can be adjusted to suit the needs of each client.
(d) Service deployment: Cloud implementation refers to enabling solutions for SaaS (software as a service), PaaS (platform as a service) or IaaS (infrastructure as a service) which can be accessed by end users or customers on demand. A model for cloud implementation refers to the kind of cloud infrastructure framework on which a cloud solution would be deployed. The implementation of the cloud requires all the installation and configuration measures needed to be carried out before user provisioning can occur.
(e) Cloud resource management: A central feature of any man-made system is the control of resources. A cloud is a dynamic structure, subject to unexpected demands and influenced by external events that it cannot manage, with a very large number of shared resources. For multi-objective optimization, cloud resource management requires complicated policies and decisions. A cloud is a dynamic system subject to unexpected requests with a very large number of shared resources and influenced by external events that it cannot monitor. For multi-objective optimization, cloud resource management involves complicated policies and decisions.
(f) Virtual desktop infrastructure: Virtual desktop infrastructure, or VDI, is a technology that refers to the provision and management of virtual desktops using virtual machines. On a centralized server, virtual desktop infrastructure (VDI) hosts virtual environments and deploys them on request to end users. In VDI, a hypervisor segments servers into virtual machines which, in turn, host virtual desktops accessed remotely by users from their devices. Users from any device

or location can access these virtual desktops and all processing is done on the host server. Users relate to their desktop instances through a link broker, which is a software-based portal that acts as an intermediary between the user and the server.

(g) Server virtualization: Virtualization of servers is used to hide server resources from users of the server. This could include the number of operating systems, processors and individual physical servers and their identity. It is the method of separating a physical server by means of a software application into several specific and separate virtual servers. Each virtual server is able to independently operate its own operating systems.

(h) Storage virtualization: Virtualization of storage is the physical data storage services abstracting technology to make them look as though they were a centralized resource. Virtualization hides the memory, networks, servers and storage complexities of managing resources. Virtualization of data runs on different storage devices, making them look as if they were a single pool of information. It is likely that pooled storage devices come from various manufacturers and networks.

(i) Network virtualization: Network virtualization (NV) refers to the abstraction of network services typically distributed to software in hardware. NV may combine several physical networks into a single virtual network based on software or split a single physical network into different virtual networks that are independent. Network virtualization software allows network managers, without reconfiguring the network, to transfer virtual machines around various domains.

Now the remaining structure of this chapter is organized as follows: Sect. 2 describes about the evolution of machine learning techniques. Further, Sect. 3 tells about the motivation behind this chapter. Section 4 describes about the various IoT and cloud applications. Then, Sect. 5 briefly describes about the comparison of scope of machine learning pre-COVID-19 and post-COVID-19 era. Section 6 describes different machine learning techniques for IoT-based cloud applications. Section 7 tells about the IoTs and cloud computing in post-COVID-19 era, i.e. it applications and challenges. Further Sect. 8 gives short view about future research directions for machine learning towards IoT-based cloud applications. In the end, Sect. 9 concludes this work in brief with including several interesting remarks for the future. Section 10 gives the information about acknowledgements, Sect. 11 talks about conflicts of interest, Sect. 12 discusses about disclosures and Sect. 13 briefs about the scope of this work. Note that in this work IoT-based applications or cloud-based IoT networking or cloud-based smart applications have been used interchangeably.

2 Evolution of Machine Learning Techniques

As computerized social information have gotten progressively universal, many have directed their concentration towards bridling these monstrous informational indexes so as to create purportedly increasingly precise and complete understandings of social procedures and have given it a term of "big data" [6]. Information volumes will proceed to increment and relocate to the cloud. Later, this information is compiled or analysed for making many sectors profitable, for example, in retail companies can target required customers based on their habits, interests, etc. But, refining this vast amount of data via traditional techniques (in current machine or deep learning), we face several serious concerns. Several issues and challenges during implementation of machine learning in big data have been discussed in [7–9]. There are billions of smart devices/gadgets connected together, which make, gather, and offer an abundance of IoT information examination consistently throghout the world [10, 11].

As ventures gain the chance to store and dissect colossal volumes of information, they will get the chance to make and oversee 60% of enormous information sooner rather than later. Specialists accept that the capacity of computers using their cloud and various ways to gain information will improve extensively because of further developed unaided calculations, more profound personalization and subjective administrations. We see noteworthy potential in large information as they have a multifaceted nature of social and spatial procedures. In the current world, large information implies working with immense datasets that are regularly unstructured. The datasets being worked with will be a blend of restrictive in-house information and freely accessible information. Working with these datasets will require devices that permit the unstructured information to be worked with and can likewise deal with the huge volumes. Information security and protection have consistently been squeezing issues, demonstrating an enormous snowballing potential. Ever-developing information volumes make extra difficulties in shielding it from interruptions and cyberattacks, as the degrees of information insurance can't stay aware of the information development rates.

Figure 1 shows the evolution of machine learning (year-wise) in detail. Only a couple of months before NASA has uncovered the image of the dark opening, however there is an enormous play of information science behind the disclosure of black hole. Data sciences and information preparing are the key variables in the creation of streamlined self-governing vehicles, that is, self-driving vehicles. To supplement this innovation further, dark data will advance and imprint the eventual fate of enormous information in the coming years or even months. Security gap which was brought about by an absence of instruction and preparing openings and the advancement of cyberattacks where the dangers are utilized by programmers are advancing and turn out to be progressively improving too. Note the serious issues with IoT or smart gadgets like breaching privacy, security of data protection, lack of standardization, etc. in [10, 12].

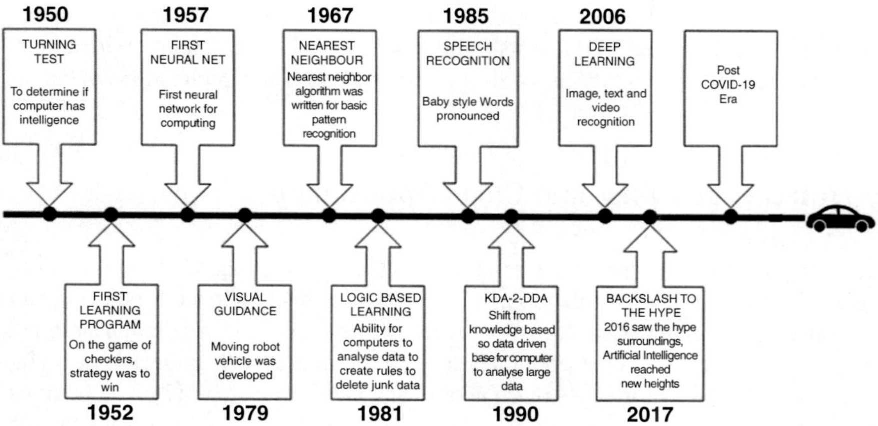

Fig. 1 The evolution of machine learning techniques (year-wise)

3 Motivation

Today we can feel or see major development in field of technology. Technology has changed the way of living life and been used almost in every sector like media and entertainment, healthcare, agriculture, defence, education, manufacturing, etc. Machines are performing several functions in parallel, IoT devices are creating a smartest environment (which save nature, e.g. IoTs are used for generating energy smartly), and also these IoTs are making a network of billions of devices and function together and generate a large amount of data (called big data) in motion/data in flight and data at rest via communicating with other devices. Note that nowadays this large amount of big data is stored at a cloud (automatically by IoTs) and can be accessed by user/consumer at any time, from anywhere. This big data is accessible from a remote location also. Later, this data is analysed by data scientist or a team of experts for generating valuable information of data, for example, for weather forecasting, experts refer data for the last 50–100 years and provide nearby (or approximate) information for the future weather. Many things are depending on such predictions in many sectors, for example, prediction of demand at early stages provides high profits to the industries. These predictions come with a day of analysis and high cost (huge investment in IoT and energy), also with several serious concerns like breaching of privacy, security of data protection, mistrust among service provider, lack of standardization of IoT, etc. Together this, cloud is also facing security, privacy, fault tolerance, etc. issue in it.

Hence, we need to provide such serious concerns and required/existing solutions for the same issues. We need to write a work which can provide information to readers about role of emerging machine learning techniques for IoT-based cloud applications. We want an environment where users can access any services from anywhere without hassle or breaches and delay. Also, user's data need to be protected in decentralized manner and must follow the CIA (confidentiality, integrity and availability) property. Hence, this section discusses about the motivation to work in the field of machine learning and its importance for IoT-based cloud applications. Now, the next section will discuss IoT applications and cloud applications in detail.

4 Internet of Things and Cloud Applications

Transformation is an ever-present trend which in today's fast-paced world is becoming an absolute necessity of the hour. There is a great deal of space for storing and manipulating the data with technology churning every bit of information in a refined new format. When the roost continues to be governed by cell phones and social media, a lot of discussion takes place about what is to come next [13]. The obvious response for the hour is the Internet of Things, or IoT. There has been a big change towards using it as a computing platform for both individuals and enterprises with the advent of the cloud. The use of cloud computing to make data available remotely puts a lot of stress on this, considering the scalability and data complexity. The basic idea behind IoT and cloud storage is to make day-to-day operations more efficient without compromising the quality of the stored or transmitted data. Since the relation is mutual, both services complement one another effectively. Although the cloud becomes the ultimate destination for storing it, the IoT becomes the database.

The question remains, however, how are the devices going to stay connected all the way through? The solution lies in the Internet of Things networking of cloud providers. The increased use of IoT in the cloud has served as a catalyst for the creation and implementation of scalable Internet of Things applications and business models. Two very closely related future Internet innovations have been cloud computing and IoT, with one providing a platform for success with the other [2]. There are many benefits derived from the convergence of IoT and cloud computing, for example:

- Increased scalability
- Increased performance
- Pay as you go

Now, we will discuss IoT and cloud applications in current era in the next subsections.

4.1 IoT Applications

IoT apps aim to add tremendous value to our lives. With newer wireless networks, superior sensors and innovative computing capabilities, the Internet of Things for its wallet share could be the next frontier in the race [13]. IoT technologies are supposed to provide the communication and knowledge for billions of everyday objects. It is already widely deployed, in different domains, namely:

- Internet of Medical Things (IoMT): The Internet of Medical Things (IoMT) is an amalgamation of medical devices and applications that can connect to healthcare information technology networks using networking technologies. By connecting patients to their physicians and enabling medical data to be transmitted over a secure network, this will reduce unnecessary hospital visits and the pressure on healthcare facilities. The IoMT market consists of smart devices for use only on the body, at home or in community, clinic or hospital environments, such as wearables and medical/vital monitors, and related real-time location, telehealth and other services.
- Internet of Everything (IoE): The Internet of Everything (IoE) "is putting people, systems, data, and items together to make networked interactions more important and meaningful than ever before, turning information into actions that generate new capacities, richer experiences, and unparalleled economic opportunities for companies, individuals, and countries".
- IoT applications in healthcare: IoT technologies are capable of converting reactive medical systems into positive wellness-focused systems. Essential real-world knowledge is missing in the tools which current medical research uses. It uses for medical examination mostly leftover info, managed environments and volunteers. IoT opens avenues to a sea of useful data through research, real-time field data and testing.

The Internet of Things also enhances the capacity, precision and functionality of existing devices. IoT focuses on designing structures rather than just equipment.

- IoT applications in the supply chain: IoT systems have revolutionized supply chain administration (SCM). Comprehending where items are, how they are handled and when they can be expected at a particular location is much simpler.

Internet of Things applications, in the supply chain, are an important way of monitoring and authenticating goods and shipments using GPS and other technologies. They can also track product storage conditions which improve quality control across the supply chain.

- IoT applications in the automotive industry: The Internet of Things, or computers that are linked over a network, is no longer a sophisticated technology. It is here and the way we live is evolving rapidly. IoT has allowed greater transportation efficiency and management capabilities in the automotive sector and leads us to a future of smart, autonomous vehicles.
- IoT applications in smart retail: Smart home technologies are commonly used, and more and more smart devices are being introduced daily. In this field, the Internet of Things technology has gone up to the top. A lot of people choose smart home systems because of their high-security measures.

With the Internet of Things technology, smart retail technology has established several technologies. Many companies have been reorganized, thanks to those inventions. As the job in many business fields is linked in a clustered manner, the rate of labour and error decreases.

- IoT applications smart cities: The thing about the idea of a smart city is that it's really city specific. The problems Mumbai faces are very different from those Delhi faces. Hong Kong's issues vary from those in New York. Also global problems, such as scarce safe drinking water, declining air quality and growing urban density, emerge through cities at various intensities. Therefore, each city is affected differently.

Government and engineers may use IoT to evaluate the often complex city-specific variables of urban planning. The use of IoT technologies will help in areas such as water control, waste management and emergency situations.

- IoT applications in smart grids: One of the most important IoT implementations is the smart grid (SG). SG is a data exchange network that is integrated with the power grid to gather and analyse data collected from transmission lines, distribution substations and customers. It addresses some IoT architectures in SG, specifications for using IoT in SG, IoT applications and SG services and challenges and potential work.
- IoT wearables: Wearable technology is a staple of IoT applications and is undoubtedly one of the first industries to use the IoT. Everywhere we see these days, we see appropriate pieces, heart rate monitors and smartwatches. One of the lesser-known wearables is part of the guardian glucose monitoring system. The technology is being created to assist sufferers with diabetes. Using a tiny electrode called a glucose sensor mounted under the skin, it measures glucose levels in the body and relays the information through radio frequency to a monitoring device. The readers can know more about wearable devices in [11, 14].

4.2 Cloud Applications

A cloud application is an Internet-based software that performs some, or all, of the computing logic and data storage in the cloud. The user interacts with the application through a web browser or mobile application, and a combination of the local computer and a cloud storage system handles the data processing:

- IoT as a Service (IoTaaS): The Internet of Things and Services allows the creation of networks that incorporate the entire manufacturing process, converting factories into a smart environment. The cloud allows for the creation of new services by a global infrastructure, allowing anyone to create content and applications for global users [3]. IoTaaS will provide a better cost-to-value comparison for existing assets with companies as it will reduce all upfront costs including hardware costs. For stakeholders, it will be easier, less expensive and far more practical to get all IoT services bundled from a single player.
- Sensing as a Service: Sensing as a Service (SaS) is an advanced type of distributed computing that takes advantage of IoT's worldwide resources and facilitates the creation of a shared sensor network that is available as a consumer service.

The organizations and developers leverage these services, thus providing users with an opportunity to monetize the data using existing infrastructure [4].

- Anything as a Service (XaaS): XaaS is a common, collective concept that refers to the delivery of something as a service. It acknowledges the large number of goods, software and technology that vendors are now providing to customers as a network service, generally the Internet, rather than local or on-site distribution within a company. There are countless XaaS examples, but the three general cloud computing models are the most common, Data as a Service (DaaS), SaaS, PaaS and IaaS infrastructure:

  - Data as a Service (DaaS): Data as a Service (DaaS) is a data management approach that utilizes the cloud via a network link to provide data storage, integration, processing and/or analytics services. DaaS is a cloud computing technique similar to software as a service, or SaaS, which includes distributing end-user applications over the network, rather than making them run applications on their computers locally.
  - Software as a Service (SaaS): SaaS is a form of software distribution that enables data to be accessed by any user with an Internet connection and a web browser. Software vendors host and manage the servers, databases and code that make up an application in this web-based model. Now that more than 60% of information seekers calling Software Advice want only web-based products, i.e. less than 2% directly request on-premise software, the cloud-based model is so popular.
  - Platform as a Service (PaaS): In the Platform-as-a-Service (PaaS) model, developers basically rent everything they need to create an application, depending on a cloud supplier for development resources, infrastructure, and operating systems. That's one of three models for cloud computing services. The creation of web applications is greatly simplified by PaaS; all backend management takes place behind the scenes from the perspective of the developer. PaaS can be accessed through any Internet connection, allowing a web browser to be integrated into the entire application. Since the development environment is not hosted locally, developers can work on the application from anywhere in the world.
  - Infrastructure as a Service (IaaS): IaaS (Infrastructure as a Service) is a type of cloud computing that provides virtualized computing services online. In an IaaS model, a cloud provider, including servers, storage and networking hardware and the virtualization or hypervisor layer, handles the infrastructure elements that are normally present in an on-site data centre. The IaaS provider also offers a variety of services to support those infrastructure components. This may include comprehensive billing, tracking, log access, protection, load balancing and clustering as well as resilience to storage, such as backup, replication and recovery.

- Network as a Service (NaaS): Network as a Service (NaaS) is the selling of third-party network services to clients who don't want to create their own networking infrastructure.

NaaS offers networking tools, software and applications as a commodity that can be bought by a number of customers, usually for a period of time contracted. It may include services such as wide area networking (WAN) connectivity, connectivity to the data centre, on-demand bandwidth (bandwidth on demand), security services and others.

- Health Cloud: Health Cloud allows for a seamless experience across the entire participant process, from recruiting and onboarding to dedication and retention. Deliver the form of personalized service that creates lasting relationships with customers. First, Health Cloud tests the leads to verify that the values in the fields System Name, Source Code and Medical Record Number have no duplicates and are not correlated with an existing account record. If there are duplicate values, the records can be fused by selecting a Lead record as master and choosing the fields you want to hold. After Health Cloud verifies that no duplicate values exist, you can delegate the patient to a care coordinator. That is facultative. You may also opt to appoint a care planner within the Care Plan case record portion of the case management department.

Hence, this section discusses about Internet applications and cloud applications used in this smart era. Now, the next section will present a comparison of scope or importance of machine learning techniques in detail with respect to pre-COVID-19 and post-COVID-19.

5 Comparison of Scope of Machine Learning Techniques Pre-COVID-19 and Post-COVID-19 Era

The artificial intelligence's sub-field is machine learning. It helps develop automated systems that can learn on their own. The machine then improves its efficiency by learning from experience without any interference by humans. This lets the machines make decisions which are data-driven. Whatever the machines learn from past experience, they use it to make predictions. In fostering global initiatives to solve COVID-19, developments in data science and artificial intelligence/machine learning (AI/ML) play a crucial role. The versatility of AI/ML technology helps scientists and technologists overcome an impressively large range of technological, epidemiological and socio-economic challenges. Before the COVID, i.e. pre-COVID-19 era, we have numerous scopes of machine learning, such as:

- Automotive Industry: One of the fields where machine learning excels by changing the definition of "healthy" driving is the automotive industry. There are a few big companies like Google, Tesla, Mercedes Benz, Nissan and so on that have invested heavily in machine learning to come up with new technologies. Yet Tesla's self-driving car is the industry's best. These self-driving cars are constructed using machine learning, IoT sensors, HD cameras, voice recognition systems, etc.

- Robotics: Robotics is one of the areas that often attract both the attention of researchers and the general interest. George Devol invented the first programmable robot in 1954, which was called Unimate. After this, Hanson Robotics developed the first AI robot, Sophia, in the twenty-first century. With the aid of machine learning and artificial intelligence, such inventions became possible.
- Quantum Computing: In the field of machine learning, we are still at an infant level. There are several advances to accomplish in this region. Quantum computing is one of the ones that can push machine learning to the next level. It is a method of computing that uses quantum mechanical phenomena like entanglement and superposition. By using the superposition quantic principle, we can construct systems (quantum systems) that can simultaneously exhibit multiple states. On the other hand, the theory of entanglement is when two distinct states can be related one to another. It helps to explain the connection between a quantum system's proprieties.
- Computer Vision: Computer vision, as the name implies, provides a view of a computer or machine. Here comes to mind what Google's Head of AI, Jeff Dean, once said, "The improvement we made from an error of 26% in 2011 to an error of 3% in 2016 is incredibly impactful. The way I like to think is now computers have evolved working eyes". The aim of computer vision is to give a machine the capacity to recognize and analyse images, videos, graphics, etc. Progress in the field of artificial intelligence and machine learning has allowed faster achievement of the computer vision target.

Now, we will discuss about scope of machine learning post-COVID-19 in brief.
Officials around the world are using different outbreak prediction models for COVID-19 to make informed decisions and implement appropriate control measures. Among the traditional COVID-19 global pandemic prediction models, simple epidemiological and statistical models have drawn more attention from authorities and are popular in the media [15]. In the face of the global public health crisis, data scientists and AI/ML innovators may be inclined to ask: Are we ready for this? Can we find a responsible path to ethically and safely exercise our technological effectiveness? In what follows, I argue that these crossroads do not need to cause confusion as to how we should go, and in fact, to find the right way forward, they give us straightforward signage [15]. In the fields of diagnostics, prognostics, genomics, drug creation, epidemiology and mobile health monitoring, biomedical AI/ML innovations have also made major strides in assisting physicians and researchers when pressed into action for the public good. And although the great promise of helping healthcare professionals to solve COVID-19 is held by each of these growth areas, they also present substantial ethical hazards. Actionable ways of coping with these are what we need now. In order to advance higher-performance models for long-term prediction, future research should be devoted to comparative studies on various ML models for individual countries. Due to the fundamental differences between the outbreaks in different countries, the creation of global models with generalization potential would not be feasible.

In summary, in post-COVID-19 era, we have shifted more towards IoT-based cloud applications and preferring things which can work automatically, i.e. we rely on programmed machines. But, as discussed above, these machines or gadgets or solutions come with several negative impacts also, which have been discussed in [10, 11].

6 Machine Learning Techniques for IoT-Based Cloud Applications

In Fig. 1, we can see the evolution of machine learning techniques since 1950. In the past few decades, machine learning use was limited, and due to having fewer amounts of data, we were unable to use machine learning techniques with its full capability. Slowly, we moved into this smart era, and then we started using machine learning in various applications like media and entertainment, education, baking finance, aerospace, retails, logistics, etc. Now, each and every application is discussed (in detail) below as:

(a) For Agriculture: IoT-based smart agriculture has intrigued many researchers among the vast array of IoT applications and has used machine learning (ML) and IoT technologies to perform groundbreaking research. By preparing production costs, minimizing losses and using resources more effectively, IoT-based data-driven farm management techniques can help increase agricultural yields. In the form of voltage values for images and actuator states for robot positions, IoT-based agriculture has resulted in a variety of data generated from various sources on and around agricultural farms. Quality data contributes to knowledge that is quality and correct. You won't be able to build predictive models using ML algorithms without having quality data. It allows for better analysis and precise predictions to apply ML algorithms to these improved datasets. Several research papers have implemented systems for IoT-based agriculture [16, 17]. Each sensor mote typically includes a microcontroller, multiple sensor types (from basic temperature sensors to cameras), actuators and wireless interfaces in an IoT solution. The IoT is capable of efficiently storing and handling vast volumes of sensor data and incorporating cloud computing resources such as agricultural maps and cloud storage. Real-time access to data from anywhere at any time provides real-time tracking and end-to-end communication across all parties.

(b) For Healthcare: The triumphant use of information mining in incredibly apparent zones like exchange, trade and e-business has coordinated to its application in another industry. The ailments are still information rich yet data low. There is a plenitude of data doable inside the clinical practices. All things considered, there is a deficiency of basic examination instruments to perceive concealed patterns and connections in information. Numerous analysts have applied data mining techniques for the forecast and determination of a few sicknesses. AI techniques have been extensively used in the visualization of various illnesses towards the early phases. The ebb and flow decade has watched an anomalous improvement in the assortment and volume of electronic information related with the turn of events and examination, tolerant self-following and wellbeing records together proposed to as big data [18]. One of the most basic interminable social insurance issues is diabetes. In the near future, this issue may harm eyes, heart, kidneys and nerves of diabetes understanding if ill-advised prescription is done which additionally prompts demise. An extensive report is done on diabetes dataset with random forest (RF), SVM, KNN, CART and LDA calculations. The accomplished outcomes show that RF is giving progressively exact expectations which contrasted with different calculations [19]. Feelings assume a critical job by the way we settle on a choice, arranging, thinking and other human mental states.

The acknowledgement of these feelings is turning into an essential assignment for e-medicinal services frameworks. Utilizing biosensors, for example, electroencephalogram (EEG), to perceive the psychological condition of patients that could require an uncommon consideration offers a significant criticism for Ambient Assisted Living (AAL) [20].

More recently, the COVID pandemic has been making the rounds. The discovery procedure was actualized on stomach computed tomography (CT) pictures. The master radiologists recognized from CT pictures that COVID-19 shows various practices from other viral pneumonia. Along these lines, the clinical specialists indicate that COVİD-19 infection should be analysed in beginning stage. For recognition of the COVID-19, 4 diverse datasets were shaped by taking patches estimated as 16 × 16, 32 × 32, 48 × 48 and 64 × 64 from 150 CT pictures. The component extraction process was applied to patches to expand the characterization execution. We can use machine learning to detect how coronavirus spreads in different bodies and how DNA strands get changed and can make comparisons from how it started to when it leaves the body. For example, IBM's Watson research group cooperated with the US Veterans Administration to build up a clinical thinking model known as the Electronic Medical Record Analyzer (EMRA). This primer innovation is intended to utilize AI methods to process patients' electronic clinical records and consequently distinguish and rank their most basic medical issues.

(c) For Education: Education is crucial to helping a person to become a good human being. Several research initiatives are underway in developing countries to strengthen their education policies and techniques in the education field. Therefore, there is a need for a great deal of effort to upgrade individual learning materials for teachers and students in developing countries. An integral aspect of educational evaluation is assessing student focus. New instruments and evaluation methods are also needed as new learning styles evolve. By logging their behaviour and analysing the resulting multimedia data using machine learning algorithms, the learning habits of learners attending remote video lectures are assessed. For the smart assessment of student learning experience, an attention-scoring algorithm, its workflow and the mathematical formulation are created.

(d) For Entertainment/Multimedia: "Data demand remains vast, says Friederike Schüür, Fast Forward Labs Senior Data Scientist and Researcher". Organizations are seeking to make their data work for them across sectors. Machine learning and data science provide them with the means to do so. Inside media and entertainment, the majority of today's data work is devoted to knowing the audience: who reads, listens and watches the material of our media and entertainment? From corporate strategies to marketing and content development, consumer observations help industries in making them profitable. Recommendation systems provide readers, listeners and watchers with relevant content. Embedding-based recommender systems can autosuggest image assets for posts, for example, as creators create new content, or help surface image assets without copyright limitations provided a target image. Sequence-to-sequence learning can not only be translated from German to English but also from one writing style to another, making it easier for the material to reach separate audiences.

(e) For Banking and Finance: There are many businesses claiming to provide banks and financial institutions with AI solutions. Artificial intelligence helps banks grant credit to those who pass machine tests more comfortably. For this function, systems and algorithms evaluate all available information about a potential borrower, research their credit background, adjust their salary levels and assess the client's efficiency and the security of the loan on this basis. In addition, Chinese banks have already gone further and have chosen not to restrict themselves solely to analysing the results. We have found that these strategies are structured to assist banking and finance companies with at least one of the following market issues:

- Detecting fraud
- Developing products
- The processes

Diebold Nixdorf, probably best known for its ATM technology, offers ATM maintenance software, allegedly enabling banks to be alerted when an ATM is due for maintenance before it fails. C3.AI supports similar applications [21]. Arimo offers fraud detection applications, a common application of machine learning in finance. Arimo says, for example, that its retail solution makes predictions based partly on eCommerce activity data from users, which can see a record of each time a customer clicks on a product link or other feature of the web. Their industrial and manufacturing approach enables the tuning process to be automated on machinery products, and Arimo says that this is achieved with a refrigerator by one of its clients.

(f) For Supply Chain Management/Logistics: For mass-scale laundry services, the authors propose a creative Internet of Things (IoT)-based cloud laundry e-commerce business model. Big data analytics, intelligent management of logistics and machine learning techniques are used in the model. It determines the best transport path using GPS and real-time update of big data and easily and simultaneously updates and reroutes the logistic terminals. Intelligently and dynamically, cloud laundry provides the best laundry solutions based on the current state of the laundry terminal spaces through user requirements and thus offers simple, reliable and clear laundry services to local hotel customers. Cloud laundry uses mobile terminal control and big data models to preserve the security needs of consumers, taking advantage of the exponential growth of the big data industry, consumer interest modelling and information security and privacy considerations. Cloud laundry companies have higher capital turnover, more liquidity and greater profitability, unlike the conventional laundry industry. Therefore, not only academic researchers but also e-commerce entrepreneurs could be interested in this new generation of smart laundry business model.

(g) For Other Sectors: The blossoming time of large information is affecting the procedure businesses enormously, giving remarkable chances to accomplish shrewd assembling. This sort of assembling expects machines to not exclusively be fit for mitigating people from serious physical work, yet in addition be powerful in taking on scholarly work and in any event, creating advancements all alone. To accomplish this objective, information examination and AI are key. Advanced analytics is used to identify the failure in the industries and is used to reduce the possibilities of error and make it error-free. The data from the past is collected and examined in a broad sense giving the creators an early idea of the possibilities of problem and how to rectify the same. We have the potential to modify the current worldview significantly. The traffic designs, endorser gear and supporters' profiles are another sector where machine learning helps in controlling traffic types, remote gadgets and endorsers differently dependent on the systems. Besides, the remote traffic load is becoming quicker than the limit, and the system administrators are confronting extreme difficulties to build arrange limit cost-successfully. While we consider the application of VANET (Vehicular Adhoc Network), the vehicle can be controlled without the presence of a driver being physically present, to make sense of the most equipped approach to cut the system and traffic, i.e. the quantity of cuts, parting traffic across cuts and so forth, which rely upon the kind of traffic and how it changes after some time and space [15]. Utilizing standards of cutting-edge large information examination and AI ideas, producing organizations can catch sensor information from shop floor instruments and hardware to take an undeniably granular and venture wide way to deal with quality control. What's more, producers will likewise have the option to recognize abandons, reveal the main driver of issues, lessen the danger of transportation non-adjusting parts, empower designing upgrades and figure out which components, procedures and work processes sway quality.

7 Internet of Things and Cloud Computing in Post-COVID-19 Era: Applications and Challenges

In the past decade, we have seen major developments in technology and price fall of smart gadgets. Smart devices are generally called as Internet of Things, i.e. the things which run through Internet or access services through Internet. These smart objects are also called Internet-connected things. Similarly, cloud uses have emerged in the past decade, and at a very high rate, its demand has also increased. Storage and energy are two major problems of computing. Storage problem is solved by cloud computing, by storing data at edge or at cloud's edge, and this data is accessible from anywhere, at any time. Hence, these two important technologies have been used at a rapid rate in post-COVID-19 era. This section will discuss various challenges faced during implementation of IoT and IoT-based cloud applications in solving real-world problems.

7.1 Challenges in Internet of Things (IoT)-Based Cloud Applications

As of late, Internet of Things (IoT) and distributed computing have been generally contemplated and applied in numerous fields, as they can give another strategy to savvy observation and association from M2M (counting man-to-man, man-to-machine and machine-to-machine), and on-request use and proficient sharing of assets, individually. Specifically, the accessibility of information at up to this point unheard of scales and worldly longitudes combined with another age of wise preparing calculations is posing as a challenge for the Internet of Things-based cloud applications. Information put away in the cloud is similarly delicate as different pieces of the IoT biological system. The foundation ought to have the option to secure information put away in the cloud. Insurance measures incorporate proper encryption, get to control, etc. Security vulnerabilities consistently exist regardless of how much endeavours you pay to upgrade your item code and equipment. For this situation, we should initially have an arrangement to fix mistakes and rapidly discharge patches, rather than leaving the blunders unfixed for an extensive stretch. Next is the necessity of furnishing clients with an immediate and secure technique to fix blunders. Right now, it is well known to refresh online gadgets over the air, yet you should guarantee that the above strategy itself would not become security vulnerability. IoT gadgets are frequently situated in open fields and are unattended and not genuinely ensured. We should guarantee that they would not be malevolently messed with by horrible association, penetrated by programmers or worked utilizing a level head screwdriver. Likewise, we should ensure information that gets put away on the gadgets in any structure. Despite the fact that it is expensive to insert a security assurance part on each IoT gadget, it is as yet critical to scramble information on these gadgets:

(a) *Privacy:* In light of the previously mentioned security defects, numerous other security and protection issues present themselves in Internet of Things. IoT utilizes Internet as a major framework used in interconnecting various topographically expanded IoT hubs, and hence cloud is utilized as a key backend supporting foundation. In the writing, the assortment of the IoT hubs and the cloud is all in all called as an IoT cloud. Tragically, the IoT cloud experiences different disadvantages, for example, colossal system inactivity as the volume of information which is being prepared inside the framework increments. To lighten this issue, the idea of haze registering is presented, in which fog computing is situated between the IoT hubs and the cloud framework to locally process a lot of local information. Contrasted with the first IoT cloud, the correspondence inertness just as the overhead at the backend cloud foundation could be essentially decreased in the haze figuring upheld IoT cloud, which we will allude as IoT mist. Hence, a few significant attempts were hard to be conveyed by the conventional IoT-based cloud. A couple of them are listed below as:

- Burglary of delicate data like bank secret phrase.
- Simple openness to individual subtleties like contact address, contact number and so on.
- It might prompt open access to classified data like money-related status of an establishment.
- An assault on any one gadget may bargain the respectability of the various associated gadgets. In this manner the interconnectivity has an immense disadvantage as a solitary security disappointment can disturb a whole system of gadgets.
- The dependence on the Internet makes the whole IoT engineering powerless to infection assault.

  A noteworthy forward jump in beating any boundary among virtual and physical universes began from the vision of the Web of Things (WoT), which uses open web quantities in achieving information sharing and articles interoperability. Social Web of Things (SWoT) further loosens up WoT to join sharp articles with casual associations and supposedly connects among physical and virtual universes just as support continued with participation between physical devices and human.

(b) *Security:* The security issues of the Internet of Things (IoT) are straightforwardly identified with the wide utilization of its framework. Starting with presenting the engineering and highlights of IoT security, among these wellbeing estimates concerned, the ones about recognition layer are especially expounded, including key administration and calculation, security steering convention, information combination innovation, just as validation and access control and so forth. A huge segment of the executed information between IoT gadgets is private data, which must not at all be listened stealthily on or altered. Security in IoT gadgets is in this manner of principal significance for additional advancement of the innovation. Such gadgets ordinarily have constrained region and vitality assets, which utilize great cryptography restrictively costly. Physically unclonable functions (PUFs) are a class of novel equipment security natives that guarantee a change in outlook in numerous security applications; their generally straightforward engineering can answer a large number of the security difficulties of vitality compelled IoT gadgets. RFID is one of the empowering innovations of the Internet of Things. RFID can possibly empower machines to recognize objects, comprehend their status and convey and make a move if vital, to make "ongoing mindfulness". The inescapability of RFID innovation has offered ascend to various significant issues including security and protection concerns.

(c) *Trust:* Trust the executives assume a significant job in IoT for dependable information combination and mining, qualified administrations with setting mindfulness and upgraded client protection and data security. It assists individuals with conquering impression of vulnerability and chance and participates in client acknowledgement and utilization on IoT administrations and applications. In any case, not many explorations about trust system for Internet of Things (IoT) could be found in the writing; however we contend that impressive need is held for applying trust component to IoT. We decay the IoT into three layers, which are sensor layer, focus layer and application layer, from parts of framework synthesis of IoT. Each layer is obliged by trust the board for specific explanation: self-created, loaded with feeling coordinating and multi-organization independently. Likewise, an official decision creation is performed by organization requester as shown by the assembled trust information similarly as requester's system. Finally, we use an ordinary semantics-based and fluffy set speculation to see all above trust part, the result of which gives a general framework to the headway of trust models of IoT. The assets ought to be doled out to gadgets as per the framework strategy, which relies upon the data given by Industrial Internet of Things (IIoT) gadgets. In the event that there are any asset requesting gadgets, they can report controlled vindictive data for their own enthusiasm to get more assets. That is, the brilliant assembling framework might be defenceless because of narrow-minded shrewd assembling gadgets' practices. This lessens the effectiveness of the whole framework and besides stops the plant-wide procedure. Many issues have been identified towards trust (inlcuding commitments) for targeting a large number of audience/researchers in this smart era.

Hence, we can say that privacy, security and trust are major issues in IoT-based cloud environments/applications. Industries are investing billions of dollars for providing users a hassle-free/secure environment to access their service. In other words, industries are spending a lot of money in avoiding any kind of cyberattacks on their databases or businesses; any breaching on data reflect trust of consumer directly.

7.2 *Internet of Things Challenges and Solutions*

Protection of IoTs in the post-COVID-19 era: In the fight against COVID-19, technological advancements are continuously making a difference to healthcare systems. The speed of progress in the Internet of Things (IoT) systems, in particular disease tracking, human movement, the identification of potential carriers and the remote monitoring of health conditions, is now being used and developed worldwide. According to Forrester Research Analyst Chris Sherman [21], two US hospitals have already been targeted via virtual care systems, after a hacker exploited a loophole in a medical IoT device (specifically, a remote patient monitoring sensor) and obtained access to hospital patient databases. And in another type of attack, the Fresenius Group, a maker of medical devices and the largest private hospital operator in Europe, has been hit by ransomware:

(a) IoTs privacy in the post-COVID-19 era: Most workers worked from offices in the pre-COVID-19 era, where the local area network (LAN) as well as the desktops/laptops were adequately protected. Sophisticated technologies could defend against cyberattacks that mainly originated from the Internet and targeted the network of enterprises. Only support staff or those who need direct system/hardware access, e.g. testing laboratories, direct consoles, unique printing machines in the banking environment, etc., operate from the office in the post-COVID-19 period. In contrast to those at the workplace, the majority of the workforce is working from home and exposed to more insecure networks.
(b) Performance of IoT applications: Elasticity, which refers to the ability to add or subtract resources according to the needs of the application or service, is one of the most essential cloud computing services. The use of this facility is especially applicable to the Internet of Things (IoTs) scope, as IoT requires a middleware that should be capable of handling high data volumes in real time.
(c) Ethical issues in IoT applications: The underlying information security issue is that there are communicator-coupled sensors in several IoT modules. For example, a camera, microphone or other sensor collects environmental data and is a connected communicator that transmits the data to a remote location, such as a cloud or other proprietary server. Although many of the components have some security features, such as passwords that limit user access, a large part of the IoT security problem is that those passwords are typically set to a factory default and have never been changed [22]. However, the default passwords are widely accessible online [23]. The responsibility of the consumer is another problem. For starters, if changing passwords was simple enough, we might think that this is a core user duty [24]. It would be similar to seat belts in vehicles in this regard; it is the obligation of the driver to put them on. But that assumes that the modifications are simple; in other words, users have the time and cyber-literacy needed to handle safe passwords.
(d) Sustainability of IoT infrastructure: The Internet of Things is regarded by business and organizations as an opportunity for the present and the future that will digitize many operations and offer enormous benefits. Yet it also has the

potential to help battle and protect the world from climate change. This is how the Internet of Things in various fields, such as water use and energy efficiency, will affect the planet's sustainability. There's a long way to go for IoT technology compatible with the SDGs. The World Economic Forum is currently estimating that 75% of IoT ventures are small- and medium-sized, focusing on business, urban energy conservation, renewable energy, health and responsible consumption. Public-private investments are crucial for projects to expand, Guerrero says. This will generalize IoT use, minimize costs and extend its use. "Training, raising awareness in society of the importance of sustainability and demystifying technology are important for the large-scale implementation of IoT sustainable development projects", he says.

(e) Scalability in IoT: The fundamental property of an IoT system is that it will provide exclusive recognition for the advancement and creation of usefulness and applications of each "thing" and its virtual personification. Scalability is a very critical component of the modern and revolutionary developments in technology that occur every day [25]. As more and more devices or "things" are linked to the Internet, the various problems that occur as a result of this are a matter of utmost concern. For scalability in the IoT, there are countless things that need to be held in mind.

(f) IoT applications maintenance: In the coming years, the IoT will be instrumental in improving productivity and performance everywhere. This can lead to more efficient power management at plants and factories by automatically adjusting environmental control systems to reduce energy use when it is not necessary. The IoT will then change the way assets are handled in seven respects, but there are certainly several more:

- Greater adoption of predictive maintenance
- Real-time data analysis
- Accurate performance metrics
- Automatic software upgrades
- Recommended repair actions
- Tighter parts and inventory control
- Remote assets

7.3 Cloud Computing Challenges and Solutions

Cloud reflects the concept of a distributed infrastructure consisting of a series of virtual machines that can be dynamically provisioned to meet the varying resource requirements of a client [1], and the entire basis of this cloud-customer relationship is governed by the service-level agreement (SLA). The National Institute of Standards and Technology (NIST) defines the cloud as a model that allows convenient on-demand network access to a common configurable resource pool of computers, such as network, storage, etc. Cloud relieves the user of the overhead of the physical installation and maintenance of their device, which automatically reduces the overall cost and enhances the system's reliability. The use of cloud-based services results in an abstract layer being added between the physical storage or servers and the users whose data or services are stored in the cloud. The current situation is such that the cloud user's data or service owner must rely entirely on the cloud service provider (CSP) for their privacy and information security. The concept of mutual trust is reached to some degree by negotiating the SLA, but a good number of cloud-specific security issues are still going to need to be addressed by either the CSP or the user itself.

When it comes to IT security concerns, data occupies the top spot, regardless of the infrastructure being used. Cloud computing is no exception to this, and it focuses on extra security problems because of its distributed nature and multi-tenant architecture. Its generation, storage, use, delivery and destruction constitute the data life cycle. All these stages in the data life cycle should be assisted by each CSP with sufficient protection mechanisms:

(a) Cloud computing security: One of the main driving factors behind a user's decision to switch into a cloud system or stick with the legacy system that trusts on the cloud service provider (CSP) and their services. Confidence is focused on the evaluation of whether all threats have been addressed by a supplier, including data protection areas, VM protection as well as other government and regulatory issues. Confidentiality, honesty and availability (CIA) are the three factors that have been considered here for the cloud system security assessment. And the domain is a convention commonly used by the CIA to assess the security issues of a traditional information system.
(b) Cloud computing privacy issues: Distributed computing is a growing innovation that encourages the sharing of resources, such as software and hardware and Internet servers. But there are some problems in data cloud protection and privacy that are not as reliable as opposed to conventional IT operations, protection patching in the cloud is much easier, enforcement is more difficult to show in the cloud, data loss is less in clouds, and more control power would boost security. There is a profound need to securely store, supervise, exchange and analyse enormous complex knowledge steps, to establish examples and trends that take into account the ultimate aim of enhancing the quality of human services, better protecting the country and investigating elective vitality. It is important that fogs are safe as a result of the unpredictable technique for the applications.
(c) Reliability in cloud computing: Reliability is defined as the ability of a system or component to perform its necessary functions for a specified period of time under stated conditions. There are different types of disappointments that can place pressure on a cloud administration's unwavering dominance, including excess, timeout, missing data asset, missing asset calculation, programming dissatisfaction, database annoyance, discontent with hardware and network failure.

(d) Quality of service in cloud computing: One of the issues posed by cloud computing is the quality of service. In making cloud services appropriate to consumers, this problem plays an important role in denoting the standards of performance, reliability and availability provided by cloud services. Many recent implementations have been mentioned/discussed in the literature in order to achieve better performance and meet the needs of producers and customers and calculate and ensure QoS in cloud computing systems.
(e) Cloud performance management: Applications for data management implemented in IaaS cloud environments must aim to reduce costs and provide good performance at the same time. Balancing these two priorities involves difficult decision-making on a variety of axes: resource provisioning, location of queries and scheduling of queries.
(f) Scalability in cloud computing: Measuring and testing the performance of cloud-based software services is critically important in the context of rapid cloud computing growth. Scalability, elasticity and coherence are interrelated aspects of the performance of cloud-based computing services. The performance evaluation and testing of cloud-based computing services is critically essential in order to support the service-level agreement (SLA) compliant quality of delivery of these services, particularly in the context of the rapid expansion of the amount of service delivery. Scalability is the ability of the cloud layer to increase the delivery potential of the information service by expanding the amount of information service offered. This concept focuses on the technological side of cloud-based computing services, but we note that the literature frequently uses alternate, utility-oriented (i.e. economic cost/benefit focused) approaches.
(g) Cloud interoperability and portability: An underlying evolution of cloud computing is interconnected cloud computing. The academic and industry sectors have been attracted by various advantages offered by connecting clouds. Just like everyone else does, interconnected clouds have their own set of challenges, such as protection, control, authorization and identity management, vendor lock-in and so on. The latest evolution faces challenges. Since it is difficult to find a cloud provider that can satisfy all customer criteria, customers attempt to merge services or switch from one provider to another. But one provider's customization of services does not allow it to be done easily. It takes a great deal of effort and cost, even though it is finished. This scenario is often referred to as vendor lock-in.
(h) Cloud compliance: There is an ever rising need for businesses to comply with a variety of laws, regulations and standards in the current market environment. Companies will need to be consistent in showing that they are in compliance, and this will demonstrate because of the existence of such cloud vulnerabilities. There is a strong incentive for businesses to be able to clearly show complete compliance, considering the possible severity of penalties for noncompliance.

Hence, this section discusses IoT challenges and its solutions, as well as cloud challenges and solutions for the same. Now, the next section will discuss several future research gaps or possibility in (with) the respective technology in the next decade.

8 Future Research Directions for Machine Learning Towards IoT-Based Cloud Applications

Currently, a large amount of picture, audio and video and metadata created by users, such as network and user activity information, is being transferred to the cloud. This is due to the availability of comparatively inexpensive storage of information and cloud backup. Continuous research is underway to apply machine learning to applications for speech/audio recognition; text, image and video processing; and language translation. In view of the broad computational requirements, conventional machine learning algorithms were limited to execution on large clusters. Furthermore, in order to perform complex learning tasks without incurring large monetary costs, APIs and software libraries are now available, for example, TensorFlow and Nervana Cloud from Google. The availability in cloud environments of hardware accelerators, such as GPUs, has reduced the computation time on large quantities of data for machine learning algorithms. The industry's interest in this field is due to the ability of predictive analytics to include deep learning.

The availability in cloud environments of hardware accelerators, such as GPUs, has reduced the computation time on large quantities of data for machine learning algorithms. The industry's interest in this field is due to the potential for deep learning in predictive analytics. Cognitive computing is a closely related avenue in the sense of future clouds. Cognitive systems in this visionary model would rely on machine learning algorithms and data generated to continuously generate knowledge. The obvious advantages of using fog computing include reducing device latency and enhancing users' quality of service (QoS) and experience (QoE) while exploiting hierarchical networking and tapping into tools that are not typically used for general computing purposes. Fog computing is therefore anticipated to allow the Internet of Things (IoT) vision to be realized.

8.1 Advanced Analytics with Machine Leaning: A Way to Forward

Our main focus is to be at the front line of giving customized arrangements dependent on joining of machine learning strategy, programming equipment elite processing and preparing for a wide scope of information concentrated registering, information handling and progressed examination for logical and business applications, utilizing both open-source and in-house R&D. With AI we can move towards:

- *Artificial Intelligence (AI) for Cybersecurity:* Intrusion detection automatically by machine and artificial intelligence: In the near future, AI-based solutions may find vulnerabilities over web and reduce the load of cybersecurity professional. Note that hackers may use AI-based solutions to perform a serious attack on cyber-physical systems or any other computing systems or networking.
- *AI for Industry 4.0:* AI will provide full automation in the near future to industry and intelligence too. AI-based manufacturing or logistics are few examples in the near future, i.e. timely and error-free delivery, increased productivity, etc. Machine intelligence in medical imaging, machine learning and AI for penetration testing and machine learning in chemical sciences are also other examples of AI for Industry 4.0.
- *AI for Data Science:* AI-based data science will be helpful (in trend in the near future) in retail, banking, finance, weather forecasting or stock market prediction.
- *AI-IoTs Integration for Other Sectors:* AI can be used in many other sectors like software development and cloud computing/fog computing. Hence, few more possibilities of AI-IoT are:
  - IoT is higher education systems.
  - Impact of IoT is education now and in the future.
  - A learning management system enhanced with Internet of Things.

Hence, this section lists various future research directions for machine learning-based IoTs – cloud applications. The readers/researchers are suggested to read work [9, 10, 12, 14] for knowing more machine learning, deep learning and their importance in the next decade. Further in [26], authors provide various future research gaps and opportunities in the future towards Internet of Things and Internet of Everything. Also, readers can find out several issues in various computing platforms in [26]. Now, the next section will conclude this work in brief with including several interesting future remarks for researchers/scientists.

9 Conclusion

Cloud and Internet of Things are two developed concepts and both are made for each other. Cloud is something which stores information which can be accessed by users/consumer remotely anytime. This requirement is completed by IoT devices and their communication via completing everyday task. The large amount of data is generating every day by these IoT devices, especially post-COVID-19. This work has discussed IoT-based cloud applications based on pre-COVID-19 and post-COVID-19. In summary, we get to know that in post-COVID-19 era, we have shifted more towards IoT-based cloud applications and preferring doing the tasks with machines/things which can work automatically, i.e. we depend on programmed machines. But, as discussed above, these machines or gadgets or solutions come with several negative impacts like security, privacy, trust, standardization, etc. For which, solutions are yet to provide. Today, many other technologies, like artificial intelligence, blockchain technology, deep learning and their integration like integration of IoT blockchain, etc., will be more useful for making IoT-based cloud applications successful.

This work provides maximum information to its readers (from academia and industry) and to learn more and more about machine learning and IoT-based cloud field. IoT-based cloud applications are in trend nowadays and will be the more adaptable by all industries (and sectors) in the near future. Success of an application depends on its trust for its user/consumer; if we provide sufficient level of security using innovative techniques, then we can build a higher trust among user and service provider. As discussed previously, dependency on these applications or on technology comes with several negative impacts (or serious concerns), which we need to solve yet. We require (expect) innovative solutions or mechanisms from academia/industry researchers (from around the globe) for raised concerns in IoT and IoT-based cloud environment.

Acknowledgement This research work is funded by the Anumit Academy's Research and Innovation Network (AARIN), India. The authors would like to thank AARIN, India, a research network for supporting the project through its financial assistance.

Conflict of Interest The authors have declared that they do not have any conflict with respect to publication of this work.

Contribution/Disclosure Both authors of this chapter are the original participants and have contributed equally; also the first author was the meditation instructor for making/providing useful information in this work for future readers/researchers/scientists.

Scope of the Work This work provides in-depth information to its readers (from academia and industry) and to learn more and more about machine learning and IoT-based cloud environment.

References

1. Mahdavinejad, M. S., et al. (2018, August). Machine learning for internet of things data analysis: A survey. *Digital Communications and Networks, 4*(3), 161–175.
2. Gadhi, S. (2020). Stonefly. https://stonefly.com/blog/role-cloud-computing-internet-things
3. Edureka. https://www.edureka.co/blog/iot-applications/
4. Asir, R. *IoT as a service.* Research Gate. Published in March 2016.
5. *How digital infrastructure can help us through the COVID-19 crisis*. World Economic Forum.
6. Tyagi, A. K., & Rekha, G. (2019, March 20). Machine learning with big data. In *Proceedings of international conference on sustainable computing in science, technology and management (SUSCOM), Amity University Rajasthan, Jaipur, India, February 26–28, 2019.*
7. Tyagi, A. K., & Chahal, P. (2020). Artificial intelligence and machine learning algorithms. In *Challenges and applications for implementing machine learning in computer vision.* IGI Global.
8. Tyagi, A. K., & Rekha, G. (2020). Challenges of applying deep learning in real-world applications. In *Challenges and applications for implementing machine learning in computer vision* (pp. 92–118). IGI Global. https://doi.org/10.4018/978-1-7998-0182-5.ch004

9. Pramod, A., Naicker, H. S., & Tyagi, A. K. (2020). Machine learning and deep learning: Open issues and future research directions for next ten years. In *Computational analysis and understanding of deep learning for medical care: principles, methods, and applications*. Wiley Scrivener. https://hdsr.mitpress.mit.edu/pub/as1p81um/release/3
10. Tyagi, A. K., Rekha, G., & Sreenath, N. (2020). Beyond the hype: Internet of Things concepts, security and privacy concerns. In S. Satapathy, K. Raju, K. Shyamala, D. Krishna, & M. Favorskaya (Eds.), *Advances in decision sciences, image processing, security and computer vision. ICETE 2019. Learning and analytics in intelligent systems* (Vol. 3). Springer.
11. Jayaraman, P. P., Yavari, A., Georgakopoulos, D., Morshed, A., & Zaslavsky, A. (2016). Internet of Things Platform for Smart Farming: Experiences and Lessons Learnt. *Sensors*, 16, 1884.
12. Reddy, K. S., Agarwal, K., & Tyagi, A. K. (2021). Beyond things: A systematic study of Internet of Everything. In A. Abraham, M. Panda, S. Pradhan, L. Garcia-Hernandez, & K. Ma (Eds.), *Innovations in bio-inspired computing and applications. IBICA 2019. Advances in intelligent systems and computing* (Vol. 1180). Springer.
13. Idexcel. https://www.idexcel.com/blog/tag/iot-and-cloud-computing/
14. Nair, M. M., Tyagi, A. K., & Goyal, R. (2019). Medical cyber physical systems and its issues. *Procedia Computer Science, 165*, 647–655. ISSN 1877-0509.
15. Sampath Kumar, Y. R., & Champa, H. N. (2019). An extensive review on sensing as a service paradigm in IoT: Architecture, research challenges, lessons learned and future directions. *International Journal of Applied Engineering Research, 14*(6), 1220–1243. ISSN 0973-4562.
16. Ardabili, S. F., Mosavi, A., Ghamisi, P., Ferdinand, F., Varkonyi-Koczy, A. R., Reuter, U., Rabczuk, T., & Atkinson, P. M. (2020). COVID-19 Outbreak Prediction with Machine Learning. *Algorithms, 13*, 249.
17. Leslie, D. (2020). Tackling COVID-19 through responsible AI innovation: Five steps in the right direction. Published on: May 27, 2020, *Harvard Data Science Review*.
18. Kamilaris, A., et al. (2016). *Agri-IoT: A semantic framework for internet of things-enabled smart farming applications*. European Union.
19. Jayaraman, P. P., et al. (2018). Internet of Things platform for smart farming: Experiences and lessons learnt. *Sensors (Basel), 16*(11), 1884.
20. https://emerj.com/ai-sector-overviews/ai-and-iot-in-banking-and-finance-current-applications/
21. https://www.cxotoday.com/news-analysis/iot-security-in-the-era-of-covid-19/
22. Kan, M. (2016, October 4). *IoT botnet highlights the dangers of default passwords*. InfoWorld. Available at https://www.infoworld.com/article/3127167/password-security/iot-botnet-highlights-the-dangers-of-default-passwords.html
23. Hiner,J.(2018,March7).*Newresearch:MostIoTdevicescanbehackedintobotnets*.TechRepublic. Available at https://www.techrepublic.com/article/new-research-most-iot-devices-can-be-hacked-into-botnets/
24. van de Poel, I., & Robaey, Z. (2017). Safe-by-design: From safety to responsibility. *NanoEthics, 11*(3), 297–306.
25. Gupta, A., et al. (2017). Scalability in Internet of Things: Features, techniques and research challenges. *International Journal of Computational Intelligence Research, 13*(7), 1617–1627. ISSN 0973-1873. Research India Publications. *15*(2020), 153–177.
26. Tyagi, A. K., Nair, M. M., Niladhuri, S., & Abraham, A. (2020). Security, privacy research issues in various computing platforms: A survey and the road ahead. *Journal of Information Assurance & Security, 15*(1), 1–16. 16p.

Further Reading

1. Computer Vision.
2. Fog Computing.
3. Edge computing.
4. Blockchain Technology.
5. Deep Learning.
6. Intelligent Automation.
7. Advanced Analytics.
8. Internet of Things-Blockchain based Applications.

Deep Learning Frameworks for Internet of Things

Dristi Datta and Nurul I. Sarkar

1 Introduction

In recent times, deep learning (DL) is receiving a lot of consideration today. The idea of artificial intelligence (AI) came into existence and the term AI was first coined in 1956 [1]. The concept is very old; however, it gains its popularity recently. Prior we had an exceptionally limited quantity of data and it is not enough to predict accurate results. However, in recent years, there is a remarkable growth in the number of record volumes. "Statistics suggest that by 2020 the accumulated volume of data will increase from 4.4 ZB to around 44 ZB or 44 trillion GBs of data" [2]. To keep pace with the increasing amount of data, improved algorithms and higher computer storage are required.

Computer-based intelligence is a procedure that empowers the machine to mirror human conduct. AI is cultivated by concentrating on how the human cerebrum thinks and how the human mind learns, chooses, and works while attempting to take care of the issue. The results of this investigation are utilized as a reason for creating keen programming and frameworks. With AI innovation, it is conceivable to train models from the past and present experience. The machines alter their reactions dependent on new data and in this manner performing human-like errands. Simulated intelligence can be prepared to achieve explicit assignments by handling a lot of information and perceiving designs on them. Man-made intelligence framework includes structuring various segments, parts, and learning calculation, never know

D. Datta (✉)
Department of Electrical & Electronic Engineering, Varendra University, Rajshahi, Bangladesh

N. I. Sarkar
Department of IT and Software Engineering, Auckland University of Technology, Auckland, New Zealand
e-mail: nurul.sarkar@aut.ac.nz

S. Pal et al. (eds.), *Artificial Intelligence-based Internet of Things Systems*, Internet of Things, https://doi.org/10.1007/978-3-030-87059-1_5

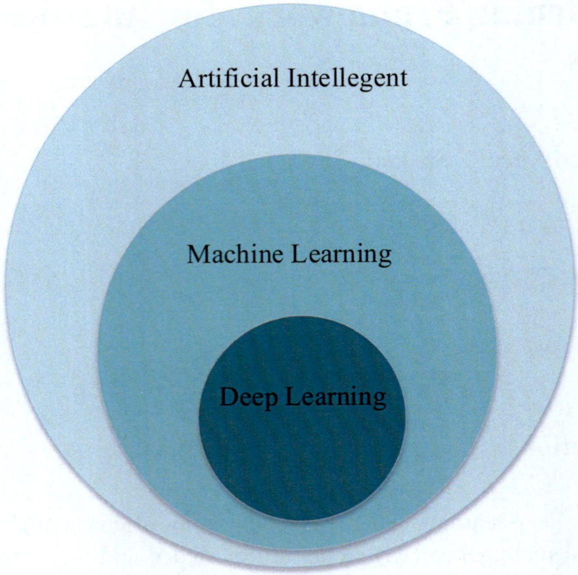

Fig. 1 Illustrating the concept of AI, ML, and DL

the conclusive outcome. In this way, the principal target of AI is to have frameworks or programming that can reflect human conduct. To accomplish AI, some innovation thinks of each other. Both ML and DL are a subset of AI that appeared in Fig. 1 [3].

ML came into existence in the early 1990s [4]. The ML is adopted in statistics fields to proficiently prepare enormous complex models. In software engineering, ML is utilized to prepare a more vigorous variant of the AI framework and furthermore in neuroscience. This ML approach is utilized to plan operational models of the cerebrum. Due to these issues, the interest in ML is increasing day by day. ML is acquired from AI and moved toward the strategies and models obtained from statistic and probability theory. Thus, ML is a subset of AI that utilizes factual techniques to empower machines to improve understanding. These algorithms and calculations are structured such that they can learn and improve our time when presented with new information.

The main impediment of ML is the high dimensionality of the information that is produced is immense in size; subsequently we have countless data sources and outputs. Due to that ML algorithms fail. In this manner, the ML cannot manage the high dimensionality of information. Another issue of ML is it cannot take care of the essential AI issues, for example, regular language handling and picture recognization. Another huge problem with conventional ML is characteristic extraction. For instance, object recognition as well as handwriting recognition turns into enormous problems for ML calculations to settle. On the other hand, DL models are proficient to concentrate on the correct features without anyone else with a little direction

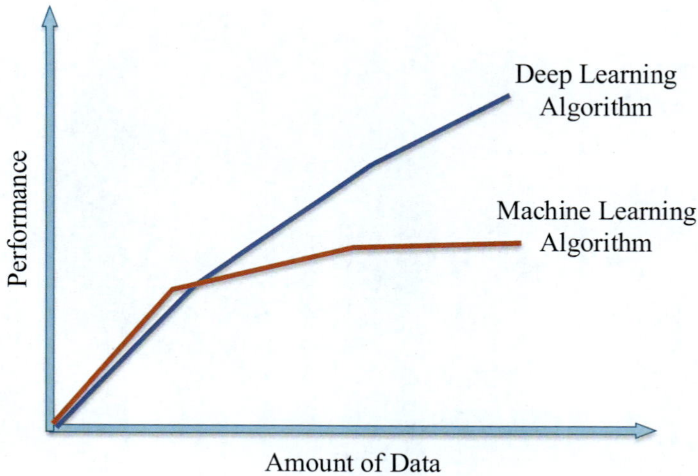

Fig. 2 Performance vs the amount of data [5]

from the software engineer. Hence the machine can also generate the features that are needed for the right output. These models also partially solve the dimensionality problem. If we have a large number of data, then the DL will be the best choice [5].

In the lower number of data, both ML and DL perform similarly; however, when the volume of the data is increased, DL performs significantly better than ML as shown in Fig. 2. In terms of hardware dependency, the DL calculation is exceptionally subject to machines, while the ML calculation can chip away at low machines also. This is on the grounds that the prerequisite of the DL calculation incorporates GPUs which is a necessary piece of work. DL calculation includes countless framework augmentation activities, and this must be improved by utilizing GPU [6].

Highlight designing is a procedure of putting the area information to decrease the complicated nature of information and make the example more noticeable to learning algorithms. This procedure is troublesome and costly regarding time and aptitude. On account of ML, the vast majority of the highlights are expected to be recognized by a specialist and afterward hand-coded according to the area and data type. For instance, highlights can be pixel esteems, shapes, surfaces, positions, direction, or anything. Figure 3 shows the spectrum of the different data types. Data can be categorized into three types: image data, text data, and sequential data. Video data involves the image data and must be maintained in sequential frames as each of the video frame is correlated with one another. Additionally, for natural language processing (NLP), it combines the feature of text recognization and sequential data. Similarly, to detect the optical character recognization (OCR), it combines the feature of image and text. Finally, human-computer interaction (HCI) for building a cognitive system requires a combination of all three types of data.

The execution of the greater part of the ML algorithm relies upon how precisely the features are recognized and extricated. Though on account of DL algorithms, it

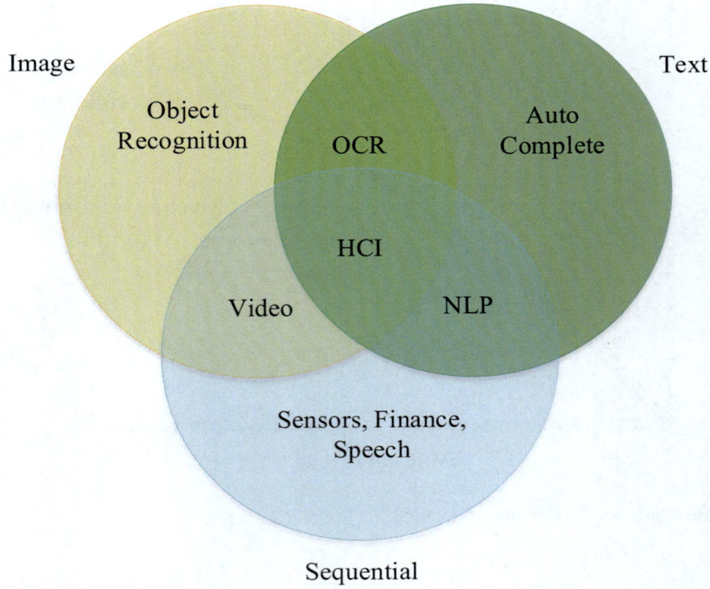

Fig. 3 The spectrum of data types

attempts to take in elevated levels features from the data. This is the important feature of DL that makes it ahead of traditional ML. DL minimizes the errand of another element extractor for each issue. Like on account of the CNN calculation, it first attempts to learn low-level features of the picture, for example, ages and lines, and afterward, it continues the pieces of appearances of individuals and lastly the significant-level description of the face.

To solve a problem using the ML algorithm, it is mostly suggested that we first separate the issue into various subparts and unravel them independently and lastly consolidate them to get the ideal outcomes. This is the manner by which the ML algorithm handles the problems. Then again, the DL algorithm tackles the issue from start to finish. In the case of multiple object detection, if we pass an image with objects, it will detect the object and also the type of an object at the same time. Hence, DL performs faster than the ML algorithm [7]. In contrast, DL methods require longer training time as it requires a huge number of parameters and data to train, whereas for ML it takes relatively less time to train. However, the execution time is converse with regard to testing data [7].

DL is the ML technique; more precisely it is the subsequent evaluation of ML. DL understands features and errands straightforwardly from data. The utilization of DL has risen throughout the most recent years essentially because of three elements. Firstly, DL strategies are currently more precise than individuals categorizing pictures. Secondly, with the update of the technology, now it is possible to train a deep network in a reduced amount of time. Lastly, a lot of leveled data required for DL has grown significantly in recent years.

The integration of the IoT platform with the DL algorithm provides a wider range of facilities to control and handle the system more effectively by the interaction of humans with the physical objects. It also allows the remote wireless control of all the IoT devices and sensors and updates data continuously. By getting experience from the past and by evaluating the present value, the prediction of the future is also possible with the analysis of big data.

IoT consists of multiple sensor systems, and data is collected throughout the IoT devices in real-time series. It can wirelessly communicate in the cloud and can do it continuously and automatically. Therefore, the IoT platform requires a new structure to communicate multiple sensor data. To learn and provide decisions from the sensor data, many authors proposed DeepSense with the DL framework [8]. The implementation of DNN with IoT devices requires high resources demand to train this neural network. It also requires a long training time. In paper [9], the author proposed DeepIoT that can significantly compress the structure of DNN; hence, execution time, device storage, and energy consumption are noticeably reduced. Additionally, reliability and prediction are considered one of the major issues for cyber-physical IoT devices. For accurate and well-structured prediction, RDeepSense is proposed in [10].

The rest of the chapter is organized as follows. The architecture for DL is described in Sect. 2. Section 3 presents the working frameworks for the deep neural network. The deep reinforcement learning approaches are discussed in Sect. 4. The applications of DL in IoT environments are highlighted in Sect. 5. The challenges and future research directions are discussed in Sect. 6, and a brief conclusion in Sect. 7 ends the chapter.

2 Architecture for Deep Neural Network

The fundamental structure of DNN comprises of an input layer, some hidden layers, and output layers. There are some basic ways of implementing DNN that are discussed below.

2.1 Convolution Neural Networks

The convolution neural networks (CNN) mainly carried out two operations: one is a convolution operation and another is a pooling operation. The fundamental layout of CNN is shown in Fig. 4. It consists of an input image that is fed throughout a convolution layer supported by a pooling layer and again repetition of a convolution and pool layer. After that data is fed into multiple fully connected layers, and finally the image is classified through the softmax unit [11].

In convolution, we require three inputs and they are filter size, stride, and padding. Let us consider that we have taken as an input image of size 6/6 and a filter

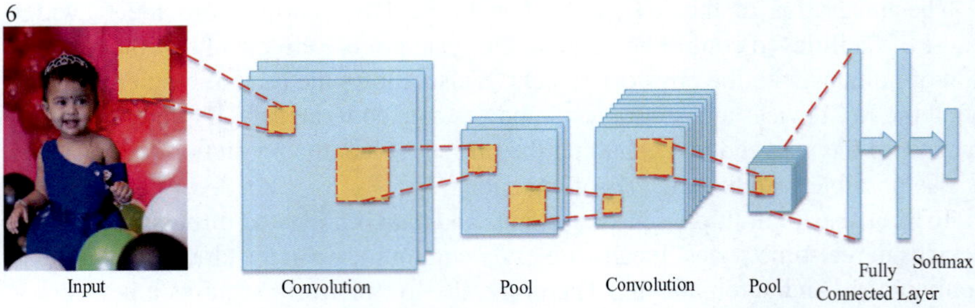

Fig. 4 Typical structure of CNN

kernel of 3/3. Now multiplication is performed with the corresponding input values with the filter values, and finally, the value of three separate equations is added to obtain a final corresponding value −12 (Fig. 5a). In this problem, we have considered the kernel size f = 3; it means that it is a 3/3 kernel. As the stride value is s = 1, it will shift by one unit on the right side to calculate the result of the second position (Fig. 5b). Similarly, when the results of the first row are completed, then the red box is shifted down to one unit to calculate the value of −6 (Fig. 5c). The process will continue until we obtained the final resultant matrix that is shown in Fig. 5d. Therefore, we obtained our final resultant matrix of 4/4 order.

The height of the resultant matrix is calculated from Eq. 1.

$$n'_H = (n_H - f + 1) \tag{1}$$

where

n'_H is the height of the resultant matrix
n_H is the height of the input matrix
f is the kernel size.

In this type of convolution, if we consider the padding, p = 0 (Fig. 5), then the output matrix is reduced to 4/4 from the given input of the 6/6 matrix. It is seen that by applying more stages of convolution, then the size of the image becomes too small to use. However, if we add padding (p = 1) with 0, then the output matrix remains the same size as the input matrix as shown in Fig. 6, and the height of the final matrix can also be justified from Eq. 2.

$$n'_H = (n_H - f + 2p + 1) \tag{2}$$

Now if we consider the stride, s = 2, it means the red box will shift by two units and significantly reduce the height of the resulting image (Fig. 7). The effect of stride is given in Eq. 3.

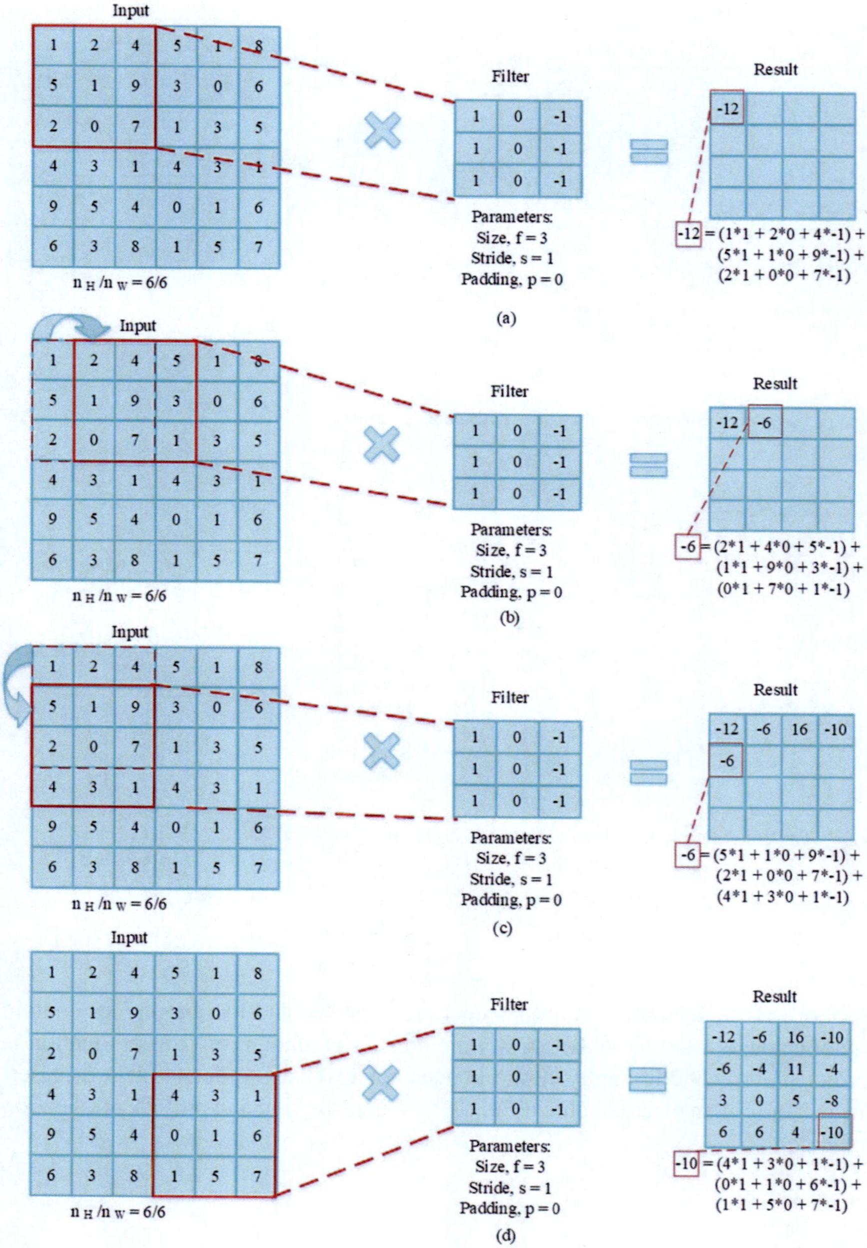

Fig. 5 Various algorithms of convolution neural networks

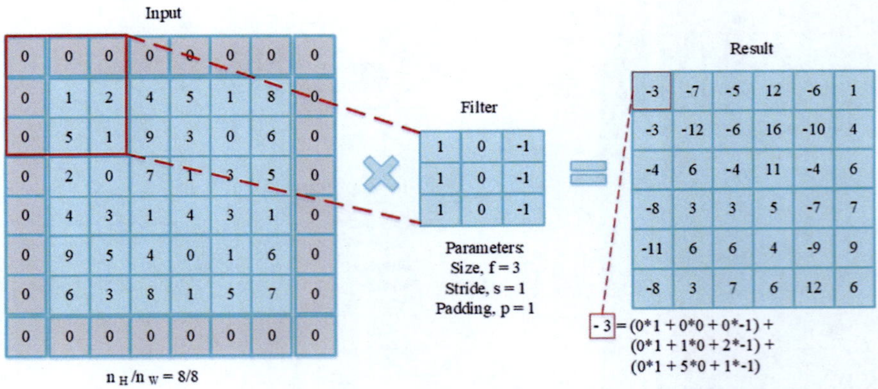

Fig. 6 The convolution neural networks with padding effect

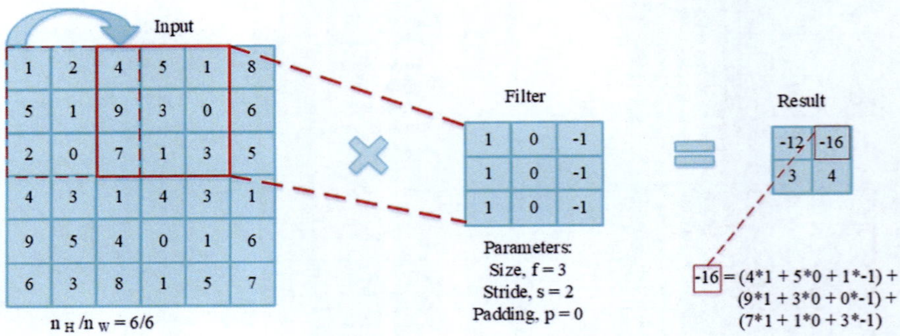

Fig. 7 The convolution neural networks with stride effect

$$n_H^{'} = \frac{(n_H + 2p - f)}{s} + 1 \tag{3}$$

Modern IoT devices such as drones, electric vehicles, and mobile phones contain digital cameras used to analyze CNN methods. The amount of crop production is predicted easily. Additionally, drone images are also used to detect plant diseases. The cameras of an electric vehicle can detect the traffic sign and provide notification to the driver.

2.2 *Recurrent Neural Network*

The CNN suffers mainly from two limitations. The first one is CNN works with the fixed grids (images) where the inputs and outputs are fixed and can only work with images with the fixed size. It cannot work with images of varying lengths. Another

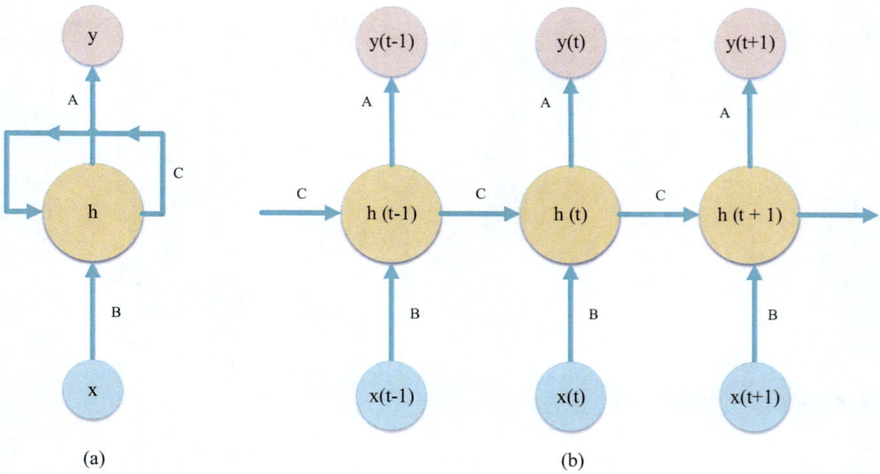

Fig. 8 Schematic diagram of RNN

problem of CNN is it cannot handle the sequential data, i.e., its output values do not depend on past values. Therefore it fails to correlate the output value with the past value (e.g., text or speech). To solve these problems, recurrent neural networks (RNN) may be a promising solution [12].

The RNN structure is designed in a way that its present input directly depends on the past input. In the hidden neuron layer, it has memory feedback as shown in Fig. 8a. The input of this network depends on direct current input with a past observed sample. Therefore, the output of $y(t)$ directly depends on the input of $x(t)$ and the feedback input of past hidden layer output $h(t-1)$ that is given in Fig. 8b.

The network can be divided into four types (Fig. 9). One-to-one network (Fig. 9a) is a basic type of RNN used for regular ML problems. One-to-many network (Fig. 9b) generates the sequence of outputs and is highly used in image captioning. Numerous-to-one system (Fig. 9c) takes in an arrangement of sources of data, for instance, an opinion investigation where a given sentence can be delegated communicating positive or negative feelings. For language translation, many-to-many networks are used as it takes in a succession of sources of data and produces a series of outputs (Fig. 9d).

2.3 Autoencoders (AEs)

Autoencoders are mostly used in data compression which has a wide scope of utilization in computer vision and computer networks. AEs are unsupervised neural networks that use ML to do this compression. AEs mean to get familiar with a packed dispersed portrayal for given information, ordinarily with the end goal of

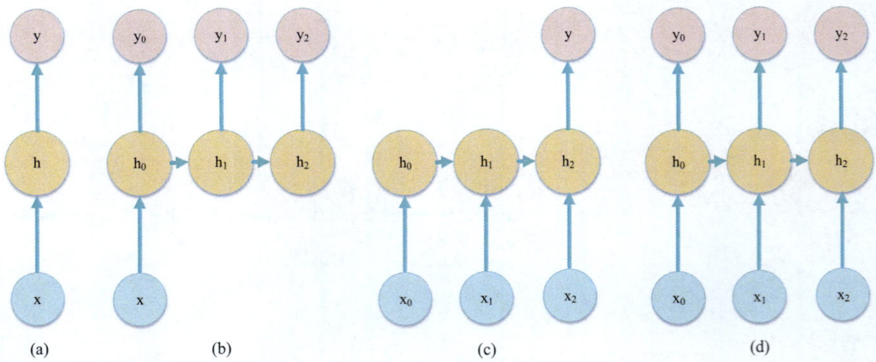

Fig. 9 Different types of RNN

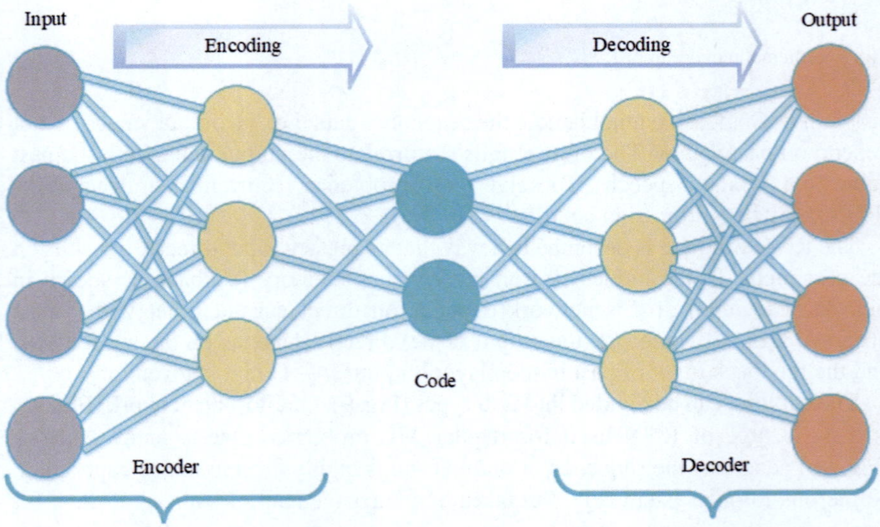

Fig. 10 The architecture of AEs

dimensionality decrease. Additionally, AEs are also used to reduce noise from the images.

AEs are a simple learning network that aims to transform inputs into outputs with the minimum possible error that applies backpropagation, setting the objective qualities to be equivalent to the sources of info. In industrial IoT applications, AEs are mostly used in anomaly detection and to trace the location of fault [13].

AEs are consisting of three components that are encoder, code, and decoder as shown in Fig. 10. In the part of the encoder, the data is compressed into a latent space representation or typically a code interpretation. The encoder layer encodes the input picture as a compacted portrayal in a diminished measurement. The next component represents the latent space. The code is the piece of the system that signifies to the packed input that is taken care of to the decoder. The decoder interprets the compacted picture back to the first position.

2.4 Generative Adversarial Networks (GANs)

The GANs are DL-based reproductive versions that are utilized for unsupervised learning. It is generally a system where two competing neuron networks are competing with each other to create or generate a variation in the data. It was first proposed by Goodfellow in 2014 [14]. The GANs architecture consists of mainly two sub-models known as the generator model and discriminator model. The generator system takes an example and produces an example of information. On the other hand, the discriminator system chooses whether the information is created or taken from the genuine example utilizing a binary characterization issue with the assistance of a sigmoid function that gives the output from 0 to 1.

The working procedure of GANs is given in Fig. 11. The generator system receipts the example and produces the example information, and after this, the discriminator network chooses whether the information is created or taken from the

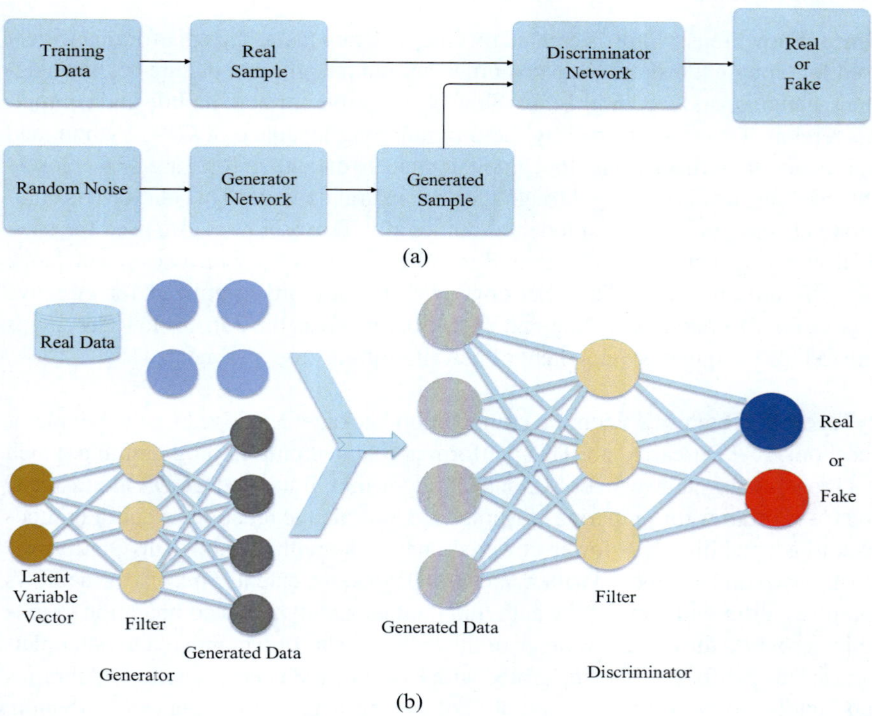

Fig. 11 (**a**) Working principle and (**b**) architecture of GANs

genuine example utilizing a binary grouping issue with an assistance of a sigmoid function that gives the output 0 to 1. The generative model analyzed the distribution of data in such a way that after the tanning phase, the probability of the discriminator committing an error amplifies. On the other hand, the discriminator network guesses the probability that the sample is coming from genuine data or the generator.

There is a wide scope of utilization of GNNs in the field of IoT. Sensors can create something new decision by analyzing the present data and can also classify real and fake characters. This architecture is also useful to convert images or text to sound. Additionally, it can detect the anomaly from different images and also identify the fake images.

3 Framework for Deep Neural Network

The growing interest to use DL in a wide field of applications including IoT influences a lot to invent and develop different DL frameworks in recent days. Each DL framework consists of its architecture, programming language, and algorithms. These different frameworks are used for research purposes to train DNN. Among them, some of the popular working frameworks for DNN are discussed below, and a comparison study is given in Table 1.

TensorFlow TensorFlow is created by Google Brain team. This platform is widely used for Google translate, NLP, text order, speech recognition, picture acknowledgment, gauging, and labeling. TensorFlow is accessible for both mobile and computers version. This is supported by the programming language of C++, Python, and R. It generally utilizes a dataflow graph to manage data; therefore, users can closely observe how data is flowing through the deep neural network. This platform is comparatively easy to build ML models. Additionally, TensorFlow is also used for powerful experimentation for research. TensorFlow comes up with two useful tools: TensorBoard and TensorFlow Serving. The TensorBoard is utilized for effective data vision of system modeling and execution, whereas the TensorFlow Serving is utilized for the quick development of new algorithms or experiments [15].

PyTorch PyTorch is a scientific computation package developed by the people of Facebook's AI Research lab. This platform is founded on Lua programming which is a processing framework for ML and DL algorithm. It utilized CUDA alongside C and C++ libraries for preparing was intended to scale the creation of building models and adaptability. It is lower powered and works with Python. This platform is mostly used on Facebook, Twitter, and Google to some extent. This platform allows automatic differentiation by the autograd module and dynamic computation graphs which provides the extra advantage of the PyTorch platform compared to other platforms. This platform is also integrated with Python which permits common libraries and bundles to be utilized for rapidly composing neural system layers in Python. PyTorch is engaged with the vast community and lots of developers are engaged to develop it. This is a close competitor in terms of popularity with TensorFlow [16].

Table 1 Comparison of various DL frameworks

Frameworks	Developer	Release date	Support language	Feature	Applications
TensorFlow	Google brain team	November 2015	C++, python, R	TensorBoard and TensorFlow serving tools make easy use of TensorFlow	Google translate, Google search, Google maps, AIRBUS, twitter, IBM
PyTorch	Adam Paszke, Sam gross, Soumith Chintala, Gregory Chanan	October 2016	C, C++, python	Automatic differentiation and dynamic computation graphs	Facebook, twitter, Google
Keras	François Chollet	March 2015	Python	Runs on top of TensorFlow, Theano, CNTK	Microsoft, CERN, NETFLIX, NASA
Theano	University of Montreal	2007	Python	Used for fast numerical computation, can run both CPU and GPU	Graph structures, traversing the graph, automatic differentiation
DL4J	Alex D. black, Adam Gibson, Vyacheslav Kokorin	October 2017	Java, Scala	Parallel training for both GPUs and CPUs	NLP, text or image recognition
MXNET	Apache software foundation	February 2020	C++, python, R, Java, Julia, JavaScript, Scala, go, Perl	Wide range of programming language, flexible operation	Speech recognition, forecasting, NLP
Caffe	Yangqing Jia	April 2017	C++	Fast in training and prediction	Vision recognition
CNTK	Microsoft Research	January 2016	C++	Multiple machines scalability, better than Theano or TensorFlow	Handwriting recognition, speech recognition
Chainer	Seiya Tokui	June 2015	Python	Run on top of Numpy and CuPy python libraries	Machine translation, speech recognition, sentiment analysis

Keras Keras is considered a superior level of neural system application programming interface (API) that supports python language. It supports both CNN and RNN that are fit for running the head of TensorFlow, Theano, and CNTK. Therefore, this framework is considered the fastest-growing DL platform. This is an open-source platform and developed by contributors across the world. It also offers high-performing quick prototyping used to specify and train differentiable programs. The Keras interface is easy to use as it offers straightforward APIs and gives clear and significant criticism upon the client blunder. It likewise gives seclusion as an arrangement or a diagram of independent, completely configurable modules that can be stopped along with as not many limitations as could be expected under the circumstances. This is effectively extensible as new modules are easy to include. This element makes Keras reasonable for cutting-edge research [17].

Theano The open-source Theano platform is designed with Python and has an integration with the NVIDIA CUDA library; therefore, it can run on GPUs. This framework also allows parallelism with CPUs. The Python library of Theano permits us to characterize, enhance, and assess scientific articulations including multi-dimensional exhibits effectively. Additionally, Theano has tight reconciliation with NumPy for data computations. The uses of GPUs to perform information escalated calculations which are a lot quicker than on a CPU. It has a broad unit test and self-confirmation that can distinguish and analyze numerous kinds of mistakes [18].

DL4J DL for Java (DL4J) was contributed to Eclipse Foundation, incorporated with Hadoop and Apache Spark. It provides parallel training to the distributed CPUs and GPUs. The programming language is Java and Scala which allows the programmer to code more easily compared to the other platform. It gives a circulated computing system as training with DL4J happens in clusters. It incorporates an n-dimensional exhibit class utilizing ND4J that permits logical figuring in Java and Scala. It additionally offers a vector space demonstrating and subject displaying toolbox that is intended to deal with enormous content sets and perform NLP. This platform is mostly used for picture acknowledgment, front location, text mining, and ways of discourse labeling. DL4J is supported by restricted Boltzmann machine (RBM) and also DBN. Additionally, it is also supported by the LSTM network, CNN, and RNN. This platform is highly efficient compared to Python. In multiple GPU systems, it also works as fast as Caffe [19].

MXNET The MXNET platform provides a variety of programming language facilities to code for the programmer. Among them, Python, C++, R, Julia, and Scala are the most popular programming language. MXNET is intended for high effectiveness, high productivity, and lots of adaptability. It consists of long-term short-term memory which is also known as LSTM networks along with CNN and RNN. This platform is highly used for dialogue response, forecasting, and natural language processing (NLP) [20].

Caffe The "Convolutional Architecture for Fast Feature Embedding" or Caffe is another DL framework that is created in C++ and supports the interface of Python [21]. This is popularly used for vision recognition, image detection, and classification. However, this system does not include fine granularity network layers that are noticed in TensorFlow or CNTK. It requires low-level language code. This is an open-source platform utilized in scholarly exploration ventures, startup models, and huge scope mechanical applications in vision, discourse, and sight and sound. It additionally bolsters GPU- and CPU-based speeding up computational part libraries, for example, NVIDIA, cuDNN, and IntelMLK. The Caffe is getting famous for its speed. It can process 60 million pictures with a simple single NVIDIA K40 GPU. The presence of the Caffe model zoo which is a deep network pertained model makes the use of the Caffe platform easier.

CNTK The CNTK is a popular framework by Microsoft. It is also known as the Microsoft Cognitive Toolkit [22]. This system is upheld by interfaces, for example, C++ and Python. CNTK underpins both RNN and CNN sort of neural model. It is set to provide high salability in terms of tanning a convolution neural network for images, speech, or any text-based data. When working on multiple machines, scalability is better than Theano or TensorFlow to some extent. This platform is also used for handwriting recognition and speech recognition. Microsoft provides lots of open sources to develop the model of CNTK. However, because of the absence of help from the ARM engineering, the capability on mobile is very constrained. This platform is intended for speed and effectiveness; CNTK scales well underway utilizing GPUs yet has restricted help from the organization.

Chainer The Chainer is a Python-based DL framework developed by preferred networks in collaboration with IBM, Intel, Microsoft, and NVIDIA [23]. This platform is completely written in Python and can run on the head of Numpy and CuPy Python libraries. This additionally gives various expanded libraries, for example, Chainer MN, Chainer RL, Chainer CV, etc. It also upholds CUDA computation. It additionally runs on various GPUs with little exertion. It also runs on multiple GPUs with little effort. The Chainer is becoming popular for its high-performance capabilities for many of the domains. The coding units of this framework are comparatively easy. This system is interrogative. It provides good dynamism and helps to understand the flow of code better. This is also considered a good machine translation. The Chainer is broadly utilized in speech recognition and sentiment examination.

4 Deep Reinforcement Learning Approaches

This section describes deep reinforcement learning which is the arrangement of DL and the community of reinforcement learning.

At a high level, reinforcement learning provides us a set of mathematical tools and methods for teaching agents how to go to perceiving and to figure out how to optimally act with a certain response.

Comparing with the other learning methods, supervised learning is the most commonly used learning techniques where the data is labeled and is used for general classification problems. On the other hand, the data is unleveled for

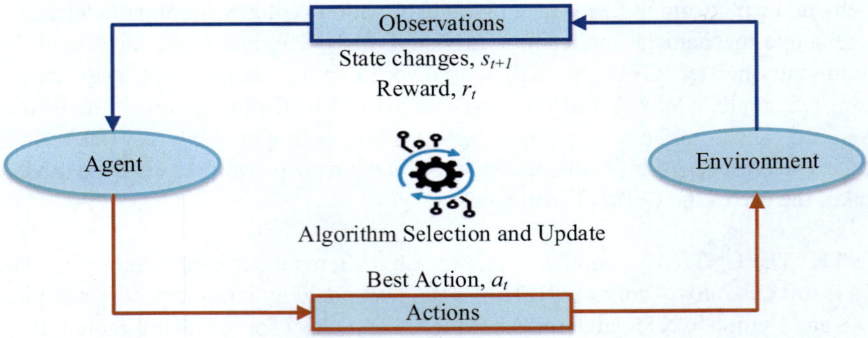

Fig. 12 Deep reinforcement learning

unsupervised learning and aims to predict the outputs by analyzing the behavior of the training data. However, reinforcement learning permits the product specialists and machines to consequently decide the perfect conduct with a particular setting to expand its performance. These learning models interact with both the environments and the learning agents as shown in Fig. 12.

Here, an agent can send an action to the environment, and action is the possible move that makes the agent. The action that executes the agent in time t that denoted as a_t can usually be chosen from a discrete subset of all possible action A. After taking an action, the agent receives an observation from the environment. The observation defines how the agent cooperates with the environment and also includes the state changes, s_{t+1}. Finally the reward, r_t, is given from the environment by analyzing the activities that take the agent. The reward is the feedback that measures the success or failure of a given action considering the environment. For a given system, an agent senses an output in the form of actions to the environment, and the environment returns the agent's new state that is associated with the reward. The environment rewards the agent for the correct actions, and the agent enhances the environment knowledge to choose the next best action [24].

Hence, the total reward, R_t, that an agent obtains the total time t can be presented as

$$R_t = \sum_{\infty}^{i=t} r_i = r_t + r_{t+1} + \ldots + r_{t+n} + \ldots \tag{4}$$

However, if we are going to infinity than the summation, R_t is also going to infinity. Therefore, to handle this issue, the discount factor, γ, is considered, and the discounted total reward is presented in Eq. 5. The term γ signifies the more weight that the reward obtained in the near future and less weight that the reward placed in the long term from the current state.

$$R_t = \sum_{\infty}^{i=t} \gamma^i r_i = \gamma^t r_t + \gamma^{t+1} r_{t+1} + \ldots + \gamma^{t+n} r_{t+n} + \ldots \tag{5}$$

In general, the discounted reward, R_t, is the discounted sum of all rewards obtained from time t that is the summation of the discounted reward in the future shown in Eq. 6.

$$R_t = r_t + \gamma r_{t+1} + \gamma^2 r_{t+2} + \dots \tag{6}$$

Now the Q-function can be defined considering the inputs of state and action that the agent can execute at that certain state. The Q-function represents the expected total discounted future reward for an agent in the state, s, can be achieved by executing an action, in the future.

$$Q(s,a) = \mathrm{E}[R_t] \tag{7}$$

In this case, an agent needs a policy function, $\pi(s)$, to infer the best action to take at its state, s. The policy should choose an action that maximizes the future reward to attain the maximum Q value shown in Eq. 8.

$$\pi^*(s) = \arg_a \max Q(s,a) \tag{8}$$

In DL, there are two main approaches that we can learn a policy function. The first way is value learning and another way is policy learning which is discussed below.

Value Learning Deep Q network (DQN) is used to model the Q function and estimate the Q value in the DL environment. In Fig. 13a, there are two parameters (state and action) that are considered as input, and the output is just the Q value. The deep neural network (DNN) is predicting the estimated and expected total reward that can be obtained given the state in a particular action. The problem with this approach is that if we want to use our policy now and we want our agent to act, we have to feed through the network with a whole bunch of different actions in every time step to find the optimal Q value. Another alternative that is shown in Fig. 13b can eliminate the above problems. In this case, we only input the state value, and the DNN computes the Q value for each of the possible actions. In most of the DL cases where the actions are fixed and with a certain state, it can compute the maximum Q value associated with the best action.

However, this value approach faces some complexities which are given below:

- Cannot handle continuous action spaces.
- Can model scenarios where the action space is discrete and small.

Another approach, i.e., policy gradient learning, is used to overcome these problems.

Policy Learning Policy learning is slightly different in comparison to value learning. In policy learning, we take state, s, as input, and a probability distribution over a possible action is taken as output shown in Fig. 14.

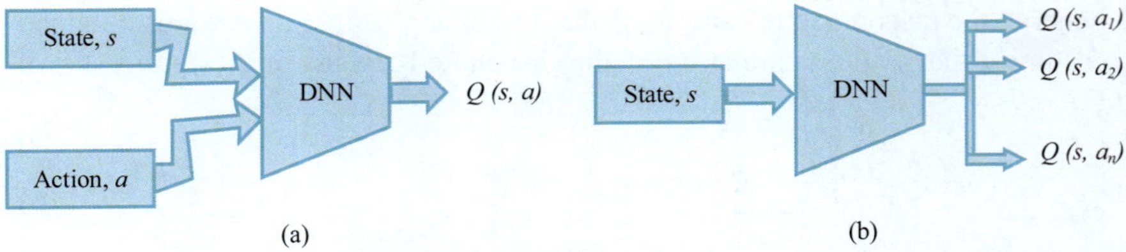

Fig. 13 Value learning approach

Fig. 14 Policy learning approach

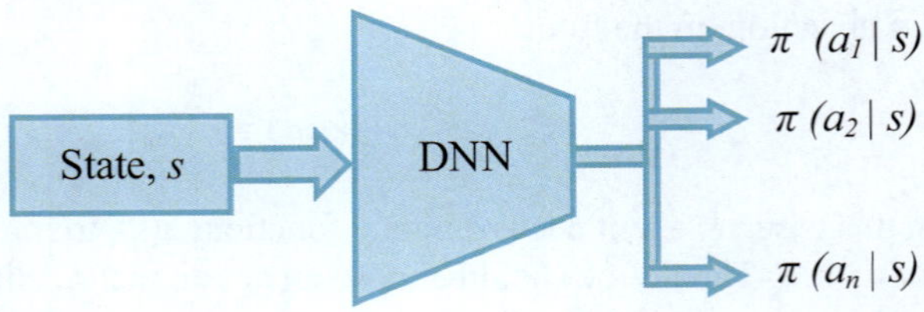

Here, $\pi\,(a_i)$ is the probability that we should execute an action a_i given the state, s, as input and science as it is a probability distribution, and all these outputs add up to 1 that is shown in Eq. 9.

$$\sum_{a_i \varepsilon A} \pi\left(a_i s\right) = 1 \tag{9}$$

Therefore, in policy learning, we can learn directly the policy, and it illustrates which policy will be the best to take action and just execute the correct action.

5 Applications of Deep Learning in IoT Scenarios

The DL algorithm is widely used in many real-life applications because of its self-learning nature. The algorithm of the machine is automatically updated with past experience. The learning environment is developed with more advanced learning algorithms and big data. In this segment, we summarize some effective applications of DL algorithms in the IoT environment. Some applications of DL such as road sign detection for a smart car and real-time distance measuring in the plane for IoTs are significant.

The detailed DL applications in IoT scenarios are briefly discussed below.

- **Image Detection and Recognition:** In most IoT applications, image recognition and detection become an important issue. Many of the services such as roads, smart homes, industries, and campuses are incorporated with the intelligence cameras that take the input as an image or videos. Therefore, the DL algorithm is used to analyze, detect, and categorize the images. The model is trained in such a way that it can handle the images and provide the decision depending upon the situation.

- **Speech Recognition:** Another important application of the DL is voice or speech recognition. Many of the IoT-based smart devices have a feature of voice analysis. In a typical voice recognition device that uses DL neural network algorithm, it takes voice as raw input data, and the data is processed in the hidden layer, and a processed voice is presented in the output layer.
- **Indoor Position Localization:** Indoor localization is becoming a vital application field for DL in the IoT domain such as smart homes, hospitals, and modern campuses. The raw input data is produced from different sources such as light communication, WiFi, ultrasound, infrared, and Bluetooth. With the advancement of the DL algorithm, the indoor position is located with high precision.
- **Pose Detection:** Many of the DL applications incorporated with IoT applications require body pose detection to understand the psychology of the human including emotions and activity to provide their accurate services. For example, the cameras installed in smart homes, smart cars, entertainment clubs or discos, gyms, and hospitals are transferring the footage into DNN and take automatic decisions depending on the body language of the customers.
- **Power Management of Smart Grid:** The smart grid provides the opportunity to communicate both ways from energy providers to consumers and vice versa. Therefore, consumers can be able to take advantage of used maximum energy in the off-peak hours to avoid higher electric bills. Different DL algorithms are involved in effective power management of the smart grid. However, the extra demand for electric power can also be predicted, and minor power system faults can automatically be cleared.
- **Road Traffic Management:** The DL algorithms are used in effective traffic management on the road. Traffic congestions are also predicted with DL algorithms such as AEs, RNN, RBN, and LSTM. Therefore, effective road traffic control is possible by analyzing the big data that is generated from the different smart IoT devices.
- **Agricultural Issues:** The DL algorithms are developed to estimate crop production and also to detect the various plant diseases that can ensure the production of healthy crops. Additionally, DL algorithms also are used in remote sensing for crop and land detection and classifications in large-scale agricultural production.
- **Educational Purpose:** The DL algorithm can be applied to make the study more attractive and interesting from kindergarten school to the university level. Mobile devices are significant in terms of gathering learners' information and progress reports which are equipped with DL algorithms. Students can take advantage of language translation and text summarization software and make the study more relaxing and comfortable.
- **Industrial Sector:** Industry can use the DL algorithms in many applications that increase the production and maintenance efficiency and significantly reduce cost. Therefore, DL can play a vital role in industrial applications. Automatic product inspection is one of the major applications in the field of industrial appli-

cations. Additionally, feature extraction and fault detection can be performed by analyzing the image of smart cameras.

- **Government Sector:** With the development of DL algorithms and IoT, the government can also take a wide range of advantages for effective monitoring of the city. Different types of natural disasters such as floods, landslides, and fires in the forest can be predicted, and proper steps are taken timely. DNN algorithms are also applied in the damage detection in the infrastructure including buildings, water pipelines, and roads.
- **Online Shopping:** Most of the online shopping apps are using the DL algorithm widely to promote and boost up their business. Customers can easily find their choice list on their screen because the DL algorithm shows the results by analyzing the past experience of the customers and recommends products depending on their choice and taste.
- **Security Issues:** The IoT-based applications are highly porn to get attack by various cyber-threats, and among them, false data injection is a common problem. The DL network is trained in such a way that it can be able to identify false data and provide significant security and privacy to the customers.

6 Challenges and Future Research Directions

The IoT digitizes the physical resources like sensors, gadgets, machines, gateways, and networks in real time. It associates individuals with things and things to things continuously. A commonplace IoT system can develop quickly and bring about an exponential increment in assortment, speed, and general volume of information. Therefore, the prime difficulties of the IoT-coordinated framework are the means by which to break down the enormous volume of data got from all sources and make a move continuously. In this section, first, we discuss some of the challenges of IoT-integrated DL networks and then list out some of the future research directions.

6.1 Challenges

- **Availability of Practical Dataset:** To train DL models, it requires a large volume of the realistic dataset, and the accuracy of the model directly depends on the amount of training dataset. Sometimes, it requires several stages of training to obtain the desired output. Although IoT devices are generating a large volume of data, sometimes, these data are not suitable or convenient to train DL models. Additionally, the security issue becomes a major concern to get suitable training dataset. For example, data related to education, healthcare, banking, and utility are highly restricted to access without authorized permission. This also makes a barrier to the train DL algorithm. Some training dataset is given by Wikipedia to train an experimental DL model [25]; however, these are not enough at all.

- **Data Pre-processing:** Per-processed and filtered data are needed to train a DL model which is becoming a vital challenge for IoT applications since data that are obtained from IoT devices are consisting of noises. Additionally, different IoT devices produce data in a different format, different types, and different lengths. Therefore, a unique data format is needed to prepare before training the DL model. For example, in image processing, CNN architecture performs well with normalized, fixed pixel range data.
- **Security over Anomaly Data:** IoT devices are directly connected with the IoT cloud. Hence, data that is generated from the IoT devices fed into the cloud, and this IoT cloud may have access to the unauthorized person. Additionally, the IoT cloud is highly porn to cyberattack. Data can be hacked or modified; therefore, when to train a DL model, there may be a chance of training with anomaly data. On the other hand, there are many data found in IoT servers that are an unnecessary or invalid form of data for training DL models which makes the training of the DL model more difficult.
- **Training Model with a High Velocity of Data:** IoT devices generate a huge volume of data at a very high speed. Data also come from different sources and different formats. Therefore, the DL model needs to train within a very short time with a high volume of data. This becomes challenging. Additionally, some data are leveled and others are unleveled. Although DNN can handle the heterogeneous data and gather experience from the unleveled data, it is not sure to what extent the DL model can handle and works accurately in the presence of a large volume of data.
- **Developing DL for IoT:** Developing a new DL framework that is suitable in the IoT environment is also becoming challenging as the data size is growing every day. DNN is needed to handle a resource-constrained architecture.
- **DL Self-Limitation:** Despite many good applications of DL models in a wide range of fields, it suffers from some limitations. Some limitations of DL approaches are listed below.
- One of the biggest challenges of DNN is that it works like a black box. Why the DNN comes up with this specific output is a very difficult task to understand. On the other hand, if a human is responsible to solve a task, in case of any mistake, it is understandable why this mistake occurs. However, some algorithms, for example, decision trees, are easy to understand. In many applications, such as in the banking sector, interpretability is important. Whether a person is eligible for getting alone or not is a matter of explanation because the client may ask the bank what is the reason behind this particular decision.
- Another challenge to work with DNN is the optimization of hyperparameters. In DL, hyperparameters are such kinds of parameters, with a little modification of this value has a significant impact on the execution of the model in terms of training speed and quality. Relying totally on the default pre-set value is not a wise decision. However, there are only a few numbers of hyperparameters in a model that is hand-tuning. Proper guidelines and research are needed to optimize the value of different hyperparameters.

- DNN architecture suffers from flexibility and multitasking. When DNN is successfully trained, it works efficiently with high a range of accuracy. However, the overall system needs to recalibrate and retrain to solve a similar type of problem. Therefore, the DNN model fails to work with a little different problem without fully reconstructed architecture and lose flexibility in operation.
- Sometimes DNN shows false confidence in image recognition [26], and it also responds with foolish examples that are completely unrecognized by humans. Additionally, DNN also suffers from classification and regression problems. In IoT application, weather forecasting and load forecasting are the common scenario to handle. In [27], the author tried to improve the regression capability by proposing DBN and SVR methods. However, a more clear investigation is required to overcome the challenge.

6.2 Future Research Directions

The following research activities are suggested as future work.

- **Mobile IoT Data:** Mobile phone is getting a big platform to generate IoT data. All the mobiles are directly connected with the cloud so necessary IoT data can be obtained from mobile and apply them to train the DL model particularly in the field of smart city applications. In the article [28], the author explains the opportunity of mobile big data in DL training using a distributed DL framework. However, more research and investigation are required in this field to justify the efficient use of mobile big data in the IoT platform.
- **Adding Background with IoT Data:**. As IoT data is obtained from different IoT devices, it sometimes becomes critical to find out the source of generated IoT data. Therefore, additional contextual information is required to easily understand the data, and it will help to conveniently train the DNN. The supportive information with the IoT data makes the training process accelerate and helps to respond quickly. For example, some smart cameras can detect facial recognition and pose for the purpose of security in a smart city or self-driving car assistance applications. In this scenario, adding supportive information on the IoT data makes the DL model come up with a more accurate solution. In this field, more research is needed.
- **Online IoT Data Provision:** The IoT devices produce data and send it to the IoT cloud. As the IoT data is streaming in nature, knowing the overall size and sequence of the data in advance is a matter of challenge. In this scenario, an advanced algorithm is needed to design that can stream the IoT data significantly without having prior knowledge of the data stream. The study in [29] shows the online auction architecture considering the cloud resources in IoT applications. However, a wide range of research is mandatory in this field.
- **Implementation of Semi-supervised Training Frameworks:** In DL algorithms, most of them are supervised learning process which uses leveled data. On

the contrary, most of the data generated from different IoT sources are unleveled. To define this data is a time-consuming as well as expensive task. Therefore, research is required to train DL models with a lower number of leveled data and most of the unleveled data. Advanced DL frameworks need to be designed to attain their required output with good accuracy.

- **Protection from Threats:** IoT is a cyber-physical system and highly porn to a cyberattack. Most of the time, we rely on the data that are generated from the IoT devices to train the DL model. Hence, it is important to provide data safe from any kind of cyber-threats. In this scenario, DL models can be used to trace specious data from the huge amount of data. It is also possible to determine the weak zone where there is a high possibility to get an attack. To make the overall system more robust, research is needed in this area to provide multilayer protection against cyber-threats.
- **Self-Maintenance of Communication Networks:** The large number of IoT gadgets, their communication protocols, and the architecture of the network become more complex to maintain with the traditional ML models. However, DL approaches can play a vital role to make the system self-service by utilizing the quality of self-healing, self-configuration, self-optimization, and self-load balancing [30]. However, complete guidelines and research are needed to make the IoT system more easily controllable with the DL approach.

7 Conclusion

Because of the advancement of the newly developed technology in IoT and DL, lives have become smooth and comfortable day by day. IoT devices produce raw data that are connected with the cloud, and these data are fetched to train the convolution neural network. Several working frameworks have been developed that are specialized in image recognition, voice translation, anomaly detection, forecasting, and many others. It is a significant achievement in the recent era that machines can work more perfectly with a high range of accuracy than humans do to some extent and the machines also never get tired. In this chapter, we explored DL framework and applications for IoTs. The various challenges to implement DL in IoT scenarios are discussed. Finally, future research directions are discussed which might help network researchers to contribute more in the field toward the development of next-generation IoTs.

References

1. Kaplan, A., & Haenlein, M. (2019). Siri, Siri, in my hand: Who's the fairest in the land? On the interpretations, illustrations, and implications of artificial intelligence. *Business Horizons, 62*(1), 15–25.

2. IDC. (2014). *Executive summary: Data growth, business opportunities, and the IT imperatives*. International Data Corporation.
3. Quick, D., & Choo, K. K. R. (2014). Impacts of increasing volume of digital forensic data: A survey and future research challenges. *Digital Investigation, 11*(4), 273–294.
4. Esparza Tortosa, R. (2020). *Dataset for artificial intelligence*.
5. Machine Learning Mastery. https://machinelearningmastery.com/what-is-deep-learning/
6. Wu, R., Yan, S., Shan, Y., Dang, Q., & Sun, G. (2015). Deep image: Scaling up image recognition. *arXiv preprint arXiv:1501.02876, 7*(8).
7. O'Mahony, N., Campbell, S., Carvalho, A., Harapanahalli, S., Hernandez, G. V., Krpalkova, L., Riordan, D., & Walsh, J. (2019). Deep learning vs. traditional computer vision. In *Science and information conference* (pp. 128–144). Springer.
8. Yao, S., Hu, S., Zhao, Y., Zhang, A., & Abdelzaher, T. (2017, April). Deepsense: A unified deep learning framework for time-series mobile sensing data processing. In *Proceedings of the 26th international conference on world wide web* (pp. 351–360).
9. Yao, S., Zhao, Y., Zhang, A., Su, L., & Abdelzaher, T. (2017, November). Deepiot: Compressing deep neural network structures for sensing systems with a compressor-critic framework. In *Proceedings of the 15th ACM conference on embedded network sensor systems* (pp. 1–14).
10. Yao, S., Zhao, Y., Shao, H., Zhang, A., Zhang, C., Li, S., & Abdelzaher, T. (2018). Rdeepsense: Reliable deep mobile computing models with uncertainty estimations. *Proceedings of the ACM on Interactive, Mobile, Wearable and Ubiquitous Technologies, 1*(4), 1–26.
11. O'Shea, K., & Nash, R. (2015). An introduction to convolutional neural networks. *arXiv preprint arXiv:1511.08458*.
12. Hermans, M., & Schrauwen, B. (2013). Training and analysing deep recurrent neural networks. *Advances in Neural Information Processing Systems, 26*, 190–198.
13. Baldi, P. (2012, June). Autoencoders, unsupervised learning, and deep architectures. In *Proceedings of ICML workshop on unsupervised and transfer learning* (pp. 37–49).
14. Goodfellow, I., Pouget-Abadie, J., Mirza, M., Xu, B., Warde-Farley, D., Ozair, S., Courville, A., & Bengio, Y. (2014). Generative adversarial nets. *Advances in Neural Information Processing Systems, 27*, 2672–2680.
15. Abadi, M., Agarwal, A., Barham, P., Brevdo, E., Chen, Z., Citro, C., Corrado, G.S., Davis, A., Dean, J., Devin, M., & Ghemawat, S. (2016). Tensorflow: Large-scale machine learning on heterogeneous distributed systems. *arXiv preprint arXiv:1603.04467*.
16. Collobert, R., Kavukcuoglu, K., & Farabet, C. (2011). Torch7: A matlab-like environment for machine learning. In *BigLearn, NIPS workshop* (No. CONF).
17. Keras – Wikipedia. https://en.wikipedia.org/wiki/Keras
18. Bastien, F., Lamblin, P., Pascanu, R., Bergstra, J., Goodfellow, I., Bergeron, A., Bouchard, N., Warde-Farley, D., & Bengio, Y. (2012). Theano: New features and speed improvements. *arXiv preprint arXiv:1211.5590*.
19. Deeplearning4j – Wikipedia. https://en.wikipedia.org/wiki/Deeplearning4j
20. Apache MXNet – Wikipedia. https://en.wikipedia.org/wiki/Apache_MXNet
21. Jia, Y., Shelhamer, E., Donahue, J., Karayev, S., Long, J., Girshick, R., Guadarrama, S., & Darrell, T. (2014, November). Caffe: Convolutional architecture for fast feature embedding. In *Proceedings of the 22nd ACM international conference on multimedia* (pp. 675–678).
22. Microsoft releases beta of Microsoft Cognitive Toolkit for deep learning advances. https://blogs.microsoft.com/ai/microsoft-releases-beta-microsoft-cognitive-toolkit-deep-learning-advances/
23. Chainer. https://en.wikipedia.org/wiki/Chainer
24. Ayanzadeh, R., Halem, M., & Finin, T. (2020). Reinforcement quantum annealing: A hybrid quantum learning automata. *Scientific Reports, 10*(1), 1–11.
25. List of datasets for machine-learning research - Wikipedia. https://en.wikipedia.org/wiki/List_of_datasets_for_machine-learning_research
26. Nguyen, A., Yosinski, J., & Clune, J., 2015. Deep neural networks are easily fooled: High confidence predictions for unrecognizable images. In *Proceedings of the IEEE conference on computer vision and pattern recognition* (pp. 427–436).

27. Qiu, X., Zhang, L., Ren, Y., Suganthan, P. N., & Amaratunga, G. (2014, December). Ensemble deep learning for regression and time series forecasting. In *2014 IEEE symposium on computational intelligence in ensemble learning (CIEL)* (pp. 1–6). IEEE.
28. Alsheikh, M. A., Niyato, D., Lin, S., Tan, H. P., & Han, Z. (2016). Mobile big data analytics using deep learning and apache spark. *IEEE Network, 30*(3), 22–29.
29. Gharaibeh, A., Khreishah, A., Mohammadi, M., Al-Fuqaha, A., Khalil, I., & Rayes, A. (2017). Online auction of cloud resources in support of the internet of things. *IEEE Internet of Things Journal, 4*(5), 1583–1596.
30. Klaine, P. V., Imran, M. A., Onireti, O., & Souza, R. D. (2017). A survey of machine learning techniques applied to self-organizing cellular networks. *IEEE Communications Surveys & Tutorials, 19*(4), 2392–2431.

Fog-Cloud Enabled Internet of Things Using Extended Classifier System (XCS)

A. S. Gowri, P. ShanthiBala, and Immanuel Zion Ramdinthara

1 Introduction

Nowadays computers, sensors, and actuator-embedded handheld devices have made our life much smarter than ever before. These handheld devices connect through the Internet, thus making the Internet of Things (IoT) a practical solution for everyday life. The increase in smart devices has enabled education, health care, transportation, and even food court as online digital services [1]. The variety and voluminous data conceived by these digital services are accumulating at a phenomenal rate. Hence, there is a need for appropriate resources and technology to manage the magnitude of data for further processing and analytics.

Although the cloud is known for its scalable processing capacity, it suffers from poor performance when it comes to time-sensitive IoT applications [2]. Besides, the industrial IoT protocol needs IP translation that incurs further interoperability costs. With the growth of IoT-enabled online services, the heavy transmission load in the network requires huge bandwidth, lack of which escalates congestion and delay in the cloud. Hence, a technology like fog computing that facilitates the services of the cloud near the data producing/consuming devices is most preferred.[1]

Fog computing is a decentralized ubiquitous system that affords compute and storage facilities to a large number of clients in the edge network, thus saving bandwidth and congestion issues. The geographically distributed fog device not only assures the least delay for time-sensitive applications but caters to the stochastic needs of IoT far faster than that of cloud [3, 4]. Unlike the cloud in which data are stored in remote locations, the nearness of the fog devices to the data source

[1] Cisco- Fog Computing, 2015.

A. S. Gowri (✉) · P. ShanthiBala · I. Z. Ramdinthara
Department of Computer Science, School of Engineering and Technology,
Pondicherry University, Pondicherry, India

S. Pal et al. (eds.), *Artificial Intelligence-based Internet of Things Systems*,
Internet of Things, https://doi.org/10.1007/978-3-030-87059-1_6

guarantees high security. Though fog serves best for IoT needs, it cannot scale the compute requirements of complex applications like big data analytics. With pros and cons on either side, the federation of fog and cloud complements each other, thus providing everything as service [5].

With the proliferation of online services, the amount of workload that arrives from heterogeneous IoT devices is highly stochastic. Allocation of fog and cloud resources to the incoming workload is a type of real-world problem that keeps changing its requirement continuously. Real-world problems are solved by a system of interconnected components. These systems as a whole exhibit a property, but often, the individual interacting components are not clearly defined, thus turning the system into a complex one. When such complex systems are equipped with the capacity to learn and change their behavior by experience, they are termed as complex adaptive systems (CAS) [6]. Our human immune system, ecosystem, weather forecasting system, traffic control system, search engine, and autonomic Web services are some of the examples of CAS.

Depending upon the instantaneous requirement, CAS prefers an evolving set of solutions rather than a single best solution. A common way to represent CAS is through rule-based agents. An agent is a single individual module, whose interaction with other modules forms the collection of rules [7]. CAS requires problem-solving techniques like learning classifier system (LCS) that can adapt to the environment through interaction. LCS is a rule-based methodology that models an intelligent decision-maker agent. LCS is used to solve continuous decision-making problems whose behavior is stochastic [8]. The saga of LCS is best described in the Fig. 1.

The fog-cloud enabled IoT services are a CAS that needs to learn and evolve decision-making rules instantaneously depending upon the situation. Though there are several techniques available, XCS a variant of LCS is one of the options that assure the best result. Unlike other approaches, XCS has got the uniqueness of learning and evolving capability that is far essential for the stochastic requirements of IoT.

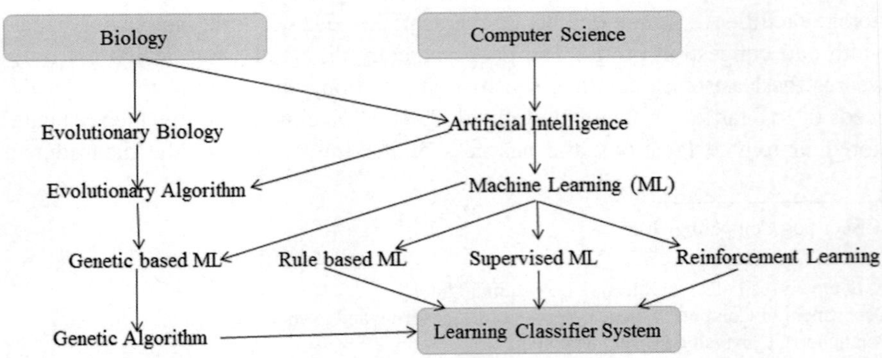

Fig. 1 Chronicle of LCS [6]

The chapter is organized with a literature review of existing works in Sect. 2 that relates resource allocation (RA) in fog-cloud frameworks, LCS, and XCS in specific. Then the architecture and characteristics of LCS that are inherited by XCS are discussed in Sect. 3. Section 4 elaborates on the driving mechanism of LCS. The section also differentiates the variants of LCS with their specialty. Section 5 focuses exclusively on the description of XCS with its importance in the RA problem. A case study on optimal RA in fog-cloud for IoT applications through XCS is discussed in Sect. 6. The chapter concludes with the benefits of using XCS for continuous decision-making problem and scatters a glimpse of the open issues of IoT that could be addressed through XCS in the future.

2 Literature Review

[1] substantiates that the sense, process, and actuate characteristic of IoT differ from one application to the other, thus making it difficult to define a unilateral reference architecture for IoT. The functional views of IoT architectures that are adopted by giant vendors like Microsoft, Intel, and SAP were compared. The author [2] compared the advantage of fog computing over the cloud from the perspective of IoT applications. The constraints of IoT, cloud, and fog in handling time-sensitive applications are discussed in detail.

[3] detailed the taxonomy of fog computing and presented the research gaps in it. Standards and regulations to adopt fog devices as the resource to process Web applications were recommended by [4]. The book chapter [5] explained the implementation of RL for RA in a fog-cloud environment for IoT applications. A novel framework was introduced that assured an efficient RA model for time-sensitive applications of IoT.

[6] explained the historical perspective of the LCS family in terms of the bucket brigade algorithm, Sutton's temporal difference, and Q-learning methods. The article tabulated the various works carried out using LCS. GA was majorly applied in the population set for problems like classification, navigation, and robotic. But for problems like data mining, non-Markov environment, function approximation, and RL problems, GA was applied both on the action set and population set. Certain works implemented GA in the match set too. The author [7] explained the difference in choosing strength and accuracy for fitness computation.

[8] presented a survey on LCS which describes its historical evolution. The role of MDP involved in RL and the steps involved in rule evolution are discussed. The author explained how the single rule set in Michigan LCS style differs from multiple competing rule set of Pittsburgh style. A review of optimization techniques for RA in the cloud, based on load balancing is presented in [9]. Features like pricing, monitoring, processing, and management of service level agreement (SLA) were dealt in SaaS (software as a service) layer, whereas the PaaS (platform as a service) layer handled the scheduler and load balancer. Works related to loading balances

were evaluated based on CPU utilization, SLA violation, energy consumption, and cost parameter, respectively.

With the growing number of infrastructure as a service (IaaS) providers, not only does it require expertise but is time-consuming for the clients (consumer of cloud services) to select the provider who provides efficient service [10]. The algorithm follows the Technique for Order of Preference by Similarity to Ideal Solution (TOPSIS) in which the available IaaS resources are ranked by their similarity index concerning the application requirements. Finally, the top IaaS resource was allocated to the corresponding application.

A novel approach for workflow scheduling in an uncertain environment was suggested in [11]. The unexpected failure of hardware/software resources and stochastic change in the number of objectives were treated as uncertainty in the work. The evolutionary approach called neural network-based non-dominated sorting genetic algorithm (NN-DNSGA) was employed to solve the problem.

A context-sensitive multitier fog architecture for IoT task placement was recommended in [12]. Location identification, current network condition, type of service (sense/actuate/compute/storage), and QoS requirements are considered as context for task placement. The multitier fog architecture comprises IoT device (edge) layer, a fog orchestration layer, a fog colony consisting of fog nodes, a neighbor colony, and the cloud layer in the respective order. Tasks are placed in either of the nodes for processing depending upon the deadline requirement.

[13] proposed a novel approach where RA in fog is dealt with market equilibrium (ME)-based solution. The requested services are considered as buyers of fog resources where they are processed. The work aimed to maximize service utility. The utility with finite and infinite demands was evaluated under Eisenberg-Gale (EG) scheme, generalized EG scheme, proportional sharing scheme, and maximum fairness scheme.

[14] substantiated the utilization of radio access network (RAN) as a compute and storage resource in the fog layer, called Fog-RAN. The RA problem in Fog-RAN was formulated as a Markov decision process (MDP) which allocated fog's resources to the IoT tasks. The performance of the system was evaluated through various RL methods like Monte Carlo, Q-learning, Sarsa, and E-Sarsa. [15] employed RL for the dynamic allocation of compute nodes to serve the arriving tasks. A cost mapping table was used to store the cost of deploying data chunks (tasks) in computation nodes. The RL-based RA algorithm resulted in an optimal solution than its counterparts.

Butz and Martin[2] elaborated the application of LCS in detail for problems like data mining, robot arm controller problem, and optimal path finding in the maze environment. The author differentiated XCS from XCSF and other behavior learning techniques. [16] elucidated XCS through pseudo-code and explanations. The fundamentals of XCS, its type, and how and when XCS is more suitable are discussed. The author explained the idea behind an environment, the parameter settings required in the XCS for the appropriate problem domain.

---

[2] Butz, Martin V. 47.Learning Classifier Systems.

The author [17] described the intricate concepts of XCS. The uses of Moyenne Adaptive Modifier (MAM) technique, Widrow Hoff or the Least Mean Square (LMS) technique, the need for generalization and value estimation were explained with appropriate examples. The book on LCS [18] deals with generalization, scalability, parallelism, complexity, and other issues of the classifier system. [19] explained the fundamentals of GA, its types, fitness sharing, penalty function, multi-objective optimization, and ML for dimensionality reduction.

[20] discussed the tournament/wheel selection techniques to select classifiers for reproduction and deletion. The 20-bit multiplexer problem experimented with various parameter settings of the learning rate (β). The resource management and scalability problem of XCSF was discussed in [21]. It explained how XCSF scales optimally by population size. The author revealed the problem involved in the local linear learning method. [22] described how RL is used for automated RA problems. The different classifications of RA and scheduling problems in the perspective of Monte Carlo, temporal difference, and Q-learning were discussed.

The best classifier memory-based XCS (BCM-XCS) algorithm estimated the number of workloads that are to be processed in the fog layer. The algorithm predicted the amount of workload such that the system minimizes the cost of delay and power. Green energy-equipped battery is recommended as the renewable energy source to fulfill the scarcity of fuel energy in the future [23].

[24] presented the open challenges related to energy-efficient resource allocation (EERA). Articles were broadly categorized as power-aware allocation and thermal aware allocation. Static power management was dealt with, at the circuit level, logic level, and architecture level, and dynamic power management provided solutions through service migration and service shutdown techniques. [25] implemented an energy-efficient RA framework that guaranteed energy efficiency in power usage effectiveness and datacenter infrastructure efficiency.

[26] suggested dynamic energy-aware cloudlet-based mobile cloud computing (MCC) framework to reduce the energy consumption in wireless networks through cloudlets for offloading and processing the data. Cloudlets are the compute/storage devices found in the edge network closer to the edge devices that requested services. The total energy consumption of all the cloudlet was computed as the product of the performance level of each cloudlet by per unit energy cost. [27] focused on different types of RL algorithms with their core concept, theoretical properties, and limitations. The author claimed how partial feedback about the learner's prediction in supervised learning differentiates it from RL. [28] presented the practical implication of value functions, policies in model-based, and model-free RL. A detailed study on different versions of LCS, their analysis, and their application to classical tasks of accuracy-based LCS is presented by [29].

LCS with extended memory to solve non-Markov decision is presented by [30]. The concept paves the way to incorporate memory-enabled XCS for Markov decision problems. This helps to boost the processing speed of XCS. [31] suggested the concept of how to incorporate anything as a service (XaaS) in the cloud. Self-management of cloud services that lessen the burden of cloud administrators was also recommended.

3 Architecture of LCS

Fundamentally, LCS comprises a population of classifiers, also called a finite set of rules. Each classifier constitutes a condition part, an action part, a payoff prediction (reward), and other parameters [18]. The population of classifiers represents the current observation (state) of the environment for which the LCS is implemented. LCS is made up of four main components, namely, an environment, a performance component, a reinforcement learning component, and an evolutionary/discovery component [6]. The architecture of LCS is shown in Fig. 2.

- The first and last contact of LCS is with the environment through the detectors and effectors, respectively. A detector senses the current state of the environment and encodes it into formatted input data. The effector ultimately executes the action decided by the LCS algorithm.

The input data is either a Boolean value or nominal value. While the Boolean input is of form {0,1,#}, the nominal inputs are specified within an interval range along with # symbol. The wildcard symbol # is considered equivalent to 0 or 1 depending upon the situation. The input data is considered equivalent to a classifier in the population if the non # value in the condition part of the population classifier matches the corresponding value of the input data. In the case of nominal values, the input data range is compared with the specified interval range in the population classifier.

- The performance component gears the interface between the environment and the classifier population. Population set, match set, prediction array (a utility that evaluates the action), and action set forms the performance component.

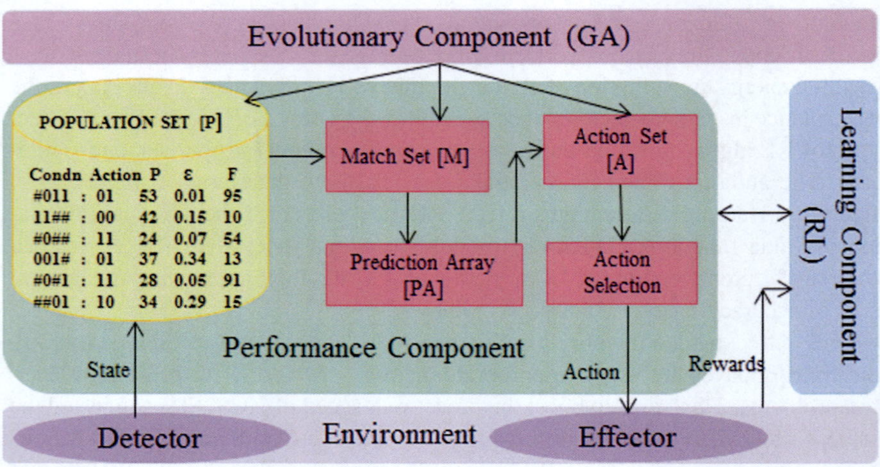

Fig. 2 Architecture of LCS[2]

The problem instance (input) is in the form of a condition part of the classifier. If the input coincides with any of the classifiers in the population set, such classifiers are grouped into the match set. For every possible action in the match set, the prediction payoff is computed as the fitness weighted average of the prediction of all classifiers in the match set. These prediction payoffs are stored in the prediction array. Depending upon the values in the prediction array, one action is chosen for execution. After selecting an action for execution, an action set is created with those classifiers in the match set that specifies the chosen action.

- The reinforcement learning component or the credit assignment component.

Once the chosen action is executed, a scalar reward is generated and the next problem instance is perceived. Meanwhile, the action set is updated by computing the Q-value which is the summation of the reward received and the discounted maximum value in the prediction array. The updated classifier parameters like the prediction payoff p_j, error e_j, and fitness f_j in the action set are distributed to the population set. Now, the population set is renewed by the discovery component.

- The evolutionary component uses different genetic operators (selection, crossover, mutation) that discover better rules or improve the existing rules.

Whenever the classifier parameters are updated in the population set, the evolutionary component checks if the population size does not exceed the maximum specified size. Then the genetic algorithm is executed in the action set and the population set. Some versions of LCS apply GA in the match set too.

Characteristics of LCS

The efficiency of LCS is characterized by certain properties like adaptability, generalization, performance, scalability, and speed [18].

- *Generalization*: In a population, a classifier (rule) probably matches more than one input vector. The different problem instances with similar consequences in the environment are recognized equivalent which leads to generalization. In short, generalization influences the reduction of population size.
- *Adaptability*: The capability of the LCS to generalize a classifier even when the accuracy criterion is rigid. It is also defined as the tolerance level of LCS for prediction error.
- *Performance*: The performance of LCS is gauged by the quality of classifiers that evolve in a population. A classifier population is said to be good when it provides less chance to add new rules. This indicates that the existing classifiers are more experienced (aged) and effective in deriving the required action.
- *Scalability*: It is the ability of the LCS to quickly arrive at the solution even with the increase in size and complexity of the population.
- *Speed*: The speed is the actual time taken by the LCS, to build a consistent classifier population, which needs no further insertion or deletion.

Correctness, completeness, and compactness are the other characteristics that accelerate the quality of LCS. These characteristics enable LCS to be successful in the application domains like classification, decision-making, data mining, RL problems, regression, cognitive mapping, and robot navigation problems. LCS is also seen as a model for cognitive abilities with the goal to design an agent-based intelligent decision-maker [7]. The two mechanisms that drive LCS toward its goal are the learning and evolutionary process.

4 Driving Mechanisms of LCS

Learning

Learning is the improvement in performance, by the experience and knowledge acquired through the interaction with the environment. The process of learning depends on the way the information is received from the environment [8]. It can be made either offline or online. In offline, the learning model is trained with a batch of training instances simultaneously, whose outcome is a single rule set. The outcome is very much static and does not change with respect to time. Offline training is suitable for data mining problems, which analyze the history of the dataset.

On the contrary, online learning, also referred to as incremental learning, trains the learning model with each problem instance one step at a time. The outcome is a rule set that changes with respect to time, to every additional problem instance, observed from the environment. A robot navigation controller or an autonomous service provider that receives a stream of continuous input from the environment is an example of online learning.

Artificial intelligence differentiates the type of learning according to the feedback received from the environment[3]. Supervised learning is the process of learning through training from a labeled set of data. Each problem instance consists of an input and the desired output. The input is the description of the situation and the output is the correct action that the learner system is expected to apply in that situation. The goal of learning is to generalize a rule so as to act appropriately in situations that were not presented during the training.

Unsupervised learning is about learning from a group of unlabeled data. The target is to find the hidden pattern present in the collection of unlabeled data. But, RL varies from supervised and unsupervised learning. RL is the concept of learning and deriving a policy through continuous interaction with the environment [25]. It suits more for the interactive sort of problems which is not the competence of supervised or unsupervised learning.

In RL, the learner agent is supplied with observed data (state) from the environment. The agent learns to decide by interacting with the environment through the trial and error method which is quite different from supervised and unsupervised learning. Unlike other learning techniques, RL derives policy and possesses a reward signal that is maximized in the long run [14]. The basic idea of RL is shown in Fig. 3. Each type of learning suits a specific problem domain. As LCS has been successfully applied for classification, regression, and RL type of problems, the discussion of this chapter sticks to RL methodology.

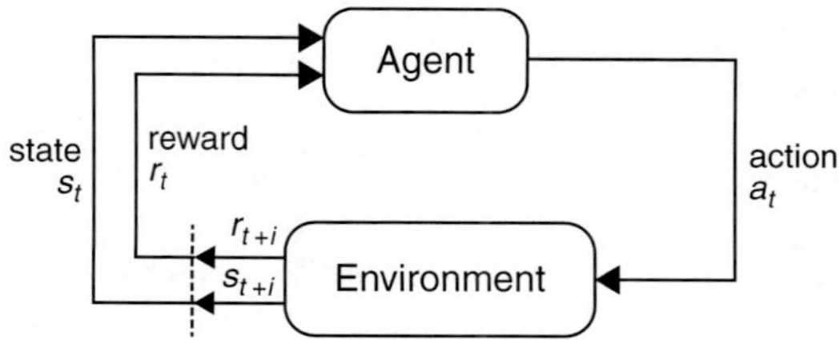

Fig. 3 Reinforcement learning. (Sutton, R. S., & Barto, A. G. Reinforcement Learning: An Introduction)

Markov Decision Process for RL

The decision-making problems which are solved by learning and interacting with the environment are called RL problems. Decisions made in such problems can be framed as a tree structure. The vertices of the tree represent the system states (s_1, s_2, s_3,.....s_t∊S: state space) and the edges denote the individual decision or action (a_1, a_2, a_3,...........a_t∊A) of the agent, respectively.

In principle, an RL problem is epitomized in the form of an environment. The environment is epitomized by a set of states (S), and for each state s∊S, there is a set of allowable action a∊A. Depending upon the particular state "s," the agent takes an action which results in a transition to a new state s'. Based on the quality of the action, the agent receives some reinforcement or reward that is scalar value either positive or negative. The aim of the agent is to maximize this reward in the long run [27]. The components environment, agent, state, action, and reward are the building blocks of RL as described in Table 1.

Markov decision process (MDP) helps to formulate the RL problem mathematically. For this, one has to understand the definition of Markov property and whether a given problem is Markov or not. MDP is a sequence of random processes.

Markov property states "the future is independent of the past given its present," which means that the current (present) state is adequate to predict the next state. So it is not necessary to track or store the past states [28]. Hence, those problems in which the future state is predicted given the current state is said to possess Markov property. MDP is defined as an ordered list (**S, A, P, R, γ**) where:

S is a finite set of random states s_t, s_{t+1},... at every discrete time step "t"
A is a finite set of discrete actions
P: SxA→π(S), P is the transition function, π(S) probability distribution function, i.e., the probability the state will change from one state to another depending upon an action.
R: SxA→R, a reward function that obtains a scalar reward for each pair (s_t, a_t) that the agent receives from the environment.
γ is the discount factor ranging between zero and one. It indicates the amount of prominence given to the immediate reward and future reward for non-episodic tasks.

An episodic task is one with a finite number of states that possess a terminal state. On the other side, the non-episodic tasks are continuous tasks with an infinite number of states. While it is easy to compute the cumulative reward from the start to the end state for episodic tasks, it is complex to compute the cumulative rewards

Table 1 Building blocks of RL

Components of RL	Description
Environment	It is the depiction of the RL problem that needs to be solved for its different states at each discrete time step
Agent	The learner or the decision-making program of the system that takes action on the environment and receives a reward for it
State	A state denotes the current situation of the environment at a specific time step
Action	It is the decision that the agent takes on the environment resulting in the next state
Reward	It is the reinforcement or the feedback received by the agent for the action taken

for the infinite states of non-episodic tasks [27]. Hence, a discount factor between zero and one is used to calibrate the value of immediate reward and future reward, respectively. The cumulative reward (returns) computed for episodic and continuous (non-episodic) tasks are given in Eqs. (1) and (2).

Returns for episodic tasks:

$$G_t = r_{t+1} + r_{t+2} + r_{t+3} \ldots\ldots\ldots\ldots = \sum_{n}^{k=1} r_{t+k} \tag{1}$$

Returns for continuous tasks:

$$G_t = \gamma^0 r_{t+1} + \gamma^1 r_{t+2} + \gamma^2 r_{t+3} + \ldots\ldots\ldots = \sum_{\infty}^{k=0} \gamma^k r_{t+k+1} \tag{2}$$

In RL, the agent strives to maximize the cumulative reward. The agent prefers the particular action that produced profitable rewards in the past (exploitation). Nevertheless, the discovery of that prudent action is not possible without several trial and error attempts (exploration).

The tradeoff between exploration and exploitation is a unique feature of RL. Through trial and error, the agent discovers a rule/strategy to choose an action for a particular observation of the environment. The rule that the agent follows to take any action is called policy (π). Mathematically, a policy is a decision-making rule π:S→A that denotes state-action mapping.

A policy can be deterministic or non-deterministic depending upon the problem [28]. Deterministic policy denoted by $\pi(s)$ holds a single possible action for every state. When there is more than one possible action from a state, the probability of executing an action is non-deterministic, denoted by $\pi(s, a)$. A policy is indicated as $\pi(a/s) = P[A_t = a/S_t = s]$ which is the probability that an agent takes an action aϵA on each states s ϵS.

The policy is subject to change with experience. Hence, it is essential to find an optimal policy that yields the best value of a function. An optimal value function is the one that pays maximum value compared to all other value functions. Bellman equation helps to find optimal value function.

The Bellman expectation for state-action value function $q_\pi(s, a)$ when the agent exercises action "a" on the state "s" by following policy π is shown in Eq. (3). It is defined as the expected sum of immediate reward and the discounted state-action value for the successor state s_{t+1}, with respect to the action a_{t+1} that the agent will take from that state.

$$q_\pi(s,a) = E\left[r_{t+1} + \gamma q_\pi(s_{t+1}, a_{t+1}) / s_t = s, a_t = a\right] \tag{3}$$

The Bellman optimality equation for state-action $q^*(s, a)$ is the sum of the probability of the immediate reward and the discounted state-action value as shown in Eq. (4). $p(s', r/s, a)$ is the probability to transit from state "s"to the next state s' when action "a" was taken with reward r.

$$q^*(s,a) = \sum_{s',r} p(s', r / s, a)\left[r + \gamma \max_{a'} q^*(s', a')\right] \tag{4}$$

Evolutionary Process

The evolutionary process refers to the discovery of new rules or classifiers that are not present in the population. New classifiers are intended to make good decisions than the existing ones. In practice, genetic algorithm (GA) is used to create such new rules. GA is a computational search technique that generates a population of classifiers each signifying a prospective solution to a given problem [20]. Some of the terminologies used in GA are genome, phenome, and genetic operator.

- A genome is also known as genotype/code that represents the condition and action part of the classifier.
- A phenome or phenotype is the solution built from the offspring.
- The genetic operators' selection, crossover, and mutation modify the population by reproducing a well-performing classifier and deleting ill-performing classifiers.
- The well-performing classifiers are the one that possesses high fitness value and contributes to the better performance of the intended system.
- The ill-performing genomes usually possess low fitness value and do not contribute to the improvement of the system.
- Single-point crossover disintegrates parent genomes into two equal or unequal parts and builds a new genome by reversing the subparts from each parent.
- The multipoint operator fragments the parent genomes into several parts at more than one place and builds new genomes.

In GA, the selection process usually extracts the well-performing genomes (classifiers/rules) for reproduction. Tournament selection and roulette wheel selection are some of the commonly referred selection techniques used. The same selection technique is used for the deletion of ill-performing classifiers also [20].

The crossover operator creates new genomes by recombining the subparts of two or more parent genomes. The crossover operator can be single-point or multi-point. Followed by the selection and crossover, mutation takes place that randomly alters an individual attribute of the new genome. The procedure to implement the GA is described by the following steps [19]:

(i) Initialize the size of the population.
(ii) Declare the number of iterations/generations for which the following steps are repeated.
(iii) Initialize the population with random instances that cover the possible search space.
(iv) Compute the fitness of every individual classifier in the population.
(v) Select the parent genome from the population that possesses the highest fitness.
(vi) Crossover the selected genomes to derive new offspring.
(vii) Mutate the new classifier and add them to the next-generation population.
(viii) Repeat the above steps until the population is refined with classifiers of high fitness.

Significance of Reinforcement Learning and Genetic Algorithm
Besides RL, the state-action value function can be computed through other techniques like dynamic programming (DP) or deep neural network (DNN). But unlike RL, these techniques are mostly opaque to human computation analysis and require prior knowledge of the system.[3]

Further, RL involves the concept of Q-tables (look-up tables) that comprises the tuple <state-s, action-a, prediction reward-p>. The size of the Q-table keeps growing with problem size, thereby increasing space complexity. Hence, RL faces a setback for large MDP problems. To overcome this scalability issue, the evolutionary computing paradigm like GA is used.

GA comprises the tuple <condition-c, action-a, prediction value-p>, where c refers to a cluster of states, instead of a single state. A cluster is a group of conditions that represent a common state [8]. The cluster approach saves memory by representing various forms of a state in a single notation, thereby reducing the space complexity. At the same time, GA alone is not sufficient for addressing MDP problems because of the following reasons:

1. GA does not use the fact like a policy that maps the state-action pair.
2. GA neither notices nor keeps track of the states that an individual problem instance passes through.
3. GA does not take into account the action type that it selects.

As GA ignores such useful structures in complex adaptive problems, RL is employed to bridge the gap [22]. RL helps the evolutionary algorithm to generate a better population of classifiers or improve the existing ones. Thus the combination

[3] https://pythonhosted.org/xcs

of RL and GA enriches the capacity of each other, thereby conceiving another powerful concept called LCS.

Different Models of LCS

Over the years, different versions of LCS have evolved each of which holds good, depending upon the probability requirement. Zeroth level classifier (ZCS), extended classifier system (XCS), anticipatory classifier system (ACS), supervised classifier system (UCS), and XCS for function approximation (XCSF) are the different types of LCS found in the literature [6].

The ZCS is the initial version of LCS architecture. In ZCS, the fitness of the classifier depends on the accumulated reward got by executing the classifier. The greater the accumulated reward, the higher is the fitness strength. The classifiers with less reward are considered as ill-performing and more likely chosen for deletion.

The concept of the extended classifier system (XCS) is that the fitness of a classifier is based on the accuracy of the prediction of reward rather than the value of the reward. It happens that a classifier with less reward value possesses high fitness, while a classifier with larger reward holds low fitness. In XCS, the population classifiers that survive every generation are those whose reward is predicted accurately rather than those with higher reward value[2].

The anticipatory classifier system (ACS) is quite different from other versions of the LCS family. In ACS, the classifiers are denoted by <condition, action, effect> rather than <condition, action, reward>. The effect component indicates the impact or consequence of the action. It reflects the possibility of the next state caused due to the action specified in the current situation. Since they learn a sequence of transition models, ACS is considered as a model-based RL architecture [7].

XCSF is a function approximation oriented XCS that accepts real value inputs for the condition part instead of Boolean values. It approximates continuous real-valued function specified in the form of an interval. XCSF computes reward prediction using a linear approximation technique like recursive least squares (RLS) [21].

The supervised classifier system (UCS) is exclusively designed for episodic problems like classification and data mining where delay in reward prediction is tolerable. While other versions of LCS use RL for learning, UCS adopts supervised learning. In UCS, GA is exerted both on the match set and population. UCS prioritizes the accuracy of the reward prediction to gauge the classifier. In large space problems, UCS proves effective generalization and consistent knowledge representation [6].

5 Accuracy-Based Extended Learning Classifier System (XCS)

XCS was conceived by Wilson in 1995. It evolves accurate and maximized general classifier for a given problem. The key feature of XCS is not only it derives optimal input-output mapping but also produces a minimal set of classifiers (rules) that

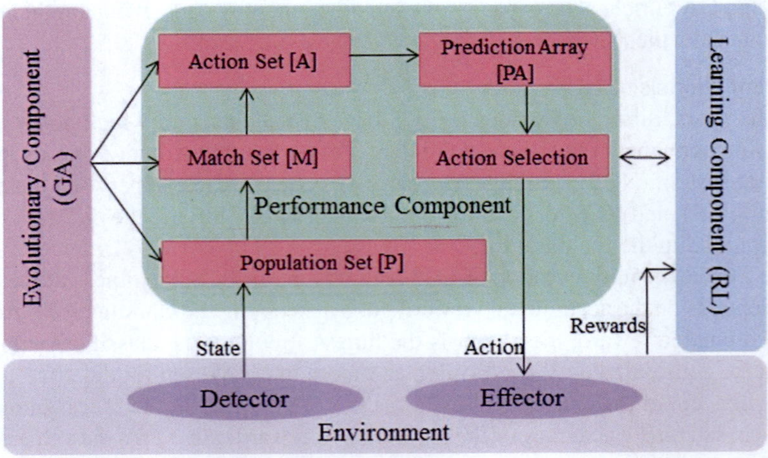

Fig. 4 Components of XCS [30]

explain the mapping. This property distinguishes XCS from other ML algorithms whose model is opaque to human computational analysis. The different components of XCS that reflect LCS are shown in Fig. 4.

In short, XCS generates a feasible set of solutions, instead of a single best solution to a given problem. With its niche GA evolution and fitness description, XCS attempts to estimate the Q-table with an accurate and maximally generalized classifier [29].

Terminologies and Parameters of XCS

Situation: A situation is a sensory input received from the environment through the detector. The input is a sequence of bits received at a time, usually denoted by $\sigma(t) \in \{0,1\}^L$, where L refers to the number of bits in each input. XCS accepts a fixed-length string of bits as its input. It refers to the condition part of a classifier[16].

Scenario: A series of situations for which the XCS algorithm executes an appropriate action to maximize the reward.

Action: An action is an output executed by the XCS algorithm in response to a situation. The action is denoted by $\alpha(t) \in A$ where A is a list of actions ordered as $\{a_1, a_2, a_3, \ldots\ldots\ldots\ldots a_n\}$. XCS selects the best action from the list of action choices depending upon the situation.

RL Problem: The given RL problem can be either a single-step problem (episodic) or multi-step (continuous). In the case of a single-step problem, the action performed on a situation has no impact on its successor. Execution of action results in a reward that indicates the end of the task. Boolean multiplexer and data mining-based classification problems are examples of single-step RL problems [27].

In a multi-step problem, the action performed in each state has a serious impact on its successor. In other words, the current state is an outcome of the action taken at the previous time step. Also, the reward is not obtained for every state-action pair.

Table 2 Attributes of a classifier

Attributes	Description
c	The condition c ϵ $\{0, 1, \#\}^L$ that specifies the sensor input, otherwise called a situation
a	The action a ϵ{ a_1, a_2, a_3, ... a_n } itemize the action list, otherwise called the output
p	The prediction estimate p is the payoff expected for the action taken
ε	The error is made in predicting the reward. It is the difference between the target prediction and the existing prediction of the classifier
f	The fitness of the classifier
exp	The experience denotes the number of times that the classifier has been a member of the action set
ts	The timestamp stipulates the last time step at which the GA was invoked
as	The size of the action set [A], to which the classifier is included
n	The numerosity denotes the number of times that the classifier is subsumed by an experienced classifier

The maze problem, regression problem, or sequential decision-making problems are some of the multi-step problems.

Classifier/Rule: XCS maintains a set of classifiers that represent the state of the environment [16]. Each classifier “cl” is a tuple defined by the letter **x** ϵ **{c, a, p, ε, f, exp, ts, as, n}.** Table 2 portrays the attributes of a classifier.

Population Set: It is the collection of all rules or classifiers present in the set [P] at a time “t.” The attribute “N” enumerate the maximum size of the population.

Match Set: At every time step, the group of classifiers in the population [P] that matches the current situation (input) are moved to a set called match set [M] [8].

Prediction Array: For every possible action in the match set, the fitness weighted prediction average is computed and stored in the prediction array. The values in [PA] assist to decide the most promising action to be chosen.

Action Set: An action set [A] is a subset of match set [M]. The action set [A] includes those classifiers from the match set whose action is chosen for execution. The process of choosing the appropriate action depends on the computation performed in the prediction array [PA].

Reward: A reward is a positive or negative scalar value which is the outcome from the environment in XCS. RL always attempts to maximize the cumulative reward in the long term. The type of reward can be categorized into three types [27].

(a) *Immediate reward*: It is the reward value that is returned by the environment in response to the action at every time step. It is denoted by “*r*.”
(b) *Expected future reward*: The estimated reward for the successive states, especially excluding the current one. It is denoted by $q(s_{t+1}, a_{t+1})$.
(c) *The payoff or the combined reward*: It is the sum of immediate reward and the discounted expected future reward. It is denoted by $\{r + \gamma q(s_{t+1}, a_{t+1})\}$.

The following sequence of steps explains the XCS algorithm with a detailed block diagram in Fig. 5 [16]:

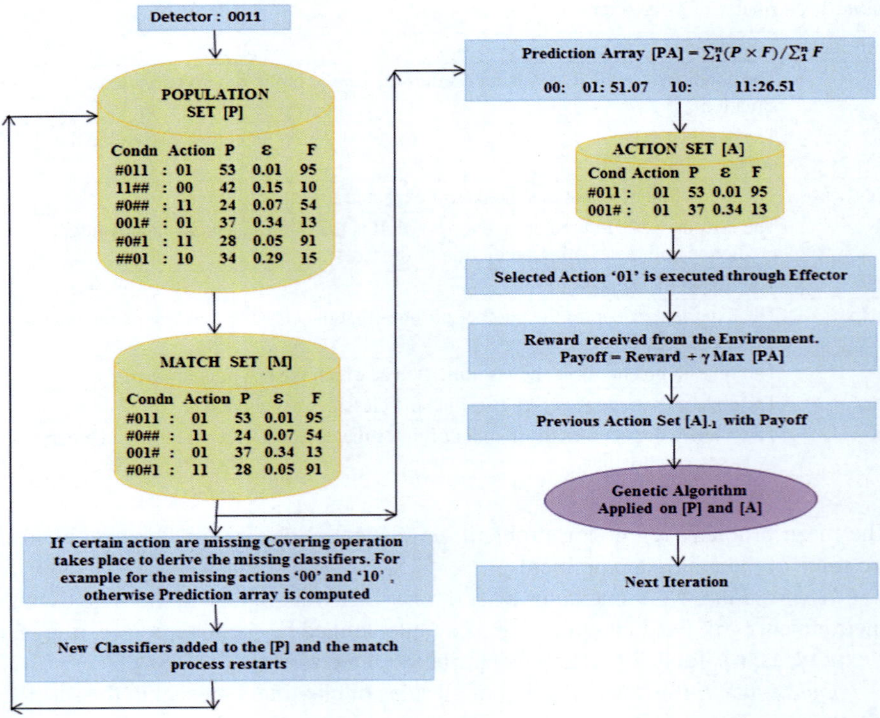

Fig. 5 Flow diagram of XCS algorithm [6]

(i) Define the environment, agent, and initialization of population set [P].
(ii) Formation of the match set with covering operation as and when required.
(iii) Deriving the prediction array and choosing an action.
(iv) Creation of action set [A] and executing the action.
(v) Reward prediction estimate.
(vi) Updating the classifier parameters of the action set.
(vii) Rule evolution by genetic algorithm and action subsumption.
(viii) Iteration repeats for the next time step.

Environment, Agent, and Initialization of Population Set

The XCS algorithm begins when the environment is instantiated. In the RA problem of our case study, the edge devices that generate the workload, and the fog/cloud nodes that process them constitute the environment. The amount of workload at every time step forms the input. The population set [P] is initialized with classifiers composing of arbitrary values for conditions, action, prediction, and other parameters. With the classifier denoted by the letter "cl," pseudo-code for the [P] initialization is illustrated with the following procedure [16]:

```
classifier_initialization( )
{Initialize the classifier's condition with length [situation]
for each bit b in cl.c            // cl.c refers to the condition part of the classifier cl.
{ if random [0,1] < P#)           // P# denotes the probability of including '#' as a bit in condition part of the classifier
        b ← #
  else
        b ← the corresponding bit in situation
}
cl.A ← random action from the action set [A]
cl.p ← pi                         // reward prediction, prediction error, and fitness are initialized with an arbitrary value
cl.ϵ ← ϵi
cl.F ← Fi
cl.exp ← 0
cl.ts ← actual time
cl.as ← 1
cl.n ← 1
return cl
}
```

Match Set Formation and Covering Operation

The match set is initialized as an empty set. The XCS algorithm compares the current situation with the condition part of each classifier in the population set [P]. Those classifiers in [P] whose condition part coincides with the situation (input) are moved to a set called match set. Thus, the match set [M] is created as a subset of [P].

The matching process though sounds trivial; yet, it involves a character-wise comparison between the input and the condition part of each classifier in [P]. In general, the condition in the classifier is a combination of care and don't care [#] symbols. A care symbol is an element of [0, 1] values in XCS. Whereas "#" in the situation (input) indicates don't care symbol. When the comparison is performed, each input symbol should match with the exact value of the condition part [29].

The match set [M] thus created verifies whether it encompasses all types of actions from the action list { a_1, a_2, a_3, ... a_n }. It checks if the count of action types in [M] is equal to θ_{mna} (maximum number of action types). Usually, the count of action types found in [M] will not exceed θ_{mna}. If it is less than θ_{mna}, then it indicates that some action types are absent in [M] for the given condition. Hence, a covering operation is invoked to include those missing actions [16].

The covering operation creates a new classifier by inserting a wildcard symbol (#) in the condition part with probability $P_{\#}$ at each position. It is ensured that the condition part of the new classifier matches the current situation. The action part of the new classifier is assigned the value which is not present in the match set. The new classifier thus created is included in the population set [P], as it differs from all other classifiers. Then matching operation repeats resulting in a new match set [29].

Prediction Array Derivation and Action Selection

When the match set is formed, the prediction array [PA] is constructed for every type of action $\mathbf{a}_i$ in the [M]. It is the fitness weighted prediction average as given in Eq. (5).

$$\text{Prediction Array}\left[\text{PA}\right] = \sum_{n}^{1}\left(P \times F\right) / \sum_{n}^{1} F \tag{5}$$

where n is the number of classifiers that advocate the action in the match set. The variables P and F are the corresponding prediction value and fitness of those classifiers with the action $\mathbf{a}_{i.}$ Out of the values in the prediction array, an action is selected for the execution. The choice of the action is either by pure explore style (random action selection) or by exploitation (choosing the best action). In general, the action with the highest prediction value or highest fitness or the action that yielded maximum fitness weighted average is considered as the best action [16].

Creation of an Action Set and Execution
An action set [A] is a subset of the match set [M]. Those classifiers whose action in the [M] matches the chosen action are grouped as the action set [A]. The procedure behind the formation of the action set is coded as a function called FORM_ACTION_SET().

```
FORM_ACTION_SET[M], chosen_action)
{
for each classifier cl in [M]
 if (cl.A == chosen_action)
    [A] = [A]+cl                    // update the action set with the classifier whose action matches with chosen_action
}
```

Once an action set is obtained, the chosen action is implemented through the effector on the environment. Then, the environment generates a reward prediction **P** as feedback.

Reward Prediction Estimate
The Q-learning estimation is performed in the RL part of the XCS algorithm [18]. The reward prediction P depends upon the type of RL problem encountered. For a multi-step problem (continuous task), the classifier's prediction is calculated as the sum of immediate reward and the discounted rate of maximum reward expected from the future states in the successive time step. For a single-step problem (episodic task), the prediction P is equivalent to the immediate reward (feedback) given by the environment. The computation of the reward prediction for the two types of tasks is shown in Eqs. (6) and (7).

$$P = r + \gamma.\max\left(\left[\mathrm{PA}\right]\right) \tag{6}$$

$$P = r \tag{7}$$

Classifier Parameter Update
Every time a classifier is added to the action set [A], its parameters like prediction (p), error (ϵ), fitness (F), experience (exp), size of the action set (as), and the numerosity (n) are updated. Among all, experience parameter exp is incremented to indicate that the corresponding classifier is encountered in the [A] yet another time. The procedure to update the classifier parameter follows the Q-learning method of RL [27]. The Q-value update is performed using Eq. (8)

$$Q(s_t,a_t) = Q(s_t,a_t) + \beta\left[\left\{r_{t+1} + \gamma.\max_{a+1} Q(s_{t+1},a_{t+1})\right\} - Q(s_t,a_t)\right] \quad (8)$$

where $Q(s_t, a_t)$ denotes the existing payoff which is to be updated, β signifies the learning rate, $(r_{t+1} + \gamma.\ \mathrm{Max}_{a+1}\ Q(s_{t+1}, a_{t+1}))$ indicates the learned payoff, γ is the discount rate, and r_{t+1} is the immediate reward received for the action a_t taken on the state s_t. The value $(\mathrm{Max}_{a+1}\ Q(s_{t+1}, a_{t+1}))$ represents the largest reward expected from the future states in the successive time.

The parameter prediction (p), error (ϵ), and action set size (as) are updated using Moyenne adaptive modified (MAM) method for the first few iterations followed by least mean square (LMS) for the rest of the iterations [16]. According to MAM, until 1/β times, a parameter is updated by an average of the current and previous value. Once the classifier is updated to a minimum of 1/β times, the LMS strategy is applied [18]. The update computations behind the two techniques are described by the pseudo-code using arbitrary variables x and y.

```
if(cl.exp< 1/β)
     x= x+(y-x)/cl.exp    // MAM technique
else
     x= x+β(y-x)          // LMS technique
```

The MAM technique enables the movement of early values of parameters more quickly to their "true" average value in order to avoid the parameters taking arbitrary values. It means that the MAM technique quickly gets rough approximation. However, the LMS technique used after 1/ β updates refines the estimate from approximation to accurate values [17].

The value of β is usually set within 0.05–0.2 which is a recommended range. The value of β adjusts the sensitivity of the updated parameter with respect to the changes in the population. The β value is used as the step size in the updates of prediction, error, fitness, and action set size [17].

Reward Prediction Update

The reward prediction (P) is computed as explained in Eqs. (6) and (7) depending upon the problem whether it is episodic or continuous. The existing reward prediction is updated by the MAM and LMS technique based on the classifier's experience. The procedure for the reward prediction update is given by the following pseudo-code.

```
if (cl.exp< 1/β)
    cl.p=cl.p + (P-cl.p)/cl.exp      // (P - cl.p) is the error in the predicted reward
else
    cl.p=cl.p + β(P-cl.p)
```

where cl.p is the classifier's existing reward prediction that is updated and P is the learned reward prediction (target).

Prediction Error Update

The prediction error indicates the difference between the target/learned reward prediction error and the existing error. The difference |(P-cl.p)| is the learned reward prediction error, and cl.є is the existing reward prediction error. The difference between the two errors results in the current error that is updated. The error update procedure is coded as shown below.

```
If (cl.exp<1/β)
   cl.є = cl.є+ (|P-cl.p|-cl.є)/cl.exp          // MAM technique
else
   cl.є = cl.є+β(|P-cl.p|-cl.є)                 // LMS technique
```

In practice, it is always better to compute the update of prediction error (*cl.*є) first, followed by the prediction update. If the prediction update is computed first, the chance of prediction error becoming zero is high. This enables faster learning and is suitable only for simple problems. But in the case of complex problems, zero error in the very early stage goes misleading. So updating the prediction error (*cl.*є) followed by a reward prediction update is best recommended.

Fitness Update

The fitness of the classifier indicates the accuracy level of the reward prediction. The accuracy of the reward prediction is reciprocal of the reward prediction error[2]. The lesser the error, the greater the accuracy is considered. Hence, the accuracy of the *i*th classifier (k_i) is calculated as the reverse of the classifier's error by Eq. (9):

$$\left.\begin{array}{ll} k_i = \mathbf{1} & \text{if } e_i < e_0 \\ k_i = \alpha\left[\left(e_i - e_0\right)/e_0\right]^{-v}, & \text{otherwise} \end{array}\right\} \tag{9}$$

where the parameter e_0 ($0<e_0<1$) is the error threshold under which a classifier is measured as most accurate. So if the error of the *i*th classifier e_i is less than the threshold error, the accuracy of k_i is set to one. The values $v > 0$ and $\propto$ within the range [0, 1] are the constants that control the degree of diminishing inaccuracy if the reward prediction of the classifier is inaccurate. Next, the relative prediction accuracy (k_i') over other classifiers is estimated by Eq. (10):

$$\mathbf{k}_i' = \frac{\mathbf{k}_i \times \textbf{numerosity of ith classifier}}{\sum_{x \in [A]} \mathbf{k}_x \times \textbf{numerosity of } (\mathbf{x})} \tag{10}$$

where k_i is the accuracy of the ith classifier and k_x is classifier x in the [A]. Eventually, the fitness of each classifier is updated through the LMS method, by Eq. (11):

$$f = f + \beta\left(k_i' - f\right) \tag{11}$$

where *f* represent the existing fitness value which is updated and k_i' represents the learned fitness[2].

GA Rule Evolution and Action Subsumption

The rule evolution part of the XCS determines the insertion of well-performing classifiers and deletion of ill-performing classifiers with respect to the scenario, as the time proceeds. While the action set [A] is used for choosing the more effective classifier for reproduction, the population set [P] is meant for the insertion of a new classifier or deletion of the existing classifier [6]. Before invoking GA, it is verified whether it is possible to apply GA at that particular time. Given the current [A], it is possible to invoke GA, only when the average time (ts) is greater than the GA threshold (θ_{GA}) for all classifiers in the [A].

The parent classifiers are chosen from [A] based on their fitness value. The chosen classifier then undergoes crossover and mutation to generate new child classifier(s). The parameters of the children classifier are initialized with the average of their parent values [20]. Usually, the offspring classifiers thus created are added to the [P]. But if the subsumption procedure is involved in the system, then the addition of the new classifier to the [P] is reconsidered.

GA subsumption is a process that accelerates the generalization of the classifier. A classifier is treated more general if its error ($cl.\ e_i$) is less than the threshold error (e_0) and its experience (cl.exp) is greater than the subsumption threshold (θ_{sub}) [16]. The subsumption process searches for an equivalent classifier in the [P] that is more generally such that the new classifier could be generated from it or even could be replaced.

The idea is that the new classifier that does not satisfy the generalization condition is considered incompetent. They do not contribute much to the improvement of the system since everything is accomplished by the parent (general) classifier itself. When such a general classifier is encountered in [P], the new classifier is subsumed by incrementing the numerosity of the parent classifier and eliminating the new offspring.

Independent of GA subsumption, action set subsumption is solely meant to perform in the action set. The process searches the most general classifier that is maximally general and accurate in the [A] instead of [P]. If such a classifier exists, its numerosity is increased, followed by the elimination of the subsumed classifier from [A] and [P].

Resource Allocation and the Significance of XCS

The revolution of availing anything as a service (XaaS) has led the industries and enterprises to host their applications on the Web from where it provides online service to the consumers [31]. Hosting the application and its supporting packages requires enough compute/storage facilities like fog computing that resides near to clients providing service at low latency. The allocation of fog/cloud resources to the workload (requests from IoT) for processing is referred to as fog-cloud resource allocation [10].

Moreover, the exponential growth of the IoT device makes it the biggest consumer of power. Power consumed per unit of time is estimated as energy. Hence, energy conservation is an inevitable factor in the performance estimation of the RA model [24]. The success of any RA framework lies in the efficient algorithm that is deployed to handle the computing resources. The choice of the algorithm depends on its capability to manage the application requirement with minimum delay, energy, and budget.

The proliferation of IoT application exhibits complexity in terms of heterogeneity, mobility, and context awareness [5]. Hence an algorithm that learns and adapts according to the changing requirements of the IoT-fog-cloud framework is required. Amid other approaches, XCS not only learns and adapts the optimal I/O mapping but evolves a minimal set of rules that describe those mapping, thus making it prominent among data scientist gadget.[4]

6 Case Study

An optimal RA in fog-cloud for IoT applications through XCS is the problem considered as a case study. The workload request from various IoT devices and the resource pool (fog and cloud) that supply the compute nodes forms the environment. The agent is the program that encapsulates the detector, population set [P], match set [M], prediction array [PA], action set [A], the GA, and the effector.

At every time step, the amount of workload is given as the input. The goal of the proposed framework is to estimate the optimal number of workload that is to be processed in fog with minimum delay and energy consumption. The fog-cloud RA framework balances the quantity of workload that is distributed among the fog layer and the cloud.

System Model

The proposed RA framework considers every time period "t" as 10 seconds. At every time period, the amount of user requests ($\sigma(t)$) from their edge devices to the base station is termed as workload. It is represented as a discrete-time signal to be in the range $[0, \sigma_{max}]$. The value of σ_{max} is set to 100, which is the maximum workload that the base station receives per second. Hence with every "t" set as 10 s, the amount of workload ranges between 0 and 1000 requests per second.

The fog layer encompasses a fog controller (base transceiver station) and a collection of fog nodes that are geographically located near to each other. The fog controller node distributes the incoming workload between the fog nodes and the cloud. The number of active fog nodes at every time period "t" is denoted by $fn(t) \in [0, fn_{max}]$, where fn_{max} is the maximum amount of fog nodes available in the fog layer.

The job of the agent program hosted in the fog controller is to estimate the number of workloads ($\delta(t)$) that are to be processed in the fog layer. Therefore $\delta(t)$ is

[4] http://hosford42.github.io/xcs/

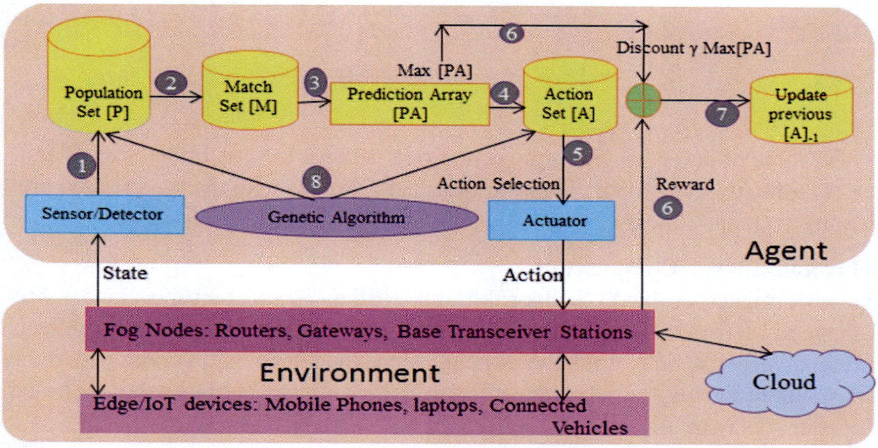

Fig. 6 XCS-based fog-cloud framework [23]

supposed to be less than or equal to σ(*t*). Hence the residual workload (σ(*t*)−δ(*t*)) is transmitted to the cloud. At any time period "t," the estimation of δ(*t*) is influenced by various factors like the amount of workload, number of active fog nodes, the processing capacity of each fog node, and the battery capacity of the active fog nodes.

The fog-cloud framework for RA with respect to XCS comprises environment and agent as shown in Fig. 6. At any instant of the time period "t," the fog controller is expected to receive 0–100 requests per second. The workload first reaches the fog controller from where it is partially distributed to the fog layer or to the cloud for processing. The fog controller hosts the agent program that estimates δ(t). Finally, the performance of the fog-cloud RA model is evaluated in terms of delay and energy consumption cost.

XCS Solution

The RA framework modeled as XCS contributes to the estimation of the optimal number of workloads to be processed in fog (δ(*t*)). The actual input to the system at every time period is the condition part of the classifier. The specific combination of bits in the classifier's condition represents the amount of incoming workload σ(t), battery capacity of the fog nodes b(*t*), and the magnitude of network congestion h(t). The action δ(*t*) which is the optimum amount of workload to be processed in the fog layer depends on these input parameters and the processing capacity (k) of each fog node.

The pseudo-code to compute δ(*t*) given below first checks if the available battery capacity is less than the operating power (P_{op}) required to handle the σ(t) workload. If true, it indicates that the fog layer no longer has enough power to process and all the workload is offloaded to the cloud which makes $\delta(t) = 0$. Otherwise, $\delta(t)$ is computed as the product of the active number of fog nodes $fn(t)$, the processing capacity of the fog node (k), and the unit of one time period (t).

```
if b(t) ≤ P_op(σ(t)) then
    δ(t) = 0
else
    δ(t) = fn(t) × k × 10secs
```

The value of δ(*t*) is stored as the action part, against the condition of the classifier. As time passes, the population of classifiers evolves with their associated prediction, error, and fitness values.

Delay and Energy Computation

The performance of the XCS algorithm that enables an efficient fog-cloud RA is evaluated in terms of delay and energy consumption. The delay parameter reflects the actual response time of the RA system. Hence the total delay is computed as the sum of three delays, namely, transmission delay (d_t), processing delay (d_p), and offloading delay (d_{off}).

The transmission delay represents the time consumed for propagation on the wireless network. The processing delay is the time consumed to compute the amount of workload (δ(*t*),) in the *fn*(*t*) amount of fog nodes, each with "k" processing capacity. Hence the processing delay in the fog layer for the time period "t" is given by $d_p = \frac{\delta(t)}{fn(t)\times k - \delta(t)}$. The offloading delay ($d_{\text{off}}$) is the time taken to transmit the residual workload (σ(*t*) – δ(*t*)) on to the cloud. Thus, the offloading delay depends on the amount of residual workload and the intensity of congestion (h(t)) in the network. It is calculated as $d_{\text{off}} = (\sigma(t) - \delta(t))\cdot h(t)$. Then the total delay is computed as shown in Eq. (12):

$$d_{\textbf{total}} = d_t + d_p + d_{\textbf{off}},$$
$$d_{\textbf{total}} = d_t + \frac{\delta(t)}{fn(t)\times k - \delta(t)} + (\sigma(t) - \delta(t))\cdot h(t) \quad (12)$$

With the trend moving toward online services, the massive scale of IoT devices that enable these services has become the largest consumer of power nowadays. The power consumed per unit of time which is referred to as energy has a direct effect on the cost of the RA framework. Hence, the calibration of energy is inevitable in evaluating the performance of the RA framework. The magnitude of energy consumption is estimated as the sum of operating energy and computing energy.

The operating energy is estimated as the sum of three types of power consumption as shown in Eq. (13)

$$e_{\textbf{op}} = e_{\textbf{on}} + e_{\textbf{maint}} + e_{\textbf{trans}} \quad (13)$$

where (e_{on}) is the power consumed by the fog nodes to be in on-state, (e_{maint}) is the power consumed by the maintenance equipment like UPS and cooling systems, and (e_{trans}) is the power consumed for transmitting the workload to the cloud.

The computing energy refers to the power consumed by the active server/nodes ($fn(t)$) to process the workload ($\sigma(t)$) in terms of computing and storage. It is defined as a function of $e_{cp}(fn(t),\sigma(t))$ as in Eq. (14)

$$e_{cp}\left(fn\left(t\right),\sigma\left(t\right)\right)=fn\left(t\right)\times\sigma\left(t\right)\times\textbf{energy cost / unit} \tag{14}$$

The overall energy consumption of RA is then estimated as the sum of operating energy (e_{op}) and computation energy (e_{cp}) as given in Eq. (15)

$$e_{\mathbf{RA}} = e_{\mathbf{op}} + e_{\mathbf{cp}} \tag{15}$$

Thus the computation technique for delay and energy consumption, incorporated in the fog-cloud resource allocation framework through the XCS algorithm, increases the efficiency of the model. The crust of the proposed work lies in deriving the optimum number of workload that is to be processed in the fog layer. Ultimately, the efficiency of RA depends on the correct distribution of workload among the fog and the cloud layer to balance energy consumption with the minimum delay.

7 Conclusion and Future Enhancements

The chapter reveals how XCS is used to solve the RL problem that involves continuous decision-making tasks. The fog-cloud resource allocation for IoT applications dealt through XCS consumes minimum energy and delay than its rival works. Though contemporary techniques like XCS were found to be efficient to solve the modern IoT-based RA problem, still there are research areas that are to be explored. Lack of sufficient IoT-based training dataset, constraints found with the ML models for IoT, cost involved in training, and generalizing the model for IoT are some of the open issues that need to be addressed in the future.

References

1. Di Martino, B., Rak, M., Ficco, M., Esposito, A., Maisto, S. A., & Nacchia, S. (2018, September). Internet of things reference architectures, security and interoperability: A survey. *Internet of Things, 1–2*, 99–112.
2. Atlam, H., Walters, R., & Wills, G. (2018, April). Fog Computing and the Internet of Things: A Review. *Big Data and Cognitive Computing, 2*(2), 10.
3. Naha, R. K., et al. (2018). Fog computing: Survey of trends, architectures, requirements, and research directions. *IEEE Access, 6*, 47980–48009.
4. Iorga, M., Feldman, L., Barton, R., Martin, M. J., Goren, N., & Mahmoudi, C. (2018, March). *Fog computing conceptual model*. In National Institute of Standards and Technology, Gaithersburg, MD, NIST SP 500-325.

5. Gowri, A. S., & Shanthi Bala, P. (2020). *Architecture and security issues in fog computing applications* (Ed. S. Goundar). IGI Global.
6. Urbanowicz, R. J., & Moore, J. H. (2009). Learning classifier systems: A complete introduction, review, and roadmap. *Journal of Artificial Evolution and Applications, 2009*, 1–25.
7. Holland, J. H., et al. (1813). What is a learning classifier system? In P. L. Lanzi, W. Stolzmann, & S. W. Wilson (Eds.), *Learning classifier systems* (Vol. 2000, pp. 3–32). Springer.
8. Sigaud, O., & Wilson, S. W. (2007, May). Learning classifier systems: A survey. *Soft Computing, 11*(11), 1065–1078.
9. Kaur, A., Kaur, B., & Singh, D. (2017, January). Optimization techniques for resource provisioning and load balancing in cloud environment: A review. *IJIEEB, 9*(1), 28–35.
10. Soltani, S., Martin, P., & Elgazzar, K. (2018, December). A hybrid approach to automatic IaaS service selection. *Journal of Cloud Computing, 7*(1), 12.
11. Ismayilov, G., & Topcuoglu, H. R. (2020, January). Neural network based multi-objective evolutionary algorithm for dynamic workflow scheduling in cloud computing. *Future Generation Computer Systems, 102*, 307–322.
12. Tran, M.-Q., Nguyen, D. T., Le, V. A., Nguyen, D. H., & Pham, T. V. (2019, January). Task placement on fog computing made efficient for IoT application provision. *Wireless Communications and Mobile Computing, 2019*, 1–17.
13. Nguyen, D. T., Le, L. B., & Bhargava, V. K. (2019, June). A market-based framework for multi-resource allocation in fog computing. *IEEE/ACM Transactions on Networking, 27*(3), 1151–1164.
14. Nassar, A., & Yilmaz, Y. (2019). Reinforcement learning for adaptive resource allocation in fog RAN for IoT with heterogeneous latency requirements. *IEEE Access, 7*, 128014–128025.
15. Gai, K., & Qiu, M. (2018, September). Optimal resource allocation using reinforcement learning for IoT content-centric services. *Applied Soft Computing, 70*, 12–21.
16. Butz, M. V., & Wilson, S. W. (2002, June). An algorithmic description of XCS. *Soft Computing – A Fusion of Foundations, Methodologies and Applications, 6*(3–4), 144–153.
17. Kovacs, T. (2004). *Strength or accuracy: Credit assignment in learning classifier systems*. Springer.
18. Lanzi, P. L., Stolzmann, W., & Wilson, S. W. (Eds.). (2000). *Learning classifier systems: From foundations to applications*. Springer.
19. Kramer, O. (2017). *Genetic algorithm essentials* (Vol. 679). Springer.
20. Kharbat, F., Bull, L., & Odeh, M. (2008). *Revisiting genetic selection in the XCS learning classifier system*. In 2005 IEEE Congress on evolutionary computation, Edinburgh, Scotland, UK (Vol. 3, pp. 2061–2068).
21. Stalph, P. O., Llorà, X., Goldberg, D. E., & Butz, M. V. (2012, March). Resource management and scalability of the XCSF learning classifier system. *Theoretical Computer Science, 425*, 126–141.
22. Schwind, M. (2007). *Dynamic pricing and automated resource allocation for complex information services: Reinforcement learning and combinatorial auctions*. Springer.
23. Abbasi, M., Yaghoobikia, M., Rafiee, M., Jolfaei, A., & Khosravi, M. R. (2020). Efficient resource management and workload allocation in fog–cloud computing paradigm in IoT using learning classifier systems. *Computer Communications*, 13.
24. Hameed, A., et al. (2016, July). A survey and taxonomy on energy efficient resource allocation techniques for cloud computing systems. *Computing, 98*(7), 751–774.
25. Thein, T., Myo, M. M., Parvin, S., & Gawanmeh, A. (2018, November). Reinforcement learning based methodology for energy-efficient resource allocation in cloud data centers. *Journal of King Saud University – Computer and Information Sciences*, S1319157818306554.
26. Gai, K., Qiu, M., Zhao, H., Tao, L., & Zong, Z. (2016, January). Dynamic energy-aware cloudlet-based mobile cloud computing model for green computing. *Journal of Network and Computer Applications, 59*, 46–54.
27. Szepesvári, C. (2010, January). Algorithms for reinforcement learning. *Synthesis Lectures on Artificial Intelligence and Machine Learning, 4*(1), 1–103.

28. Maia, T. V. (2009, December). Reinforcement learning, conditioning, and the brain: Successes and challenges. *Cognitive, Affective, & Behavioral Neuroscience, 9*(4), 343–364.
29. Bernadó-Mansilla, E., & Garrell-Guiu, J. M. (2003, September). Accuracy-based learning classifier systems: Models, analysis and applications to classification tasks. *Evolutionary Computation, 11*(3), 209–238.
30. Zang, Z., Li, D., & Wang, J. (2015, June). Learning classifier systems with memory condition to solve non-Markov problems. *Soft Computing, 19*(6), 1679–1699.
31. Al-Shara, Z., Alvares, F., Bruneliere, H., Lejeune, J., Prud'Homme, C., & Ledoux, T. (2018, September). CoMe4ACloud: An end-to-end framework for autonomic cloud systems. *Future Generation Computer Systems, 86*, 339–354.

Convolutional Neural Network (CNN)-Based Signature Verification via Cloud-Enabled Raspberry Pi System

Iqraq Kamal, Hwa Jen Yap, Sivadas Chandra Sekaran, and Kan Ern Liew

1 Introduction

Based on the Global Economic Crime and Fraud Survey 2018 [1] '46% of Southeast Asian companies reported experiencing economic crime in the last two years, up from 26% in 2016'. One of the misconducts that are occurring is signature forgery, even though paper-based form of authorisation is expected to be consigned to history by 2018. Many argued, however, that paper-based forms of payment such as cheques should not be phased out until suitable alternatives are introduced. Furthermore, any non-contextual document that contains signature can be easily forged. This could cause multiple fraudulence in any industry despite the size of the organisation. A more proactive action needs to be taken to ensure that the cheque book can be verified in a more accurate manner to reduce the monetarily loss of the people.

Many problems that require high accuracy classification has adopted machine learning. This is including a signature verification system to identify forged signature and real signature. Many previous works utilise CNN to develop an offline signature verification system. CNN is the state-of-the-art for deep learning computer vision [2]. The convolutional layer feature extraction process is flexible in recognising the signature despite its inconsistent position in a picture. This key feature makes them a viable deep learning architecture to verify signature. However, most of them focus more on developing a writer-independent signature verification system. The algorithm is trained based on available online dataset to find the pattern

I. Kamal · S. Chandra Sekaran · K. E. Liew
Aerospace Malaysia Innovation Centre, Kajang, Malaysia
e-mail: iqraq@amic.my; sivadas@amic.my; liew@amic.my

H. J. Yap (✉)
University of Malaya, Kuala Lumpur, Malaysia
e-mail: hjyap737@um.edu.my

S. Pal et al. (eds.), *Artificial Intelligence-based Internet of Things Systems*,
Internet of Things, https://doi.org/10.1007/978-3-030-87059-1_7

between the forged signature and the real signature. Even so, for security sake, it is important not to just classify signature into forged and real signature but to also identify the owner of the signature itself.

To make this possible, a writer-dependent signature verification process needs to be developed. The system must be portable and reliable. The portability is going to help verify signatures of important documents and cheque. Having this system is expected to reduce the amount of fraudulence and scam. This is especially true if the banks can implement the system to validate the signature of the person on the cheque with high accuracy and repeatability. Furthermore, the rapid development of IoT enhances the accessibility of the data with a remote device. To add, with the presence of cloud storage, the data is safer and can be shared globally. This is very useful for the implementation of machine learning due to its data-hungry nature. The purpose of the research is to develop a portable product that can verify signatures. The three main objectives to attain are as follows:

1. To develop a suitable offline writer-dependent signature verification using CNN.
2. To integrate the verification process to an IoT architecture.
3. To maintain the security and the accuracy of the verification process.

2 Literature Review

This chapter provides an overview of previous research on CNN and Raspberry Pi. It introduces the framework for the case study that comprises the focus of the research described in this paper.

It is important to set the context of the literature review work by first providing:

1. An explanation of its specific purpose for this case study.
2. Comments on the previous treatment on the broad topic of CNN on handwritten signature and the usage of Raspberry Pi.
3. An indication of scope of the work presented in this chapter.

An appreciation of previous work in this area served two further purposes. Firstly, working the findings from past literature into a formal review helped to maintain the topics' perspective throughout the study. Past literature can be used as a comparison point of the implementation of tools such as cloud and Raspberry Pi. Secondly, having a complete literature review will raise the opportunities to articulate a more critical analysis on the application of CNN on hand signature verification. To add, results from previous models in past literatures can be used as a benchmark of the optimisation level of the model we are currently using. By comparing the result, we can know the best architecture to implement for verifying handwritten signature.

2.1 Background of Signature Verification

Signature verification is validating the identity of an individual by analysing his/her signature. The process intends to differentiate a genuine signature from a forgery [3]. Unlike any handwriting process, a signature contains unique individual features and is usually a combination of personal symbols that are far more complicated to be validated.

A survey by Pal [4] stated that the common approach towards this problem is separated into two classes: Class 1 is a genuine set of signature and Class 2 is forged signature. To calculate the performance of each test, two types of errors are considered which are the false rejection and the false acceptance. Based on that, there are two types of error rate which are the false rejection rate (FRR) and the false acceptance rate (FAR).

Based on Justino [5], there are three common types of common forgeries as listed below:

1. Random forgery. It is not done by intent, the forger might not have any initial idea of the genuine signature or even the author's name. It might include the forger's own signature.
2. Simple forgery. The author's name is known, but the forger does not have any access to the sample of the genuine signature. The signature produced is in his/her own style.
3. Skilled forgery. The forger knows the author's name and they have access to the sample of genuine signature. They can recreate the signature in a similar fashion as with the original.

The skilled forgery can be divided into 4 four kill levels which are:

1. Forged signature that is produced with the genuine author's name only.
2. Forged signature that is produced by an inexperienced forger without the knowledge of the exact spelling but rather by observation.
3. Forged signature that is produced after some unrestricted practice by non-professional forgers.
4. Forged signature that is produced by a professional impostor or person who has experience in copying signatures.

One of the most challenging tasks in classifying a signature is in finding a significant distinction between each signature. As stated by Zhang [6], feature extraction plays a very important role as it should able to extract features that help to increase interpersonal distance between each signature example of different persons while reducing interpersonal difference for those belonging to the same individual.

2.2 Convolutional Neural Network (CNN)

In the work of Fukushima [7], CNN is inspired by Hubel [8] study on receptive fields in the visual cortex. Fukushima [6] introduced two basic types of layers in CNN, which is the convolutional layer and the downsampling layer. Later, Weng [9] introduced a method called max pooling where a downsampling unit computes the maximum of the activation of the units in its patch. Max pooling is commonly used in modern CNN. CNN is heavily used in image classification and is more reliable in handling image data compared to the common artificial neural network (ANN). 'One of the largest limitations of traditional forms of ANN is that they tend to struggle with the computational complexity required to compute image data' [10].

2.3 Standard Components of CNN

Basic CNN are composed of three types of layers [11] :

1. Convolutional layer
   a. Focuses on the use of learnable kernel. The kernel glides through the input and the scalar product are calculated for each value in that kernel. The network will learn which kernels are activated when there are specific features at a given position of input [10]. The process is also known as sparse interaction and needs less parameters to be stored which will reduce the memory requirement of a model and improve statistical efficiency compared to ANN which uses the traditional matrix multiplication method [12].
2. Pooling layer
   a. The purpose of the pooling layers is to reduce the spatial resolution of the feature maps and thus achieve spatial invariance to input distortions and translations [13]. As stated by [12], pooling is crucial for handling input of varying sizes. It is done by varying the size of an offset between pooling regions ensuring the same number of summary statistic is received by the classification layer.
3. Fully connected layer
   a. The fully connected layer contains neurons of which are directly connected to the neurons in the two adjacent layers, without being connected to any layers within them. This is analogous to the way that neurons are arranged in traditional forms of multilayer perceptron. The arrangement for a fully connected layer could dictate the performance of the whole architecture. Based on Basha [14], higher number of nodes in the fully connected layer is needed in a shallow architecture to obtain better result, while deeper architectures require a smaller number of nodes irrespective of dataset type.

2.4 Activation Function

Activation function serves neural networks into two primary manners, which are to help take into account the interaction effect between the features and to help the model accommodate the non-linear effects.

2.5 Rectified Linear Units (ReLU) Activation Function

ReLU is one of the non-linear activation function. The units are easy to optimise because they are similar to linear units. The only difference that ReLU has is that half of its domain provides zero as the output. This makes the derivatives through a ReLU stay large and consistent. In short, ReLU can be presented as follows:

$$f(x) = \max(0,x)$$

Due to its simplistic nature, the ReLU function should be able to accelerate the training speed of deep neural network compared to the traditional activation function. However, one drawback using the ReLU is that there will be no learning via gradient-based method whenever their activation is zero. All the input that has less than zero value will be rendered meaningless [12].

2.6 CNN Against Traditional Image Classifier

In the year of 2019, there were at least 500 million of photos uploaded online. This vast amount of data created leads to the need to advance the field of computer vision especially in image classification. Image classification is highly useful to help users to recognise data and pattern. The advancement has introduced multiple technologies that are highly accurate to handle different variation of image classification. This includes the development of CNN, a state-of-the-art technology for image classification in deep learning. With more hardware capacity available, a deeper and more complex architecture can be developed and deployed. This leads to a more accurate and general model to be created. However, there are still many other image classifiers that are available. The CNN is expected to perform better based on its ability to learn and extract features from the image, to provide a more accurate prediction and for the overall generality of the prediction compared to another image classifier.

There are many other traditional image classifiers developed without the implementation of machine learning. Thakur [15] conducted a review between different types of image classification. The review compared five different image classification methods which are parallelepiped techniques, minimum distance classifier, maximum likelihood classifier, ANN and support vector machine. Based on the comparison, the ANN can deal with noisy input efficiently and has a high computation rate. Due to the nature of it being data driven, a typical neural network should be able to have a better generality in predictions compared to other image classifiers. However, being data driven means it is also exposed to some other disadvantages such as heavy time consumption for data training, is considered as semantically poor and will occasionally encounter the problem of overfitting. Even though the architecture of the ANN is different compared to CNN, it is comparable since CNN commonly uses fully dense multilayer perceptron to classify the image extracted from the convolutional layer.

In another study conducted by Lee [16], a deep neural network is compared to scale-invariant feature transform, speeded-up robust features and binary robust independent elementary feature. The comparison is performed using a webcam attached to a turtle bot. It is found that the deep neural network works 20% better compared to other classification methods. However, there are certain trade-offs when using a deep learning method. The time taken to train the CNN to achieve satisfying accuracy is significantly longer. The comparison shows that for a mobile robot, traditional vision classification only requires between 30 s and 1 minutes to achieve 70% accuracy. This shows that, for supervised learning, CNN will require a lot of data and time to train to achieve an acceptable level of accuracy. Even so, for a mobile robotic application, the user can still utilise traditional computer vision since it does not require the level of accuracy a CNN could offer. This is supported in a study conducted by O'Mahony [17] where they figured, despite the accuracy that the CNN offers, it could be an overkill for certain applications. They found that deep learning excels at solving a closed-ended problem, but not in problems such as robotic, augmented reality, virtual reality, and 3D modelling where traditional techniques flourish and are more efficient. To add, deep learning requires highly powered GPU and TPU to be trained on.

The CNN has been proven to perform better based on its ability to learn based on the image data, capability to provide a more accurate prediction and the overall generality of the prediction compared to traditional image classifier. Nevertheless, there are many limitations that come with the perks that it offers, which requires a high volume of data and high GPU computation for training and is less dynamic when it comes to robotic and 3D modelling-related tasks. As society embraces the inclusions of the IoT in everyday life, more portable devices that can give a real-time feedback will be deployed; hence, it is very important to overcome the limitations that CNN have to make it future-proof and increase its overall dynamic so that it can be vastly used later.

2.7 CNN-Based Signature Verification Process

Signature verification process usually includes two different data extraction model. The first one being online and the latter one being offline. For comparison sake, we will look in the online verification system. Online signature verification will record more variables compared to offline verification. In a study conducted by Julita [18], it is suggested to take account three different variables which are the speed, pressure and orientation during the signature capturing process. The paper utilises support vector machine as its learning model. The result shows a 100% rate of accuracy to predict forgery and non-forgery. To add, Sae-Bae [19] found that, by taking account the number and length of strokes of each signature, they managed to get an average of 3% error rate. The signature data used is taken from mobile devices. Despite being a promising technology, this will only work if the whole signature system in every industry is managed online and digitally and the user has an accurate touch-screen pad for data collection.

For practicality, offline (static) signature still offers a more pragmatic approach towards our current situation in the society. Offline signature verification is done using pictures or images of signature as its data. Based on the literature we reviewed and compared, we are expecting CNN to provide a good classification and prediction when it comes to verifying a signature. In a research by Khalajzadeh [20], a 99.86% accuracy is achieved from predicting a Persian signature from 22 persons. Based on the paper, it is stated that having a convolutional layer as a feature extraction increases the robustness about signature location change and scale variation. From that, we assume that it is likely that CNN will be able to predict forgery in the signature. This is proven in a study conducted by Mohapatra [21] where the model is trained to differentiate between the forged and original signature. Their model can outperform other models on the same dataset. Even so, we would like to focus on the fact that they are using NVIDIA Tesla GPU which is made specifically for deep learning. The same case is true for Hafemann [22], where they used Nvidia Tesla to train their dataset. The need to have a very good GPU to accelerate the training time is supported by Cozzens et al. [23] as they ran their CNN on a CPU instead of a GPU. The time taken to train a large dataset was significantly different, and, in their case, due to the time constraint, a smaller size of data is used which then caused the model to overfit and affect its overall accuracy. Finding the right ratio between amount of data and complexity can be challenging. However, the finding is obvious that larger dataset led to a better model accuracy as more variation of data being fed will help to generalise the model from being overfitted. This is shown in the study conducted by Dey [24] as they stated that a possible reason for the lower performance on the GPDS300 is the lesser number of signatures for each signature style as they are not able to outperform the state-of-the-art model for that dataset.

2.8 Writer-Dependent and Writer-Independent Signature

The difference between writer-dependent and writer-independent signature is in the utilisation of convolutional layer as a feature extractor. The amount of data available to be processed can be hugely impacted by the type of signature we are processing. Most of the state-of-the-art models for signature verification utilise writer-independent method to study the correlation between the CNN model and the accuracy it can provide. This can be seen in the study conducted by Sronothara [25] where they deploy two main convolutional architectures which are LeNet and AlexNet to be compared on the same dataset, and in the research conducted by Cozzens [23], the paper utilises SigComp2011 dataset to test the accuracy of the model that they are developing. Despite the number of the signature available online, the outcome of the paper is limited only to classify between two classes, which are the original and forged signatures. However, in a real-life scenario, classification that is directly related to the end user is much more important. The model must be able to classify the owner and the authenticity of the signature.

To overcome the limitation of getting sufficient data for writer-dependent signature verification, Hafemann [22] proposed a two-stage approach, which is to use the convolutional layer to extract feature from writer-independent signature and then utilise the weight on a writer-dependent-based classifier. They managed to achieve a very low equal error rate. However, the model shows a very high FAR and a low FRR. This technique is promising to overcome the lack of data to verify writer-dependent signature. However, the limitation is in the inter-relation between the writer-independent dataset and writer-dependent dataset. Based on Hafemann [22], the technique is not stable as it shows varying rates of accuracy across multiple users and dataset and a more defining user-specific threshold to verify writer-dependent signature needs to be studied.

The use of CNN to verify a signature is proven to be viable and promising due to its ability to extract feature and finding inter-relation between pixels despite the position and orientation of the signature in the image. Based on previous studies, across multiple architectures and dataset, the results show that with the right implementation, CNN could achieve an average of 85% accuracy rate. However, there are several obstacles that need to be addressed which are its capability to handle a small dataset of unique signature, the computation power it requires and the time taken to train the dataset. To implement a CNN-based signature verification in IoT, we will require a more rigid user-specific threshold to be trained as data to increase its accuracy. Furthermore, its IoT compatibility can be improved either by deploying the model on a single-board computer or utilising cloud as a training centre.

2.9 IoT and Machine Learning

It is expected that by the year of 2025, there will be 21 billon IoT devices deployed around the globe. The massive increment of fully connected devices will accelerate the amount of data produced exponentially. Meaningful data collected helps the advancement of many other industries. A user is expected to have more insight of their day-to-day operation and gained a better perspective to increase the efficiency of their own organisation. This is especially true to help implement machine learning or any other prediction models. This is demonstrated by Kumar [26] where machine learning is implemented for early detection of heart diseases. The device is first used to monitor real-time heart conditions and is connected to the hospital to alert them if the user is facing any cardiac problem. On a bigger scale, machine learning is also implemented in a surveillance system. A computer vision is integrated with IoT to automate the use of IP cameras. The system is used to detect any movement based on the sequencing of the picture with a 100% success rate. Machine learning has a very promising application in the IoT; the presence of huge amount of data enhances the accuracy of a predictive model. However, computation power is still a main issue to implement machine learning. Even in the paper by Rajan [27], they implemented fog computing for the IoT instead of the conventional wireless transmitting method in healthcare. It is one of the ways to process the data received from the sensors.

Based on the trends that come from previous studies, it is obvious that most of the novel system developed is for large-scale application. The topology will usually require a direct connection from the sensor head to the processor or cloud for data processing. The sensor must be able to measure and record data accurately, and the size must be compatible to accommodate its objective. However, arguably, this arrangement is not suitable for all applications. The suitability of the application is influenced heavily by the number of sensors used in a single system.

For a single sensor system like signature verification, the topology is not cost- and time-effective. When it comes to applying a signature verification system, most of previous use cases propose an online signature verification solution rather than offline [18, 27]. Despite being able to overcome the rigidity of the system by utilising day-to-day smartphones and PDAs, the verification system is limited to digital system and cannot be implemented on a paper-based signature. There are only small areas covering the usage of single-board computers (SBC) as a sensor and processing unit. We could argue that this is because of two main reasons. The first one being most of the problems that were solved using machine learning on IoT devices require a real-time data feedback. Especially from Kumar [25] and Rajan [26], both are utilising clinical data that comes from IoT sensors worn by humans. Since the data is uploaded in real time and will only be meaningful after a certain pattern is discovered, it would be more efficient to directly train those datasets through cloud computing. The second point is that the previous use case does not require a physical human-machine interaction (HMI)). The sensors are used directly to feed the data into the data cloud storage. On SBC, multiple peripherals can be utilised, and the data can be directly processed before being sent out to the cloud. To add, with SBC, the user can have computation power to run a prediction model without the need of cloud computation. SBC can solve a more unique problem such as signature identification and verification due to its dynamic nature to portability and flexibility.

2.10 Single-Board Computer (SBC)

Small but compact, SBC is a device that has the capability of a normal computer but with a fraction of its computational power. Moving towards IR 4.0, SBC was deemed to be one of the most important tools that can be deployed to complement an IoT system. With the portability its offering, it opens multiple possibilities in different fields such as automation and robotics. To add, with the computational power that is sufficient to run a normal program, SBC offers portability to run a machine learning model. With the use of a more sophisticated machine learning model, there are multiple reasons to choose single-board computer over cloud computing such as having real-time human-machine interface capacity through extra peripherals that it provides and the capability to act as an independent server for cloud storage that could enhance data security.

As stated by Maksimović [28], Raspberry Pi has a similar capacity of a personal computer to the domain sensor network. It is the perfect platform for interfacing with wide variety of external peripherals. Having Wi-Fi and Internet access, it is one of the most viable devices for remote communication and cloud application. Raspberry Pi is also relatively cheap and is more readily accessible to the public.

2.11 Deep Learning on Raspberry Pi

It is normal to assume that any machine learning and deep learning model should run on a highly powered computer. However, with the presence of SBC such as Raspberry Pi, it is not necessary. Despite the limited computing power that Raspberry Pi offers, it has a built-in software such as Scratch which enables users to program and design animations, games or interesting videos. In addition, programmers can also develop script or program using Python language. It is still able to run a simple yet crucial deep learning architecture. The implications of such possibilities are huge as more classification task can be done on a more remote area and in turn will promote the development of both SBC and deep learning model such as CNN.

Training data on a Raspberry Pi has issues. However, there are several ways that can be executed to help train the model. In the paper written by Dürr [29], the dataset was trained separately on a NVIDIA GTX 780 GPU for 40 minutes. The trained model then is transferred into Raspberry Pi. To verify the capability of Raspberry Pi in handling a pre-trained model, the model is compared with the commonly used OPENCV. The model managed to outperform it in terms of accuracy and speed, even though the computational power in Raspberry Pi is significantly less compared to standard gaming or tensor laptops. By utilising a pre-trained network, it is still able to achieve the accuracy of the model without compromising any other factor. In addition, it is possible to implement a more complex architecture with this method such as VGG 16. The model can be pruned before the training occurs with only a small drop in accuracy as trade-off.

The rationality of having a lightweight CNN model can be seen from a CNN-embedded Raspberry Pi to monitor structural health. Monteiro [30] trains the whole dataset of the model on Raspberry Pi. The architecture that they are using is five layers of CNN and a max dropout layer in between. Despite being able to achieve 100% accuracy and beating the state-of-the-art algorithm in structural health monitoring, the model relies on a very simple pixelated image as its dataset. This is not applicable for our use case as a signature is a highly complex object. In order to overcome that, Nikouei [31] introduced a depthwise separable convolution to reduce the computational cost of the convolution layer itself. The L-CNN file size is only 5% from the typical VGG-16 architecture, but it is still be able to produce a remarkable accuracy to detect human presence in motion. Despite being very light, the whole architecture of the model is built to detect presence of human-based object in a video. The objective does not require a high-level accuracy; hence, this application is not the most suitable for signature verification.

Raspberry Pi is a viable device to deploy CNN algorithm for various applications. However, the most crucial matter here is to recognise the need of the application and its relation to the limitations of Raspberry Pi. Various low feature tasks can be carried out by allowing the training to happen directly on Raspberry Pi. However, to perform a highly accurate verification process, a deeper layer of network needs to be deployed. For the time being, this process is more feasible to be done on a compatible computer. In the same context, a longer time is required to do hyperparameter tuning in Raspberry Pi, hence affecting the optimisation process of the model. On a final note, to have a well-built deployable model straight on a Raspberry Pi, it is preferable to have a strong base model trained in a personal computer. Having a strong benchmarking model will help to indicate the sufficiency of the customised model.

2.12 *Raspberry Pi-Enabled Cloud System*

The key point to all IoT application is the remote accessibility towards the data available in any system. With the capability that Raspberry Pi offers, the user will be able to connect it to larger inter-device system through the Internet. In the case of a signature verification system, all the data collected through Raspberry Pi can be stored in the cloud. The user can access the signatures collected through Raspberry Pi on other devices for verification process monitoring and model training purposes.

One of the ways of setting up cloud is by utilising an open-source software such as the ownCloud. By installing ownCloud, the accessibility of the files in the cloud is not restricted locally. This can be seen in a paper written by Princy [32] where they set up a cloud storage for the Raspberry Pi to accommodate real-time data storing system. The data is then stored in the cloud for future use. By having a private cloud storage, the data stored are more secure and are personalised to the user only. However, the accessibility is limited if the network is restricted by firework and blocked ports. To overcome this, Pimple [33] implemented Weaved which provides a unique IP by utilising tunnel protocol on the physical device.

On the other hand, for IoT application, Raspberry Pi can be connected to well-known cloud storage providers such as Google Drive or Dropbox. The provided service usually comes with multiple other perks such as free storage and high accessibility without the need to personally maintain the server. It is an advantage if the user is not able to provide enough storage in their drive to be used as cloud storage. This can be seen in a paper written by Ikuomola [34] where Raspberry Pi is connected to Google Drive for a smart surveillance system to overcome the need for having a physical hard drive.

Raspberry Pi has enough capability from hardware and a software standpoint to be connected directly to the cloud. The open-source software such as ownCloud and the readily available Python library make it easier for Raspberry Pi to transmit and receive any data via the Internet. Therefore, the integration of IoT with signature verification process is important to ensure data accessibility and security through cloud implementation. In addition, by having a push notification that informs the end user regarding the signature verification progress enhances the overall security and the reliability of the system. For our current use case, we will be using push bullet as the mean of notifying end user.

3 Methodology

The purpose of this chapter is to describe the methods used to create the CNN-based signature verification on a cloud-enabled Raspberry Pi and to assess the system's ability to meet the aim and objectives of the paper. This chapter will explain the hardware setup, application design, data processing, user experience and the limitations that come with it. This will be followed by a discussion on the full setup of the whole system on the Raspberry Pi 4 including the advantages and disadvantages of the tools chosen. It concludes with problems encountered during the research and future suggestion to improve the whole system (Fig. 1).

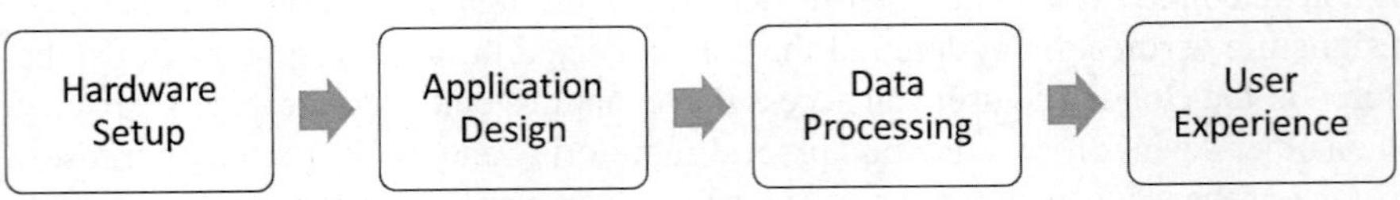

Fig. 1 Research approach overview

3.1 Hardware and Environment Setup

We are dividing the product development system into two parts which are the development of our CNN as our classification model for signature verification and implementation in SBC and the integration of the IoT system to make our system portable and able to function remotely. The development of the CNN is done in our laptop due to the availability of more computational power. By using our laptop, the CNN can be trained and optimised faster compared to a SBC. To ensure our model training can be done in the laptop, we are enabling CUDA for our NVIDIA GPU. CUDA is a parallel computing platform and is commonly used to train deep learning using GPU. Table 1 shows the specs of the laptop that we are using.

We are using Raspberry Pi 4 as our main device to capture images, verify signature and upload images to the cloud. The main reason we are using Raspberry Pi is because of its size and compactness. Having a small signature verification device is going to help make the product portable and can be easily transported everywhere. Plus, Raspberry Pi has a built-in camera slot that is useful to capture signature image and a Wi-Fi adapter making it easy to connect it to the Internet and to upload data to our cloud storage. For the Raspberry Pi camera, we are using the standard 5 MP 1080 camera module board. The camera weighs only 3g with the dimensions of 25 mm × 20 mm × 9 mm which makes it suitable for mobile application. The development of the CNN is done using Python 3.7.7 on PyCharm IDE. We utilised the TensorFlow and Keras library to configure our deep learning model and data processing application such as image augmentation and model evaluation. For Raspberry Pi, we are using Python 3.7 on Thonny IDE to capture image and run the pre-trained deep learning model for the verification process. Unlike in the PC, the Raspberry Pi is running TensorFlow Lite. The TensorFlow Lite enables on-device machine learning with low latency.

3.2 Cloud Storage Setup

One of our main objectives is to create a signature verification system that is portable and accessible through cloud. The Raspberry Pi is expected to upload the captured signature image to the cloud, thus making it available for it to be accessed

Table 1 Dell G5 specifications

Dell G5 Tech Specs
Intel® Core™ i5-8300H CPU @ 2.30GHz
8192MB RAM
NVIDIA GeForce GTX 1050Ti

using a highly powered computer for future model training. Instead of using the common cloud provider service like Google Drive or Dropbox, we are using ownCloud, a software used to create and utilise file hosting services. ownCloud is free, thus allowing any capable user to install and operate without any charges. Since it is running on a user private server, the user can avoid limited quotas on storage space and the number of connected clients. The capacity of the storage is depending on the physical capabilities of the server. In our case, ownCloud utilises the 64GB space of micro-SD installed on the Raspberry Pi.

Furthermore, since the data is trained outside of Raspberry Pi, it is crucial to transfer the data to the training PC to improve the accuracy of the training model progressively. By automatically feeding the captured image of the signature to the server, the user can download the data as a client. This data pipeline from Raspberry Pi to cloud will be able to cover the limitation to train the data directly on the Raspberry Pi. Users can access it anywhere and utilise a more powerful computer to train the dataset to have a better prediction model in the future. For our use case, having a private storage increases the safety and the security of the data we own. A private server plays an important role to counter any kind of identity theft or misconduct towards stored data.

To implement the ownCloud, we are using PHP and NGINX to run the cloud software. NGINX is an open-source http server. It is well known for its high performance, simple configuration and low resource consumption. Since NGINX is an event-driven architecture, users will only need small and predictable amounts of memory under load. To add, with the presence of SSL certificate on NGINX, the cloud connection is fully encrypted and secured to be used.

3.3 Pushbullet

To notify the user of the status of the program, the paper utilises third-party app called Pushbullet. The app can be used to easily transfer data from your phone to your PC. In our case, we are using it to transmit data directly from Raspberry Pi to our smartphone. The data transmission is not limited locally as it is going through cloud. By using this app, the user is notified with the progress and the result of the signature verification on the Raspberry Pi. The push notification is useful to solidify the signature verification system by keeping the end user notified with the usage of their signature from time to time. The coding of the push notification is integrated into the main coding flow in Raspberry Pi.

The Pushbullet package is installed through pip pushbullet.py in the Raspberry Pi terminal. The package includes all the crucial function to notify the user through the push notification system. The unique API key is available for any signed-up user. It is used to synchronise all devices that are registered within the account. Once the user runs the program, the function will send the notification to all selected devices as shown in Fig. 2.

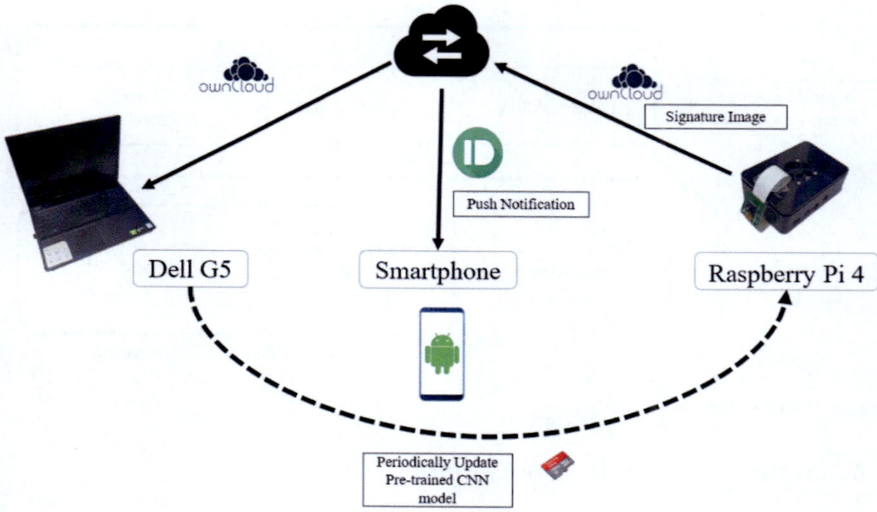

Fig. 2 Inter-connection between devices

3.4 Application Design

The process of the whole signature verification system starts with the development of deep learning model in our Dell G5. The development is explained under the data processing chapter. Once the deep learning model is fully developed and tested, it is uploaded manually into the Raspberry Pi through the SD card. The signature verification process is done in the Raspberry Pi. It captures the signature image while at the same time produces the prediction based on the pre-trained CNN that we developed in the PC. While the verification process is executing, it will notify the users' smartphone through the third-party push notification app. Once the verification is done, the captured image will be uploaded to the ownCloud for it to be accessed by the PC. Having this loop between the PC and Raspberry Pi can ensure the execution of periodic progression of the deep learning model by using newer user signature.

3.5 Data Handling

In this subsection, we are going to explain the development of the CNN used in our system. This includes our sampling strategy, data pre-processing and explanation of the CNN architecture we are using. It is then followed with our analysis on the model performance based on the hyperparameter tuning we did (Fig. 3).

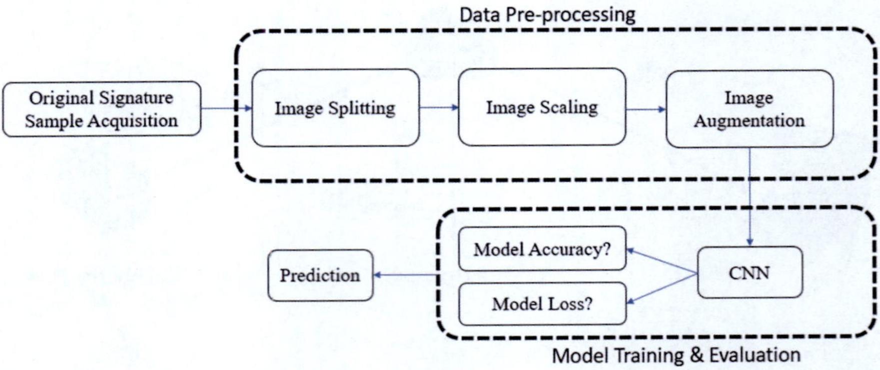

Fig. 3 Data handling sequence

3.6 Signature Sample Acquisition

For the purposes of this use case, the sample acquisition will be in the form of gathering real signatures and forged signatures. 659 signatures are collected in total, from 2 users. The initial idea is to have the signature that comes from multiple people. However, since we will require a model to predict the class of signature, the number of signature classes is reduced to 2 to improve its overall accuracy. The data is collected using 3x3 cm grid drawn on a paper. The users are then asked to sign in the space provided. To get the forgery signature, the writer collects the data from four random persons. They were instructed to copy the original user's signature. In this case, even though the four forgers have never seen the sample signature previously, they are still categorised as skilled forger since they are exposed with real forgeries and are not within time constraint while copying the original signature (Fig. 4).

Once the users and forger finished signing on the 3 × 3 grid, the paper is then scanned for us to obtain the image of signatures. We then chose the acceptable signature to be used as training and validation image for our model. Once the signatures are chosen, the grid is then split using Windows photo editor.

The images are then divided into four different folders based on four different classes consisting of user 1 real signatures, user 2 real signatures, user 1 forged signatures and user 2 forged signatures. 70% of the signatures in each class are being used as training set and 30% of the signatures are used for validation set (Fig. 5).

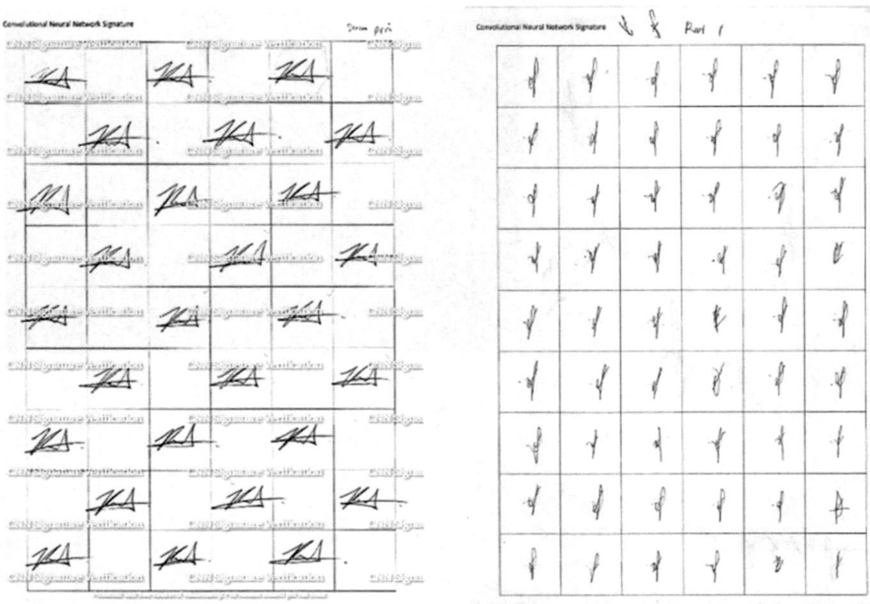

Fig. 4 Sample signatures gathered on 3 × 3 cm grid

Fig. 5 Post-processed signature image

3.7 *Image Augmentation*

Since the number of signature images that we have is only 659 signatures, we are using image augmentation to fully utilise each one of them. Image augmentation is one of the many ways to increase the variety of images used to train a model without acquiring new data/images. When it comes to computer vision, it is best to have at least 1000 images per class. Having a large dataset is very crucial as it will improve the performance of the model significantly and prevent overfitting from happening. For our use case, generating image data manually is time-consuming and tedious.

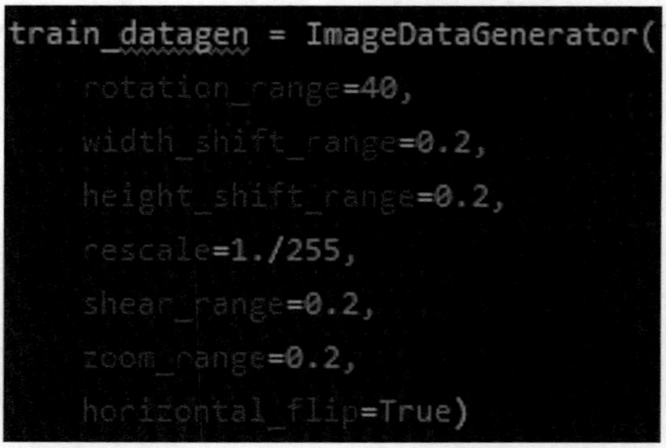

```
train_datagen = ImageDataGenerator(
    rotation_range=40,
    width_shift_range=0.2,
    height_shift_range=0.2,
    rescale=1./255,
    shear_range=0.2,
    zoom_range=0.2,
    horizontal_flip=True)
```

Fig. 6 Image augmentation parameter

Fig. 7 Sample of augmented images

The best solution is to apply different techniques to process the images such as rotation and flipping. To augment the image that we have, we are utilising the function in TensorFlow package called ImageDataGenerator. The function will automatically augment the images based on the parameters we have set (Fig. 6).

Even though image augmentation does not increase the amount of raw data that we have, the process is expected to produce enough variability to prevent our model from overfitting without losing meaning and relevant data of the signature and to increase the overall robustness of our CNN. Below is the sample of the augmented signature we have managed to obtain (Fig. 7).

3.8 Convolutional Neural Network (CNN)

For this paper, we are using a customised CNN model architecture. The configuration of this model is used based on the amount of data that we have and the problem that we are trying to solve which is to identify a writer-dependent signature. Since there were not many past literatures focusing on offline writer-dependent signature verification, we are proposing this configuration as our pilot model. In addition, we are not able to utilise the signature data available online because we are verifying the signature and classifying them to their user rather than just differentiating

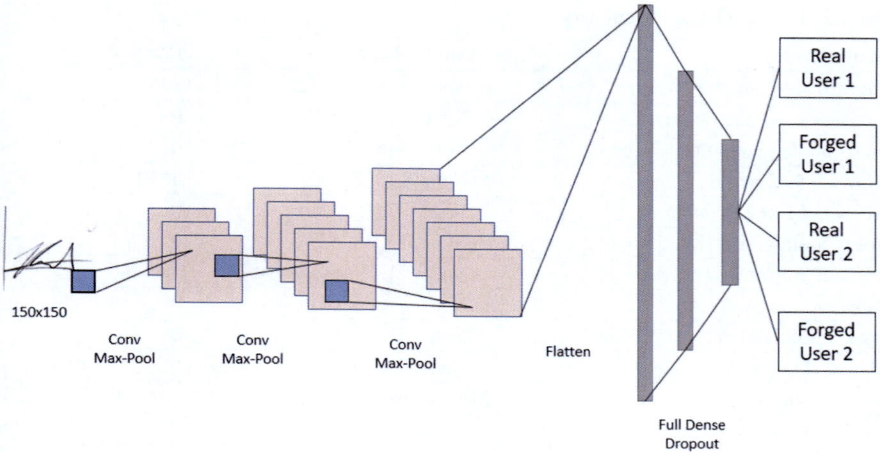

Fig. 8 Pilot model convolutional layer

between a real signature and a forged signature. To add, most of state-of-the-art CNN architectures for image classification require huge amounts of data and are therefore less suitable for our use case (Fig. 8).

Table 2 shows the summary of the architectures. It includes the output shape of kernel for each layer of parameter it has. The model that we are using consists of three convolutional layers complimented with three max pooling layers. Having three convolutional layers is expected to cover all the important parameters to learn and classify the complexity of any signature. The activation function used at each layer is the ReLU. As stated in the previous chapter, ReLU is used to improve the training speed of the model. As for the classification layer, we are using the dense neural network with the SoftMax activation function. To prevent overfitting, we implemented dropout before the final classification layers. The dropout is used to randomly ignore some number of layer output which in turn provides a different view of the configured layer despite multiple iterations. The architecture has over 2,461,476 parameters.

3.9 Training Result

To review our model performance, we are using matplotlib library in Python to visualise the training and testing progress over time. The accuracy metric is determined after the model parameters, and it is the measure of how accurate the models' prediction is compared to true data. On the other hand, loss function is used to optimise the model. The loss is measured based on training and validation process. It is the sum of the difference between predicted images and the real class of the image. Since this is a multi-class classification problem, we prioritised the accuracy of the

Table 2 Pilot model layer summary

Layer (type)	Output shape	Param #
conv2d_4 (Conv2D)	(None, 148, 148, 32)	896
activation_6 (Activation)	(None, 148, 148, 32)	0
max_pooling2d_4 (MaxPooling2	(None, 146, 72, 32)	0
conv2d_5 (Conv2D)	(None, 146, 72, 32)	4640
activation_7 (Activation)	(None, 73, 36, 32)	0
max_pooling2d_5 (MaxPooling2	(None, 73, 36, 32)	0
conv2d_6 (Conv2D)	(None, 71, 34, 64)	18496
activation_8 (Activation)	(None, 71, 34, 64)	0
max_pooling2d_6 (MaxPooling2	(None, 35, 17, 64)	0
flatten_2 (Flatten)	(None, 38080)	0
dense_3 (Dense)	(None, 64)	2437184
activation_9 (Activation)	(None, 64)	0
dropout_2 (Dropout)	(None, 64)	0
dense_4 (Dense)	(None, 4)	260
activation_10 (Activation)	(None, 4)	0
Total params: 2,461,476 Trainable params: 2,461,476 Non-trainable params: 0		

model more than its loss. To plot the graph of performance for our model, we plot two graphs which are the model accuracy graph and the model loss graph for every parameter tuning we did with the model. For the model accuracy, the y-axis is the accuracy of the model and on the x-axis is the epoch. For the loss accuracy graph, the y-axis shows the loss and the x-axis shows the epoch. The graphs provide meaningful indication of the model such as the speed of convergence and the status of the model as to whether it may be overlearning the training data.

There are few hyperparameters that we tune throughout the training process, which are the epochs, batch size, steps per epoch and dropout ratio. We run through multiple parameter setups to find the highest point of accuracy while having lowest point for model loss. Here we only include three different pairs of graphs that are significantly different from each other based on the hyperparameter tuning that we did.

One of the first hyperparameter configurations we tested was by setting the epoch at 300. As shown in Fig. 9, both graphs fail to converge. Even though we manage to achieve 100% result on the train accuracy, the test accuracy started to decline after the 50th epoch. The split of the graph explains that our model is overfitted and it is impacted by the number of epochs we set, which is 300. As we allow the model to keep on iterating after point of convergence which is around the 100th epoch, the separation becomes more obvious. This is supported by the trend on the model loss. The projection of the test set increases after each iteration. The increment could mean that there are too many parameters being introduced in this model for such a limited amount of data, which leads us to reduce the epoch and alter the dropout ratio to prevent overfitting of the data.

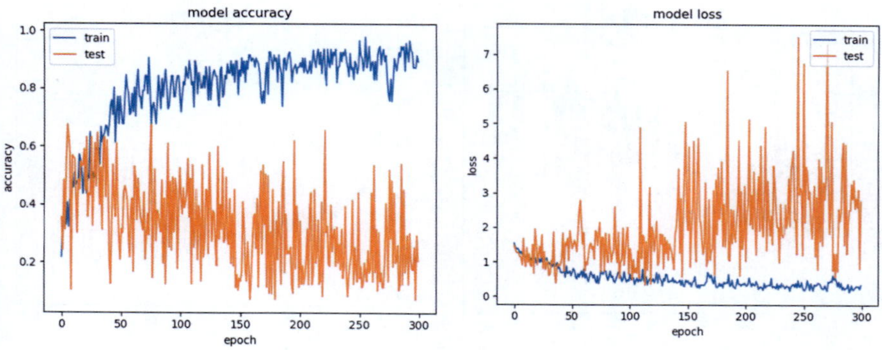

Fig. 9 Model accuracy and model loss at 300 epochs

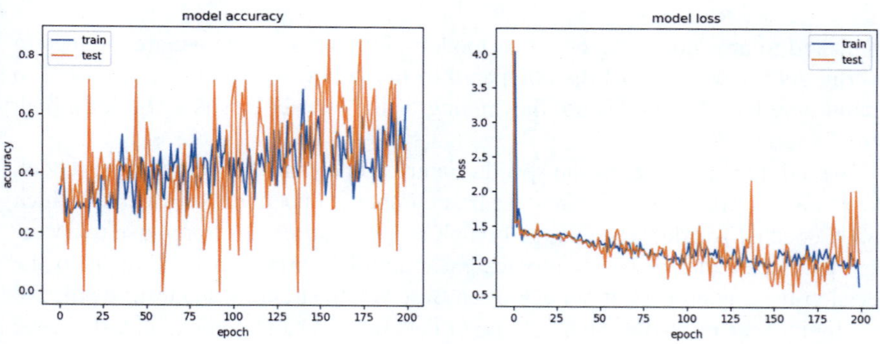

Fig. 10 Model accuracy and model loss at 200 epochs with batch size of 64 and dropout ratio of 0.2

As for the second major hyperparameter tuning, we reduced the number of epoch to 200, increased the batch size to 64 and increased the dropout ratio to 0.2. For the model accuracy basing on Fig. 10, we are able to maintain the convergence of both train and test set. This is possibly because the model is not overfitted and is still able to generalise well due to the lower number of epoch. Even so, the graph is a bit flat. This was clearly due to the increment of the batch size. This leads the model to have only 63% average accuracy. On the other hand, the model loss converges well despite the presence of spikes in the graph. This is possibly due to the increment of the dropout ratio preventing the model from overfitting. This pair of graphs highlights the evidence that the epoch and dropout ratio have a correlation to the number of data available. Having lower epoch and high dropout ratio could increase the overall accuracy and prevent the model from overfitting.

To further increase the accuracy of our model, we increase the dropout ratio from 0.25 to 0.1 while maintaining other parameters. The implication can be seen in the model accuracy where we managed to obtain 83% accuracy of the model. As shown in Fig. 11, even though the test accuracy dropped at 160th epoch, it manages to converge back at the 200th epoch. This is due to the higher number of dropout

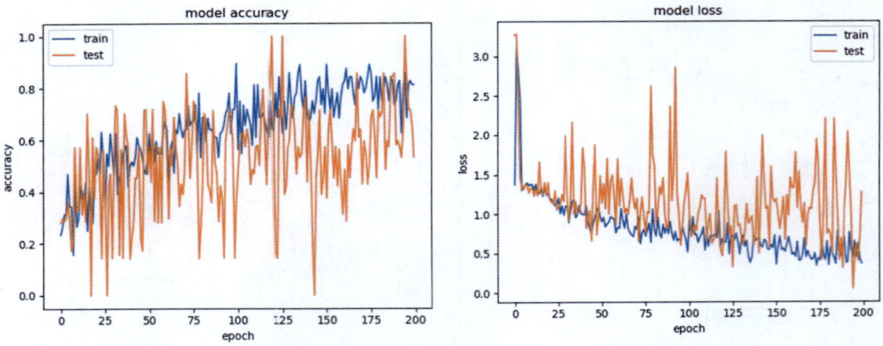

Fig. 11 Final model used for signature verification. The model epoch is at 200 with 0.1 dropout ratio

compared to previous parameter. Our models are more accurate but are inclined to overfit. As for the model loss, compared to the previous graphs, we managed to obtain only 0.49 loss. Even so, the spiking of the test set loss is a sign of a high dropout ratio.

Overall, the graphs show the speed of convergence and the status of the model. Based on that, we are able to know the best number of epoch for our use case, which is 200 epochs. To add, setting a really high dropout ratio helps to remove unnecessary parameter that does not provide meaningful information that can help the decision-making process of our classifier. Based on the training and testing process, the significance of having high volume of data becomes more apparent to improve the accuracy and at the same time increase the overall generality of the model. In the context of signature verification, the problem can be more complicated, since signatures are more prone to have a lower consistency per user. To add, the complexity increases when we are trying to classify a writer-dependent signature. This due to the limited amount of signature that we can obtain from a single user (Fig. 12).

Once the trained model is transferred to the Raspberry Pi, the user will be able to execute the signature verification and classification process through the Raspberry Pi terminal. The system is going to update the user every time a new picture is taken by the program and once the prediction result is out.

Based on Fig. 13, the picture on the right is sent before the authentication process, and image on the left is the slide-down notification on the prediction result. For our example in the figure, Class 1 means the signature belongs to real user 1. This is true and to test it further, we repeat it with seven original signatures that belong to user 1 and four fake signatures of real user 1.

Based on Table 3, the model predicted three signatures wrongly making the accuracy to fall at only 70%. This means that 20% of the time the system can falsely declare an original signature as forgery and 10% of the time it can falsely claim fake signature to be correct. The latter is considered as more dangerous error where it let forged signature to slip by. However, the rate at which this happens within an

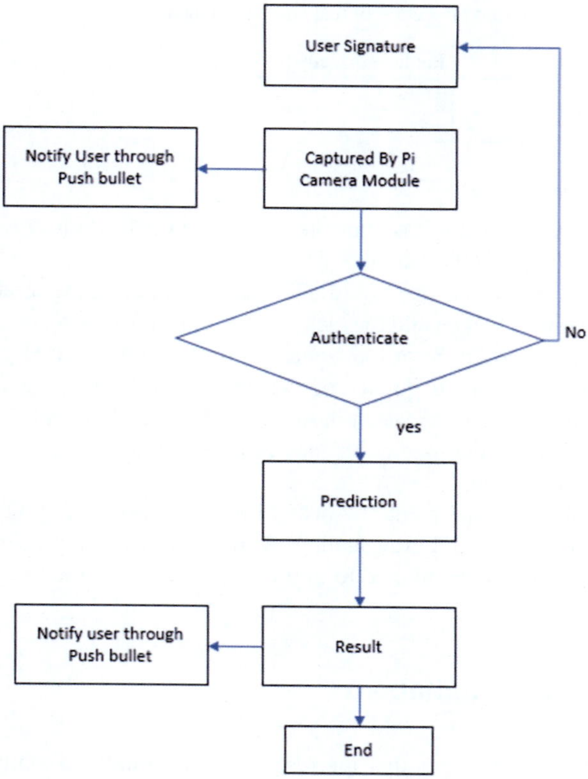

Fig. 12 User experience flowchart

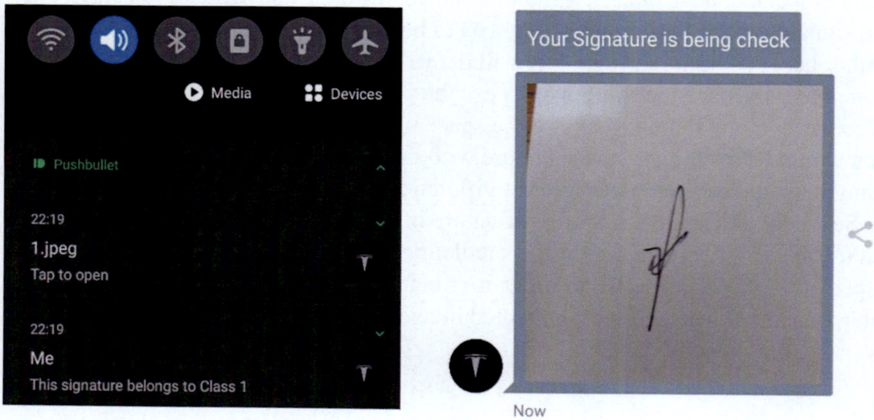

Fig. 13 Notifications from Pushbullet during program execution in Raspberry Pi

Table 3 Confusion matrix on prediction of real and fake signature

N = 1	Predict correctly	Predict wrongly
Real signature	3	2
Fake signature	4	1

organisation that does not possess a certified signature verification system is reduced by 90% with the help of this system.

There are several reasons that could affect the accuracy of the model which are the quality of the image taken and the lack of image pre-processing in the Raspberry Pi. While capturing the image of the signature, there is no standard procedure set. This could affect the overall quality of the image taken since we used scanned images as training and testing dataset during the CNN development. The difference of quality could impact the prediction accuracy. Therefore, it is best to train based on its real-life application.

To add, we did not implement any heavy image pre-processing of the captured image other than scaling and grey scaling. A more refined pre-processing method such as sharpening could be applied to improve the quality of the image.

4 Problems and Limitations

There were several challenges that the researchers encountered while conducting the experiment for this paper. The first challenge was to get a good number of signatures to be used as trainable data. Unlike recognising other common or typical daily object like dogs or cats, signatures are very limited and difficult to get. To add, the signature used to train the model has to be consistent. Having a very high variability between each real signature will defeat the purpose of the signature verification process. Hence the solution is to use the grid prepared on paper so that the size is consistent, and the user can use the space provided as a guidance. Another solution would be to compare the signature with the available accepted signature before using it as training data to ensure the difference is within an acceptable tolerance.

Secondly, due to the inconsistent nature of the signature and the volume of the data, it is very likely to overfit the predicting model. One of the solutions in this paper is to augment the data before it is being fed into the model. However, the interconnection between the augmented image is still present, and it is not as effective as getting a good raw data.

5 Conclusion

A CNN-based signature verification system on a cloud-enabled Raspberry Pi has been successfully developed. The paper has acknowledged past literature on signature verification, CNN, Raspberry Pi and the usage of cloud. It is followed by detailed explanation of training the CNN and the execution process on Raspberry Pi. The product is supposed to help the end user to detect and prevent fraudulent in workplace. Coming back to our research objectives which are:

1. To develop a suitable offline writer-dependent signature verification using CNN.
2. To integrate the verification process in an IoT architecture.

We managed to create a pilot CNN model for our use case. However, improvements are needed in terms of preparing high-quality real-life application image of the signature and reducing the complexity of the CNN model to prevent overfitting improvement as we only managed to achieve 70% accuracy on real-life application. We did manage to integrate the process in an IoT architecture with the implementation of Raspberry Pi, Pushbullet notification and ownCloud service. Although there are certain limitations that need to be addressed, the premise of having a portable signature verification tool is promising.

6 Future Research

There are many ways to improve the current system produced in this paper. First, the CNN could be implemented directly into the Raspberry Pi by using a lighter or pruned version model. By doing this, user will be able to update the model in real time without having to export the data to an external computer to be trained again. However, the data is not more secured compared to being at the server.

Secondly, having a larger dataset of user signatures will significantly help to optimise the model and to prevent overfitting from happening. Since we are only limiting the signatures to only two people, in the future, more user signatures should be introduced to the model to improve its usability in a real scenario.

Thirdly, a better camera model could be used to produce a higher resolution image of the signature. This research did not compare the specification of the hardware or the quality of the image against the accuracy of the model. Future research could develop or propose a baseline or minimum specification of imaging hardware and quality to be used for the signature verification process.

Acknowledgement The authors would like to thank Aerospace Malaysia Innovation Centre for the hospitality to conduct the research and the University of Malaya for guiding throughout the completion of this paper.

References

1. PricewaterhouseCoopers (PWC). https://www.pwc.com/my/en/publications/2018/2018-gecfs-sea.html
2. Sharma, N., Jain, V., & Mishra, A. (2018). An analysis of convolutional neural networks for image classification. *Procedia Computer Science, 132*(Iccids), 377–384. https://doi.org/10.1016/j.procs.2018.05.198
3. Rivard, D., Granger, E., & Sabourin, R. (2013). Multi-feature extraction and selection in writer-independent off-line signature verification. *International Journal on Document Analysis and Recognition, 16*(1), 83–103. https://doi.org/10.1007/s10032-011-0180-6
4. Pal, S., Blumenstein, M., & Pal, U. (2011). *Off-line signature verification systems: A survey*. In International conference & workshop on emerging trends in technology 2011, ICWET 2011 – Conference proceedings (Vol. 1, pp. 652–657). https://doi.org/10.1145/1980022.1980163.
5. Justino, E. J. R., Bortolozzi, F., & Sabourin, R. (2005). A comparison of SVM and HMM classifiers in the off-line signature verification. *Pattern Recognition Letters, 26*(9), 1377–1385. https://doi.org/10.1016/j.patrec.2004.11.015
6. Zhang, Z., Wang, K., & Wang, Y. (2011). *A survey of on-line signature verification. Lecture notes in computer sciences (including subseries lecture notes in artificial intelligence lecture notes bioinformatics)*, 7098 LNCS (pp. 141–149). https://doi.org/10.1007/978-3-642-25449-9_18.
7. Fukushima, K. (1980). Neocognitron: A self-organizing neural network model for a mechanism of pattern recognition unaffected by shift in position. *Biological Cybernetics, 36*(4), 193–202. https://doi.org/10.1007/BF00344251
8. Hubel, D. H., & Wiesel, T. N. (1962). Receptive fields, binocular interaction and functional architecture in the cat's visual cortex. *Journal of Physiology, 160*(1), 106–154. https://doi.org/10.1113/jphysiol.1962.sp006837
9. Weng, J., Ahuja, N., & Huang, T. S. (1997). Learning recognition and segmentation using the cresceptron. *International Journal of Computer Vision, 25*(2), 109–143. https://doi.org/10.1023/A:1007967800668
10. O'Shea, K., & Nash, R. (2015). *An introduction to convolutional neural networks* (pp. 1–11) http://arxiv.org/abs/1511.08458.
11. Venkatesan, R., & Li, B. (2018). *Convolutional neural networks in visual computing: A concise guide*. CRC Press.
12. Goodfellow, I., Bengio, Y., & Courville, A. (2016). *Deep learning*. MIT Press.
13. Rawat, W., & Wang, Z. (2017). Deep convolutional neural networks for image classification: A comprehensive review. *Neural Computation, 29*(9), 2352–2449. https://doi.org/10.1162/neco_a_00990
14. Basha, S. H. S., Dubey, S. R., Pulabaigari, V., & Mukherjee, S. (2020). Impact of fully connected layers on performance of convolutional neural networks for image classification. *Neurocomputing, 378*(April), 112–119. https://doi.org/10.1016/j.neucom.2019.10.008
15. Thakur, N., & Maheshwari, D. (2017). A review of image classification techniques. *International Research Journal of Engineering and Technology, 4*(11), 1588–1591.
16. Lee, A. (2015). *Comparing deep neural networks and traditional vision algorithms in mobile robotics*. Pdfs.Semanticscholar.Org. https://pdfs.semanticscholar.org/1b6f/569b79721037425fca034c7ae47904fb9276.pdf.
17. O'Mahony, N., et al. (2020). Deep learning vs. Traditional computer vision. *Advances in Intelligent Systems and Computing, 943*(Cv), 128–144. https://doi.org/10.1007/978-3-030-17795-9_10
18. Julita, A., Fauziyah, S., Azlina, O., Mardiana, B., Hazura, H., & Zahariah, A. M. (2009). *Online signature verification system*. In Proceedings of the 2009 5th International colloquium on signal processing and its applications. CSPA 2009, April 2015 (pp. 8–12). https://doi.org/10.1109/CSPA.2009.5069177.

19. Sae-Bae, N., & Memon, N. (2014). Online signature verification on mobile devices. *IEEE Transactions on Information Forensics and Security, 9*(6), 933–947. https://doi.org/10.1109/TIFS.2014.2316472
20. Khalajzadeh, H., Mansouri, M., & Teshnehlab, M., (2017). *Persian signature verification using fully convolutional networks* (Vol. 1, issue 2, pp. 7–12).
21. Mohapatra, R. K., Shaswat, K., & Kedia, S., (2019, November). *Offline handwritten signature verification using CNN inspired by inception V1 architecture*. In Proceedings of the IEEE International conference on image information processing (pp. 263–267). https://doi.org/10.1109/ICIIP47207.2019.8985925.
22. Hafemann, L. G., Sabourin, R., & Oliveira, L. S. (2017). Learning features for offline handwritten signature verification using deep convolutional neural networks. *Pattern Recognition, 70*, 163–176. https://doi.org/10.1016/j.patcog.2017.05.012
23. Cozzens, B., et al. (2019). *Signature verification using convolutional neural network*. In 2019 IEEE International conference on robotics, automation, artificial-intelligence and internet-of-things, RAAICON 2019 (pp. 35–39). https://doi.org/10.1109/RAAICON48939.2019.19.
24. Dey, S., Dutta, A., Toledo, J. I., Ghosh, S. K., Llados, J., & Pal, U. (2017). *SigNet: Convolutional Siamese Network for writer independent offline signature verification* (Vol. 1, pp.1–7), http://arxiv.org/abs/1707.02131.
25. Sronothara, A. B., & Hanmandlu, M. (2018). Off-line signature verification using CNN. *International Journal on Future Revolution in Computer Science & Communication Engineering, 28*(12), 1407–1414. https://doi.org/10.1016/j.patrec.2007.02.016
26. Kumar, P. M., & Devi Gandhi, U. (2018). A novel three-tier Internet of Things architecture with machine learning algorithm for early detection of heart diseases. *Computers & Electrical Engineering, 65*, 222–235. https://doi.org/10.1016/j.compeleceng.2017.09.001
27. Rajan, J. P., Rajan, S. E., Martis, R. J., & Panigrahi, B. K. (2020). Fog computing employed computer aided cancer classification system using deep neural network in Internet of Things based healthcare system. *Journal of Medical Systems, 44*(2). https://doi.org/10.1007/s10916-019-1500-5
28. Maksimović, M., Vujović, V., Davidović, N., Milošević, V., & Perišić, B. (2014). Raspberry Pi as Internet of Things hardware: Performances and constraints. *Des. Issues, 3*(June), 8.
29. Dürr, O., Pauchard, Y., Browarnik, D., Axthelm, R., & Loeser, M. (2015, January). *Deep learning on a Raspberry Pi for real time face recognition*. In Eurographics (p. 4), https://doi.org/10.2312/egp.20151036.
30. Monteiro, A., De Oliveira, M., De Oliveira, R., & Da Silva, T. (2018). Embedded application of convolutional neural networks on Raspberry Pi for SHM. *Electronics Letters, 54*(11), 680–682. https://doi.org/10.1049/el.2018.0877
31. Nikouei, S. Y., Chen, Y., Song, S., Xu, R., Choi, B. Y., & Faughnan, T. R. (2018). *Real-time human detection as an edge service enabled by a lightweight CNN*. In Proceedings of the – 2018 IEEE international conference edge computing EDGE 2018 – Part 2018 IEEE World Congress on services (pp.125–129). https://doi.org/10.1109/EDGE.2018.00025.
32. Princy, S. E., & Nigel, K. G. J. (2016). *Implementation of cloud server for real time data storage using Raspberry Pi*. In IC-GET 2015 – Proceedings of the. 2015 online international conference green engineering and technologies (p. 3). https://doi.org/10.1109/GET.2015.7453790.
33. K. Pimple, J. Shubham, R. Gaurav, G. Pooja, and D. Gaurav, 2017. Deploying and extending on-premise cloud storage based on own cloud. *International Journal of Computer Science Trends and Technology*, 5(2), 440–442, doi: 8.
34. Ikuomola, A. J. (2019). An embedded cloud-based video surveillance system. *Computing, Information Systems & Development Informatics Journal, 10*(1), 1–6. https://doi.org/10.22624/aims/cisdi/v10n1p1.

Machine to Machine (M2M), Radio-frequency Identification (RFID), and Software-Defined Networking (SDN): Facilitators of the Internet of Things

S. Sharmila and S. Vijayarani

1 Introduction

In the current era of networking, seamless connectivity and ubiquitous computing are not a great challenge. Human and machine communication was restricted at the initial stage of Internet growth, and the advancement in Internet technology has made communication between everything as in the IoT. It is the latest technology that creates a universal network of devices and machines embedded with software that is capable of exchanging and communicating information with each other via the Internet [1]. The significant aspect of the IoT is that it can renovate an object into an intelligent smart object by providing actuating, communicating, sensing, and computing ability to the object.

The IoT generates data from the connected object, and it is used for further analysis and decision-making. The exponential escalation in Internet connectivity and the gadgets has necessitated the IoT that bridges the gap among the objects and services [2]. In the IoT, an object determines everything that can communicate or not and the flow of an event, as well as data created by the interconnection of smart things, facilitating the process of management, control, decision-making, tracking, and coordination. Bind of heterogeneous techniques and significance has made the IoT as a great success [3]. Figure 1 shows the significance of the IoT.

IoT paradigms are framed with the assistance of M2M, RFID, and SDN technologies. The flaws met in the design of applications are rectified using these technologies. The functionality of the industrial, healthcare, and data transmission applications is accomplished by the facilitators, namely, M2M, RFID, and SDN. Each technology has a unique feature and is applied based on the requirement

S. Sharmila (✉) · S. Vijayarani
Department of Computer Science, Bharathiar University, Coimbatore, Tamilnadu, India
e-mail: vijayarani@buc.edu.in

S. Pal et al. (eds.), *Artificial Intelligence-based Internet of Things Systems*,
Internet of Things, https://doi.org/10.1007/978-3-030-87059-1_8

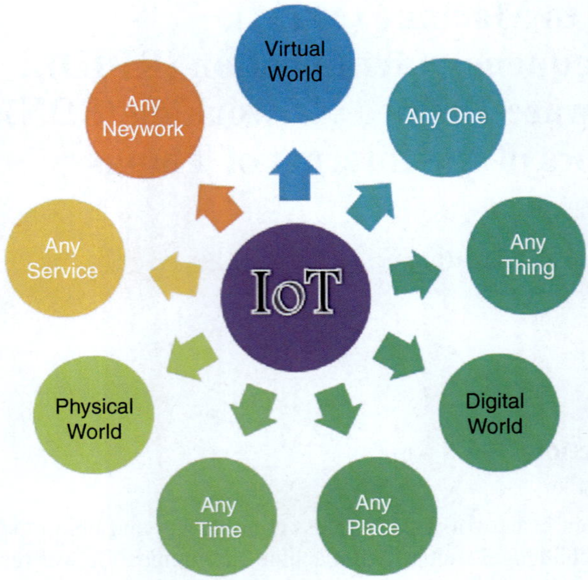

Fig. 1 Important features and significance of the IoT

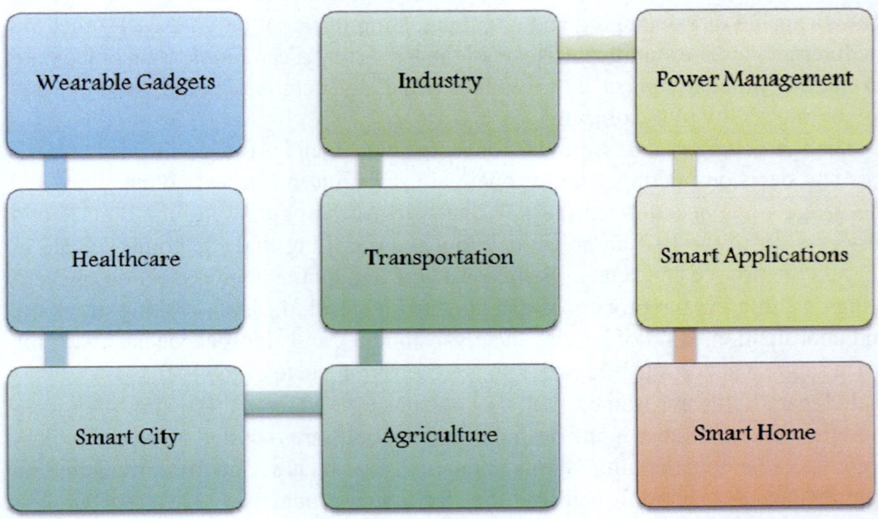

Fig. 2 Applications of the IoT

of the application. The IoT is widely used in smart home, industry, automation, and healthcare [4], and it is shown in Fig. 2.

M2M is a combination of communication system and information technology with machines that provide the data transmission facility with reduced human intervention. In the context of M2M, the transmission of information is attained by the network of communication. The functionality of the entire device is accomplished by the information exchanged among the devices. The information is generated and exchanged among the devices that provide significant inference and for the formulation of a conclusion. M2M achieves a ubiquitous communication system with complete mechanical automation. The framework of M2M is widely used in the applications of e-healthcare, smart grids, industrial automation, intelligent transportation system, environmental monitoring, and smart cities [5, 6].

RFID is a data transmission approach, which is based on wireless technology, and it captures the data from varied identification attributes. RFID is a standardized scheme, and it is used in various economic and tracking sectors [7]. It is also employed in the field of access control, traceability, and logistics. The key advantages of RFID are identification, unitary, low cost of tags, and wireless communication that provide the application with significant practical benefit and progression in terms of applications and concept [8].

SDN is an innovative paradigm based on a wireless network that has changed the control and management aspects of the networking system by developing them simpler and flexible. The main intent of SDN is to segregate the management and control plane from the data plane by initiating the necessary protocol into the system. The scalability and the performance of the network are enriched by the features of SDN to attain better transmission. It provides effective transmission architecture and widely used in the data transmission paradigm [9, 10].

The IoT is a combination of intelligent objects, and the sensed raw data through the objects are transmitted to the cloud for further processing and decision-making. The processed outcome that will be displayed at the user terminal is represented in Fig. 3.

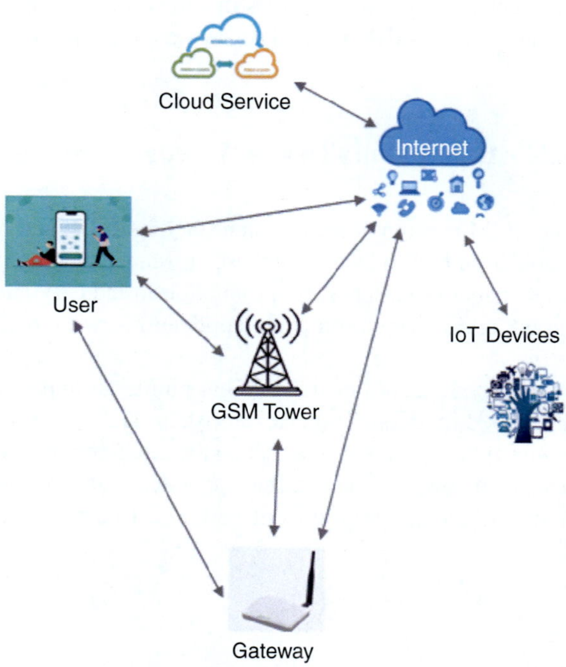

Fig. 3 Overall framework of the IoT

This chapter is organized as follows: Sect. 2 illustrates the facilitators of the IoT, Sect. 3 discusses the Machine 2 Machine communication, Sect. 4 illustrates the Radio-Frequency Identification (RDIF), Sect. 5 describes the Software-Defined Networking (SDN), Sect. 6 discusses the issues and challenges, and Sect. 7 concludes the facilitators of the IoT.

2 Facilitators of the IoT

The interrelated electronic devices, digital machines, and software in the IoT can transmit or exchange information over the connected network without the requirement of human to the computer or human to human interaction. The transmission is achieved with the assistance of prominent technologies, namely, machine to machine (M2M), radio-frequency identification (RFID), and software-defined networking (SDN). The significance of these technologies and their importance in the application are discussed in this section.

3 Machine to Machine (M2M)

The autonomous communication system among two electronic devices without the involvement of human is determined as machine to machine (M2M) communication. The rapid adoption and emergence of wireless approach, the pervasiveness of the electronic system, and the software system complexity have attracted the M2M system that is widely used in academia and industry [11].

3.1 Autonomous Device Management System via M2M

The M2M communication system facilitates the autonomous and ubiquitous communication facility, which embarks mechanical automation with potential features. M2M network enriches the production in industry and has shown effectiveness in varied areas. The promising applications with M2M technology are outlined in Fig. 4.

As illustrated in Fig. 4, M2M has numerous applications, and the devices require diversified functions and characteristics. The nature of nodes differs for every individual application that is simple to the intelligent node. The M2M nodes are categorized from low, mid, and end sensor nodes in accordance with the hardware capacity. According to the requirement and based on the variety of nodes as well as the

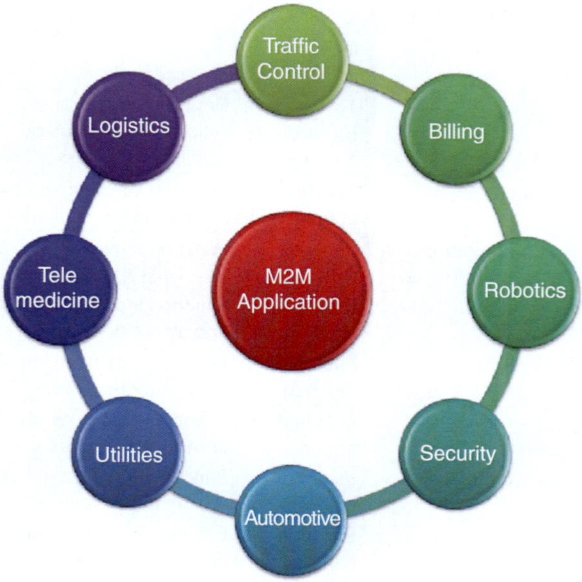

Fig. 4 Applications of M2M communication

capability, the necessity of function will be determined. The M2M node category is summarized and their functions are categorized in Table 1.

In the paradigm of machine automation, the machines perform the operations such as

- Optimizing the parameters that are relevant to the consumption of power, monitoring, and the distribution of multimedia at homes.
- Weather monitoring and production of huge products in the giant industry.
- Maintenance time of a certain variety of products and developers.

The M2M communication system can enrich the performance and satisfaction of customers, and the productivity of the industry is also enriched. The variation between the M2M and IoT is illustrated in Table 2.

3.2 *Collaborative Infrastructure Formulated by M2M*

The system model of M2M is composed of M2M device domain, M2M network area domain, and M2M user or admin domain, which are shown in Fig. 3. The domains in the M2M model are interlinked and they work together.

Table 1 Categories of M2M senor nodes

Category of node		High-end node	Mid-end node	Low-end node
Application service		Biomedical, Military	Home network, automation in industry, management of asset	Environment protection
Function	Data aggregation	Function not required	Function may be required or may not be required	Function required
	Autoconfiguration	Function required	Function required	Function required
	Power saving	Function required	Function may be required or may not be required	Function required
	Localization	Function required	Function may be required or may not be required	Function not required
	Quality of service support	Function required	Function may be required or may not be required	Function not required
	Localization	Function required	Function may be required or may not be required	Function not required
	Transmission control protocol/Internet protocol	Function required	Function may be required or may not be required	Function not required
	Power control	Function required	Function may be required or may not be required	Function not required
	Traffic control	Function required	Function may be required or may not be required	Function not required
Characteristics	Density	Low	Mid	High
	Complexity	High	Mid	Low
	Energy efficiency	Low	Mid	High
	Cost	High	Mid	Low
	Scalability	Low	Mid	High
	Internet protocol	IP	IP	Non-IP
	Mobility	Mobility	Low mobility	Low, almost static
	Hop count	Low	Mid	High
	Intelligence	High	Mid	Low

3.2.1 M2M Device Domain

It is the collaboration of a huge number of devices, electronic components (sensors, smart meters, and actuators), and gateways (concentrators and point of data aggregation). The sensory information is collected from the atmospheric condition and from the varied parts of the M2M area domain that collaboratively formulate intelligent decisions to broadcast the data to the gateway. The gateway receives and manages the received sensed data that acts as an intelligent device. The data transmission is attained through a single- or multihop channel through network area domain to the back-end user/admin server. The device instilled in the M2M varies based on the type of application.

Table 2 Differentiation of M2M and the IoT

Machine to machine communication (M2M)	Internet of Things (IoT)
M2M permits machines to relay or transmit information based on the necessity with the assistance of protocol	The object in the environment is connected using the IoT. The objects connected to the Internet without any intelligence brought into the IoT through smartphones, which act as a gateway between objects and the Internet
M2M works on the point to point transmission that is an embedded module of hardware or mobile communication system	The IoT depends on the network with IP that bridges device data to middleware or cloud platform
Big data is handled by the hardware reliance in M2M	Embedded software dependence makes the IoT to handle big data in an easy way
The generation and utilization of data only from the electronic device	Sensor data and devices are integrated into the system
Example: Data transmission with machines and electronic devices, namely, washing machine, refrigerator, and smart meters via Internet protocol (IP) over a wireless or wired medium	Example: Near-field communication (NFC) enables the facility of interaction with advertisements in the newspaper through movie posters or shortcodes
M2M is denoted as plumbing due to the limited scalability	The IoT utilizes various standards with high scalability, and it is universal enablers

3.2.2 M2M Network Area Domain

The M2M network area domain acts as an interface between the M2M device domain and the M2M user/admin domain. The network protocols are employed to provide reliable and cost-effective device with extensive coverage areas that transmit the sensory information from the M2M device to the user/admin domain.

3.2.3 M2M User or Admin Domain

The user/admin domain is composed of clients in the M2M application and a back-end server (BS). The key component of BS is the M2M system, and it is the integration point of sensory information storage that is propagated from the device domain. This domain also offers the real-time observation of data from numerous client applications for the remote monitoring management (RMM) system. In smart grids, the control center works as the BS, and the M2M health observation server acts as the BS in healthcare application. The BS will vary based on the nature of the application and requirement.

The M2M communication scenario is categorized as a client/server model and the peer-to-peer model. The client/server model establishes communication between the M2M devices with one or more M2M servers. In the scenario of the client/server model, the electronic devices rely on the M2M device domain and servers placed within the application domain. It is mostly used in environmental monitoring, smart

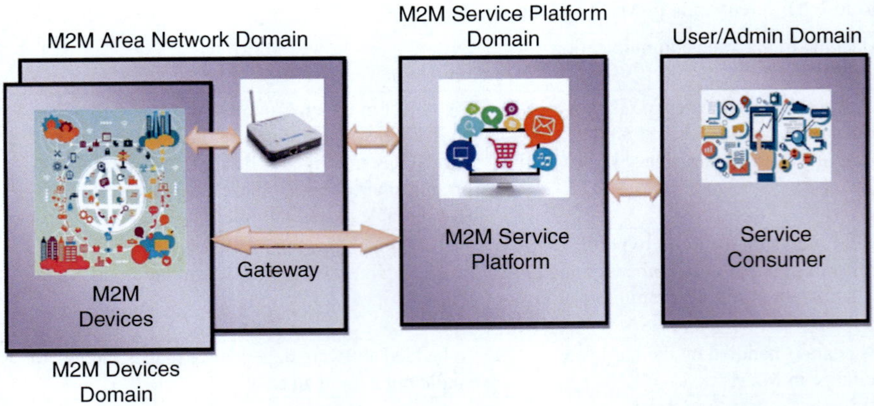

Fig. 5 Collaborative infrastructure of M2M communication

Table 3 M2M network key technology

Standards	Data rate	Used in	Application
Bluetooth	Low	(Personal area network) PAN	Sharing of music and video files
802.15.6	Low	(Body area network) BAN	e-Healthcare
Zigbee	Low	PAN	Automatic control
Ultra-wideband	High	PAN	Live video streamlining
Wi-Fi	High	(Local area network) LAN	Laptop connectivity and smart metering

electrical power grid, and home automation. Other communication paradigms use the peer-to-peer model, whereby the devices transmit directly among them. This type of communication is signified as inter-M2M device transmission that can be either via an ad hoc or by the mobile network mode [12]. The collaborative infrastructure is shown in Fig. 5, and the key technology of M2M is shown in Table 3.

3.3 Significance of M2M in the IoT

M2M communication system utilizes a licensed spectrum and long-distance communication technology, which has numerous benefits to industry, academia, and user community. The network deployment cost is low and network coverage is better in M2M [13, 14]. The significance of M2M in the IoT is described below:

- The communications among the electronic devices are standardized using the 3GPP technology, and it facilitates the machine-type communication (MTC).
- M2M plays a vital role in many applications and business; the operational efficiency is enriched.

- The installation of M2M technology is simple and easy to maintain.
- The establishment of M2M-based devices is highly scalable, reliable, and cost-effective.
- M2M offers higher throughput, minimum latency, less energy utilization, and a high range of compatibility.
- M2M facilitates high security and privacy features.
- M2M offers massive concurrent device communication and reduces bursty traffic.
- The priority scheduling and mechanism of access control is available in the M2M communication system.

Due to the significance of the M2M technology, it has been widely used in many commercial applications, and their characteristics are given in Table 4.

Table 4 M2M and their commercial platforms

Platform	Application protocol	Characteristics	Application interface	Feature
Cosm	Hypertext Transfer Protocol	Web application, mobile application, device sharing, user management, data analysis	Representational state transfer	Real-time data, open-source application, analyze automatic control and monitoring
ThingSpeak	Hypertext Transfer Protocol	Mobile application, device sharing, user management, data analysis	Representational state transfer	Real-time data, open-source application, analyze automatic control and monitoring
Nimbits	Hypertext Transfer Protocol	Mobile application, device sharing, user management	Representational state transfer	ioBridge, sharing data points and Google Cloud
EVERYTHING	Hypertext Transfer Protocol	Web application, mobile application, device sharing, user management	Representational state transfer	Active digital identification

(continued)

Table 4 (continued)

Platform	Application protocol	Characteristics	Application interface	Feature
Sensinode	Hypertext Transfer Protocol/Constrained Application Protocol	Web application, user management, M2M area network support	Representational state transfer	Nano stack, Zigbee
One Platform	Hypertext Transfer Protocol	Mobile application, user management, data analysis, M2M area network support	Representational state transfer	Cloud-based service
Axeda	Hypertext Transfer Protocol Secure	Web application, mobile application, user management, data analysis, M2M area network support	Representational state transfer/ Simple Object Access Protocol	Axeda Wireless Protocol, cloud-based service
SensorCloud	Hypertext Transfer Protocol Secure	Mobile application, user management, data analysis, M2M area network support	Representational state transfer	MicroStrain Sensor, Math Engine
BugSwarm	Hypertext Transfer Protocol	Web application, device sharing	Representational state transfer	Linux support
NeuAer	Hypertext Transfer Protocol	Mobile application, device sharing	Representational state transfer	Rules and Tag
iDigi	Hypertext Transfer Protocol	Web application, mobile application, data analysis, M2M area network support	Representational state transfer	iDigi product and Zigbee

4 Radio-frequency Identification (RFID)

Radio-frequency identification (RFID) system is based on the wireless communication technology, and it captures the information that is linked with varied recognition attributes of entities having RFID labels [15]. The process of data collection and transmission is attained by electromagnetic waves between RFID tags and readers (interrogators). The labeling granularity is accomplished by automatic identification and data capture (Auto-ID). RFID allocates diverse identification codes for related items, and better visibility is provided with the various levels of identification in the process of manufacturing and logistical. The effectiveness of the tracking and processing system is attained by the RFID technology [16]. Various applications of RFID technologies are illustrated in Fig. 6. The advancement of RFID facilitates the digital data embedded in the tags to read and capture via radio waves. The different kinds of tags and their features are given in Table 5 and Fig. 7.

4.1 Framework of RFID

The RFID is used in many application sectors, and it works on the 3C concept. The facets of RFID mechanism include context, capture, and control, whereby the process of tracking and monitoring is achieved. In the context, the RFID integration in the environmental condition is explored and the RFID tags will work. The constraint in communication and processing environments such as obstacles, reflection,

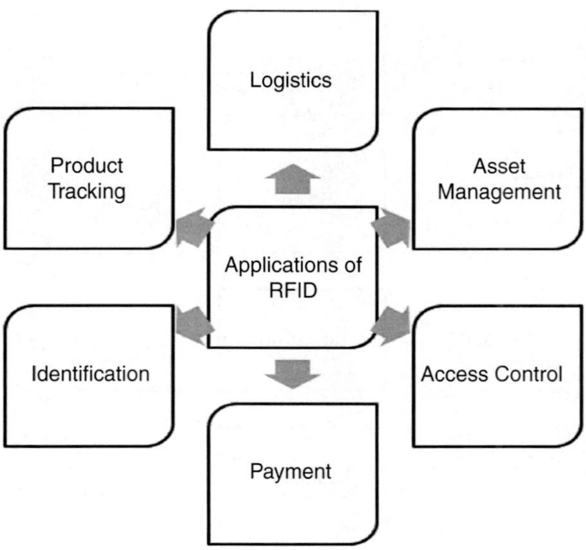

Fig. 6 Applications of RFID

Table 5 Features of RFID tags

Kind of tag	Features		
	Passive tag	Active tag	Semi-passive tag
Internal power source	No	Yes	Yes
The carrier wave backscattered from the reader that is attained by signal	Yes	No	Yes
Response	Weaker	Stronger	Stronger
Size	Small	Big	Medium
Cost	Less expensive	More expensive	Less expensive
Potential shell life	Longer	Shorter	Longer
Range	10 cm to few mts	100 mts	100 mts
Sensors	No	Yes	Yes

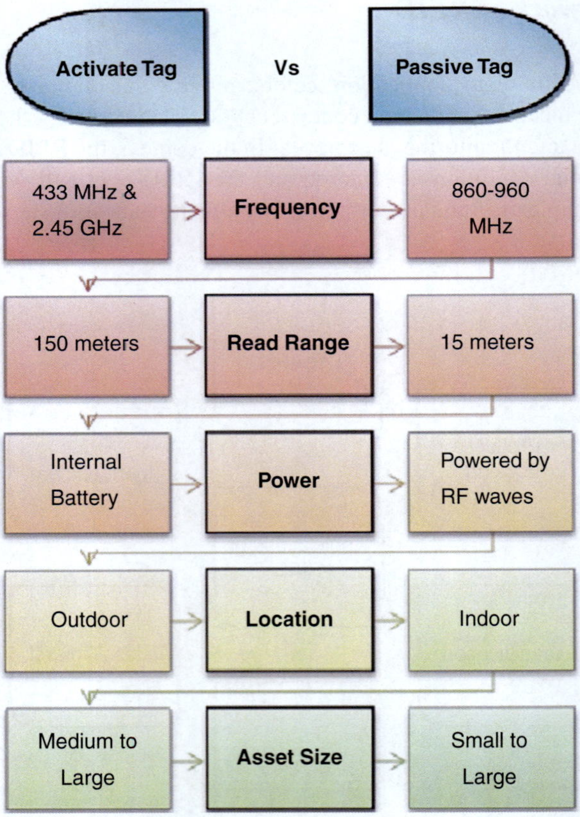

Fig. 7 Active RFID vs passive RFID

and interface is investigated. The environmental situation is identified and the processing is adjusted for the situation.

The election of equipment in RFID (readers and tags) is done in the capture process, whereas accurate identification of data is attained by the process of exploration and environment. In this phenomenon, the necessary conditions such as reading range of RFID tags, location of antenna, the privacy of data, security issues, power control, and operation frequencies are adjusted based on the necessity of zone. The real-time control system and the implementation of the business rule are administrated [17, 18]. The middleware is correlated with the enterprise system, graphical user interface, and database in the capture facet. Generally, the following components are incorporated into the RFID system:

- Unique Electronic Product Code (EPC) per firm will be assigned and the RFID tags are fixed to the item.
- Databases and networked RFID readers are instilled in real time.
- The information is exchanged via RFID antennas among the middleware, wireless networks with tag, and control platforms.

The RFID tags are categorized as active and passive that is powered by the energy source from the electromagnetic waves, and it is radiated by the backscattering method from the reader antennas of RFID. The storage and communication ability are limited, and the transmission is also not possible on its own in the passive tag. The data reading is possible only between the ranges of 0.6 to 3 m. The on-board long-life system of battery provides power supply for active tag, and the energy supply is sufficient in this system, which permits the independent data transmission within an immense range (90 meters approximately). The transmission frequency range determines the benefit of RFID technology. The RFID technology is categorized based on the frequency range, and their benefit also varies for every range. The frequency range and their benefits are explained in Table 6.

The communication is developed by the antennas, and there are numerous kinds of antennas in an active and passive RFID system. Based on the necessity of the beam width, antennas are selected. Figure 8 displays the framework of an RFID system. The RFID tags sense the atmospheric condition and transmit the sensory information. The RFID reader reads the data and the information is processed by the local software and at the backend system.

Table 6 Various kinds of RFID

Type of RFID	Benefit
Low frequency (LF)	The communication is established in the metal or water surface
High frequency (HF)	Developed based on global standards. Tags are very flat with a diameter of 1 cm
Ultrahigh frequency	Multiple tags can be read and manufactured by the ISO standard
Microwave frequency	Tags are very small in size

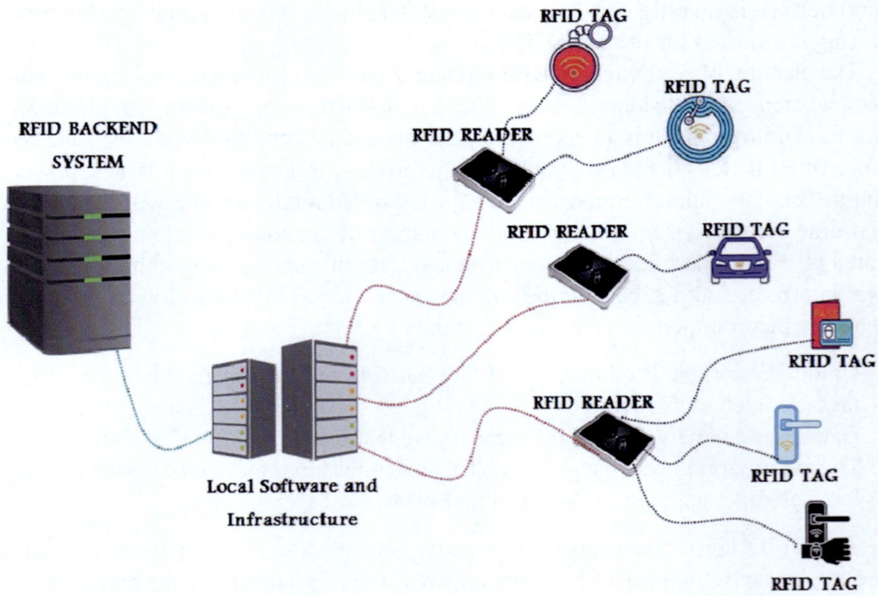

Fig. 8 Overall system design of RFID

4.2 *Significance of RFID in the IoT*

The RFID uses the electromagnetic signal to identify, capture, and track the tag attached to the object [19]. RFID has made automation in various sectors and also shows proficiency in several aspects that are explained in this section. The significance of RFID is depicted in Fig. 9.

- The data collection process in RFID technology is highly automated, whereas it minimizes the occurrence of error and effort of humans.
- RFID accepts the tag reading by item scan or line of prospect is not needed.
- Multiple RFID tags can be recognized and read by the RFID readers that increase the efficiency of RFID.
- The presence of any item with RFID tags contained in a specific range of environments can be identified instantly and matched with the relevant data in the database.
- In the case of assets, identified assets are cross-referred against the preferred location and the recorded information such as presence, relocation, and missing data are verified.
- The fixed readers and active scanners are integrated with RFID to make the complete automation of the tracking system.

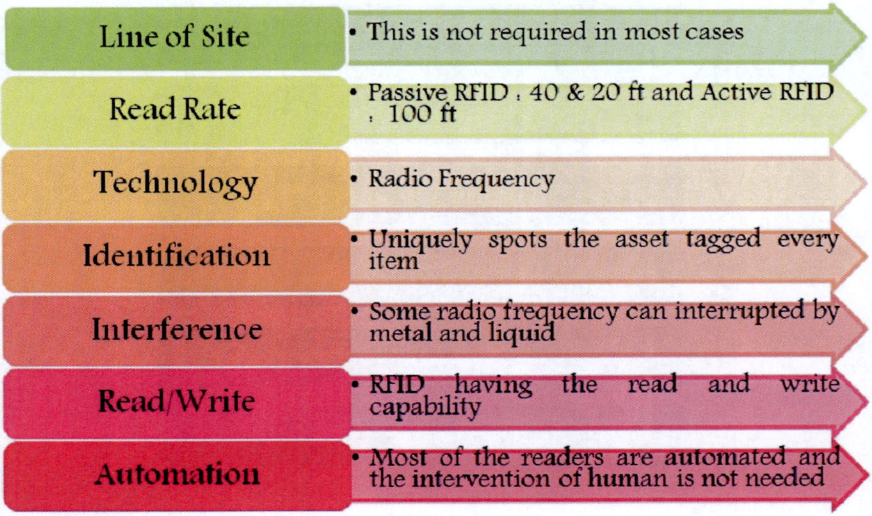

Fig. 9 Significant features of RFID

- RFID system makes the complete automation of emergency planning and arranging security solutions in the industrial and other business sectors.
- The objects and employees can be tracked and maintained with the assistance of RFID tags and readers.
- The RFID-attached asset control system enriches the security premises and also provides automation in the maintenance of products.

5 Software-Defined Networking (SDN)

Software-defined networking (SDN) is a network management technique, which facilitates the configuration of effective network programmatically and dynamically, and it enhances the performance of the network and monitoring schemes. SDN addresses the issues in static architecture, and it is generally associated with the OpenFlow protocol [20]. SDN technology is widely applied in the industrial sector and attained significance in the performance. The architecture of SDN decouples the data forwarding function and control of the network. The SDN is agile, directly programmable, open standard, managed centrally, and configured programmatically [21, 22]. Figure 10 shows various use cases of the software-defined network.

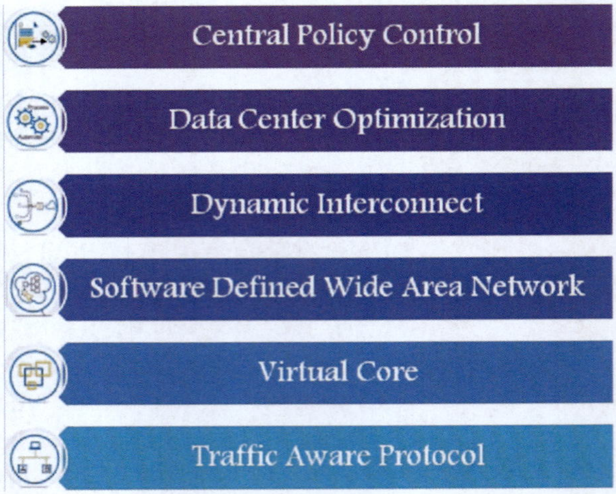

Fig. 10 SDN use cases

5.1 Three-Layered SDN Architecture

SDN is a network that has changed the control and management view of the networking system by developing them simpler and flexible. In SDN, the framework segregates the management and control plane from the data plane by initiating the necessary protocol into the system. It provides effective transmission architecture and is widely used in most cases of data transmission paradigm [23, 24]. The three-layered architecture of SDN is displayed in Fig. 11.

5.2 SDN Application Plane

The SDN application plane communicates directly to their necessity of network and determined behavior of network that controls through northbound interface (NBI). The decision-making is attained by the abstract data view, and it has one application logic as well as more than one NBI driver. The process of controlling the network is achieved by the relevant NBI agents.

5.3 SDN Control Plane

The SDN controllers lie in the mid of framework, and it is centralized logically which is in charge of the following:

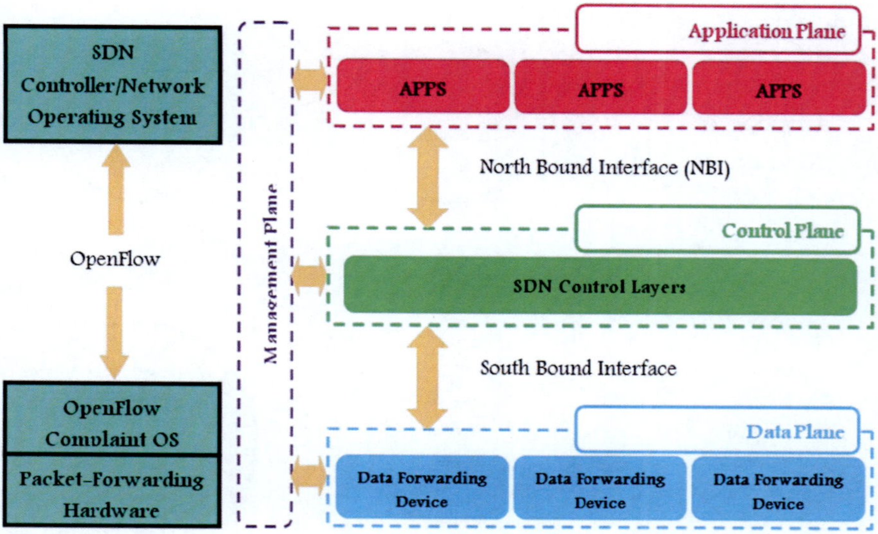

Fig. 11 Three-layered SDN framework

- The necessities are translated from the application plane to data plane.
- An abstract view of the network is provided by this plane.

It acts as an intermediary plane and controls the data flow among the planes.

5.4 SDN Data Plane

The SDN data plane permits the forwarding of data and maintains the traffic between the interfaces that reside outside the plane and the internal processing unit. This plane precludes the details of implementation such as maintenance of physical resources, slicing or visualizing the data, and mapping the logical and physical data [25].

5.5 Significance of SDN in the IoT

The software-defined network has numerous advantages in the integration of the IoT and data transmission network that is explained below [26]. Figure 12 illustrates the significance of SDN in the IoT.

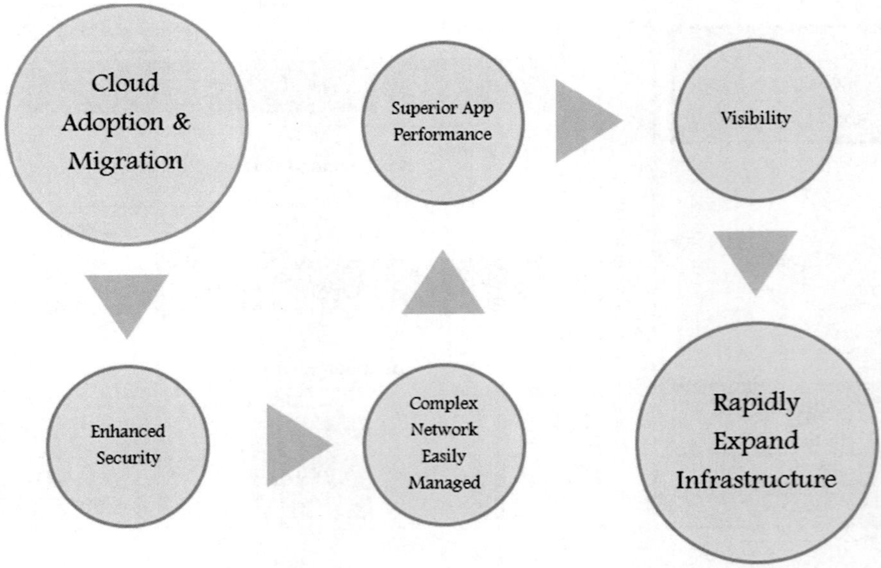

Fig. 12 Importance of SDN

5.6 Central and Standardized Policy

SDN incorporated the standard policy across the connected network. In the public and private cloud, data policy is utilized to manage the data process and management.

5.7 Single Plane of Glass

In the context of cloud, SDN act as a single plane of glass that facilitates the simple way of data transmission and processing.

5.8 Native Integration

The effective integration of native software with the cloud service and relevant software constructs is achieved.

5.9 Visibility and Analytics

SDN can perform visibility and analytics into the network and relevant applications.

5.10 Data Integration

SDN could attain the integration of the native system with the platform of cloud management.

5.11 Rapid Cloud Adoption

The integration of software, network, and cloud adoption is attained with minimum disturbance.

6 Issues and Challenges of the IoT

The Internet of Things (IoT) has embarked the connected, smart nodes, and pervasive computing that communicate autonomously and provide the needed service. Moreover, IoT nodes are gathering huge sensory information and private information that becomes a goldmine of malicious nodes. This context creates various issues and challenges in the IoT [27–30]. Figure 13 shows various issues in the IoT.

6.1 Connectivity

Data transmission across the electronic device in the IoT is achieved by wireless or wired communication technology by incorporating connectivity standards. In the IoT, there are numerous connectivity standards, and standardization is a complicated issue to meet all kinds of needs.

6.2 Security

The data transmitted over the network is confidential, and it is necessary to safeguard the privacy of the data. The hardware and connectivity security protocols are utilized to achieve the privacy and security of data over the network. The evolution of security is illustrated in Fig. 14, and the attacks as well as their nature are illustrated in Table 7.

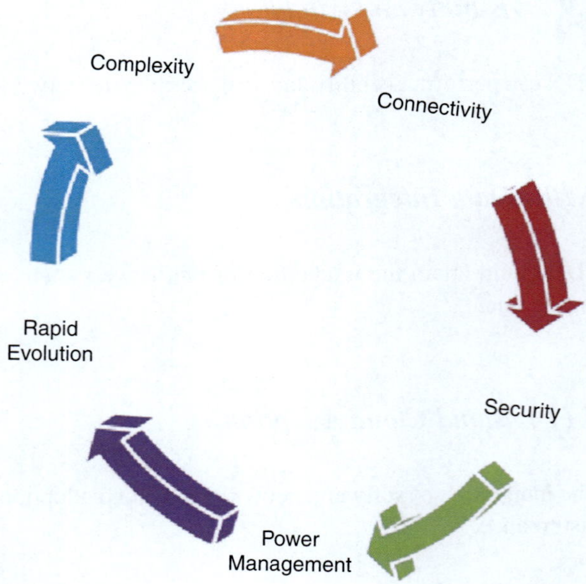

Fig. 13 Various issues in the IoT

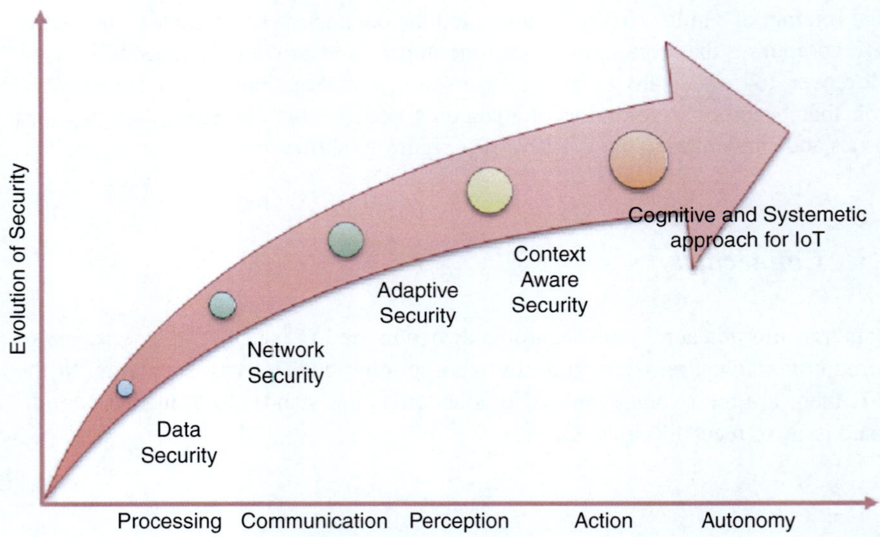

Fig. 14 Evolution of security in the IoT

Table 7 Various security attacks and their nature

Type of attack	Level of the threat	Suggested solution for the attack
Active	High	Ensuring both the confidentiality and integrity
Passive	Low	Utilizing symmetric encryption approaches
Man in the middle (MIM)	Low to medium	Data confidentiality approaches incorporated
Eavesdropping	Low to medium	Encryption approaches applied to all the devices
Privacy	High	Blind signature and ring signature approaches used
Interruption	High	Robust authorization mechanism employed
Routing diversion	High	Connectivity-based method used
Blocking	Extremely high	Anti-jamming, packet filtering, active jamming, and updated antivirus programs are installed
Fabrication	Extremely high	Authenticity mechanism in data is applied
DoS	Extremely high	Cryptography mechanism in data is applied

6.3 Power Management

The electronic component resides inside the IoT device utilizes the energy supplied by the battery or external power source. The process of replacing or recharging the power suppliers in the IoT is complicated. Hence, the proper energy maintenance mechanism is needed and the incorporation of mechanism for every individual method is a complicated process. Designing an IoT device with an effective power management approach is a huge challenge in the IoT.

6.4 Complexity

The integration of various technologies, embedded software, and device together is a complicated process. The connectivity is achieved by the development and ease of design of the device, which is a vague progression.

6.5 Rapid Evolution

In the information era, numerous devices are connected to make the automation in life, and it is a naissance in industry. The designing of IoT devices needs to consider various aspects, namely, security, user-friendly, flexible, and ease of integration. Besides the advancement in technology, the IoT has various limitations and also faces diversified challenges. In Table 8, various challenges and their benefits, as well as limitations, are listed.

Table 8 Challenges and benefits of the IoT

Challenges of the IoT	Benefits	Limitations
Big data	Innovative applications are enabled Big data analysis provides useful insights	Management of big data Centralization of data The network edge and data
Security and privacy	With the assistance of sensitive data, innovative applications are designed	Inspection of private data A new variety of attacks
Scalability	A new variety of devices are connected The access control is gained for the connected devices	The complexity of data management Quality of service (QoS) The capacity of the network
Heterogeneity	Various transmission technologies, data, and devices are integrated Varied IoT vertical silos are combined	Interoperability

7 Conclusion

This chapter outlines the overview of M2M, RFID, and SDN, whereas these technologies play a prominent role in the integration and data transmission in the IoT. Based on the necessity in every individual application, appropriate technology is incorporated into the application. The exchange of information among the electronic device is attained by M2M technology. The process of tracking and control of every individual item is carried effectively by the RFID technology. Every individual technology poses an individual role in the application area and achieved effectiveness in the process of automation. This chapter anticipates varied technologies, challenges, and prominent features in various applications.

References

1. Said, O., & Masud, M. (2013). Towards internet of things: Survey and future vision. *International Journal of Computer Networks, 5*(1), 1–17.
2. Whitmore, A., Agarwal, A., & Da Xu, L. (2015). The Internet of Things – A survey of topics and trends. *Information Systems Frontiers, 17*(2), 261–274.
3. Rayes, A., & Salam, S. (2019). Internet of things (IoT) overview. In *Internet of Things from hype to reality* (pp. 1–35). Springer.
4. Tharini, V. J., & Vijayarani, S. (2020). IoT in healthcare: Ecosystem, pillars, design challenges, applications, vulnerabilities, privacy, and security concerns. In *Incorporating the Internet of Things in healthcare applications and wearable devices* (pp. 1–22). IGI Global.
5. Park, C. W., Hwang, D., & Lee, T. J. (2014). Enhancement of IEEE 802.11 ah MAC for M2M communications. *IEEE Communications Letters, 18*(7), 1151–1154.
6. Mehmood, Y., Görg, C., Muehleisen, M., & Timm-Giel, A. (2015). Mobile M2M communication architectures, upcoming challenges, applications, and future directions. *EURASIP Journal on Wireless Communications and Networking, 2015*(1), 250.

7. Casier, H., Steyaert, M., & Van Roermund, A. H. (Eds.). (2011). Analog circuit design: robust design, sigma delta converters, RFID. .
8. Ahson, S. A., & Ilyas, M. (2017). *RFID handbook: Applications, technology, security, and privacy*. CRC Press.
9. Yousaf, F. Z., Bredel, M., Schaller, S., & Schneider, F. (2017). NFV and SDN – Key technology enablers for 5G networks. *IEEE Journal on Selected Areas in Communications, 35*(11), 2468–2478.
10. Nguyen, T. M. C., Hoang, D. B., &Chaczko, Z. (2016, December). *Can SDN technology be transported to software-defined WSN/IoT?*. In 2016 IEEE International Conference on Internet of Things (iThings) and IEEE Green Computing and Communications (GreenCom) and IEEE Cyber, Physical and Social Computing (CPSCom) and IEEE Smart Data (SmartData) (pp. 234–239). IEEE.
11. Verma, P. K., Verma, R., Prakash, A., Agrawal, A., Naik, K., Tripathi, R., ... Abogharaf, A. (2016). Machine-to-Machine (M2M) communications: A survey. *Journal of Network and Computer Applications, 66*, 83–105.
12. Datta, S. K., & Bonnet, C. (2015, April). *A lightweight framework for efficient M2M device management in one M2M architecture*. In 2015 International conference on Recent advances in Internet of Things (RIoT) (pp. 1–6). IEEE.
13. Kim, J., Lee, J., Kim, J., & Yun, J. (2013). M2M service platforms: Survey, issues, and enabling technologies. *IEEE Communications Surveys & Tutorials, 16*(1), 61–76.
14. Tuna, G., Kogias, D. G., Gungor, V. C., Gezer, C., Taşkın, E., & Ayday, E. (2017). A survey on information security threats and solutions for Machine to Machine (M2M) communications. *Journal of Parallel and Distributed Computing, 109*, 142–154.
15. Kitsos, P., & Zhang, Y. (2008). *RFID security* (Vol. 233). Springer.
16. Chetouane, F. (2015). An overview on RFID technology instruction and application. *IFAC-PapersOnLine, 48*(3), 382–387.
17. Khan, M. A., Sharma, M., & Prabhu, B. R. (2009). A survey of RFID tags. *International Journal of Recent Trends in Engineering, 1*(4), 68.
18. Wu, D. L., Ng, W. W., Yeung, D. S., & Ding, H. L. (2009, July). *A brief survey on current RFID applications*. In 2009 International conference on machine learning and cybernetics (Vol. 4, pp. 2330–2335). IEEE.
19. Das, M. L., Kumar, P., & Martin, A. (2020). Secure and privacy-preserving RFID authentication scheme for Internet of Things applications. *Wireless Personal Communications, 110*(1), 339–353.
20. Dawoud, A., Shahristani, S., & Raun, C. (2018). Deep learning and software-defined networks: Towards secure IoT architecture. *Internet of Things, 3*, 82–89.
21. Salman, O., Elhajj, I., Chehab, A., & Kayssi, A. (2018). IoT survey: An SDN and fog computing perspective. *Computer Networks, 143*, 221–246.
22. Barakabitze, A. A., Ahmad, A., Mijumbi, R., & Hines, A. (2020). 5G network slicing using SDN and NFV: A survey of taxonomy, architectures and future challenges. *Computer Networks, 167*, 106984.
23. Bannour, F., Souihi, S., & Mellouk, A. (2018). Distributed SDN control: Survey, taxonomy, and challenges. *IEEE Communications Surveys & Tutorials, 20*(1), 333–354.
24. Lin, T., Kang, J. M., Bannazadeh, H., & Leon-Garcia, A. (2014, May). *Enabling SDN applications on software-defined infrastructure*. In 2014 IEEE Network Operations and Management Symposium (NOMS) (pp. 1–7). IEEE.
25. Beckett, R., Zou, X. K., Zhang, S., Malik, S., Rexford, J., & Walker, D. (2014, August). *An assertion language for debugging SDN applications*. In Proceedings of the third workshop on hot topics in software defined networking (pp. 91–96).
26. Pham, M., & Hoang, D. B. (2016, June). *SDN applications-The intent-based Northbound Interface realisation for extended applications*. In 2016 IEEE NetSoft conference and workshops (NetSoft) (pp. 372–377). IEEE.

27. Conti, M., Dehghantanha, A., Franke, K., & Watson, S. (2018). Internet of Things security and forensics: Challenges and opportunities.
28. Ngu, A. H., Gutierrez, M., Metsis, V., Nepal, S., & Sheng, Q. Z. (2016). IoT middleware: A survey on issues and enabling technologies. *IEEE Internet of Things Journal, 4*(1), 1–20.
29. Hassan, W. H. (2019). Current research on Internet of Things (IoT) security: A survey. *Computer Networks, 148*, 283–294.
30. Ng, I. C., & Wakenshaw, S. Y. (2017). The Internet-of-Things: Review and research directions. *International Journal of Research in Marketing, 34*(1), 3–21.

Architecture, Generative Model, and Deep Reinforcement Learning for IoT Applications: Deep Learning Perspective

Shaveta Malik, Amit Kumar Tyagi, and Sameer Mahajan

1 Introduction

Along with dig data and cloud computing, artificial intelligence is growing rapidly with the advancement of deep learning especially with the neural networks with an increase in the venture in smart cities, smart healthcare, and the Internet of Things (IoT). A large amount of data need to be processed that is generated by the IoT [30]. Deep learning plays an important role in many applications with the IoT for better results, and it dominates the industry and research along with the IoT. Deep learning technology has the capability and potential to reduce the large complex dataset into the predictive and alteration output. Moreover, many companies like Microsoft, Amazon, Google, Apple, etc. [1] are seriously investing through deep learning technology. Deep learning applications are unsupervised learning, supervised learning, and reinforcement learning. It is one of the leading areas of deep learning. Supervised learning on datasets for the classification is used in many areas like image recognition and handwriting recognition [2, 3]. Unsupervised learning on datasets with no label is used for clustering. Now deep learning can be used in the field of robotics also. With the fastest growth of deep learning, the IoT is also connected with it with sensors, and all devices can communicate through sensed data. For the application of deep learning, large datasets are becoming obtainable. Potential learning for deep learning is provided by the reinforcement learning.

S. Malik (✉) · S. Mahajan
Terna Engineering College, Navi Mumbai, Maharashtra, India

A. K. Tyagi
School of Computer Science and Engineering, Vellore Institute of Technology, Chennai, Tamilnadu, India

Terna Engineering College, Navi Mumbai, Maharashtra, India

S. Pal et al. (eds.), *Artificial Intelligence-based Internet of Things Systems*, Internet of Things, https://doi.org/10.1007/978-3-030-87059-1_9

The learning algorithms of deep learning are based on the artificial neural network or layers of artificial deep learning without any supervision. The concepts behind the ANN are related to the human brain. Deep learning algorithms give better performance as compared to machine learning algorithms. There are a number of deep learning frameworks, i.e., deep sense, etc.; one of the deep learning algorithms is DeepIoT that is used to compress the structure of the deep neural network, and after compression, the model can be deployed to commodity devices. Various deep learning architectures such as convolution deep neural network (CNN), deep neural network, and recurrent neural network (RNN) have been applied on various applications such as computer vision and automatic recognition of speech. Deep learning basically covers one approach that is to build and train the neural network (NN). Neural network uses multiple hidden layers and multiple nonlinear operations which are performed on each layer. There are three deep learning algorithms, i.e., supervised, unsupervised, and reinforcement learning algorithms. A number of sensors are deployed in the IoT, and the IoT generate a homogeneous volume of data for many applications such as smart homes, smart manufacturing, and smart agriculture. Deep learning plays role in embedded and mobile deep learning, and it is feasible of making IoT applications more reliable and effective.

Various Software Libraries

There are various frameworks that support various techniques, i.e., machine learning and deep learning.

- Torch
- Theono
- TensorFlow
- Caffe

Torch

In which there are many community-driven packages. For example, image processing, video processing, computer vision, etc. It is efficient and easy to use, which is based on LnaJIT. It is a scripting language with the implementation of CUDA.

Theano

It works in an n-dimensional array to evaluate the mathematical expression, and it is a faster and efficient approach. This library is for large-scale computation.

TensorFlow

It is for the distributed computing. It is not as much as faster than the Theono. For numerical computations, it uses data flow graph.

Caffe

It gives poor performance for recurrent networks. Caffe 2 is more powerful and scalable, and it is a successor of Caffe. Apart from those libraries, some other libraries are Keras, Lasagne, etc. All these libraries are reusable and user-friendly. They all support a distributed environment.

This chapter will discuss the challenges in embedded, mobile, multimedia healthcare, and many more applications in the IoT with deep learning to process large amounts of data with advanced techniques and methods, for example, deep learning, machine learning, etc. Most of the applications include sensor data, e.g., most of the healthcare applications include sensors, i.e., detection of pulse rate, heart rate detection, automatic blood pressure detection, etc. It can be connected with smartphones for the fitness-based application in measuring a number of steps, count of the calories, count of heartbeat, etc.

Deep Learning Techniques/Methods Used in the Current Era

- Just like machine learning is considered as the subfield inside the circle of artificial intelligence, similarly, deep learning is the subfield present inside the circle of machine learning. The mechanism used in deep learning is learning deep representations which involve hidden layer processing, i.e., the involvement of learning multiple layers and abstracts from data [14]. Practically, any neural differentiable architecture is considered as deep learning until it satisfies the conditions of optimizing a differentiable objective function with the use of a variant of SGD, i.e., stochastic gradient descent. Both supervised and unsupervised learning tasks have supported the tremendous growth of neural architectures. A few major techniques are listed below:
- Multilayer Perceptron (MLP): The feed-forward neural network is the multilayer perceptron, and between the input and the output layers, it consists of multiple layers hidden. The mechanism involves the perceptron employing the arbitrary activation function and does not force compulsions on strict binary classifiers. MLPs are generally seen as nonlinearly transformed stacked layers, intending on learning hierarchical feature representations. Universal approximations are another name given to MLPs.
- Autoencoder (AE): Autoencoder is an unsupervised model that practices upon the reconstruction of input data in the output layer. The middlemost layer also known as the bottleneck layer is considered as the salient feature representation in the input data. We might witness a few variants of autoencoders, majorly denoising autoencoder, etc.
- Convolutional Neural Network (CNN): It is also one of the special kinds of feed-forward neural network which consists of convolution layers in addition to pooling operations. It is capable of enhancing the efficiency and accuracy through capturing the global as well as local features significantly. It works best on the data which show grid-like topology.
- Recurrent Neural Network (RNN): RNN focuses on modeling and processing of sequential data. It consists of loops and memory so that they remember the former computations, unlike any other feed-forward neural network. In such cases there might be a problem of vanishing gradient, variants like LSTM (long short-term memory) and GRU (gated recurrent network) are put into action to overcome such an issue.
- Restricted Boltzmann Machine (RBM): RBM generally comprises two layers, the hidden layer and the visible layer, and therefore may be referred to as a two-layered neural network at times. The restricted word here implies that there exists

no phenomenon of intra-layer communication in either of the layers of this network. It is very easy to stack it to a deep net.

- Neural Autoregressive Distribution Estimation (NADE): NADE follows unsupervised model which is built on top of an autoregressive model along with feed-forward neural nets. In order to model data distribution and densities, NADE is proved to be a quite efficient and tractable estimator.
- Adversarial Networks (AN): AN comprises a discriminator and generator which makes it a generative neural network. The two neural networks are allowed to compete within themselves in a minimax game framework by training them simultaneously.
- Attentional Models (AM): AM are considered to be differential neural architectures. Their operational implementation is established on content addressing on an input sequence. Majorly domains of computer vision and natural language processing incept such mechanisms because of them being typically ubiquitous. Even deep recommender system research is witnessing the emergence of the attentional mechanism.
- Deep Reinforcement Learning (DRL): DRL is a trial and error-based operation. The basic components which come under this paradigm are agent, actions, environment, states, and rewards (can be even penalties). DRL is the only learning process that has achieved human-level performance in the fields of gaming, self-driving cars, and many other domains. It is based on the reinforcement learning; in fact with the combination of deep neural network, without handcrafted features, and domain heuristics through deep neural nets, the agents are enabled to acquire knowledge from the raw data and come up with efficient representations [14].

Organization of This Work

Section 2 discusses the evolution of deep learning techniques in detail. Section 3 discusses our motivation, the reason behind writing this article on IoT applications with deep learning. Further, the scope of computer vision or machine learning and deep learning during COVID-19 and post COVID-19 era is presented in Sect. 4. Section 5 discusses deep learning frameworks, used in various applications (e.g., healthcare, transportation, agriculture, finance, etc.). Then, Sect.6 discusses several generative models in detail. Section 7 discusses the uses or role of deep learning applications in an Internet of Things-based environment. Further, Sect. 8 discusses various possibilities with deep learning for IoT-based applications. Section 9 discusses several opportunities and challenges in deep learning for IoT-based applications. Further, Sect. 10 discusses future research gaps in deep learning-based IoT environment for the twenty-first century. In last, Sect. 11 concludes this work in brief by including several interesting future remarks.

2 Evolution of Deep Learning Techniques

The human brain links to neural networks and relates to computer model which was established by Walter Pitts and Warren McCulloch in 1943. They called threshold logic to mimic the thought process with the combination of mathematics and algorithm. In 1960 the ongoing back model was lauded by Henry J. Kelley.

Deep learning model is CNN based on ANN (artificial neural network). In deep generative model (DGM) such as the nodes in deep belief neural network and deep Boltzmann machine [4].

In 1967 [5], Alexey Ivakhnenko and Lapa published a working learning algorithm for supervised, deep, feed-forward, and multilayer perceptrons.

In 1986 [6, 7], Rina Dechter introduced deep learning to the machine learning community. In 2000 [8, 9], Igor Aizenberg and his colleagues introduce ANN (artificial neural network) with Boolean threshold neurons.

In 1989, Yann Lecun et al. applied the backpropagation algorithm which was the reverse automatic differentiation in 1970, with the deep neural network in handwritten recognition [10–13] of ZIP codes on mails. It required 3 days to train the algorithm [14].

In 1992, cresceptron was published [15–17]; in recognition of 3D objects, it is a method in scenes of cluttering. In cresceptron, in each layer without supervision, cresceptron learned an open number of features, through back analysis. By deep neural network, max pooling is now often adapted, e.g., ImageNet tests.

In 1994 Carvalho with Mike Fairhust and David Bisset introduced or printed results of Boolean neural network with multilayer [18].

In 1995 [19–21], Brendon Frey introduced the wake-sleep algorithm which was fully connected with layer and several hundred hidden units, and it was taken over 2 days to train the network.

The way to get backpropagation to work is to use continuous activation functions; the researchers used binary neurons. That time, they didn't think about the gradient that they can train with gradient also even if they didn't have the idea of a continuous neuron.

In 1995 the field again died and the idea of the neural network got out of control but in 2010 neural nets again came out and people used neural nets in recognition of speech with dynamic performance with good improvement; in 2013, computer vision also worked with neural nets. In 2016, neural nets were used by natural language processing (NLP). From 2017 to 2020, neural nets or neural network is used in many applications, i.e., robotics and prediction. Now, machine learning and deep learning algorithms are working in many areas in managing/tracking/identifying COVID-19 patients also.

Supervised Learning

90% of deep learning applications use supervised learning. Supervised learning is a process by which, you collect a bunch of pairs of inputs and outputs, and the inputs are feed into a machine to learn the correct output. When the output is correct, you don't do anything. If the output is wrong, you tweak the parameter of the machine and correct the output toward the one you want. The trick here is how you figure out which direction and how much you tweak the parameter, and this goes back to gradient calculation and backpropagation. Supervised learning stems from perceptron and Adaline. The Adaline is based on the same architecture with weighted inputs; when it is above the threshold, it turns on and below the threshold, it turns off. The perceptron is a two-layer neuron net where the second layer is trainable and the first layer is fixed. Most of the time, the first layer is determined randomly and that's what they call associative layers.

3 Motivation

In-deep learning strategies have appreciated great achievement in the corrective and speaking society in the last few years, hitting previous methods of acoustic modeling technology, language modeling, etc. There are several factors to consider that contributed to this success. In all application backgrounds, deep learning structural skills practical presentations from unlabeled data and multidisciplinary work studies were extensively useful. In many language processing tasks and language model, integrating the space vector learned word models, smooth and compiled based on semantic and syntactic knowledge.

Content in the context of words, with repetitive structures, has led to significant improvement. Repeated neural networks have also shown promise in music processing. In acoustic modeling, the ability of deep structures to distinguish many aspects of input variables, such as diversity results that rely on the speaker in speech acoustics.

4 Scope of Machine Learning and Deep Learning During COVID-19 and Post COVID-19 Era

COVID has exposed the limitations of deep learning and machine learning. That was not the first time the technology has failed. In 2016, machine learning algorithm failed to predict the Brexit vote as well as the US presidential election. Few reasons behind the popularity of techniques in the Recent Few Years are as follows:

- Machine learning and deep learning have high computational power.
- Due to high speed of deep learning algorithms, it improves the graphics processing.
- Large data provide better training material for the deep learning algorithm as large amount of data is increasing day by day.

Rate of Increase of the Adaption of Internet of Things
As the rate of increase in the adaption of the IoT, the popularity of machine learning and deep learning increases. It also enhances the data storage capacity which leads to enhanced market growth. Before COVID-19 or a few years back, there has been a production of medical science with informatics techniques like medical imaging, genomics, etc., and now these have been transformed through advances in computer science.

4.1 Use of Machine Learning and Deep Learning to Fight Coronavirus

Deep learning has been applied in various fields, for example, healthcare, robotics, cyber-security, etc.; the more refined architecture of deep learning has made all the development possible. Machine learning and deep learning are used in many applications in COVID-19 to detect the virus. One of the technologies is thermal imaging which is used in early detection of virus and detects the person who has the virus and beginning to show the symptoms. These techniques come under the field of thermal imaging science. The camera is to detect the infrared radiations and show different colors for a different level of radiations. If the patient has a symptom of fever in the early stages of infection with the virus, then the body temperature will rise.

As shown in the image, the symptom with virus has more radiations on the upper part of the body. In Fig. 1, we can see the red colors and the more brutish color. Note that this image is captured by a computer machine.

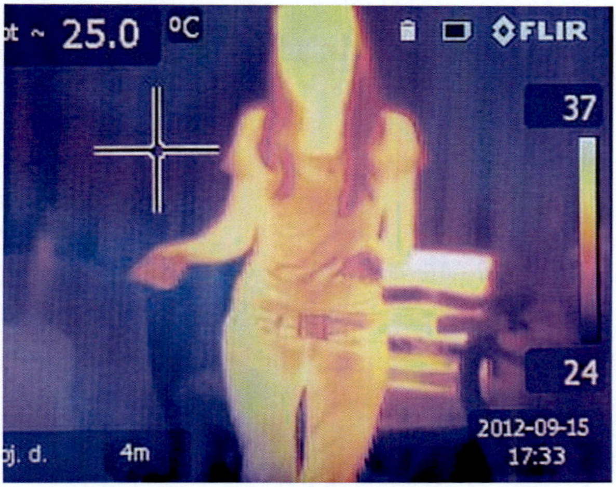

Fig. 1 Thermal imaging of human by artificial intelligence [2]

Through machine learning and deep learning algorithm, it is possible to detect the symptom or early detection of the virus. In transportation, business, and commerce, it is insufficient or expensive to test or detect the infection or virus, but by using cameras with thermal imaging and to scan individual, this would be based on machine learning or deep learning algorithm with processing of input data in real time. Many algorithms of ML and DL can be used for detecting the virus.

Some algorithms are support vector machine (SVM), K-nearest neighborhood (K-NN), random forest model, CNN, etc. Artificial intelligence (AI) is used in several applications within the COVID-19 pandemic situation, i.e., scaling customer communication, analyze the behavior of patient who has disease, understand how COVID-19 spreads, and hospitals are using mobile app to manage COVID-19. Various learning algorithms are used with the IoT, i.e., searches the patterns where people are infected or sick and analyzes the captured data through hospitals, etc.

4.2 Role of Artificial Intelligence with the Internet of Things During COVID Situation

Artificial intelligence is the most demanding field of work, and it plays an important role in managing COVID-19. The devices that helped in managing the virus are the following:

- **Connected Thermometers:** It measures the body temperature of the patient, and it analyzes the real-time patient data and sends it to the nurse or doctor for continuous monitoring.
- **IoT Button:** It is designed for the emergency or any need in the hospital. If any facility needs in the hospital related to maintenance or cleaning issue, etc.
- **COVID Voice Detector:** It detects the voice of the infected patient by evaluating the sound of their cough, even breathe, and the way they speak. Many machine learning and deep learning algorithms are used to predict or analyze speech and cough data.

4.3 Artificial Intelligence with IoT Post COVID Outbreak

Artificial intelligence will also play a dynamic role in post COVID outbreak. According to expert only by wearing mask is difficult to handle the virus.

- **Wearable Devices:** This helps in monitoring real-time data related to person whose having symptom of COVID, collecting a large amount of data on the cloud, fast decision-making, and faster response time with analysis of data. In short, it is used to collect the signs and status of health from users. Devices are in the form of bracelets and ring.

- **Telemedicine and Remote Diagnosis:** In which patient take consultancy through online and doctors diagnose the patients and prescribed the medicine to them virtually. Moreover, with the portable IoT devices, data is captured and sent to the doctors, and it helps the doctor to prescribed medicine to patients. For example, data related to body temperature, sugar and oxygen level in blood, heart rate, digital image of ear, throat, and other body parts, etc.
- **Robotics and Automation:** This is used in various hospitals by delivering drugs to the patients, doing surgery, and serving food to patients. This helps in reducing the direct contact of the doctor with patients. It also improves the efficiency of the medical staff. Camera is used in it to detect if the patient has taken medicine or not, etc.
- **Automatic Air Purifier:** It automatic purifies the air and maintains the quality levels of the air. It checks the air quality index; when the air quality index decreases, it gives alert that is also connected with the mobile application. It automatically sanitizes the hospitals, theater, and patient rooms through robots.
- **Drones:** Remotely monitoring of hospitals, any infected area, and dispensaries, and it helps the ambulance in clearing the traffic with AI-enabled cameras. It also helps to supply essential things, for example, blood, injections, etc.

5 Architecture of Deep Learning Frameworks for Various Applications (e.g., Healthcare, Transportation, etc.)

Today deep learning frameworks or deep learning architectures are used in various applications like healthcare, agriculture, etc. The details of the deep learning (DL) architectures in the section are given below and explain their underlying algorithms. There are three major types of neural networks (NN), convolutional neural networks (CNN), pretrained unsupervised networks, and recurrent or recursive neural networks (RNN), and up-to-date description will be provided.

5.1 Convolutional Neural Network (CNN)

The convolutional neural network (CNN) is one of the key neural networks (NN) for image classifications and image recognition in neural networks. Detections of objects, identification of faces, etc. are some of the places where CNNs are commonly used. As compared to conventional, a reduced pre-processing with less data pre-processing is required for image classification algorithms [22]. Parameter sharing and sparse interactions are key aspects of CNN. Parameter sharing in a specific feature map is the sharing of weights among all neurons. This aims to reduce the number of system-wide parameters and makes the computation more effective and efficient, and on the other hand using kernels or feature detectors smaller than the input image, sparse interaction, or sparse weights is implemented. Three phases are usually performed by each layer of a convolutional neural network. To create a series of linear activations, the layer performs multiple convolutions in the first step, this is followed by the generally called detector stage, and it involves running linear activations through a nonlinear activation function [23, 24].

CNN has been in development for a long time and many CNN-based architectures have been developed. Some of the best-known examples are AlexNet, ResNet, LeNet and LeNet-5, and the GoogLeNet. We will explore more about these in the subsequent sections.

5.1.1 CNN-Based Architectures

One of the pioneers of deep learning, Geoffrey Hinton and his colleagues, Alex Krizhevsky and Ilya Sutskever, introduced AlexNet as the first deep architecture. AlexNet has several layers of convolution, followed by a layer of POOL. It includes eight main layers, where convolutional layers are the first five layers and fully connected layers are the last three layers.

Designed by researchers at Google, GoogLeNet won the ImageNet 2014. Compared to AlexNet, CNN is much deeper; GoogLeNet comprises 22 layers relative to the 8 layers of AlexNet. The key contribution of GoogLeNet is to create an initiation layer that reduces network parameters. This in turn helps the network to perform better. At the top of the convolutional layers, GoogLeNet also uses the average method of pooling, which further removes parameters that lead to marginal gains inefficiency.

ResNet introduced a novel architecture with "skip connections" and heavy batch normalization features. Such skip connections are also referred to as gated units or recurrent gated units and have a strong resemblance to the successful recent elements used in RNNs. They were able to train a neural network with 152 layers thanks to this strategy, while still having less complexity than VGGNet. The layers are reformulated instead of unreferenced functions when learning residual functions. As a result, by increasing network depth, in terms of refine and to achieve extensive precision, these residual networks are easier. The residual modules are stacked one over another, similar to GoogLeNet, to form a full end-to-end network. Simonyan and Zisserman created VGGNet. VGGNet consists of 16 convolutional layers and is very appealing because of its very standardized architecture. VGG involves subsequent convolutional layers accompanied by layers of pooling. To make the layers narrower, the pooling layers are responsible. The narrower the layer, the deeper it is.

In Fig. 2, it shows the top 5 error rate percentage. AlexNet is the highest in error rate and ResNet is the lowest in error rate.

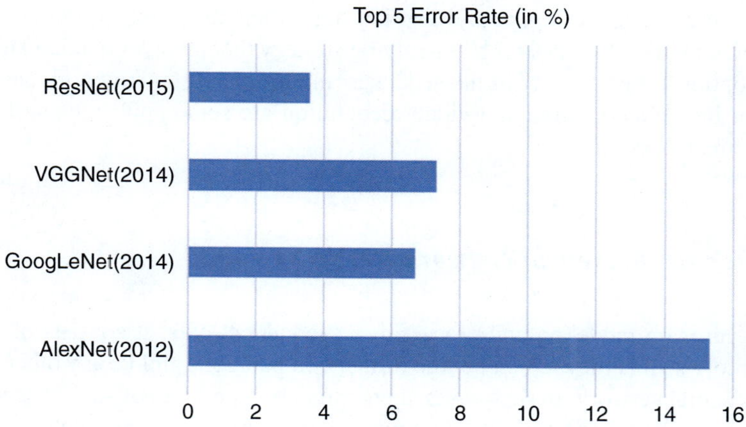

Fig. 2 Error rate

5.2 *Pretrained Unsupervised Networks*

Pretrained unsupervised networks are a deep learning (DL) model; in a neural network using unsupervised learning, each hidden layer is trained in terns to get more accurate and appropriate fit of dataset. In unsupervised learning algorithm, the previous layer is used as an input which is a trained layer and then trains the whole network in supervised network. We will be discussing some pretrained unsupervised networks such as autoencoders, generative adversarial networks (GAN), and deep belief networks in the following sections.

- Autoencoders are an unsupervised learning method in which neural networks are utilized for the purpose of representation learning. The network figures out how to pack and encode information successfully and afterward figures out how to recreate the information back to a portrayal that is as close as conceivable to the first contribution from the diminished encoded representations. The input, output, and the hidden layer constitute an autoencoder. Dimensionality reduction for data visualization and data denoising are some of the most important applications of autoencoders.
- Generative adversarial networks (GANs) are networks that use generative models and have shown great progress over the years. Its abilities to generate and manipulate images in never seen before ways have made it one of the best approaches of deep learning. Subsequent sections have a brief introduction to GANs.
- Deep belief networks are algorithms that create outputs using probabilities and unsupervised learning. They are composed of latent binary variables, and both undirected and directed layers are present. Binary variables range in 0 and 1, and the probability lies between 0 and 1. Neural networks (NN) can be linked together in a series of diverse combinations. In order to do this, a connection is formed

between each network. Greedy learning algorithms are used to pretrain deep belief networks. This is a problem-solving strategy that involves making an optimal option for each layer in the sequence and ultimately finding a global optimum. Image recognition and video recognition are some applications of deep belief networks [25].

5.3 Recurrent Neural Networks (RNNS)

RNN focuses on modeling and processing of sequential data. It consists of loops and memory so that they remember the former computations, unlike any other feed-forward neural network. In such cases, there might be a problem of vanishing gradient; variants like LSTM and GRU are put into action to overcome such an issue. RNN is a type of neural network, in which the previous layer is used as an input to the current layer.

It actively captures consecutive and time dependencies. RNNs typically add cycles that connect neighboring nodes or time steps to the traditional multilayer network architecture [27].

Networks of long short-term memory (LSTM) are a form of recurrent neural networks that can learn order dependence in case of a sequence prediction issue. In complex problem fields like machine translation, speech recognition, and much more, this is an essential behavior. If the gap between references is minimal, RNN works accurately in bits of information referring. When the gap between the referenced data is large where RNN starts to suffer, RNN cannot always link this data. As the gap between dependency increases, the error gradients disappear exponentially, and as a result, the network can be very slow or not able to process learning. To cope with this, LSTM and TBPTT are used. RNNs are extremely powerful, and this is evident by the range of their applications. Some of the major applications of RNN involve prediction problems, language modeling and generating text, machine translation, recognition of speech, generating image descriptions, tagging of video, text summarization, call center analysis, detection of face, applications as image recognition in OCR, and composition of music [28].

Hence, this section discusses several deep learning frameworks which are in trend in this era and used in many popular and critical applications. Now, the next section will discuss about generative models of deep learning in detail.

6 Generative Models

The use of probability, statistics, and artificial intelligence (AI) in applications to generate a depiction or abstraction of observational evidence or target variables that can be computed from findings is called generative modeling [29]. In unsupervised machine learning, generative modeling is used as an attempt to justify phenomena in data, allowing algorithms to consider the real world. It is possible to use this AI understanding to determine all sorts of probabilities through modeled data on a matter [29].

Generative models can be classified into two types: explicit density models and implicit density models. An explicit density function pmodel (x; ϴ) is defined by explicit density models, while implicit density models describe a stochastic operation that can produce data directly. Maximum likelihood models are a good example of explicit density models. Meanwhile, generative adversarial networks (GANs) have proven to be the best implementation of implicit density models [30]. GANs, being in active development and interest for several years now, have been described in the subsequent section.

6.1 Generative Adversarial Networks (GANs)

In 2014, Ian Goodfellow and others from the University of Montreal first proposed the idea of generative adversarial networks (GANs). In any domain, GANs are able to imitate any data distribution: images, audio, voice, and prose. GANs train two models in parallel using unsupervised learning. How they utilize a boundary check that is significantly less than average as for the volume of information on which the organization is prepared is a pivotal component of GANs. As comparable to the training dataset, the network is compelled to effectively the training data, leading to improved performance at generating data. A GAN network is made up of a parallel working discriminator (D) and a generator (G). The purpose of the generator is about being able to produce a fake output that matches a real output, through the training of the generator from its encounters with the discriminator and not with any real material [31]. With the ability to distinguish false data from real data, the generator's purpose is to create an output that is so identical to the reality which confuses the discriminator [23].

A random vector of fixed length is accepted as an input by the generator model which then generates a domain model. The vector is extracted from a Gaussian distribution completely at random. Points in this multidimensional vector space relate to points in the problem area and produce a compact representation of the dataset after training. Latent variables are those variables that are essential to a domain but not explicitly measurable. The vector space is referred to as a latent space which provides compression of the defined raw data or high-level principles, such as the arrangement of data input. The generator model adds sense to points in a specified latent space in the case of GANs so that additional points taken from the latent space can indeed be offered as input to the generator model and used to introduce additional and related examples of output. The generator model is retained and used to produce new samples after practice [33].

The discriminator model generates a validation binary decision mark by taking an instance (true or generated) from the domain as input (refer Fig. 3). The model generator produces sample, and actual examples come out from training set. A

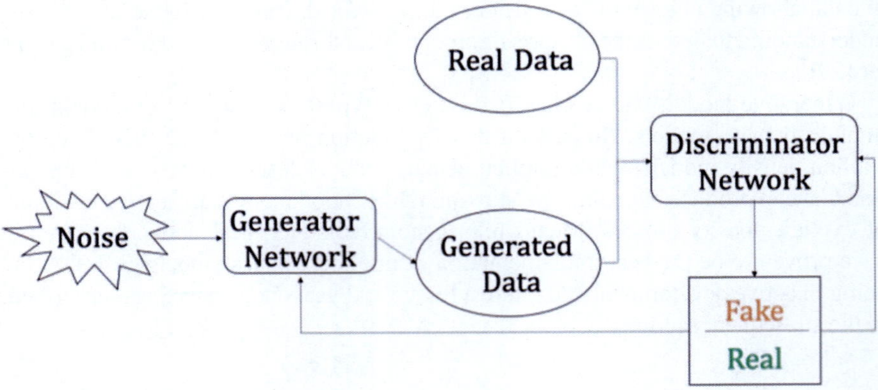

Fig. 3 Discriminator model of GAN

normal (and well-acknowledged) model of categorization is the discriminator. The model of the discriminator is eliminated after the training phase, now that we are concerned mostly in the generator. The generator can often be reused based on its learnings. Using the same or related input data, any or more of the function extraction layers may be used in transfer learning applications [33].

Generative models have the ability to be used in numerous applications and are often used to improve image resolution [32]. Another helpful application of GANs (generative adversarial networks) has become the ability to produce images based on a comprehensive caption specification [34].

7 Deep Learning Applications in the Internet of Things

The neural network model works better in some special domain, i.e., CNN (convolution neural network) model works better in vision-based application and dimensionality reduction works better in visualization [35–37]. There are many other applications of DL in which the IOT is used, i.e., human poses detection for smart home automation and automatic car assistance. This section first discusses about the foundation services of the IoT that is used by deep learning and then discusses about applications of deep learning with the IoT.

7.1 Foundation Service of the Internet of Things That Is Used by Deep Learning

a) **In General**

Image Recognition: In which the input data for the deep learning is in the form of images and videos, and it can be equipped by the mobile devices with the high-resolution cameras Intelligent and automated with high resolution cameras can be used in many applications like smart home, etc. [38].

Speech and Voice Recognition: Smart wearable devices, automatic speech recognition, and mobile devices, these are more suitable way for people to connect with their devices. The main concern in speech recognition or voice recognition is energy intensiveness when the functionality is on the resource-constrained devices. In neural network model, voice or speech data is taken as an input and then passes it through different hidden layers and then speech or voice sound of particular is presented as an output.

Indoor Localization: Indoor services can be used for the indoor navigation, and it is also used with the IoT in different sectors, i.e., smart homes, hospitals, etc. The input data used for these applications are mobile or data from mobile devices and other technologies also, i.e., RFID, Wi-Fi or Bluetooth, and ultrasound. DL models can be used along with this to predict the location [39–41].

Figure 4 shows the number of IoT applications and functional services

b) **As Applications**

- **Smart Homes**: The concept of the smart home is used in many applications connected with the IoT; nowadays, most of the appliances are connected with the IOT for enhancing the homes, to improve the life quality. This can be connected with DL models to predict or monitor the trends or health trends, etc.[42]
- **Smart City:** In which the IoT is included with a number of domains, i.e., agriculture, energy, transportation, etc. Large amount of data that come from different domains can be analyzed by the DL models which will give high-quality output [43].
- **Smart Energy:** Smart grid and consumers of energy constitute a major part of IoT big data; this is the two-way communication. Based on the real- time analytics, energy providers are dependent on energy consumption patterns and thus learnings from this analysis can be used to predict and figure out the needs of consumer and take appropriate business decision [44].
- **Intelligent Transportation System:** Many applications used related to automatic recognition of traffic signals, autonomous driving, and automatic car assistance system. Lin et al. proposed a method or concept of real-time detection of traffic signs based on CNN model with GPU [45].

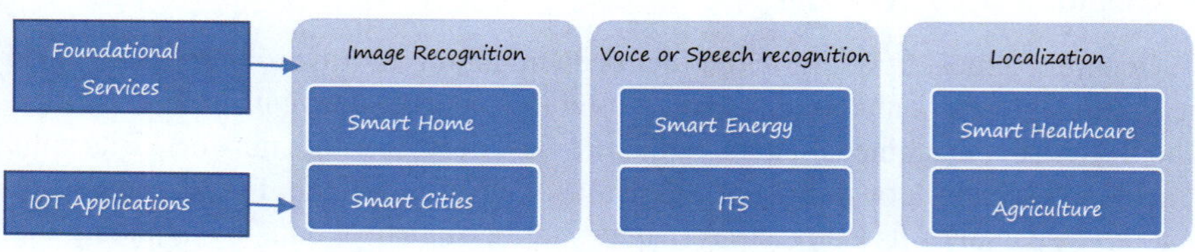

Fig. 4 IoT applications and functional services

- **Smart Healthcare:** DL models can be used for healthcare application, i.e., classification of medical images that is a very known and trending application in healthcare domain. Pereira et al. [46] detect Parkinson's disease at early stages with the idea of recognition of handwriting through CNN deep learning model.
- **Smart Agriculture:** IoT and DL models can be used in domain of agriculture, to recognize the number of crops and diseases of plants. Classification and prediction of crop disease can be done using CNN or other models of deep learning. Mobile app can be used to identify the fruit, vegetable, and plant disease. It can be helpful for farmers [47].

  – Other Applications of Deep Learning Used with the IoT

- **Education:** In which IOT and DL models can be used in education domain. Mobile devices can be used to collect the data of learners or students. DL models can be used to predict the progress of the students, for example, MOOC courses are very popular nowadays and large amount of data generated from the learners or students. It is helpful to analyze and predict the behavior of struggling student [48, 49].

8 Possibilities with Deep Learning in IoT-Based Applications

The present deep learning gives the medical care industry the capacity to examine huge measure of information at remarkable paces without settling on precision. In general, mathematical methods enable deep learning models to operate at human level cognitive ability; also it works based on hidden layers. In the near future, deep learning use will be a hot technique to use in healthcare, as well as for suggesting healthcare best tools, techniques, curing techniques for a disease, etc. It will change and make a huge in this world. We can find how many multimedia applications are used in healthcare industry.

Hence, Figure 5 discusses future uses of deep learning for multimedia applications in healthcare industry.

8.1 Possibilities with Deep Learning

- **Quantum Deep Learning**

Quantum annealers along with other deep quantum information processors like programmable photonic circuits can be used to construct deep quantum networks. The Boltzmann machine is said to be the simplest deep quantum network. Bits with tunable interactions constitute the classical Boltzmann machine. In order to ensure that the dispersal of its expressions matches to the statistical data, by adjusting the interaction of its constituting bits, Boltzmann machine will be trained. Using an

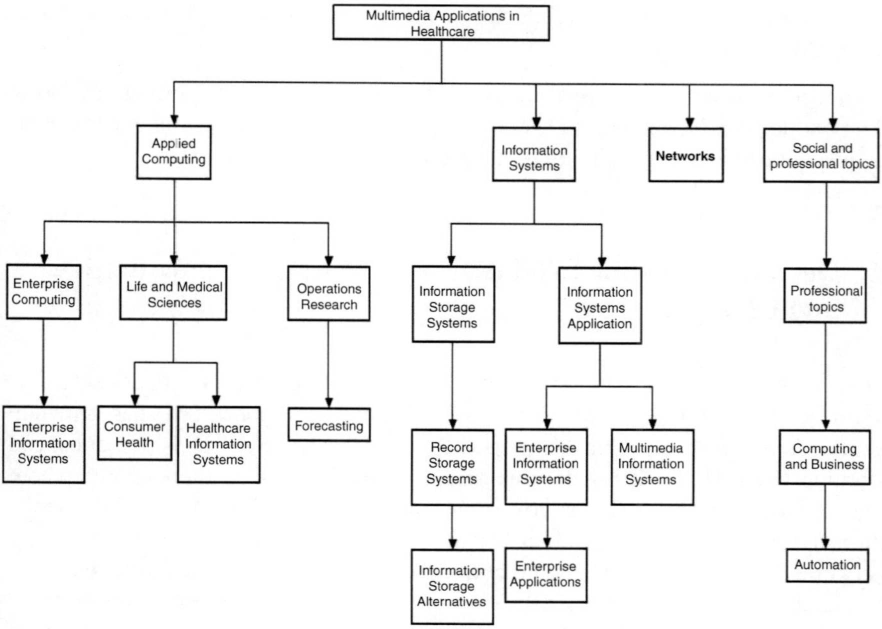

Fig. 5 Deep learning future uses for multimedia applications in healthcare industry

adjustable model corresponds with the set of interacting quantum spins if neural network (NN) is considered and it can be quantized. Thus, we can peruse out the yield qubits to get the outcome by instating the information neurons in the Boltzmann machine to a fixed state and letting the framework to get warmed up. Compared to a general-purpose quantum computer, the quantum annealing device (QAD) (quantum information processor) is much easy to construct and enlarge. D-Wave computer is one such device. The readers are suggested to refer [50] to know more about potential limitations of machine learning and deep learning.

8.1.1 Others

In the future, for fast coding, deep learning tools will incorporate simplified programming frameworks. The most popular deep learning tools, like TensorFlow, BigDL, OpenDeep, Caffe, Theano, Torch, and MXNet, will be more useful in the near future. Deep learning tools will be embedded in every design surface in almost every application. For visual development of reusable components, the deep learning toolkits will support a larger scale. Hence, few possibilities can be made with deep learning as:

- Deep learning networks will clarify computer memory.
- In building datasets for DL models, neural architecture search will play a key role.

- To search convolutional architectures, many applications will continue to use reinforcement learning.

Hence, this section discusses the uses of deep learning in the current era and in the next decade. Now, the next section will discuss several opportunities and challenges in deep learning-based IoT-based applications in detail.

9 Deep Learning for IoT-Based Applications: Opportunities and Challenges

To support reliable and low-cost communication, next-generation wireless networks are needed. Over the previous few years, artificial intelligence (AI) has grownup exponentially with machine learning (ML) in deep learning (DL) and enhanced learning (EL) which shows its importance in many types of applications, where planning or retrieval problems govern an important role. Internet of Things (IoT) infrastructure is one of the most important applications for next- generation wireless networks. Many researchers and practitioners widespread nature of this type of system leads to evaluate the use of DL techniques to make IoT smarter, more trustworthy, and more efficient. By 2020, Forbes predicts that the IoT industry will grow. Smart cities, smart grid, health, and IoT (IoT) industries are the pioneers of this major transformation and growth. The Industry 4.0 scale is taking benefit of the IoT industry.

IoT applications have a large number of disseminated devices that contain static data. Clearly, wireless solutions are particularly important in this context as evidenced by the huge growth of industrial components (usually very low-key) that facilitated the real execution of this technology. Data gathered by devices grows slowly in size and heterogeneity. To improve intelligent and efficient resource management and network management is the edge of DL, and to improve overall system performance of the basic devices is essential. IoT devices produce large packets or short packets over wireless downlink due to which communication agreements also need to be in line with the dynamic IoT status in real time. For standard wireless and wireless networks, IoT networks often need to be developed depending on their power consumption, and load balancing techniques should be designed devices for many years in a trustworthy way to use the IoT.

The diversity of IoT devices also increases the issue of interoperability from a practical point of view. The one of the biggest problems in the industry is security, and security detection to infrastructure protection from harmful network attacks, unwarranted access, and protecting user privacy is essential but at the same time challenging. For example, in order to provide the largest category of cyber threats, the malicious detection systems need to adapt to unexpected events. The smart IoT needs to face all the major challenges listed above. Deep learning (DL) resolves a wide variety of difficulties without human intervention.

10 Future Research Gaps in Deep Learning-Based IoT Environment for Twenty-First-Century Generation

The real applications with a variety of features and complex system deal with a variety of domain changes. In case of industry with the complex system, the different manufacturers produce the same system but with different sensors and sensory numbers, e.g., to monitor the condition or climate. Research into the evolution of unintelligible domains, especially those used in complex physical systems, is limited but has great potential to have an impact, especially on industrial systems.

Another idea is to adopt the simulation into the real app and use the simulation space. This approach is particularly appealing as the details will be sufficient in the source domain. Production models are widely used in computer viewing functions. Examination of the availability of samples produced in accordance with the body's compatibility is an open-ended research question. Also, controlling the production of relevant data samples by looking at body processes in the same way is an open research questionnaire. Learning enhancements have thrived especially in applications that have a broad and detailed simulation environment.

In addition, the levels of stress that RL has been used extensively in the PHM context have been equally reduced. In maintaining decision-making, the digital twins can provide support. Complex problem requires research. Moreover model can be enriched with information obtained as deep learning (DL), and expert knowledge is a highly renouncing research area. While many of the indicators are currently being traced to physics-enabled machine learning, there is no consensus or integration into the alternative methods and how they can be transmitted to industrial applications. Further research is needed to develop and integrate these approaches, which may also lead to improvements in the definition of advanced models and methods. Datasets are supposed to be the leading drivers of research and experimentation in many areas in deep learning (DL) like NLP (natural language processing) and CV (computer vision). However, in the PHM context, the lack of representative data prevented the widespread use and modification of DL methods in industrial applications.

There are many solutions to this challenge: expansion of data, data processing, and use of physics machine learning models or rather from an organizational perspective, data sharing across corporate boundaries. ML models tend to be less constructive and have an increased risk of overheating when insufficient data is available.

However, the study of adding data to time series data is limited. Investigating how data additions can be used in PHM situations, especially in time series data, is one of the potential indicators of research. Newly, many methods of producing samples or defective traits have been suggested by neural generative networks [50]. However, most studies focus on vibration data and pre-processed signals to make them look like the image. An exciting research guide is to explore the transfer of these approaches to extremely complex databases and time series data.

In the future, an additional challenge that needs to be addressed is the effective and efficient structure and selection of training data stocks. This applies specially to changing environments with very different working conditions where the training database does not fully represent the full range of expected working conditions. The decision is to be taken if the new data is added in the database then it need to train.

The research also focuses on practical learning: selective detection that will make significant improvements in algorithm performance.

Lastly, future researchers may look forward to continue their research work on black box problems, scalability, standardization, and the structure of the deep learning model.

Hence, this section discusses several research gaps in deep learning models like black box problems, scalability, the structure of the model, etc., in detail. More details about computer vision, machine learning, and deep leaning can be found in [43–45, 50, 51]; readers are suggested to read the listed articles to enrich their knowledge. Further, reader can refer [52–55] articles for knowing about several critical issues in the Internet of Things and various computing environments.

11 Conclusion

Deep learning (DL) and the Internet of Things (IoT) have been applied to a number of applications, and they have drawn lots of attention of researchers in current years. These two technologies are the chain like producer and consumers; the IOT generate the data or raw data that is analyzed by the deep learning algorithms, and deep learning (DL) generate high end level abstraction that give into the IOT system for the sufficient changes and good performance of services. The paper explained characteristics, challenges of deep learning with IOT, and architecture of deep learning. It also presented the use of artificial intelligence technologies with the IOT and deep learning or machine learning algorithms. Moreover, in this paper, the future research direction is identified in the path of deep learning for IOT-based application.

Further Reading

- Deep Belief Network
- Robotic Process Automation
- Intelligent Automation
- Automated Analytics
- Quantum Programing
- Quantum Machine learning
- Quantum Deep learning

Bibliography

1. Cloud TPU Alpha: Train and Run Machine Learning Models Faster Than Ever Before (2017) [Online]. Available: https://cloud.google.com/tpu/
2. Krizhevsky, I., Sutskever, & Hinton, G. E. (2012). ImageNet classification with deep convolutional neural networks. *Proceedings of NIPS, 1*, 1097–1105.
3. Nielsen, M. (2017). Neural Networks and Deep Learning. [Online]. Available: http://neuralnetworksanddeeplearning.com/
4. Bengio, Y. (2009). "Learning deep architectures for AI" (PDF). Foundations and Trends in Machine Learning.
5. Vakhnenko, A. G., & Lapa, V. G. (1967). *Cybernetics and forecasting techniques*. American Elsevier Publishing co..
6. Ivakhnenko, A. (1971). "Polynomial theory of complex systems" (PDF). *IEEE Transactions on Systems, Man, and Cybernetics, SMC-1*(4), 364–378. https://doi.org/10.1109/TSMC.1971.4308320
7. Fukushima, K. (1980). Neocognitron: A self-organizing neural network model for a mechanism of pattern recognition unaffected by shift in position. *Biological Cybernetics*.
8. Dechter, R. (1986). *Learning while searching in constraint-satisfaction problems*. University of California, Computer Science Department, Cognitive Systems Laboratory. Online.
9. Aizenberg, I., Aizenberg, N. N., & Vandewalle, J. P. L. (2000). *Multi-valued and universal binary neurons: Theory, learning and applications*. Springer Science & Business Media.
10. Co-evolving Recurrent Neurons Learn Deep Memory POMDPs. Proc. GECCO, Washington, D. C., pp. 1795–1802, ACM Press, New York, NY, USA, 2005.
11. Linnainmaa, S (1970). The representation of the cumulative rounding error of an algorithm as a Taylor expansion of the local rounding errors. Master's Thesis (in Finnish), Univ. Helsinki, 6–7.
12. Griewank, A. (2012). "Who invented the reverse mode of differentiation?" (PDF). Documenta Mathematica (Extra Volume ISMP).
13. Werbos, P. (1974). *Beyond regression: New tools for prediction and analysis in the behavioral sciences*. Harvard University. Retrieved 12 June 2017.
14. Werbos, P. (1982). *Applications of advances in nonlinear sensitivity analysis (PDF). System modeling and optimization* (pp. 762–777). Springer.
15. LeCun, et al. (1989). Backpropagation applied to handwritten zip code recognition. *Neural Computation, 1*, 541–551.
16. Weng, J., Ahuja, N., & Huang, T. S. (June, 1992). *Cresceptron: a self-organizing neural network which grows adaptively* (Vol. I, pp. 576–581). Proc. International Joint Conference on Neural Networks.
17. Weng, J., Ahuja, N., & Huang, T. S. (May, 1993). *Learning recognition and segmentation of 3-D objects from 2-D images* (pp. 121–128). Proc. 4th International Conf. Computer Vision.
18. Weng, J., Ahuja, N., & Huang, T. S. (Nov. 1997). Learning recognition and segmentation using the Cresceptron. *International Journal of Computer Vision, 25*(2), 105–139.
19. de Carvalho, A. C. L. F., Fairhurst, M. C., Bisset, D. (1994-08-08). "*An integrated Boolean neural network for pattern classification*". Pattern Recognition Letters.
20. Hinton, G. E., Dayan, P., Frey, B. J., Neal, R. (1995-05-26). The wake-sleep algorithm for unsupervised neural networks.
21. Hochreiter, S. (1991) "Untersuchungen zu dynamischen neuronalen Netzen," Diploma thesis. Institut f. Informatik, Technische Univ. Munich. Advisor J. Schmidhuber
22. Hosseini, M.-P. A cloud-based brain-computer interface to analyze medical big data for epileptic seizure detection. In: The 3rd Annual New Jersey Big Data Alliance (NJBDA) Symposium (2016)
23. Hosseini, M.P. Brain-computer interface for analyzing epileptic big data. Ph.D. thesis, Rutgers University-School of Graduate Studies (2018)

24. Hosseini, M., Lu, S., Kamaraj, K., Slowikowski, A., Venkatesh, H. (n.d.). Deep learning architectures.
25. "Deep Belief Networks: How They Work and What Are Their Applications – MissingLink.ai", MissingLink.ai. [Online]. Available: https://missinglink.ai/guides/neural-network-concepts/deep-belief-networks-work-applications/
26. "Introduction to Recurrent Neural Network – Geeks for Geeks", GeeksforGeeks Available: https://www.geeksforgeeks.org/introduction-to-recurrent-neural-network/. [Online].
27. Goodfellow, I., Bengio, Y., & Courville, A. (2016). *Deep learning*. MIT Press. http://www.deeplearningbook.org
28. [Online]. Available: https://iq.opengenus.org/applications-of-rnn/
29. [Online]. Available: https://searchenterpriseai.techtarget.com/definition/generative-modeling
30. [Online]. Available: https://www.analyticsvidhya.com/blog/2020/01/generative-models-gans-computer-vision/
31. Creswell, A., White, T., Dumoulin, V., Arulkumaran, K., Sengupta, B., & Bharath, A. A. (2018). Generative adversarial networks: an overview. *IEEE Signal Processing Magazine, 35*(1), 53–65.
32. Ledig, C., Theis, L., Huszár, F., Caballero, J., Cunningham, A., Acosta, A., Aitken, A., Tejani, A., Totz, J., Wang, Z., et al. (2017). *Photo-realistic single image super-resolution using a generative adversarial network* (pp. 4681–4690). Proceedings of the IEEE Conference on Computer Vision and Pattern Recognition.
33. Reed, S., Akata, Z., Yan, X., Logeswaran, L., Schiele, B., & Lee, H. (2016). Generative adversarial text to image synthesis. *arXiv preprint arXiv*, 1605.05396.
34. Available: https://machinelearningmastery.com/what-are-generative-adversarial-networks-gans/. [Online].
35. Larry, H. (2017). Voice control everywhere: Low-power special purpose chip could make speech recognition ubiquitous in electronics. [Online]. Available: http://news.mit.edu/2017/low-power-chip-speechrecognition-electronics-0213
36. Wang, X., Gao, L., Mao, S., & Pandey, S. (2015). Deepfi: Deep learning for indoor fingerprinting using channel state information. In *2015 IEEE Wireless Communications and Networking Conference (WCNC)* (pp. 1666–1671). IEEE.
37. Gu, Y., Chen, Y., Liu, J., & Jiang, X. (2015). Semi-supervised deep extreme learning machine for Wi-Fi based localization. *Neurocomputing, 166*, 282–293.
38. Zhang, W., Liu, K., Zhang, W., Zhang, Y., & Gu, J. (2016). Deep neural networks for wireless localization in indoor and outdoor environments. *Neurocomputing, 194*, 279–287.
39. Hazen, T. J. (2016). Microsoft and Liebherr collaborating on new generation of smart refrigerators.[Online]. Available http://blogs.technet.microsoft.com/machinelearning/2016/09/02/
40. Yoshida, H. (2016). The internet of things that matter: Integrating smart cities with smart agriculture. Digitalist Magazine. [Online]. Available: http://www.digitalistmag.com/iot/2016/05/19/internet-of-thingsthat-matter-integrating-smart-cities-with-smart-agriculture-04221203
41. Mocanu, D. C., Mocanu, E., Nguyen, P. H., Gibescu, M., & Liotta, A. (2016). *Big IoT data mining for real-time energy disaggregation in buildings* (pp. 9–12). Proceedings of the IEEE International Conference on Systems, Man, and Cybernetics.
42. Ma, X., Yu, H., Wang, Y., & Wang, Y. (2015). Large-scale transportation network congestion evolution prediction using deep learning theory. *PLoS One, 10*(3), e0119044.
43. Tyagi, A. K., & Rekha, G. (2020). Challenges of applying deep learning in real-world applications. In *Challenges and applications for implementing machine learning in computer vision* (pp. 92–118). IGI Global. https://doi.org/10.4018/978-1-7998-0182-5.ch004
44. Tyagi, A. K., "Prediction models", Book: Handbook of research on disease prediction through data analytics and machine learning, IGI Global 2020, pp. 50–69. https://doi.org/10.4018/978-1-7998-2742-9.ch004
45. Tyagi, A. K., Rekha, G., Machine Learning with Big Data. (March 20, 2019). Proceedings of International Conference on Sustainable Computing in Science, Technology and Management (SUSCOM), Amity University Rajasthan, Jaipur – India, February 26–28, 2019.

46. Liu, C., Cao, Y., Luo, Y., Chen, G., Vokkarane, V., Ma, Y., Chen, S., & Hou, P. (2017). *A new deep learning-based food recognition system for dietary assessment on an edge computing service infrastructure*. IEEE Transactions on Services Computing.
47. Yang, T.-Y., Brinton, C. G., Joe-Wong, C., & Chiang, M. Behavior based grade prediction for MOOCs via time series neural networks. *IEEE Journal of Selected Topics in Signal Processing, 20.*
48. Hochreiter, S., et al. (15 January 2001). Gradient flow in recurrent nets: The difficulty of learning long-term dependencies. In Kolen, J. F. Kremer, & C. Stefan (Eds.), *A field guide to dynamical recurrent networks*. John Wiley & Sons.
49. Hosseini, M., Pompili, D., Elisevich, K., & Soltanian-Zadeh, H. (2017). Optimized deep learning for EEG big data and seizure prediction BCI via the internet of things. *IEEE Transactions on Big Data, 3*(4), 392–404.
50. Pramod, A., Naicker, H. S., & Tyagi, A. K. (2020). Machine learning and deep learning: Open issues and future research directions for next ten years. In *Book: Computational analysis and understanding of deep learning for medical care: Principles, methods, and applications, 2020.* Wiley Scrivener.
51. Tyagi, A. K., & Chahal, P. (2020). Artificial intelligence and machine learning algorithms. In *Book: Challenges and applications for implementing machine learning in computer vision.* IGI Global. https://doi.org/10.4018/978-1-7998-0182-5.ch008
52. Tyagi, A. K., Rekha, G., & Sreenath, N. (2020). Beyond the hype: Internet of things concepts, security and privacy concerns. In S. Satapathy, K. Raju, K. Shyamala, D. Krishna, & M. Favorskaya (Eds.), *Advances in decision sciences, image processing, security and computer vision. ICETE 2019. Learning and analytics in intelligent systems, Vol 3*. Springer.
53. Tyagi, A. K., & Nair, M. M. (2020). Internet of Everything (IoE) and Internet of Things (IoTs): Threat analyses. *Possible Opportunities for Future, 15*(4).
54. Tyagi, A. K., Nair, M. M., Niladhuri, S., & Abraham, A. (2020). Security, privacy research issues in various computing platforms: A survey and the road ahead. *Journal of Information Assurance & Security, 15*(1), 1, 16p–16.
55. Reddy K.S., Agarwal K., Tyagi A.K. (2021) Beyond things: A systematic study of internet of everything. In: Abraham A., Panda M., Pradhan S., Garcia-Hernandez L., Ma K. (eds) Innovations in Bio-Inspired Computing and Applications. IBICA 2019. Advances in Intelligent Systems and Computing, vol 1180.

Enabling Inference and Training of Deep Learning Models for AI Applications on IoT Edge Devices

Divyasheel Sharma and Santonu Sarkar

1 Introduction

Consider an autonomous driving vehicle that must infer the road, pedestrians, vehicles, and other objects ahead and in its surroundings. The vehicle must make real-time decisions where any delay could mean a catastrophe. Similarly, consider an industrial surveillance system that tracks objects in a factory. Since the factory workers would inevitably be part of the video stream being processed, their privacy is a concern, ethically, and enforced by regulations such as GDPR [1]. In both the scenarios, if the artificial intelligence or machine learning (AI/ML) models that provide inference are deployed on the cloud, the real-time decision-making may suffer for the autonomous vehicle, and the risk of personal and proprietary-industrial data breach increases for the factory. It would be better to deploy the inference models at the edge (i.e., on the autonomous vehicle; and within the factory premises/the camera) to reduce the latency in decision-making by the model and the risk associated with data transfer to the cloud.

There are numerous applications where edge computing brings fast intelligence locally, such as air quality monitoring, people occupancy estimation, factory goods' quality control, and precision agriculture. Nearly 70% of data is being created at the end devices, and half of this data is estimated to be processed at the edge. Moreover, the edge AI software market is said to grow from $335 million in 2018 to $1152 million by 2023 [2].

The emergence of these applications that have a tight delay and privacy requirements and the rapid advancements in inferencing and training capabilities of AI/ML models on resource-constrained infrastructure has created edge AI as an attractive alternative to the cloud. However, edge AI's success depends on how AI/ML

D. Sharma · S. Sarkar (✉)
ABB Corporate Research, Bangalore, India
e-mail: divyasheelsharma@acm.org; santonus@acm.org

S. Pal et al. (eds.), *Artificial Intelligence-based Internet of Things Systems*,
Internet of Things, https://doi.org/10.1007/978-3-030-87059-1_10

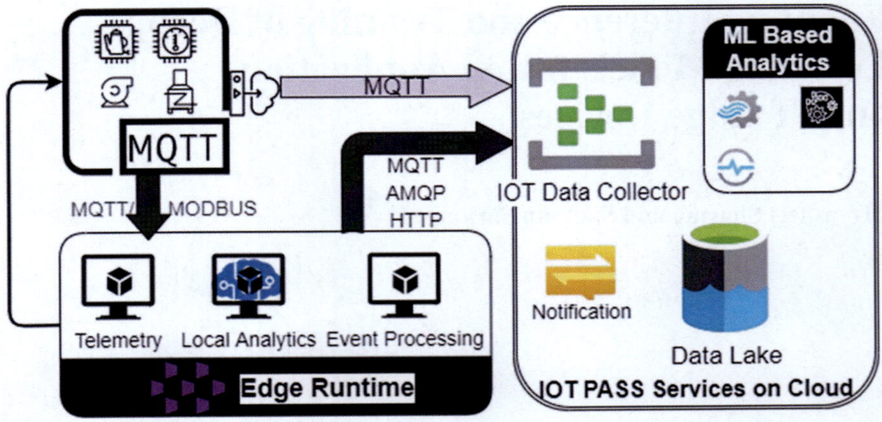

Fig. 1 Overview of AI/ML at the edge

models, especially the deep learning models that promise the state-of-the-art prediction accuracy, perform on resource-constrained edge devices.

Figure 1 shows a generic interplay between end devices, the edge computing platform, and the cloud infrastructure. Edge computing platform (also known as edge device) can have varied hardware configurations. For instance, it can very well be a small-sized ASIC device like Google Coral, a tiny computer, or a reasonably large computing infrastructure with containerized services.

An edge computing platform remains in the end devices' close vicinity to reduce the latency of data collected from these devices. The computing platform analyzes the data using AI/ML techniques and provides a real-time response back to the device. It also transmits the processed information to the cloud for large-scale data processing. Prominent cloud providers such as AWS, Google, IBM, and Microsoft Azure support building such edge platforms and provide a plethora of PAAS offerings to integrate edge devices with their cloud platforms.

Typically, the modern trained neural networks-based models are hundreds of layers deep and large with millions of model parameters and activations [3] and do not fit in the constrained memory at the IoT edge. Moreover, larger networks tend to perform a larger number of operations (FLOPS) that degrade prediction latency.

Recent literature [4] suggests that deep networks' compression and acceleration lead to an improved inference of deep learning models on the resource-constrained edge. For example, compressing the 95MB ResNet-50 model with 25.6 M parameters led to the 75% reduction of trained parameters and 50% computation time [4, 5]. Typically, there is a trade-off between model compression and inference accuracy [6, 7], and various methods of optimization such as network pruning [8, 9], model quantization [6, 7], low-rank factorization [10–15], and knowledge distillation [16] are proposed. Cheng et al. [4] have organized these methods into four categories: parameter pruning and quantization, low-rank approximation and sparsity, transferred/compact convolution filters, and knowledge distillation. Lebedev et al. [12] provide yet another overview of inference techniques that differentiate edge nodes' optimization with sufficient edge resources from edge devices with tight resource constraints. The overview includes techniques to pre-process model input to narrow the search space of models (e.g., by targeted inputs focused on regions of interest in a video frame), optimize the model structure to reduce the number of multiplication operations, model selection where the optimization function weighs both prediction accuracy and inference time, and recommendations for energy-preserving hardware. Furthermore, it describes techniques related to the segmentation of models using horizontal and vertical partitioning of the models [17] and early exit of inference [18]. Various inference frameworks that support model optimization have also been developed in the industry: TFLite [19], Nvidia TensorRT [20], and specialized hardware (ASICs), e.g., Nvidia Jetson series [21], coral from Google [22], Intel OpenVino/Movidius [23, 24], and Qualcomm's Snapdragon [25].

Another option to enable edge AI is to train the model on edge. Storing and training an AI/ML model on edge can lead to cost-savings that are otherwise associated with first, transferring data to the cloud and, second, centrally training the model on extensive computing infrastructure. Moreover, keeping the data and processing at the edge preserves privacy. Although singularly, an edge device is resource-constrained, distributed training is possible considering many edge devices' collective resources. Mostly, a technique for updating a centrally trained AI/ML model with local updates from the edge called federated learning has been successful [26]. Other possibilities are to transfer-learn a pre-trained model on the edge device [27, 28].

The rest of the chapter is organized as follows. Section 2 focuses on model inference at the edge and discusses model optimization techniques such as network pruning, model quantization, knowledge distillation, and inference frameworks. Section 3 concentrates on training/retraining at the edge and describe federated learning that leverages hybrid cloud-edge architecture and retraining at the edge device via transfer learning and weight imprinting. Finally, in Section 4, we conclude the chapter by discussing possible future directions.

2 Inference at the Edge

The inference is the process of using a trained machine learning model to make predictions on new input data. When the trained model resides at the IoT edge, inference happens at the edge.

Inference at an edge IoT device is of particular interest because it promises reduced latency and the risk associated with transferring data from the edge to the cloud, memory utilization, and dependence on constrained network bandwidth. However, the edge inference is challenging due to the typically large size of deep learning-based trained models, which provide state-of-the-art prediction accuracy. The models' large size increases prediction latency and puts pressure on limited memory resources available at the edge device. Below we discuss three approaches to optimize the models' size for inference at the edge, namely, model pruning, model quantization, and knowledge distillation.

2.1 Model Pruning

Model pruning implies removing the redundant structure from a neural network. The structure could be optimized by removing weights, neurons, filters, or channels. Typically, insignificant weights are reduced to zero, i.e., as shown in Fig. 2, structurally, the connections with those weights are removed from the neural network. This removal of neural network connections is known as connection pruning.

In [29–32], various researchers have shown that networks with too many parameters do not generalize for a fixed dataset. In contrast, networks with too few parameters may not represent the dataset adequately. Hence, a trade-off between prediction accuracy and network complexity can be explored and exploited to simplify neural network architecture, leading to improved memory footprint, energy efficiency, and model latency that are important for performing inference at the resource-constrained IoT edge.

Blalock et al. [33] have discussed various pruning algorithms in their survey article and concluded that pruning works with little or no loss of accuracy. Sometimes pruning even increases accuracy, e.g., by finding the right capacity network that avoids overfitting [8].

Several model optimization techniques for neural networks have been proposed by LeCun [32] and references therein. Zhu et al. [9] have recently experimented with more modern deep learning models, namely, deep CNNs, stacked LSTM, and seq2seq LSTM models, and shown that a 10X reduction in weight parameters is possible with minimal loss in prediction accuracy.

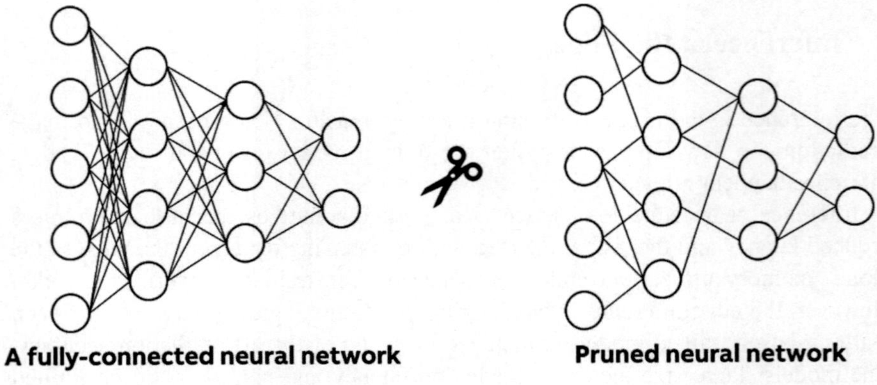

Fig. 2 A connection-pruned network

In their seminal paper on pruning, Han et al. [8] introduce pruning as an important training-time step to reduce the network size in their algorithm that learns the network weights and the essential connections. The following two-phase learning algorithm with pruning is proposed to reduce the energy required to run deep learning models on mobile devices while preserving accuracy:

1. In the first phase, the neural network architecture is trained, then all the connections with weights below a threshold are set to zero (i.e., the connections are removed, and hence, the network is pruned). This pruning results in a sparser network.
2. In the second phase, this sparse network is retrained (i.e., fine-tuned) with the remaining connections to reinforce the new network architecture and compensate for any loss inaccuracy.

As shown in Fig. 3, both phases can be iterated to simplify network architecture further while maintaining accuracy. Table 1 lists the pruning algorithm variants that are defined based on structure, scoring, scheduling, and fine-tuning [33].

Pruning is evaluated based on various goals for which it is performed, e.g., to reduce storage footprint or computational cost of inference. The overall compression ratio to the fraction of parameters pruned is evaluated to measure the impact on storage footprint. FLOPS is measured to evaluate the computational cost of inference since multiple parameters such as model input and filter size impact it. Blalock et al. [33] provide an open-source library – called ShrinkBench – for pruning with functions to evaluate pruning methods. ShrinkBench provides various magnitude-base baselines to compare pruning methods: global magnitude pruning, layer-wise magnitude pruning, global gradient magnitude pruning, layer-wise gradient magnitude pruning, and random pruning.

It has been found that for large amounts of pruning, pruning methods do better than random pruning; however, the same is not valid for small amounts of pruning. Pruning all layers of the network uniformly provides worse results than pruning selectively in different layers. Moreover, for a constant number of fine-tuning iterations, pruned models tend to perform better than the same-sparsity models trained

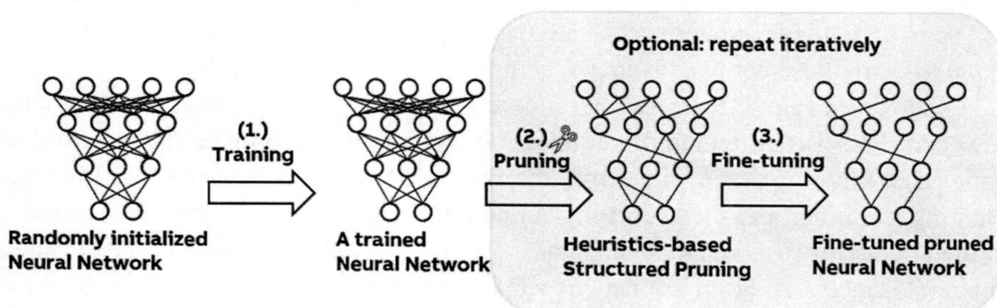

Fig. 3 Pruning neural networks with heuristic-based structured pruning with fine-tuning. For example, weights pruning with a low-magnitude heuristic below which all weight parameters are set to zero (i.e., connections are removed)

Table 1 Variants of pruning methods

Types	Variant techniques	Description	References
Structure	Unstructured pruning	Each weight parameter is pruned individually to create a sparse network	[34]
	Structure pruning	Sets of parameters are pruned together. For example, entire neurons, filters, or channels	[35]
Scoring	Local	Pruning fractionally within a sub-component of a neural network and comparing scores at the sub-component level, e.g., a layer	[8]
	Global	Comparing scores at the network level irrespective of where pruning takes place within the network	[36, 37]
Scheduling	Single-step	Prune all parameters in a single step	[38]
	Multi-step	Prune a fraction of parameters iteratively in multiple steps	[8, 39]
Fine-tuning	Original weights	Continue to train the network with original weights	
	Rewinding	Continue to fine-tune the network with weights from an earlier state	[37]
	Re-initialization	Choose new weights for the complete network	[37, 38]

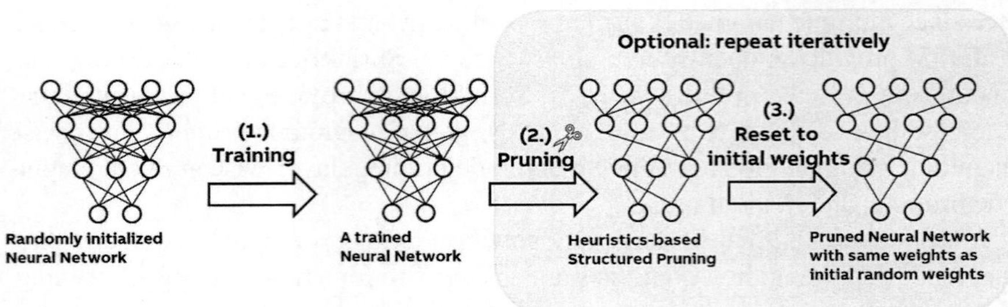

Fig. 4 The Lottery Hypothesis scheme to find equivalent sparse networks to a dense network

from scratch with random weights initialization. However, recently, the Lottery Ticket hypothesis [37] proposes that dense, trainable neural networks have equivalent sparse, trainable subnetworks with the same capacity and suggests a new technique to learn these equivalent sparse neural networks from scratch.

As shown in Fig. 4, first, train a randomly initialized neural network and prune it. Second, reset each remaining connection weights in the pruned network to the same randomly initialized value from the first step before the network was trained. Training, pruning, and then resetting remaining connections' weights can be performed iteratively till the model accuracy does not degrade significantly. The resulting pruned network is an equivalent subnetwork. Interestingly, it can be trained from scratch to achieve commensurate accuracy within commensurate iterations if the initial connection weights for the pruned structure are set to the same as the respective initial randomly initialized weights from the original large network.

However, finding the optimal subnetwork without pruning the network first is an open problem.

In practice, typically, to optimize the model, pruning is applied along with post-training quantization that we discuss next.

2.2 *Model Quantization*

Quantization means converting data to types with reduced precision. In neural networks, quantization reduces the floating-point representation of deep learning parameters (i.e., weights, activations, and gradients) to a lower-precision (typically 8-bit integers). Like pruning, research efforts in quantization have also reduced memory footprint of the deep learning models while keeping accuracy comparable to the original large deep learning models. Quantized models are memory-efficient since they take less space, lead to the faster inference, have faster downloads on networks with limited bandwidth, and consume less energy.

Figure 5 shows a scale-round approach to provide intuition on how to quantize a real number, 2.56 from [0, 10] range to an 8-bit integer in [−128, 128] range using the uniform integer quantization scheme [38]. First, the number is scaled to the target quantized range and then rounded to the nearest integer. In this example, floating-point 2.56 is quantized to 32, an 8-bit integer value. We also see that 32, when scaled back, gives 2.528, which is not quite 2.56. This loss in precision may lead to some loss of accuracy in network predictions, which is typically acceptable.

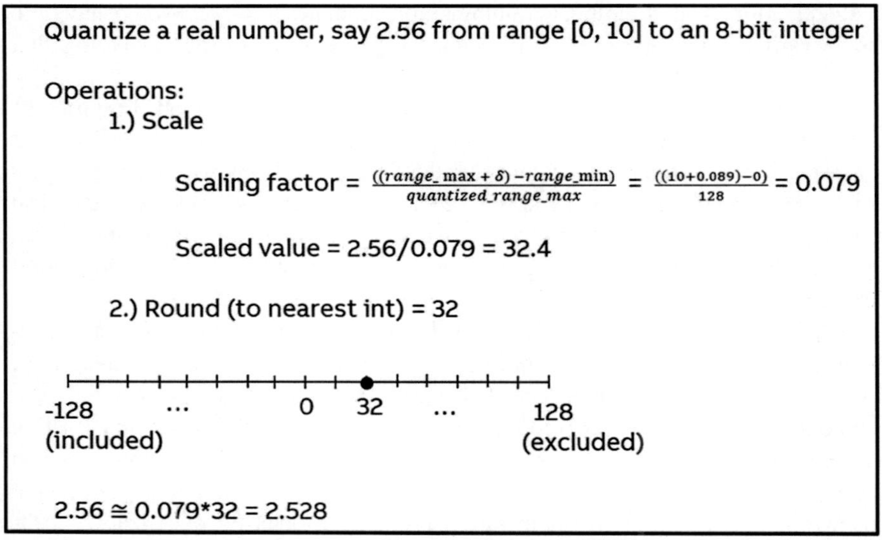

Fig. 5 A scale-round uniform integer quantization scheme to quantize a real number (i.e., 2.56) from [0,10] range to an 8-bit integer in [−128, 128] range

The technique requires the exact quantization of zero, which is achieved by providing a zero-point offset. For simplicity, the zero-point offset is omitted from the example in Fig. 5.

Given a floating-point tensor (aka multidimensional array) X_{fp}, the quantized tensor X_q can be calculated by a uniform integer quantization scale-round shift scheme:

$$X_q = round\left(\frac{X_{fp}}{S}\right) + z$$

where $S \sim (X_{q\_max} - X_{q\_min})/2^n - 1$ represents a scaling factor, $z = -X_{fp\_min}/S$ is the zero-point offset (shift), and n is the number of bits used for quantization.

Matrix multiplications are a typical operation in neural networks and involve multiplication and accumulation operations. Doing multiplication and accumulations with 32-bit floating-point numbers as operands lead to 32-bit floating-point multiplications and accumulations. Since deep neural networks involve many matrix multiplications, and floating-point operations consume more power than integer operations, quantized operands could lead to resource savings and efficient computations. If operands are quantized 8-bit integers, products become 16-bit, and accumulations are 32-bit integers. Since matrix multiplications chain in the neural networks, the operands need to be rescaled to 8-bit whenever required.

Quantizing weights and activations leads to smaller size models, whereas quantizing gradients leads to lesser communication costs in a distributed learning setting [39]. Quantization may be performed before or after training a model. When an already trained model is quantized, it is called post-training quantization. Whereas in quantization-aware training, a training graph simulates low-precision behavior in the forward pass and accounts for it as quantization error that is optimized together with the training loss.

Why quantization works in deep learning is not well understood. Empirically, quantization is an effective method of compressing models. It is also theorized that low-precision operations can be considered noise, which is known to act as a regularizer [39]. Other theories are emerging to prove that original networks' statistical properties are preserved in quantized networks even when numerically the weight (or activation) values are changed from floating-point to low-precision numbers [39].

Quantization techniques can be divided into deterministic and stochastic techniques. Further deterministic quantization can be achieved by various techniques: rounding [40], vector quantization [41], and quantization-as-optimization [42]. Rounding involves converting continuous values to discrete. Vector quantization uses cluster-based approaches to cluster real values into subgroups that are then replaced by the cluster-centroid value. Quantization-as-optimization poses the quantization problem as an optimization problem. Stochastic quantization techniques comprise random rounding and probabilistic quantization. In random rounding, the quantized weights are sampled from a discrete distribution. In probabilistic quantization, weights are sampled from a discrete prior distribution whose parameters are inferred using a learning algorithm.

2.3 Knowledge Distillation

Transfer learning [43] has shown that a trained model's essence can be preserved and used to train another network. However, transfer learned models are typically of similar neural network architecture; hence, of similar depth and the number of parameters, constraining the usage of transfer learned models on resource-constrained IoT devices. The motivation is to be still able to train a large, deep network that adequately learns a representation of the input data; however, it then transfers the generalizations from the large network to a shallow and lightweight model that can meet inference time performance constraints.

At training time, we want to extract the maximum knowledge from data. It makes sense to train large models, even ensemble models, that may even overfit individually whose predictions can be averaged at test time. However, when we want to make inferences on new data at test time, the large/ensembled trained network is accurate in predictions but inefficient under various deployment constraints such as that of an IoT edge. Large deep learning models are often over-parameterized [44], and big ensemble models have high redundancy and low knowledge per parameter. At test time, we need models that have a small memory footprint and minimum computations.

Since a trained deep learning model essentially represents a function approximation representing the training data, we have the function available to us after training the model. Can the knowledge in this learned function be transferred to a simple model?

The knowledge distillation technique [45] helps in achieving this goal. The technique differs from transfer learning in the fact that there is no weights transfer. Instead, the aim is to transfer generalizations (i.e., knowledge) from the larger trained network to a smaller, simpler model.

Knowledge distillation is a technique to create such a compressed model from a pre-trained large model. The small, compressed model preserves the knowledge from the large model and mimics its behavior. The main intuition behind knowledge transfer is as follows: Typically, a deep learning model – for example, a classification model – outputs a one-hot encoded vector representing classes with a single 1 and many 0s. However, more knowledge is embedded in the inputs (logits) to the softmax function that provides non-normalized probabilities for various classes, resulting in output as probabilities. Logits contain a more granular sense of the closeness of various classes to each other for a given input rather than a one-hot encoded vector that preserves a high-level binary output. For example, one-hot encoded vector predicting digit 7 for 0 to 9 digits is represented as [0 0 0 0 0 0 0 1 0 0] whereas logits [0.0005 0.05 0.1 0.003 0.005 0.003 0.002 0.3 0.002 0.001] may tell us that while predicting handwritten digits class a 7 is closer to a handwritten 1 or 2 rather than to 6. Hinton et al. [45] refer to the relationship between classes as dark knowledge. As discussed in the example above, the knowledge that a handwritten 1 is closer to a handwritten 7 than 6 is an example of such dark knowledge. Hinton et al. [45] call the logits output as soft targets instead of a one-hot encoded output, referred to as a hard target.

$$p_i = \frac{e^{(z_i/T)}}{\sum_j e^{(z_j/T)}} \quad (1)$$

The logits are further softened using a temperature parameter (T) in the softmax function (Eq. 1), where different temperature values can provide even more information about a class prediction's relationship with other classes. For example, increasing the temperature may lead to soft targets that may tell more about the relationships between various classes.

While training a distilled network, one could train a small network for the softened targets. However, as shown in Fig. 6, Hinton et al. [45] suggest that the best results are achieved when both soft and hard targets are considered. The cross entropy loss is minimized between the original network's soft targets and the small network's soft predictions while keeping the temperature the same as what was set for the original network to obtain the softened targets. Similarly, another cross entropy loss is minimized for hard predictions and hard targets with temperature T=1. The knowledge distillation training method has also been referred to as a teacher-student training method where the larger model is considered a teacher and the smaller model is the student [47]. The technique was first introduced in 2006 by Bucila et al. [46]. Later in 2015, the method was generalized by Hinton et al. [45].

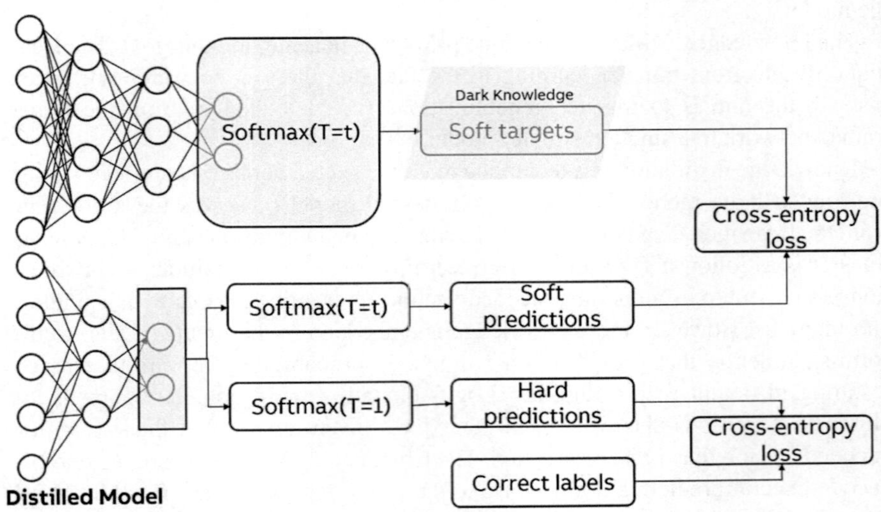

Fig. 6 Knowledge distillation

3 Training Models at the Edge

Traditionally, machine learning models are trained centrally on complete datasets. Consider training machine learning models on users' mobile data in such a setting. First, the data from each mobile device needs to be sent to a data center, and second, models need to be trained on the massive amount of collected data.

Most data, such as on a mobile device or an industrial IoT edge, are siloed and distributed, and there are data privacy and security concerns about sending data to the cloud. These challenges motivated researchers [48] to ask the question: Can machine learning models be trained locally on edge devices without sending data to the data centers? With growing computing power on the edge devices, there is an opportunity to store data locally and perform machine learning training at the edge node. This opportunity has led to the development of a new algorithm for distributed machine learning known as federated learning [48].

3.1 *Federated Learning*

Gboard is a keyboard for mobile phones that uses a successful practical implementation of federated learning. Let us consider the Gboard scenario [49].

The Gboard keyboard aims to allow mobile phone users to type on a small form factor keyboard efficiently. Since human fingers are more significant than the keys, the virtual key-presses may be imprecise. Gboard uses a machine learning model to predict a user key-press. Similarly, machine learning models predict the next words that make typing easier for users.

Since training models for mobile users centrally raise data-communication costs and privacy issues, federated machine learning is proposed. Figure 7 depicts a federated setting, where an initial model is trained centrally on a server and deployed on mobile phones. Each deployed model is then updated with training over local data. To preserve users' privacy, the local data on the mobile phone is encrypted. When users' phones are either idle or charging, they check in to the central parameter server; the server may begin new epochs for federated machine learning. Each mobile phone runs several rounds of stochastic gradient descent on the device and sends its parameter updates to the central server. Only the parameter updates (i.e., weight, bias, and gradients in the deep learning setting) are shared with the central server. Of course, not all phones are idle charging and available all the time. Hence, the central server picks a small percentage of users' phones for participation in federated learning at a time. At the central server, all the parameter updates from various mobile phones are averaged. These averages update the central model parameters. After the model is updated, the parameter updates received by the central server are deleted.

To further maintain user privacy, differential privacy-based federated learning approach is proposed [50]. Figure 8 shows the two additional steps used in this

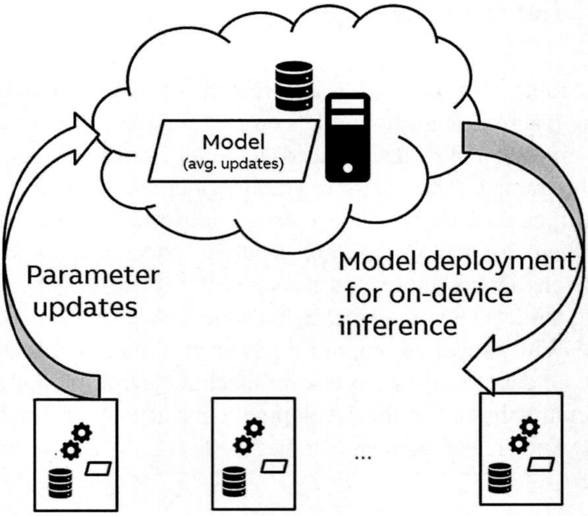

Fig. 7 Federated learning

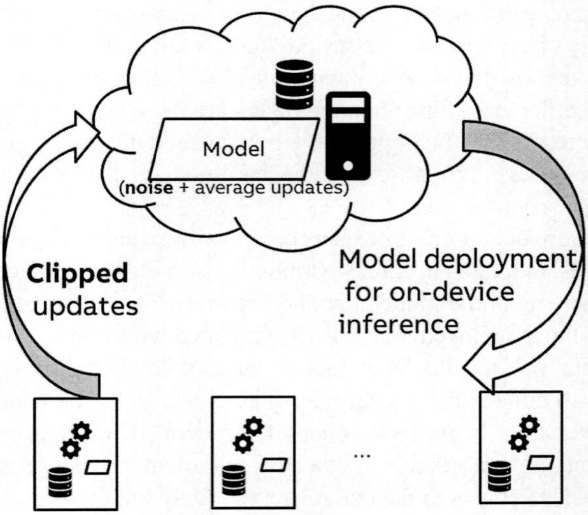

Fig. 8 Differentially private federated learning. Edge devices clip their updates, and the central server adds noise before averaging the updates

approach. First, the parameter updates to the central server are clipped below a size, and a calibrated noise is added after the model is averaged.

Unlike training in the cloud, in federated learning, the datasets are typically unbalanced and non-IID since each user's mobile phone usage behavior differs. Empirically federated learning has been shown to train models with equivalent accuracy to the centrally trained at the cloud.

FederatedSGD (FedSGD) and federated averaging (FedAvg) are two algorithms for communication-efficient updates [48]. Stochastic gradient descent (SGD) can be applied to federated optimization. Single batch gradients can be calculated per round of communication. The gradient updates are sent to the central server, and the average gradient is used to perform the central model's weight updates. The participation of devices is governed by a parameter C, where C=1 implies 100% of clients are participating.

In FedAvg, the device itself can perform multiple steps of training using gradient descent. The amount of computation is controlled by a fraction of participating clients in around (C), the number of training passes on a local dataset in each round (E), and the minibatch size (B) for edge-level updates. FedSGD is the same as FedAvg when $E = 1$ and $B = \infty$.

3.2 Adaptive Retraining

Transfer learning is the most popular method to adaptively retrain pre-trained models for related new data. Below we discuss two flavors of transfer learning that allow retraining at the resource-constrained edge device. In both cases, the last layer of the pre-trained model is retrained on the edge device.

Retraining via Backpropagation at the Edge
Backpropagation is used to update weights at each layer of a neural network. However, backpropagation can also be used to update weights only for the last fully connected layer of a pre-trained network. However, this principle is utilized to retrain a pre-trained network at the edge device with a deviation.

Typically, in transfer learning, weights at several layers are frozen except for a few (e.g., last) layers. Such a model is retrained. However, typically already compiled models are downloaded and deployed on an edge device. For example, a TensorFlow Lite model is compiled to run on edge TPU (tensor processing unit). This implies that the weights are locked and cannot be retrained on the edge device. However, in the retraining-with-back-propagation-at-the-edge technique, the last layer of the pre-trained network is removed, and then the model is compiled. The last layer is then implemented to allow retraining on the resource-constrained device. One drawback of this approach is that it requires iterative training and a more considerable amount of data.

Retraining via Weight Imprinting
The weight imprinting technique allows updating the weights of the last layer of a pre-trained neural network classifier using a small dataset on an edge device. The technique works for adding new classes to the model while preserving the network's ability to classify the pre-trained classes.

Figure 9 shows the two steps involved in retraining new models on the edge using weight imprinting. First, a base classifier is trained. Then, for a few new data points, e.g., images of a class that is not currently represented in the network, the classifier's embeddings are extracted, normalized, and added to the penultimate

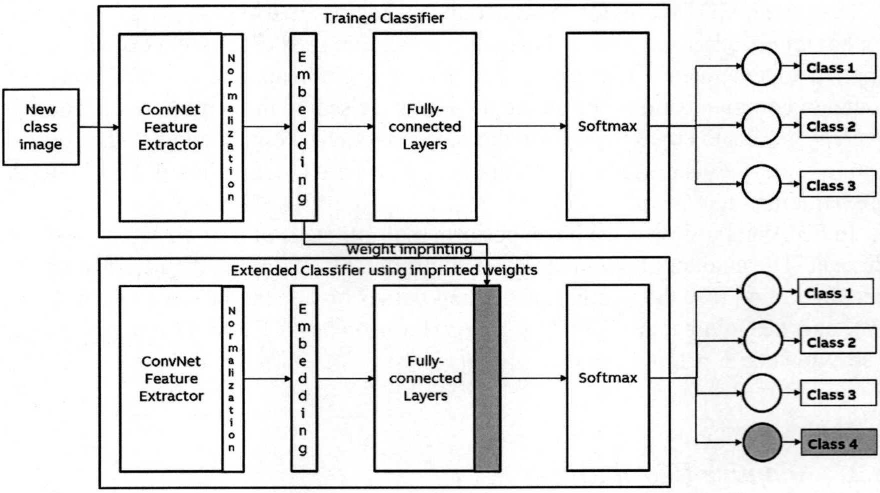

Fig. 9 Weight imprinting

layer of the base neural network as a weight update. The new extended classifier can be used to perform inference for all the existing and new classes.

Weight imprinting is based on setting proxies (a representative data point) for each class and adjusting activation vectors with L2 normalization. Essentially, using proxies and L2 normalization at the end of an embedding extractor such that the output embedding has unit length provides the mathematical trick that creates equivalence in proxy-based metrics learning and softmax-based classifiers [51], which enable learning with small amounts of data.

In contrast to retraining via backpropagation at the edge, weight imprinting does not need the model's recompilation. However, if there is a sizeable intra-class variance in the dataset, backpropagation performs better than weight imprinting. Lastly, weight imprinting needs specific engineering for each use case, whereas backpropagation at the edge is generic.

4 Conclusion

This chapter has introduced the most prevalent techniques for inference and training of neural network models at the IoT edge.

Model optimization with connection pruning leads to a sparser network with insignificant weights turned to zero. Zero weights can be ignored during inference to improve latency. Moreover, sparse networks tend to be easier to compress. Quantizing model parameters to perform inference using low-precision arithmetic produces models with smaller footprints that conserve processing power. The knowledge distillation technique generates smaller (distilled) models by targeting softer probabilities (i.e., outputs of logits) during training. Current literature has fewer references where the model optimization techniques of pruning, quantization, and knowledge distillation are combined. Combining model optimization techniques to explore higher optimization efficacies that can lead to efficient model inference at the IoT edge devices is a fertile area for further research.

Federated machine learning trains a high-quality centralized neural network with the leading training loop running over distributed local datasets at the edge devices. Only local weight updates are transferred to a centralized parameter server that averages the parameter updates to train a new model. Federated learning is a step toward privacy-preserving training and avoids large data transfers from the edge devices to cloud. One of the prominent use cases of federated learning is the smart keyboard in Android devices. However, building a successful learning model from a set of heterogeneous devices is still a big challenge. Moreover, while there are steps toward privacy-preserving federated learning, adhering to new and future regulations and explaining federated-learned models also remain open questions for future research. Lastly, we discussed adaptive retraining techniques of either retraining the last layer via backpropagation at the edge or directly imprinting weights for new classes that perform transfer learning with fewer available data and make retraining practicable at the IoT edge devices.

References

1. General Data Protection Regulation (GDPR) (https://gdpr-info.eu/)
2. Edge AI Software Market by Component (Solutions and Services), Data Source, Application (Autonomous Vehicles, Access Management, Video Surveillance, Remote Monitoring & Predictive Maintenance, Telemetry), Vertical, and Region – Global Forecast to 2023. Markets and Markets (https://www.researchandmarkets.com/reports/4752886/edge-ai-software-market-by-component-solutions)
3. Deep learning models with pre-trained weights. (https://keras.io/api/applications/)
4. Cheng, Y., Wang, D., Zhou, P., & Zhang, T. (2018). Model compression and acceleration for deep neural networks: The principles, progress, and challenges. *IEEE Signal Processing Magazine, 35(1)*, 126–136.
5. He, K., Zhang, X., Ren, S., & Sun, J. (2016). Deep residual learning for image recognition. In *IEEE Conference on Computer Vision and Pattern Recognition (CVPR)* (pp. 770–778).
6. Vanhoucke, V., Senior, A., & Mao, M. Z. (2011). Improving the speed of neural networks on CPUs. *Deep Learning and Unsupervised Feature Learning Workshop, Neural Information Processing Systems.*
7. Gupta, S., Agrawal, A., Gopalakrishnan, K., & Narayanan, P. (2015). Deep Learning with limited numerical precision. In *Proceedings of the 32nd International Conference on International Conference on Machine Learning - Volume 37* (pp. 1737–1746).
8. Han, S., Pool, J., Tran, J., and Dally, W. J, (2015). Learning both weights and connections for efficient neural networks. *Advances in Neural Information Processing Systems.*
9. Zhu, M., & Gupta, S. (2018). To prune, or not to prune: exploring the efficacy of pruning for model compression. *International Conference on Learning Representations (ICLR) Workshop.*
10. Denton, E. L., Zaremba, W., Bruna, J., LeCun, Y., & Fergus, R. (2014). Exploiting linear structure within convolutional networks for efficient evaluation. *Advances in Neural Information Processing Systems.*

11. Jaderberg, M., Vedaldi, A., & Zisserman, A. (2014). Speeding up convolutional neural networks with low-rank expansions. In *Proceedings of the British Machine Vision Conference*. BMVA Press.
12. Lebedev, V., Ganin, Y., Rakhuba, M., Oseledets, I. V., & Lempitsky, V. S. (2015). Speeding-up convolutional neural networks using fine-tuned CP-decomposition. *International Conference on Learning Representations (ICLR Poster)*.
13. Tai, C., Xiao, T., Wang, X., & Weinan, E. (2015). Convolutional neural networks with low-rank regularization. *International Conference on Learning Representations (ICLR Poster)*.
14. Denil, M., Shakibi, B., Dinh, L., Ranzato, M., & Freitas, N. D. (2013). Predicting parameters in Deep Learning. *Advances in Neural Information Processing Systems*.
15. Sainath, T. N., Kingsbury, B., Sindhwani, V., Arisoy, E., & Ramanhadran, B. (2013). Low-rank matrix factorization for deep neural network training with high-dimensional output targets. In *Proceeding of IEEE International Conference on Acoustics, Speech and Signal Processing*.
16. Gou, J., Yu, B., Maybank, S. J., & Tao, D. (2020). Knowledge distillation: A survey. In *Proceedings of the Conference on Computer Vision and Pattern Recognition (CVPR)*.
17. Kang, Y., Hauswald, J., Gao, C., et al. (2017). Neurosurgeon: collaborative intelligence between the Cloud and mobile Edge. In *Proceeding of 22nd International Conference Architecture Support Programming Language Operator System (ASPLOS)* (pp. 615–629).
18. Teerapittayanon, S., McDaniel, B., & Kung, H. (2016). Branchynet: Fast inference via early exiting from deep neural networks. *International Conference on Pattern Recognition*.
19. TFLite: ML for mobile and Edge devices. (https://www.tensorflow.org/lite)
20. NVIDIA TensorRT: Programmable inference accelerator. (https://developer.nvidia.com/tensorrt)
21. NVIDIA Jetson: The AI platform for autonomous everything (https://www.nvidia.com/en-in/autonomous-machines/embedded-systems/)
22. Google coral (https://coral.ai/)
23. Intel Open Vino (https://software.intel.com/content/www/us/en/develop/tools/openvino-toolkit.html)
24. Intel Movidius Vision Process Units (https://www.intel.com/content/www/us/en/products/processors/movidius-vpu/movidius-myriad-x.html)
25. Qualcomm's Snapdragon (https://www.qualcomm.com/news/releases/2019/04/10/qualcomm-expands-ecosystem-enable-next-gen-Edge-ai-and-machine-learning)
26. Konecný, J., McMahan, H. B., Ramage, D., & Richtárik, P. (2016). Federated Optimization: Distributed Machine Learning for On-Device Intelligence. *CoRR abs/1610.02527. arXiv:1610.02527*. (http://arxiv.org/abs/1610.02527)
27. Retrain a classification model on-device with backpropagation. (https://coral.ai/docs/Edgetpu/retrain-classification-ondevice-backprop/)
28. Qi, H., Brown, M., & Lowe, D.G. (2018). Low-shot learning with imprinted weights. *Conference on Computer Vision and Pattern recognition (CVPR)*.
29. Denker, J., Schwartz, D., Wittner, B., Solla, S. A., Howard, R., Jackel, L., & Hopfield, J. (1987). Large automatic learning, rule extraction and generalization. *Complex Systems, 1*, 877–922.
30. Baum, E. B., & Haussler, D. (1989). What size net gives valid generalization? *Neural Computation, 1*, 151–160.
31. Solla, S. A., Schwartz, D. B., Tishby, N., & Levin, E. (1990). Supervised learning: A theoretical framework. *Neural Information Processing Systems*.
32. Le Cun, Y. (1989). Generalization and network design strategies. In R. Pfeifer, Z. Schreter, F. Fogelman, & L. Steels (Eds.), *Connectionism in Perspective*. Elsevier.
33. Blalock, D., Gonzalez Ortiz, J. J., Frankle, J., & Guttag, J. (2020). What is the state of neural network pruning? In *Machine Learning and Systems (MLSys)*.
34. Konecný, J., McMahan, H. B., Yu, F. X., Richtárik, P., Suresh, A. T., & Bacon, D. (2016). Federated Learning: Strategies for Improving Communication Efficiency. *NIPS 2016 Workshop on Private Multi-Party Machine Learning*.

35. McMahan, H. B, Moore, E., Ramage, D., and Agüera-Arcas, B., (2016). Federated Learning of Deep Networks using Model Averaging. *CoRR abs/1602.05629. arXiv:1602.05629.* (http://arxiv.org/abs/1602.05629).
36. Yang, Q., Liu, Y., Chen, T., & Tong, Y. (2019). Federated machine learning: Concept and applications. *ACM Transactions on Intelligent Systems and Technology*. Article no. 12.
37. Frankle, J., & Carbin, M. (2019). The lottery ticket hypothesis: Finding sparse, trainable neural networks. In *International Conference on Learning Representations (ICLR).*
38. Jacob, B., Kligys, S., Chen, B., Zhu, M., Tang, M., Howard, A., Adam, H., & Kalenichenko, D. (2018). Quantization and training of neural networks for efficient integer-arithmetic-only inference. *Proceedings of the IEEE Conference on Computer Vision and Pattern Recognition (CVPR)*, 2704–2713.
39. Guo, Y. (2018). A survey on methods and theories on quantized neural networks. *CoRR, abs/1808.04752, arXiv:1808.04752.*
40. Courbariaux, M., Bengio, Y., & David, J.-P. (2015). Binaryconnect: Training deep neural networks with binary weights during propagations. *Advances in Neural Information Processing Systems*, 3123–3131.
41. Gong, Y., Liu, L., Yang, M., & Bourdev, L. (2014). Compressed deep convolutional networks using vector quantization. *CoRR, abs/1412.6115, arXiv:1808.04752.*
42. Rastegari, M., Ordonez, V., Redmon, J., & Farhadi, A. (2016). Xnor-net: Imagenet classification using binary convolutional neural networks. In *European Conference on Computer Vision (ECCV)* (pp. 525–542). Springer.
43. Tan, C., Sun, F., Kong, T., Zhang, W., Yang, C., & Liu, C. (2018). A survey on deep transfer learning. In *International Conference on Artificial Neural Networks (ICANN)* (pp. 270–279). Springer.
44. Brutzkus, A., & Globerson, A. (2019). Why do larger models generalize better? A theoretical perspective via the XOR problem. *International Conference on Machine Learning (ICML).*
45. Hinton, G., Vinyals, O., & Dean, J. (2015). Distilling the knowledge in a neural network. *Neural Information Processing Systems Deep Learning and Representation Learning Workshop.*
46. Bucilua, C., Caruana, R., & Niculescu-Mizil, A. (2006). Model compression. In *Proceedings of the 12th ACM SIGKDD International Conference on Knowledge Discovery and Data Mining (KDD)* (pp. 535–541). NY, USA.
47. Gou, J., Yu, B., Maybank, S. J., & Tao, D. (2020). Knowledge Distillation: a survey. *CoRR, abs/2006.05525 arXiv:2006.05525.*
48. McMahan, H. B., Moore, E., Ramage, D., Hampson, S., & Aguera-Arcas, B. (2017). Communication-efficient learning of deep networks from decentralized data. *AISTATS.*
49. Hard, A., Rao, K., Mathews, R., Ramaswamy, S., Beaufays, F., Augenstein, S., Eichner, H., Kiddon, C., & Ramage, D. (2018). Federated learning for mobile keyboard prediction. *CoRR, abs/ 1811.03604 (2018) arXiv: 1811.03604.*
50. McMahan, H. B., Ramage, D., Talwar, K., & Zhang, L. (2018). Learning differentially private recurrent language models. *International Conference on Learning Representations (ICLR).*
51. Qi, H., Brown, M., & Lowe, D. G. (2018). Low-shot learning with imprinted weights. *Conference on Computer Vision and Pattern Recognition (CVPR).*

Nonvolatile Memory-Based Internet of Things: A Survey

Ahmed Izzat Alsalibi, Mohd Khaled Yousef Shambour, Muhannad A. Abu-Hashem, Mohammad Shehab, Qusai Shambour, and Riham Muqat

1 Introduction

The IoT is a collection of smart objects that are wirelessly connected via smart sensors. The IoT consists of millions of smart devices "things" and it is considered as an integrated or extended part of the future Internet. Nowadays, the IoT dominate the research area of computer because it provides suitable solutions for various modernistic systems such as smart cities, transportation, emergency services, security, retails, automotive industries, agriculture, healthcare, and waste management [1].

A. I. Alsalibi (✉) · R. Muqat
Department of Information Technology, Faculty of Engineering and Information Technology, Israa University, Gaza, Palestine
e-mail: asalibi@israa.edu.ps

M. K. Y. Shambour
The Custodian of the Two Holy Mosques Institute for Hajj and Umrah Research, Umm Al-Qura University, Mecca, Saudi Arabia
e-mail: myshambour@uqu.edu.sa

M. A. Abu-Hashem
Department of Geomatics, Faculty of Architecture and Planning, King Abdulaziz University, Jeddah, Saudi Arabia
e-mail: Mabohasm@kau.edu.sa

M. Shehab
Department of Software Engineering, The World Islamic Science and Education University, Amman, Jordan
e-mail: mohammed.shehab@wise.edu.jo

Q. Shambour
Department of Software Engineering, Al-Ahliyya Amman University, Amman, Jordan
e-mail: q.shambour@ammanu.edu.jo

S. Pal et al. (eds.), *Artificial Intelligence-based Internet of Things Systems*,
Internet of Things, https://doi.org/10.1007/978-3-030-87059-1_11

The three main components in any IOT system are (i) processing units, (ii) main memory, and (iii) storage. Processing units are responsible for performing the logic operations, whereas the main memory is responsible for fast and direct access of data by the CPU (i.e., short-term access). Storage is responsible for delivering data to the CPU from secondary storage (i.e., long-term access), traditionally implemented by using HDD. However, nowadays, an emerging NVM type is designed to fit the requirements of evolution, such as NAND flash memory, NOR flash memory, MRAM, ReRAM, FeRAM, SCM, and STT-RAM. These types have the benefits of low-power consumption, supporting ULSI, and providing high reliability. The abovementioned benefits make NVMs the most favorable option for IOT systems and could replace the conventional storage such as HDD and volatile RAM [2].

Contributions In this chapter, we present a survey of techniques for designing and managing IoT using NVM.

The rest of the chapter is organized as follows: Section 2 mainly summarizes the characteristics and the challenges of various memory technologies.

Section 3 provides extensive detail about state-of-the-art techniques as follows: In Sect. 3.1, the IoT-based flash memory techniques are discussed. In Sect. 3.2, the IoT-based MRAM is presented. In Sect. 3.3, the IoT systems that use ReRAM are investigated. In Sects. 3.4 and 3.5, the IoT systems that use FRAM and SCM are reviewed, respectively. In Sect. 3.6, the techniques that use STT-RAM and TCAM are discussed. In Sect. 3.7 and 3.8, the IoT systems that use hybrid and others memories to improve the efficiency of IoT systems are discussed, respectively. Finally, summary and future research hints are given in Sect. 4.

Scope For the sake of a concise presentation, we limit the scope of this chapter as follows. We focus on NVM management techniques for IoT and not their circuit-level design issues. We focus on the key ideas of each work and include only selected quantitative results, since different works use disparate evaluation platforms and workloads. We hope that this chapter will be useful for computer architects, NVM designers, and researchers in the area of the IoT.[1]

---

[1] We use the following acronyms frequently in this chapter: application programming interface (API), convolution neural network (CNN), deep neural network (DNN), dynamic RAM (DRAM), embedded flash (eFlash), erasable programmable read-only memory (EPROM), error-correcting code (ECC), polyethylene glycol dimethacrylate (pEGDMA), execute in place (XIP), flash translation layer (FTL), finite impulse response filter (FIR), garbage collection (GC), hard disk drive (HDD), initiated chemical vapor deposition (iCVD), Landau-Lifshitz-Gilbert (LLG), magnetic RAM (MRAM), magnetic tunneling junction (MTJ), microcontroller unit (MCU), nonvolatile large-scale integrated (NV-LSI), one-state error recovery (OER), phase-change memory (PCM), polymer-intercalated resistive random access memory (i-RRAM), radio-frequency identification (RFID), self-write termination (SWT), silicon-oxide-nitride-oxide-silicon (SONOS), single NVM-based self-write termination (SWT1R-nvFF), single/multi/triple level cell (SLC/MLC/TLC), solid-state disk (SSD), spin-transfer torque random access memory (STT-RAM), static RAM (SRAM), storage class memory (SCM), storage memory management unit (SMMU), store energy (ES), structured query language (SQL), system on chip (SOC), ternary content-addressable mem-

2 Memory Overview

This section summarizes the characteristics and the challenges of various memory technologies. Figure 1 presents the device level properties of various memory technologies. In this figure, endurance refers to the number of erase cycles a memory block can withstand before it becomes invalid.

The latency and energy consumption of write operations are significantly higher than that of latency and energy consumption of read operations for all NVMs. The specific characteristics of different NVMs are discussed below.

Flash Memory There are two types of nonvolatile semiconductor flash memory: NOR and NAND. In NOR flash, one end of each memory cell is connected to the source line and the other end directly to a bit line resembling a NOR gate. In NAND flash memory, the memory cells are connected in series with 16 or 32 memory cells connected to the bitline and source line through two select transistors. Figure 2 depicts the schematic and circuit diagrams for NOR and NAND flash memory. NOR is a high-speed random access memory which can be programmed at byte level while NAND is similar to HDD. It is page-based and suited for storing sequential data such as pictures, audio, or PC data. Reading from NOR flash memory is

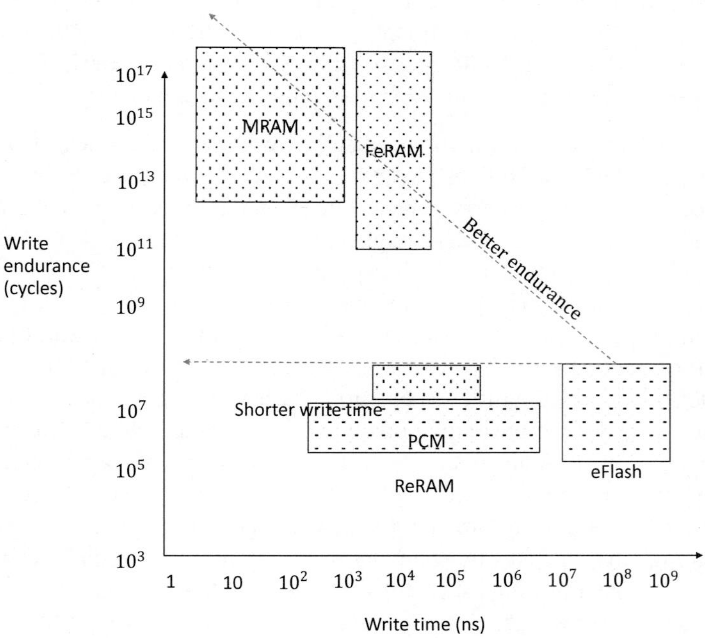

Fig. 1 Write endurance (cycles) versus write speed (ns) for different memory types

ory (TCAM), true random number generator using write speed variation of oxide-based RRAM (TRNG-UWSVOR), and ultra large-scale integration (ULSI).

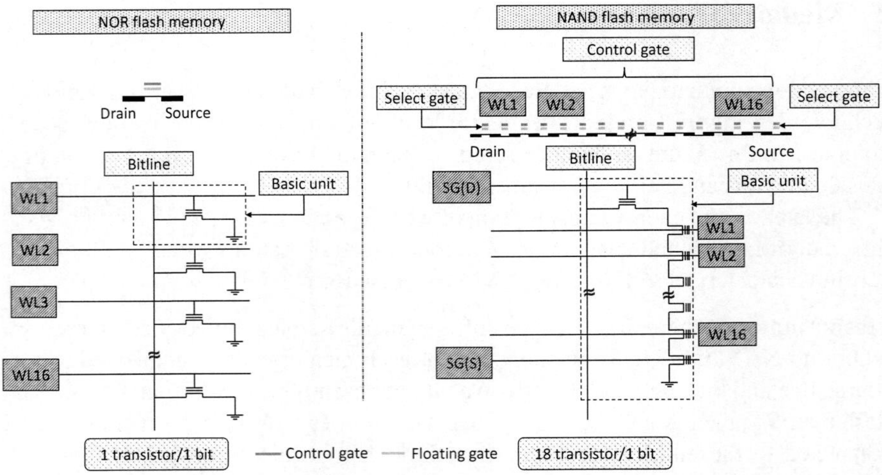

Fig. 2 Schematic cross sections and circuit diagrams for NOR and NAND flash memory (*WL* wordline, *SG(D)* select gate drain, *SG(S)* select gate source)

identical to reading from RAM. As a result, most microprocessors used NOR flash memory as XIP memory [3], which means that the program located in NOR flash memory has the ability to be directly executed from the NOR without the need to be copied into RAM before. NOR flash memory is generally used in IoT system devices like GPS and e-readers that do not require as much memory.

NAND flash memory is a high speed serial access and the speed of programming. In recent years, NAND flash memory has become popular to be used as secondary storage as it has the performance better than traditional storage. It is widely used as storage unit for data-heavy application in IoT systems, such as wearable devices, which require cheap and high-capacity storage.

ReRAM is one of the most promising memories, due to its low power consumption and high-speed operation. Panasonic started the mass production of 0.18 m ReRAM for wearable and IoT applications in 2013. The most important feature of this memory which distinguishes them from flash memory is that they are byte-addressable. IoT systems have traditionally used DRAM as a volatile memory and HDD or (SSD (i.e., flash memory)) as secondary storage. The gap in their speeds lead to large variance in their interfaces. ReRAM offer many features such as high density and high endurance comparing to flash memory. ReRAM is a form of non-volatile storage that operates by changing the resistance of a specially formulated solid dielectric material called a memristor – a contraction of "memory resistor" – whose resistance varies when different voltages are imposed across it [4].

ReRAM is based on a simple three-layer structure of a top electrode, switching medium, and bottom electrode as shown in Fig. 3. The resistance switching mechanism is based on the formation of a filament in the switching material when a voltage is applied between the two electrodes. There are different approaches to

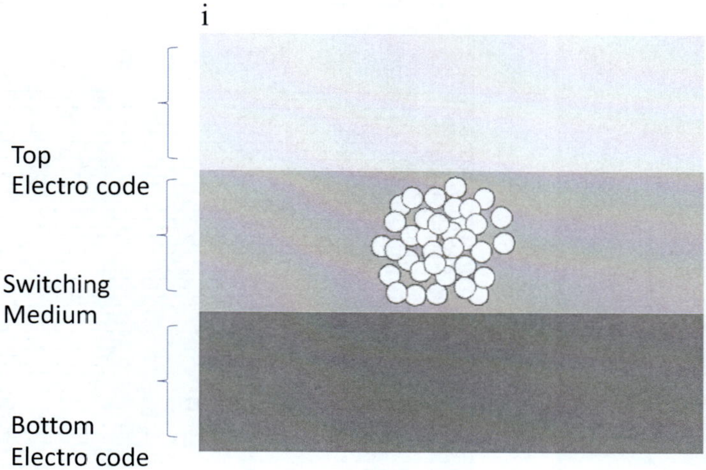

Fig. 3 ReRAM structure

implementing ReRAM, based on different switching materials and memory cell organization. ReRAM can be classified as organic or inorganic ReRAM based on the properties of the insulator. Because of its low cost, good flexibility, and simple structure, organic ReRAM is very advantageous for wearable device applications. Furthermore, ReRAM has many advantages over DRAM, PCM, and MRAM, such as water-resistant due to its internal structural simplicity and material stability [5]. For the above reasons, the ReRAM is expected to be used as universal memory technologies in IoT systems.

PCM has been widely considered as alternative to DRAM. PCM can store data by switching the GST film between two states, namely, amorphous (reset, i.e., high resistance) and crystalline (set, i.e., low resistance). Figure 4 illustrates the schematic drawing of the traditional mushroom-type PCM cell. PCM exploits the large difference between the resistivity of the amorphous and the crystalline phase in phase change materials where the electrical currents are applied to switch the material repeatedly between the two phases. The main advantages of PCM over DRAM are its nonvolatility, zero standby power, resiliency to soft errors, low read latency, and scalability. However, PCM consumes more power than DRAM, has long write latency, and has lower lifespan [6].

FeRAM the bit cell of FeRAM structure is similar to a DRAM bit cell except it uses a ferroelectric layer instead of a dielectric layer to achieve nonvolatility. Due to the high write speed and unlimited endurance of FeRAM, it is suitable for event tracking in autonomous vehicles and other industrial IoT applications.

FeRAM is significantly faster, reduces write energy, and uses nominal voltages, which make it suitable for IoT applications especially in RFID [8]. Figure 5 shows the schematic diagram of a typical FeRAM cell.

STT-RAM stores data in MTJ. As illustrated in Fig. 6, MTJ is composed of two ferromagnetic layers in addition to an isolating barrier interposed between them.

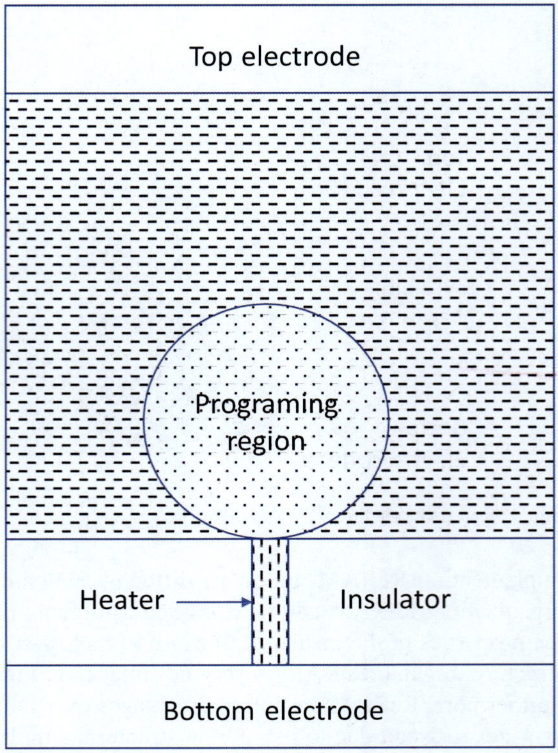

Fig. 4 PCM structure

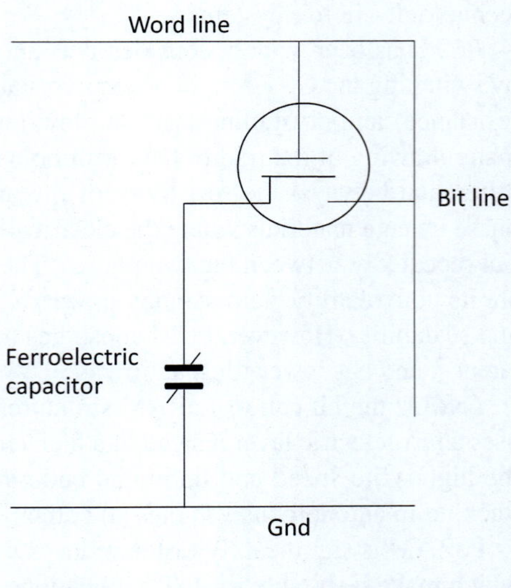

Fig. 5 FeRAM cell

This employed data storage technique has zero leakage power and can retain data without the need of any power supply for 10 years. Furthermore, STT-RAM has a comparable read speed than that of SRAM, and its single-cell structure presents

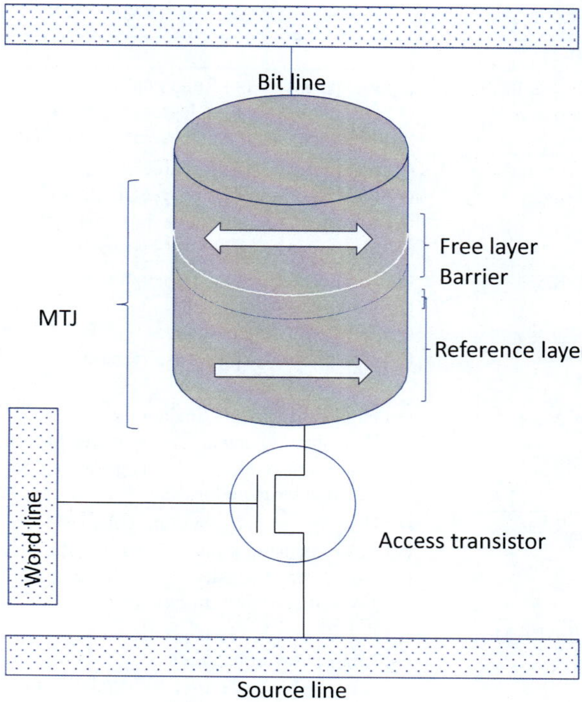

Fig. 6 STT-RAM structure [9]

much higher density than SRAM. STT-RAM is an emerging NVM technology that has better write energy, endurance, and performance than traditional NVM such as eFlash. In more details, it provides low latency (≈ ns), high lifespan (≈ 1016 cycles), and low energy consumption (≈ f J perbit). These features make it as a promising solution to replace the eFlash memory on IoTs. On the other hand, it has some disadvantages such as susceptible to data security and data privacy attacks, as well as magnetic attacks [10, 11]. Table 1 provides comparative evaluation of different NVMs in terms of pros and cons.

Table 2 classifies the previous works based on optimization metric, and it is clear that the techniques proposed are guided by multiple optimization goals which need to be carefully balanced.

3 State of the Art Techniques

The selection of tools for evaluating NVM-based IoT techniques is very important. While the simulation has flexibility to examine the designs which may currently not have any real implementation, it may be very slow to permit high exploration of the design space. In contrast, real system insemination permits more accuracy testing, and because of their high speed, it permits to execute huge number of procedures

Table 1 A comparative evaluation of different NVMs [7]

NVM	
NAND flash memory	(+) Inexpensive (+) Reliable (+) Flexible (+) High degree of density (−) Has lifespan (−) Does not support XIP
NOR flash memory	(+) Inexpensive (+) Reliable (+) Flexible (+) High degree of density (+) Ability to XIP without using too much power (−) Has lifespan (−) Very expensive
ReRAM	(+) High density (+) High endurance (+) Water-resistant (−) Limited write endurance (−) Resistance drift (−) Susceptibility to process variation (−) High write energy and latency
PCM	(+) High density (+) Scalability (+) Zero standby power (+) Low read latencies (−) Long write latency (−) High write energy consumption (−) Limited write endurance
FeRAM	(+) High speed (+) Nominal voltages (+) Power consumption (+) High endurance (−) Low storage density (−) Higher cost (−) Overall capacity limitation
STT-RAM	(+) Low latency (+) High lifespan (+) Low energy consumption (−) Susceptible to data security and data privacy attacks (−) Susceptible to magnetic attacks

Table 2 Classification based on optimization metric and evaluation platform

Category	References
Optimization metric	
Performance	[5, 12–27]
Energy consumption	[8, 11, 12, 16–24, 26, 28–32]
Accuracy	[14, 21]
Lifespan	[12, 14, 15, 18, 19, 21, 25]
Less area	[19, 20, 22, 23]
Water-resistant	[5]
Security	[11, 19, 33]
Cost	[23, 26]
Evaluation platform	
Real system/prototype	[8, 5, 11–14, 17–20, 25, 31]
Simulator	[15, 16, 21, 22, 24, 26–30, 32, 33]
Both simulator and real system	[23]

and instructions. Based on the above reasons, Table 2 also classifies the works based on their evaluation tools and platform. In the following sections, we discuss the previous techniques and mechanisms based on the type of NVM that has been used.

3.1 *Flash Memory*

Due to the explosion of data that IoT devices will collect, a huge numbers of processors and reliable embedded NVM will be required to process and manage the data from these IoT devices. Flash memory is currently the most widely used nonvolatile memory that has attracted great attention in recent years.

Balsamo et al. [12] argue that there is a need to integrate transient approaches within a general IoT programming framework such as ARM's mbed IoT device platform. The paper illustrates some current approaches in transient computing and highlights their appropriateness for operating with mbed. Following this, an assessment of the Hibemus approach on an NXP device employing mbed libraries and APIs is revealed as a validation of the transient approach with mbed. This was achieved specifically using MCU-dependent mbed APIs to access to the flash memory and MCU-independent mbed APIs to trace the allocated main SRAM memory to be saved.

Bando et al. [13] introduce three caching mechanisms that are believed to be essential parts for IoT-oriented single-level storage systems. In the first caching mechanism, the memory appears as a large main memory with the aid of SMMU where data can be accessed in a unified manner despite whether it is on the memory or on the DRAM. The second mechanism reduces the access latency to resource-limited IoT devices by caching the table of the FTL on the host DRAM.

The third one is an ad hoc device-to-device data relay where nearby IoT devices can access other's cached data by proximity-based wireless communication.

Deguchi et al. [14] propose two reliability enhancement techniques for 3D-TLC NAND flash-based SSD for DNN weight storage. The basic objective of the proposed mechanisms is to reduce the number of data bit errors by ignoring the unessential data bits to enhance memory performance. The first proposed technique, namely, OER, eliminates all of one-state errors in DNN weights when ECC fails to recover them. DNN weight data mapping, the second proposed technique, decreases errors of certain bits by assigning "0" to the reliable state of memory cells. The proposed memory has an extended data-retention lifetime by 700 times to achieve more than 10-year lifetime of IoT edge devices such as infrastructures and automobiles.

Shi et al. [15] propose an algorithm for data aggregation preprocessing called DAP to improve the efficiency of the flash memory. The proposed algorithm partitions the data into normal and hot data to decrease the number of transfers of valid page and obstruct erasures through GC. The algorithm has been validated on the flash simulation platform FlashSim. Experimental results demonstrate the usefulness of the proposed algorithm in terms of enhancing the IoT-based power grid storage system's performance, decreasing redundant write operations, reducing the number of physical block erasures, reducing the number of times of physical page read and write, and expanding the lifetime of solid-state devices.

Xu et al. [28] address the issue of reduction of power consumption in NVM by proposing two policies of cache management, which are the asymmetry-aware cache filling and the locality-aware cache flushing. The first technique was proposed to mitigate data locality and read/write asymmetry issues by triggering a new cache line filling only when all of them are occupied. Also, shadow cache, a small size SRAM, was introduced to predict data locality of a cache line. Shadow cache is looked up whenever a memory request misses in the working cache. The requested cache line will be migrated into shadow cache if it doesn't exist in it also. Locality-aware cache flushing was introduced to judge whether a cache line should be invalidated after being flushed proactively. The authors demonstrate the effectiveness of the proposed policies, using micro-benchmarks, in reducing the power consumption, and it outperforms the power-agnostic cache management policy.

3.2 MRAM

Due to the fact that nonvolatile memory-based circuits are becoming incredibly attractive for use in memory exhaustive applications such as digital FIR filters, Rao et al. [29] present a magnetic RAM-based digital FIR filter. In their paper, the authors develop a compact mathematical model for MTJ. The state of MTJ in this model is calculated in two steps: dynamic switching state and tunneling resistance. The authors employ LLG equation to accomplish the first step so that the behavior of the model can be replicated while Julliere's and Brinkman's equations are used to calculate the resistance and current flowing through the MTJ. Moreover, the paper proposes a new configuration, namely, 3T2MTJ, for MRAM that stores the data into two complimentary MTJs and designs a dual-port MRAM based on the proposed configuration. An evaluation of the performance of MRAM is conducted and compared with the classic SRAM-based system.

Senni et al. [16] investigate the employment of MRAM to design a process suitable for IoT. A validation of the rollback procedure, i.e., to completely restore a previously valid state of the processor when, for example, an error or power failure occurs, is also presented using RTL simulation. The rollback can be achieved by creating checkpoints to save the state of the system (both registers and main memory) either periodically or at strategic instants during the execution of the application. The proposed work shows that MRAM demonstrates the best performance and energy estimations of the backup/restore phases when compared with different NVMs.

Sun et al. [30] introduce a new computational architecture for CNN applications with MRAM memory. The architecture has been fabricated using the 22 nm device of STT-MRAM memory co-designed with processing-in-memory CNN accelerator. The basic operations of the proposed architecture include loading the CNN coefficients in MRAM and then, loading image into SRAM as it is faster for multiple read and write. After that, coefficients and image data are sent to CNN processing block for convolution operations. Finally, the convolution results will be sent to host processor. The MRAM CNN accelerator chip enables multiple models within one single chip, and it can be used for IoT and smart device applications.

Table 3 summarizes the advantages/disadvantages of the aforementioned techniques.

3.3 ReRAM

Due to the simple structure and low-power embedded NVM of ReRAM, it has been utilized in the IoT to enhance power consumption performance as well as stabilizing the signals by making them stronger and clearly captured. This section discusses the methods that utilize ReRAM in the field of IoT systems. As a response to resistive random access memory issues in solution-oriented process and switching mechanism uncertainty, Lee et al. [5] introduce a new organic ReRAM called i-RRAM that employs the iCVD process to deposit the polymer. In addition, pEGDMA with

Table 3 A comparative assessment of different system design and architecture techniques

Flash memory	
[12]	(+) Transient computing approach that enables computation to be sustained despite power outages (−) Limited lifespan for flash memory
[13]	(+) Introduce three caching mechanisms that are essential parts for IoT-oriented single-level storage systems
	(−) Using SLC composed of high cost
[14]	(+) Extended data-retention lifetime by 700 times to achieve more than 10-year lifetime of IoT edge devices (−) Using 3D TLC composed of low performance
[15]	(+) Enhances the IoT-based power grid storage system' performance. (+) Decreases redundant write operations. (+) Reduces the number of physical block erasures. (+) Reduces the number of times of physical page read and write. (+) Expands the lifetime of SSD. (−) Classifying hot and cold data composed of extra overhead. (−) Not considering address mapping. (−) Not considering wear leveling
[28]	(+) Reduce the power consumption (−) Flushing policy composed of extra overhead
MRAM	
[29]	(+) Reduce the power consumption by 77% as compared to SRAM-based design. (+) Reduces memory size by half by using NOR cells, add shift, and barrel shifter cells (−) Using NOR flash requires a much more complicated procedure when erasure procedure is invoked
[16]	(+) MRAM has the best performance and energy estimations of the backup/restore phases when compared with different NVMs (−) No real design (i.e., using a design kit) for MRAM is envisaged to accurately evaluate the energy, performance, and area overhead
[30]	(+) Designs MRAM CNN accelerator chip that enables multiple models within one single chip, and it can be used for IoT and smart device applications (+) Reduces the power efficiency to 9.9 TOPS/W with reliable read and write operations when compared with SRAM (−) Using CNN composed of high computational cost

a highly cross-linked structure was used as an insulator. It is shown that i-RRAM works stably even in water without any waterproof protection kit. The proposed memory presents a flexible, wearable, and waterproof organic memory as well as unambiguous switching mechanism.

Furthermore, Lo et al. [17] proposed SWT1R-nvFF for unipolar ReRAM devices. The proposed design aims to surmount the following challenges of nonvolatile nvFFs: the high waste of excessive ES in fast-switch as well as endurance and reliability degradation because of overwrite. The proposed nvFF has three modes: flip-flop, store (NVM-write), and restore (wake-up). The proposed method manages to avoid overwrite and reduces ES by combining single NVM with SWT.

In addition, Ueki et al. [18] proposed low-power 2 Mb ReRAM macro to overcome typical flash issues in data writing power consumption and reading time. The proposed ReRAM consists of a thin Ta2O5 switching layer inserted between a Ru bottom electrode and a metal cap. The function of bottom electrode is to stabilize high resistance states while the metal cap is used to control the forming voltage. The variation of off-state resistance was enhanced with a pulse modulated verify. In order to improve the high temperature retention, an interfacial layer, such as Ta2O5, was inserted on the Ru bottom electrode. This layer works as an oxygen supplier, and oxygen vacancies in the conductive filament are annihilated with the oxygen from the interfacial layer, resulting in the retention degradation.

TRNG-UWSVOR is proposed by Yang et al. [19]. The authors use reset speed variation as entropy source in their proposed circuit in order to diminish the complexity of circuit and generate more random bits. The speed variation amplifies the fluctuation of oxygen vacancy trap and de-trap which make the signal strong and has long duration to be easily and precisely captured. Self-adaptive control of write driver uses cell state detect module to sense the write end point accurately and give feedback to the arbiter module. The clock cycles during resets is used to trig a counter and then serialized into a random bit stream. The proposed work shows strength in the signals; also the signals are smoothly captured as it has long duration.

3.4 FeRAM

IoT devices are rapidly growing which triggers the need for storage tools with high capacity. Many methods have been proposed to support the IoT with strong and robust memory by utilizing and employing the capabilities of FRAM. Jeloka et al. [8] proposed charge recycling FRAM method to improve the efficiency of read and write operations. This proposed approach was introduced to address the following issues of the conventional FRAM: (1) every bit consumes at least CV2 energy to pull both the FeCap and the bitline capacitance high. (b) Moreover, most of this energy is wasted because only a small fraction of it is used to switch the FeCap polarity. The proposed method uses resonance between the FRAM array capacitance and an off-chip inductor to generate a sinusoidal clock in order to be able to read from and write to memory efficiently. Two transistors, namely, pull-up and pull-down, are employed to restore the energy consumed to write or read and maintain the amplitude of the resonating waveform.

3.5 SCM

Dinesh et al. [20] introduced a new memory interface intellectual property for interfacing IoT SoC with storage class memory or NVM. The proposed architecture employs the storage class memory as executable memory for IoT SoC while being invisible to core. In the current designs, one of the main issues is that the system has to be synchronized with a mobile device with higher memory or cloud in order to accurately process the collected data to reach conclusions and make decisions. This latency can be minimized by embedding the proposed IP interface as executable memory with parameterized cache that uses advanced caching algorithms. Table 4 summarizes the advantages/disadvantages of the aforementioned techniques.

Table 4 A comparative assessment of different system design and architecture techniques

ReRAM	
[5]	(+) Stable wearable waterproof memory with high flexibility
	(−) Using i-RRAM composed of overheating and high cost
[17]	(+) Decreases the amount of power needed to store data and keep away from overwrite operations
	(−) Using SRAM composed of low storage capacity, more complex design, and volatile (i.e., data is lost when memory is not powered)
[18]	(+) Considerable power reduction in data writing and speeds up data reading process (−) Using ReRAM composed of extra cost and difficulties in etching process
[19]	(+) Amplifies signals' strength so they would be captured clearly as they have long duration
	(−) Using TRNG not support enough flexibility because it is implemented in hardware, and verification of randomness is required
FRAM	
[8]	(+) Raises energy efficiency in read and write functions (−) Using FRAM composed of lower storage density, higher cost, and overall capacity limitation
SCM	
[20]	(+) Presents a practical solution for limited executable memory space issues (−) The support is limited for read and write functions. (−) Doesn't support write back in the IP

3.6 STTRAM and TCAM

Yu Lu [26] introduced a unified STT-MRAM-based memory system which combines both nonvolatile and volatile memories. A single memory subsystem can remove unnecessary energy-hungry transactions and dramatically improve the IOT energy efficiency. It can remove transactions that require large amount of energy and enhance IOT energy efficiency. Also, the proposed system minimizes the IoT weaknesses by providing stability and prevent anti-tearing and atomicity for secure transactions.

In IoT field, there are some major points that should be taken into account in the designing phase, for instance, decreasing power consumption, increasing security and intelligence, and nonvolatility. For that, Lai et al. [34] take the benefits of STT-MRAM to design the IoT device that improves the efficiency and achieves the stability in the processing data, as well as provides the non-power standby mode (i.e., keeps all IoT objects in the stand which leads to increase the battery life).

Replacing eFlash with SSTRAM is the main contribution of De et al. [10]. The authors utilized the SSTRAM instead of eFlash (associated with cost) to provide privacy and security for the data in the memory. In other words, STTRAM has the ability to bear the thermal and magnetic attacks which faces the IoT network architecture. STTRAM is based on the small free layer size to discover the attack and then discard it. Therefore, STTRAM helps the networks' devices to keep on executing without deadlock.

TCAM is a kind of content-addressable memory, which is characterized by its flexibility due to the mechanism of manipulating the data in the memory. In other words, TCAM uses "don't care" in the pattern matching [35].

Chang et al. [27] present the challenges in the design of the nonvolatile TCAM. These challenges are summarized as follows: (i) The tolerance of a small resistance ratio is low. (ii) The long search takes more time. (iii) The length of the word is limited. Therefore, the authors introduced a standard structure for the nonvolatile TCAM that is usable for different devices such as eFlash. Table 5 summarizes the advantages/disadvantages of the aforementioned techniques.

Table 5 A comparative assessment of different system design and architecture techniques

STTRAM	
[26]	(+) Reduce the weaknesses of the IoT devices and preventing a tearing for the safe transactions (−) Using MTJ requires extra reduction of damping and an increase of thermal stability
[11]	(+) Provides new various security and privacy challenges, like magnetic attacks. Also, reduce the number of corrupted bits (−) Partitioning EPROM into four parts or segments required extra overhead
TCAM	
[27]	(+) Improve the tolerance of a minor resistance ratio and reducing the consumption time of the searching (−) High cost (−) Each bit of storage in TCAM requires two bits: One for the value and the other for the mask

3.7 Hybrid

The great development of technologies used in our daily lives is an indicator of the rapid growth of applied science and technology area. Recently, there has been a clear trend toward improving the effectiveness and efficiency in many application technologies through inventing new techniques or improving existing techniques. This section discusses the improvement of energy efficiency of current flash memory techniques in the IoT platforms.

Cai et al. [21] used multiscale computing in several building blocks of NV-LSI circuits to examine the trade-off between saving and performance of power energy. The authors hybridized CMOS, MTJ, and NV-LSI to improve the utilization of power consumption and to reduce the sensing energy.

Chang et al. [22] used a single NVM device with bi-directional voltage-divider control scheme to develop a cell that has the ability to quick search with lesser energy consumption of search and write processes. Also, two forms of invalid-entry power consumption suppression are developed to improve the updating process of invalid-bits through adjusting the trade-off between area overhead and power consumption.

In order to reduce the execution time and power consumption of a microsystem, Chien et al. [23] designed a buffer between the MCU and flash memory using eReRAM instead of DRAM. The embedded controller in the ReRAM macro can perform the following tasks: (1) facilitate communication between the MCU and the ReRAM macro. (2) Perform built-in self-test where the whole ReRAM is checked by measuring the yield for three reference currents to calculate the best yield which is used to notify the MCU to fix the setting. (3) Perform built-in self-repair where column defects that should be repaired are recorded. (4) Compare data for the write-and-verify function. (5) Correct errors by using ECC algorithms. (6) Asymmetric coding, where data is reversed and marked by a bit flag in the case of the number of 1's in a word is greater than the number of 0's for yield enhancement. Simulation results show that eReRAM provides better performance compared to eSTT-MRAM in terms of cost, area, and reliability.

Jayakumar et al. [31] presented an energy-aware memory mapping system of program section in hybrid FRAM-SRAM micro controller devices. In the proposed technique, a program is partitioned to one or more atomic sections, namely, functions. eM-map performs a one-time characterization to find the optimal memory map for the functions that constitute a program. The function is assigned a memory map, and the eM-map performs the following operations: migration, execution, and checkpointing. Then, the total energy consumed in all three stages is calculated. If the function is executed successfully, the assigned memory map can be considered valid else eM-map iterates through all possible configurations for a function to arrive at the energy-optimal configuration. The checkpointing is performed at the end of a function in order to diminish the amount of data to be check pointed which in turn reduces the non-determinism.

Prasad et al. [24] introduced a hybrid LP8T2MTJ NV-SRAM cell through adding two nonvolatility MTJ and two additional transistors to 6T SRAM cell to reduce the energy of backup and restore operations. The proposed scheme uses MTJ to store and retrieve data held by the cell in order to accelerate the response wake times. The authors defined four operational modes for the proposed scheme including (1) normal where the proposed LP8T2MTJ cell acts as conventional 6T SRAM cell; (2) power off; (3) backup where the state of memory cell is written into the MTJ; and (4) restore mode where values stored in MTJs are reloaded into the cell nodes. The proposed architecture reduces the backup energy due to elimination of separate MTJ writing interface and additional control signals.

3.8 Others

The various benefits of NVM motivate Dou et al. [37] to present the challenges of developing NVM circuit devices for nonvolatile logics, computing-in-memory, and IoT security. NVM has several advantages including low operation voltage, compatibility with CMOS devices, and speed access. The authors define the main challenges of NVM devices including small R-ratio and large variations on cell behaviors.

Hayashikoshi et al. [32] introduced a power management technique with activity localization by rescheduling for efficient normally off computing where a task transits to power-off mode if the waiting time is extended over the breakeven time. Furthermore, autonomously standby mode transition methodology was also proposed to select the optimum standby mode of microcontrollers. The proposed NVRAM is divided into many blocks where only the target blocks are turned on and unnecessary blocks are turned off immediately.

An integration of an ultralow power transistor charge-trap embedded NVM into a 40 nm CMOS logic process is provided by Kouznetsov et al. [25]. The proposed NVM technology is based on SONOS transistor which consists of the following three components: a polysilicon gate (S), an oxide nitride oxide gate dielectric, and a silicon substrate (S). The integration of the eNVM requires only five masking layers. A typical bit cell of SONOS NVM comprises a SONOS control gate and a MOS select gate transistor connected in series. The authors developed a test chip containing an 8 Mb eNVM macro to demonstrate the capabilities and manufacture ability of the proposed SONOS eNVM technology.

Valea et al. [33] presented a secure context saving system to provide a context saving procedure for IoT devices that use NVM to store and retrieve the device state. When the power supply is going down, the system encrypts the context data and generates MAC signature and saved them into NVM. On the other side, when the power supply is becoming available, the encrypted data and the MAC signature are checked to confirm the integrity of the recovered data. Table 6 summarizes the advantages/disadvantages of the aforementioned techniques.

Table 6 A comparative assessment of different system design and architecture techniques

Hybrid	
[21]	(+) Enhances NV-LSI latency, power, and robustness. (+) dynamic power reduction (−) Using CMOS takes up more space on the chip
[22]	(+) Requires less area and enables longer WDL. (+) fast search speeds (−) Using MLC composed of extra power consumption and low lifespan
[23]	(+) Reducing of system execution time and power consumption (−) Using CMOS takes up more space on the chip
[36]	(+) Allows to benefit of both SQL-like and non-SQL DB solutions (+) Guarantees high stability and efficiency. (−) not using a real testbed
[31]	(+) Enhancing speedup of up to 2x and reducing energy consumption (−) Using hybrid FRAM-SRAM composed of complex implementation and management
[24]	(+) Reduction in backup energy. (+) reduction in restore energy (−) Not possible to refresh programs when using SRAM. (−) SRAM has more complex design
Others	
[32]	(+) Minimizing the total power consumption (−) Various sensor modules require extra management overhead
[25]	(+) Operating at the main power supply of 0.81–1.21 V with low current consumption in several operation forms (+) Affording 100 k erase/program cycles, 25 ns read access time, and data preservation for the numerous IoT applications (−) Using SONOS composed of low storage density
[33]	(+) Preventing attacks on the saved state of the IoT devices (−) Initial cost of design and development is very high for SOC

4 Conclusion and Future Outlook

In overall, the outcomes of this survey are likely to be valuable for researchers, technique makers, and NVM professionals working in the fields of IoT technologies.

To conclude, a brief summary is presented as follows:

Flash memory is appropriate to use in wearable devices, GPS, and e-card, due to its high reliability and low cost. However, limited number of erase cycles (i.e., lifespan) is the main limitation that prevents such memory to become more popular in IoT applications.

MRAM is suitable for the edge devices due its small area usage (i.e., more "cells" can be packed onto a single chip) and fast read/write times, high endurance, and strong retention. However, MRAM reliability is low because plates, which have a thickness of less than 1 nm, are complex to manufacture reliably.

ReRAM is proper for the SOC because of its lower read latency and a faster write performance. However, high cost and difficulties in etching process are the main challenges that prevent such memory to dominate in a variety of IOT applications.

FeRAM is suitable for the edge devices and MCU sensors due its low power consumption and high endurance. However, the low density and the high cost are the main challenges that prevent such memory to be widely used in IOT applications.

SCM bridges the performance and reliability gap that is acquired between DRAM and NAND in the memory hierarchy nowadays because it is using NAND as secondary storage and DRAM as main storage (i.e., a cache) for active data. However, its complex implementation and integration with DRAM and NAND flash memory are the main dilemma.

SSTRAM is with near-zero leakage power consumption and better scalability than conventional memory. However, the amount of electricity current needed to reorient the magnetization is currently too high for the most commercial IoT applications. The reduction of this electricity current density alone is the main research concern for present academic researcher in spin electronics.

Hybrid memory that exploits heterogeneity and parallelism of various NVM by combined different kinds of such memory in same IoT system is vital and could increase reliability, read and write performance, and lifespan, as well as reduce power consumption, cost, and area space, which in turn could increase the efficiency of the entire IoT system.

Future proposed models of the IoT should focus in providing syntactic interoperability along with various IoT devices. One of the most serious challenges of the syntactic interoperability among heterogeneous IoT devices is the security aspect that should be carefully considered in the near future.

References

1. Al-Turjman, F., Zahmatkesh, H., & Shahroze, R. (2019). An overview of security and privacy in smart cities' IoT communications. *Transactions on Emerging Telecommunications Technologies*, e3677. https://doi.org/10.1002/ETT.3677
2. Ghoneim, M., & Hussain, M. (2015). Review on physically flexible nonvolatile memory for internet of everything electronics. *Electronics, 4*, 424–479.
3. Rostam, A., & Abdullah, R. (2019). A proposed framework: Enhanced automated duplication algorithm for flash application. In *Intelligent and interactive computing* (pp. 377–387). Springer.
4. Mittal, S., & Vetter, J. S. (2015). A survey of software techniques for using non-volatile memories for storage and main memory systems. *IEEE Transactions on Parallel and Distributed Systems, 27*, 1537–1550.
5. Lee, B.-H., Bae, H., Seong, H., Lee, D.-I., Park, H., Choi, Y. J., Im, S.-G., Kim, S. O., & Choi, Y.-K. (2015). Direct observation of a carbon filament in water-resistant organic memory. *ACS Nano, 9*, 7306–7313.
6. Pourshirazi, B., Beigi, M. V., Zhu, Z., & Memik, G. (2019). Writeback-aware LLC management for PCM-based main memory systems. *ACM Transactions on Design Automation of Electronic Systems (TODAES), 24*, 18.
7. Mittal, S. (2018). A survey of ReRAM-based architectures for processing-in-memory and neural networks. *Machine Learning and Knowledge Extraction, 1*, 75–114.

8. Jeloka, S., Wang, Z., Xie, R., Khanna, S., Bartling, S., Sylvester, D., & Blaauw, D. (2018). Energy efficient adiabatic FRAM with 0.99 PJ/bit write for IoT applications. In *2018 IEEE symposium on VLSI circuits* (pp. 85–86). IEEE.
9. Chi, P., Li, S., Cheng, Y., Lu, Y., Kang, S. H., & Xie, Y. (2016). Architecture design with STT-RAM: Opportunities and challenges. In *2016 21st Asia and South Pacific design automation conference (ASP-DAC)* (pp. 109–114). IEEE.
10. De, A., Khan, M. N. I., & Ghosh, S. (2016). Attack resilient architecture to replace embedded flash with STTRAM in homogeneous IoTs. *arXiv* preprint arXiv:1606.00467.
11. De, A., Khan, M. N. I., Park, J., & Ghosh, S. (2017). Replacing eFlash with STTRAM in IoTs: Security challenges and solutions. *Journal of Hardware and Systems Security, 1*, 328–339.
12. Balsamo, D., Elboreini, A., Al-Hashimi, B. M., & Merrett, G. V. (2017). Exploring arm mbed support for transient computing in energy harvesting IoT systems. In *2017 7th IEEE international workshop on advances in sensors and interfaces (IWASI)* (pp. 115–120). IEEE.
13. Bando, Y., Watanabe, K., Maeda, K.-I., Kudo, H., Ishiyama, M., Kunimatsu, A., Nakai, H., Takahashi, M., & Oowaki, Y. (2015). Caching mechanisms towards single-level storage systems for internet of things. In *2015 symposium on VLSI circuits (VLSI circuits)* (pp. C132–C133). IEEE.
14. Deguchi, Y., & Takeuchi, K. (2018). 3D-NAND flash solid-state drive (SSD) for deep neural network weight storage of IoT edge devices with 700x data-retention lifetime extension. In *2018 IEEE international memory workshop (IMW)* (pp. 1–4). IEEE.
15. Shi, J., Zhang, H., Bai, Y., Han, G., & Jia, G. (2018). A novel data aggregation preprocessing algorithm in flash memory for IoT based power grid storage system. *IEEE Access, 6*, 57279–57290.
16. Senni, S., Torres, L., Sassatelli, G., & Gamatie, A. (2017). Non-volatile processor based on MRAM for ultra-low-power IoT devices. *ACM Journal on Emerging Technologies in Computing Systems (JETC), 13*, 17.
17. Lo, C.-P., Chen, W.-H., Wang, Z., Lee, A., Hsu, K.-H., Su, F., King, Y.-C., Lin, C. J., Liu, Y., Yang, H., et al. (2016). A ReRAM-based single-NVM nonvolatile flip-flop with reduced stress-time and write-power against wide distribution in write- time by using self-write-termination scheme for nonvolatile processors in IoT era. In *2016 IEEE international electron devices meeting (IEDM)* (pp. 16–13). IEEE.
18. Ueki, M., Takeuchi, K., Yamamoto, T., Tanabe, A., Ikarashi, N., Saitoh, M., Nagumo, T., Sunamura, H., Narihiro, M., Uejima, K., et al. (2015). Low-power embedded ReRAM technology for IoT applications. In *2015 symposium on VLSI technology (VLSI technology)* (pp. T108–T109). IEEE.
19. Yang, J., Lin, Y., Fu, Y., Xue, X., & Chen, B. (2017). A small area and low power true random number generator using write speed variation of oxide based RRAM for IoT security application. In *2017 IEEE international symposium on circuits and systems (ISCAS)* (pp. 1–4). IEEE.
20. Dinesh, M. K., & Bhakthavatchalu, R. (2016). Storage memory/NVM based executable memory interface IP for advanced IoT applications. In *2016 international conference on recent trends in information technology (ICRTIT)* (pp. 1–9). IEEE.
21. Cai, H., Wang, Y., de Barros Naviner, L. A., Yang, J., & Zhao, W. (2017). Exploring hybrid STT-MTJ/CMOS energy solution in near-/sub-threshold regime for IoT applications. *IEEE Transactions on Magnetics, 54*, 1–9.
22. Chang, M.-F., Lin, C.-C., Lee, A., Chiang, Y.-N., Kuo, C.-C., Yang, G.-H., Tsai, H.-J., Chen, T.-F., & Sheu, S.-S. (2017). A 3T1R nonvolatile TCAM using MLC ReRAM for frequent-off instant-on filters in IoT and big-data processing. *IEEE Journal of Solid-State Circuits, 52*, 1664–1679.
23. Chien, T.-K., Chiou, L.-Y., Sheu, S.-S., Lin, J.-C., Lee, C.-C., Ku, T.-K., Tsai, M.-J., & Wu, C.-I. (2016). Low-power MCU with embedded ReRAM buffers as sensor hub for IoT applications. *IEEE Journal on Emerging and Selected Topics in Circuits and Systems, 6*, 247–257.

24. Prasad, R. S., Chaturvedi, N., Gurunarayanan, S., et al. (2019). A low power high speed MTJ based non-volatile SRAM cell for energy harvesting based IoT applications. *Integration, 65*, 43–50.
25. Kouznetsov, I., Ramkumar, K., Prabhakar, V., Hinh, L., Shih, H., Saha, S., Govindaswamy, S., Amundson, M., Dalton, D., Phan, T., et al. (2018). 40 nm ultralow-power charge-trap embedded NVM technology for IoT applications. In *2018 IEEE international memory workshop (IMW)* (pp. 1–4). IEEE. Non-volatile memory based Internet of Things: A survey 23.
26. Lu, Y. (2016). Ultra-low-energy IoT memory architectures based on embedded STT-MRAM. In *2016 international symposium on VLSI technology, systems and application (VLSI-TSA)* (p. 1). IEEE.
27. Chang, M.-F., Chuang, C.-H., Chiang, Y.-N., Sheu, S.-S., Kuo, C.-C., Cheng, H.-Y., Sampson, J., & Irwin, M. J. (2016). Designs of emerging memory based non-volatile TCAM for Internet-of-Things (IoT) and big-data processing: A 5t2r universal cell. In *2016 IEEE international symposium on circuits and systems (ISCAS)* (pp. 1142–1145). IEEE.
28. Xu, Y., Yang, L., Hou, Z., Huo, Q., & Qiu, K. (2017). Energy-efficient cache management for NVM-based IoT systems. In *2017 IEEE international symposium on parallel and distributed processing with applications and 2017 IEEE international conference on ubiquitous computing and communications (ISPA/IUCC)* (pp. 491–493). IEEE.
29. Rao, S., Shashikanth, K., Srinivas, R., & Sunita, M. (2017). Magnetic ram based filter design for low power signal processing in IoT applications. In *2017 14th IEEE India Council international conference (INDICON)* (pp. 1–6). IEEE.
30. Sun, B., Liu, D., Yu, L., Li, J., Liu, J., Zhang, W., & Torng, T. (2018). MRAM co-designed processing-in-memory CNN accelerator for mobile and IoT applications. *arXiv* preprint arXiv:1811.12179.
31. Jayakumar, H., Raha, A., Stevens, J. R., & Raghunathan, V. (2017). Energy-aware memory mapping for hybrid FRAM-SRAM MCUS in intermittently-powered IoT devices. *ACM Transactions on Embedded Computing Systems (TECS), 16*, 65.
32. Hayashikoshi, M., Noda, H., Kawai, H., Murai, Y., Otani, S., Nii, K., Matsuda, Y., & Kondo, H. (2018). Low-power multi-sensor system with power management and nonvolatile memory access control for IoT applications. *IEEE Transactions on Multi-Scale Computing Systems, 4*, 784–792.
33. Valea, E., Da Silva, M., Di Natale, G., Flottes, M.-L., Dupuis, S., & Rouzeyre, B. (2018). Si ECCS: Secure context saving for IoT devices. In *2018 13th international conference on design & technology of integrated systems in nanoscale era (DTIS)* (pp. 1–2). IEEE.
34. Lai, H.-J., Huang, R.-Y., Huang, J.-Y., & Tsou, Y.-T. (2018). STT-MRAM application on IoT data privacy protection system. In *2018 IEEE international conference on consumer electronics-Taiwan (ICCE-TW)* (pp. 1–5). IEEE.
35. Kobayashi, M., Murase, T., & Kuriyama, A. (2000). A longest prefix match search engine for multi-gigabit IP processing. In *2000 IEEE international conference on communications. ICC 2000. Global convergence through communications. Conference record, Volume 3* (pp. 1360–1364). IEEE.
36. Fazio, M., Celesti, A., Villari, M., & Puliafito, A. (2014). The need of a hybrid storage approach for IoT in PaaS cloud federation. In *2014 28th international conference on advanced information networking and applications workshops* (pp. 779–784). IEEE.
37. Dou, C., Chen, W.-H., Chen, Y.-J., Lin, H.-T., Lin, W.-Y., Ho, M.-S., & Chang, M.-F. (2017). Challenges of emerging memory and memristor based circuits: Nonvolatile logics, IoT security, deep learning and neuromorphic computing. In *2017 IEEE 12th international conference on ASIC (ASICON)* (pp. 140–143). IEEE.

Integration of AI and IoT Approaches for Evaluating Cybersecurity Risk on Smart City

Roberto O. Andrade, Sang Guun Yoo, Luis Tello-Oquendo, Miguel Flores, and Ivan Ortiz

1 Introduction

Cities improve their decision-making capabilities by using emergent technologies such as the IoT, big data, and cloud computing. However, the hyper-connectivity product for using these technologies build new challenges for city officers because it increases the attack surface. City officers need to define the best cybersecurity strategies for minimizing the impact of cybersecurity attacks, so they require the use of a risk analysis methodology to identify the city's critical assets, the vulnerabilities of its components, and the possible attack vectors. Cities are dynamic and complex systems due to their different relationships in social, economic, and demographic axis, if they include technologies such as IoT, which are solutions with heterogeneous characteristics (several manufacturers), fast-growing (about 50 billion devices for the year 2030), and with a lack of strong security, could increase their complexity. In this context, traditional risk methodologies that are generally more static could be limited for evaluating systems and characterize the complexity,

R. O. Andrade (✉) · S. G. Yoo
Facultad de Ingeniería de Sistemas, Escuela Politécnica Nacional, Escuela Politécnica Nacional, Quito, Ecuador
e-mail: roberto.andrade@epn.edu.ec; sang.yoo@epn.edu.ec

L. Tello-Oquendo
College of Engineering, Universidad Nacional de Chimborazo, Riobamba, Ecuador
e-mail: luis.tello@unach.edu.ec

M. Flores
Department of Mathematics, Escuela Politécnica Nacional, Quito, Ecuador
e-mail: miguel.flores@epn.edu.ec

I. Ortiz
Facultad de Ingeniería y Ciencias Aplicadas, Universidad de las Américas, Quito, Ecuador
e-mail: ivan.ortiz@udla.edu.ec

S. Pal et al. (eds.), *Artificial Intelligence-based Internet of Things Systems*, Internet of Things, https://doi.org/10.1007/978-3-030-87059-1_12

uncertainty, ambiguity, and amplifying effect of cyberattacks on smart cities. To operationalize the systematic cyber-risk analysis, some researchers propose using Bayesian networks to evaluate the relationships and causes between the different components of a system in a probabilistic way. To model the cyber-risk relationships of smart cities, it is possible to abstract the city's physical features using digital twin approaches.

The IoT is a key component for building smart cities. This technology allows the generation of a digital twin smart city by obtaining data by sensing the different components of a smart city. However, its characteristics of heterogeneity and lack of advanced security expand the surface of cyberattacks; this increases the probability of affecting decisive and critical city operations. Traditional risk analysis methodologies require adapting to flexibility, dynamism, and a large amount of data from IoT ecosystems to detect threats, vulnerabilities, and risks in smart cities. In this context, Bayesian networks allow the construction of multifunction diagnostics and the development of predictive probability to face the challenges mentioned above. Therefore, the application of Bayesian networks in the field of cybersecurity brings dynamism to the traditional risk analysis methodology to evaluate IoT systems and mathematically model the smart city digital twin to simulate different types of attacks that can affect a smart city. This also allows predicting the possible impact of attacks to establish the best security strategies.

This study analyzes the use of artificial intelligence (AI) methodologies to evaluate cybersecurity risk in a smart city context due to the inclusion of emergent technologies such as big data, cloud, and the IoT. For this, we first performed an overview of cybersecurity attacks in the smart city context. Then, we analyze risk methodologies for evaluating the negative impact of cybersecurity attacks on smart cities, through the use of cognitive security techniques.

This study is structured as follows. Section 2 reviews briefly cybersecurity attacks in the context of smart cities. Section 3 introduces the risk methodology to evaluate uncertain and complex scenarios. Section 4 presents a model to assess cybersecurity risk in smart city based on Bayesian networks. Finally, Sect. 5 draws the conclusions.

2 Background

2.1 Smart City

According to the United Nations (UN), the global urban population is expected to increase to 68% by 2050. For the UN, the fastest increase is in megacities that host over 20 million inhabitants, which are located mostly in developing countries. For instance, Tokyo, New Delhi, Shanghai, Mexico City, and São Paulo are the world's most populous and largest cities with an agglomeration of 37 million, 29 million, 26 million, and 22 million inhabitants, respectively. This trend creates sustainability challenges for the cities [1]. The UN mentions that 828 million people lived in slums in 2015, lacking essential drinking water and sanitation services. Cities need to face urban growth and specific aspects like energy, pollution, sanitation services, and traffic. Cities integrate smart solutions to enhance the decision-making process and address these issues associated with urbanization. Smart cities arise as a novel dimension of urbanization due to the fusion of the city infrastructure with digital services and information [2]. A smart city can improve the power grid's resilience, prioritize road maintenance to match traffic needs, or react in events like an explosion or disease spread using integrated information [3]. A smart city is based on a sensor model that takes into account the sensing of diverse city aspects, such as automation, pollution, energy, and traffic and social synergies produced by people through diverse channels and components. For establishing the sensing of a city, smart cities use emergent technologies such as AI, big data, cloud, and the Internet of Things (IoT) [4] to embed computing and communication capabilities of physical objects of smart city [5]. Smart cities are building considering three fundamentals pillars: economic, environmental, and social (see Fig. 1).

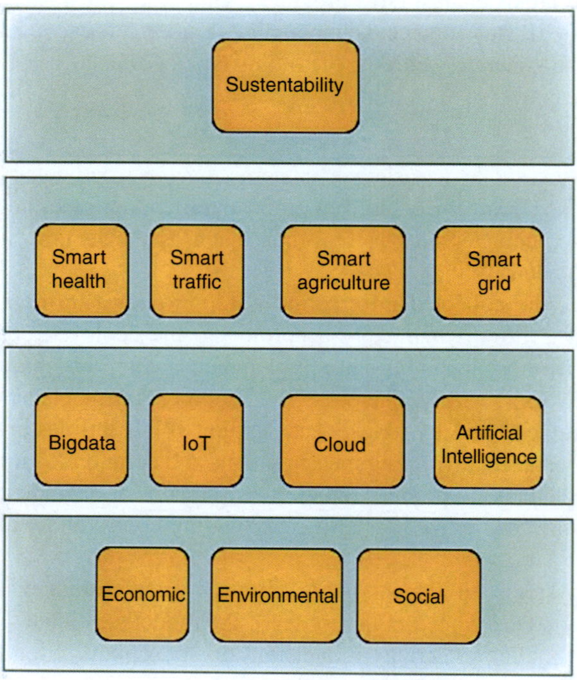

Fig. 1 Smart city components

2.2 *Smart City and Cybersecurity*

The different infrastructures and services in the smart city's domains may present vulnerabilities and expose the smart city to cyberattacks that expose confidentiality, integrity, and availability. For instance, through ransomware attacks on medical devices, attackers exploit vulnerabilities in SCADA systems.

All technology is susceptible to cybersecurity problems and being attacked by people with harmful intentions. For technologies that extend to the physical world, where there are, therefore, hybrid systems between digital and physical, the risks are higher, as in smart cities. In addition to causing severe consequences for a city's users, the attacks imply significant economic costs: the cybernetic firm IBM estimates that the average cost of a data breach is around 3.92 million dollars, an expense that could cause severe damage to any city. The systems that interact within the smart city environment are subject to a series of threats, materializing through different attack vectors, impacting the target asset or system. Security flaws are commonly exploited, whether in applications (e.g., Adobe Flash), protocols (e.g., in SSL), and devices (e.g., design flaws in smart TV), among others.

Recent cases indicate a much more alarming trend for the smart cities ecosystem: the most critical DDoS attacks are carried out using IoT devices. In October 2016, companies such as eBay, the New York Times, Netflix, PayPal, Spotify, and Twitter had serious difficulties providing their service or were inaccessible. The reason was a massive DDoS attack against DNS provider Dyn. For the most part, they used IoT elements (mainly IP surveillance cameras, digital video recorders, and home routers), managing to generate record traffic in this type of attack (the primary method of classifying its power).

Smart Traffic

In smart cities, several devices such as cameras, sensors, road detectors, and other roadside units (RSU) have been used to propose traffic control strategies that measure traffic flow characteristics in real-time, such as the length queues, vehicle speed, and traffic density.

Subsequently, the collected information is reported to a traffic controller that handles the traffic flow dynamically according to the deployed control strategy. However, most of these techniques are implemented on top of legacy traffic infrastructure, which lacks security. Therefore, the intelligent traffic management system becomes vulnerable to attack. Attackers can exploit vulnerabilities in these systems employing, for instance, disrupting measurements and communications (i.e., denial of service (DoS) attack) to produce phantom traffic jams or degrade the transportation service quality or replay attacks (i.e., providing controllers with false detection data).

Since these systems still depend on insecure sensor measurements, they must be accompanied by attack detectors that observe the model's estimates. These detectors can identify malicious activity if decisions do not support observations.

Smart Grids

Smart grids (SG) arose due to the necessity of modernizing the electricity grid, linking control and monitoring processes with green technologies known as eco-friendly or non-polluting. SGs are grid systems that integrate many computing and digital communication technologies and services into the electrical system infrastructure as described by the National Institute of Standards and Technology (NIST). Therefore, SGs go beyond the smart home and business energy meters, as bidirectional energy flows and bidirectional communication and control capabilities can bring new functionality. For example, SGs can provide a platform to maximize availability, efficiency, economic performance, reliability, and the highest security against attacks and natural power outages. These connected networks create new vulnerabilities caused by cyber intrusion and corruption, leading to devastating physical effects and considerable economic losses, meaning that the number of risks increases. For example, new nodes in the power grid create new entry points that attackers could exploit. Besides, new threats to computer systems appear day by day due to the rapid increase in sophisticated hacking tools; therefore, any telecommunications link within the electrical network represents a potentially unsafe way of operating it. Although the most obvious strategy for causing blackouts is through physical damage of generators, substations, and power lines, other activities compromise the sensors operation, communication devices, and control systems by identity theft or transmitting incorrect commands to control centers to disrupt the system. These activities cause power outages and in some circumstances, physically damages critical system components.

Smart Home

Cybersecurity attacks for their vulnerabilities have targeted smart homes. It is crucial to consider that, in most cases, users of this technology do not have in-depth technical knowledge for the proper use of the said technology. Smart homes potentially provide extra comfort, security, and more comprehensive environmental sustainability. For instance, a smart air conditioning system can employ Web-based data sources and various home sensors to perform smart operational decisions, rather than simplistic fixed-time or manual control schemes. This system can predict the house's expected occupancy by tracking location data. The air conditioner then saves energy when the house is empty or reaches the desired comfort level when it is occupied. Additionally, smart homes can help with daily tasks such as cleaning, cooking, laundry, and shopping. Nevertheless, the smart home system should be safe and reliable to exploit these benefits. In the following, some vulnerabilities that can be found in the devices of these houses are listed:

- The system hardware in smart home devices has several vulnerabilities, such as these devices' drivers. Such controllers are commonly tiny 8-bit microcontrollers with minimal computing resources and storage, limiting the capacity to implement complicated security algorithms. Another vulnerability they have in the devices comes from the factory since they contain different network standards and different software update capabilities.

– The most significant vulnerability for these devices is that the complexity of the networks lacks basic protocols; for this reason, in various situations, the owners of this type of house are not assisted by experts to give proper administration to the network.

Smart Health

The IoT is aimed at hospitals for monitoring the health status of patients. However, cyberattacks on IoT devices could cause death; for instance, an insulin overdose for a diabetic patient occurs when the blood glucose levels are below 70 mg/dL; if the patient is given insulin through an IoT device that is under attack, it could cause hypoglycemia under symptoms such as fainting, weakness, seizures, and respiratory problems, among others.

The attackers can be economically motivated for highly organized criminal syndicates to attack health and life sciences organizations to steal sensitive data and get valuable pharmaceutical and biomedical intellectual property.

In this sense, vulnerable applications and medical devices increment healthcare organizations' risk, especially healthcare providers. This risk can be attributed to more obsolete medical devices and applications that often run on antiquated and unsupported operating systems. When clinical networks are improperly segmented, there is a higher risk that IoT attacks could expand to medical systems.

Cyberattacks

There are many cyberattacks in smart city subdomains. Cyberattacks use different vector attacks, and their goal is different depending on the kind of attacks. In the following, some examples of cyberattacks on the smart city domain are presented:

– Man-in-the-middle: Attacker could intervene, interrupt, or modify the communications between two devices. For instance, the attacker could try to poison the air conditioning system by sending fake information from the thermostat and then listen to the conversation of other home's devices.
– Theft of data and identities: Attackers could access connected devices to obtain personal data because they have inadequate security. An attacker takes advantage of security lacks in IoT devices to intercept data flow and steal information to carry out possible fraudulent transactions and identity theft, among others.
– Device hijacking: The attacker takes control of a device but does not modify its functionality, but uses it as a vehicle to infect other connected devices. In this way, by infecting a smart plug, the attacker could access the smart lock and change the access PIN to open the doors.
– Distributed denial of service (DDoS): This attack attempts to stop the service to its legitimate user. Permanent denial of service (PDoS): This is the "strong" variant of DDoS since its purpose is to cause physical damage to the target devices so that they must be repaired or replaced. An example of this attack can be attacking a thermostat to provide false data and causing irreparable damage to other devices due to extreme overheating.
– Ransomware: Software whose purpose is to compromise the availability of information and systems. In most cases, a financial outlay is required by the

affected party to recover the data since the affected organization's operation can be compromised entirely, depending on the severity of the infection. As an approximation to the possible impact that ransomware could have on the smart city, in 2016, the health sector became a preferred target for cybercriminals due to their ransomware campaigns' success. In the same way, researchers have demonstrated the attack capacity of ransomware on IoT devices. For instance, on a thermostat, increasing the temperature to the maximum and then blocking it.

- Phishing: It is an attack that uses the impersonation of services and websites, misleading the user for stealing classified information, injecting malicious software, and other purposes. An example could be Anthem's case (a health insurance company), in which the use of LinkedIn made it possible to identify targets to initiate the attack, which allowed the stealing of data from 79 million people. Most of the attacks aim to steal credit card information and personal data, observing a tendency to direct this type of attack toward cloud systems due to the increase in their use. Along these lines, the theft of Office 365 credentials has been the target of recent phishing campaigns. The significant threat to smart cities focuses on the theft of credentials; this represents a gateway to access IoT devices to which users, whose data was acquired fraudulently, have authorization. In the case of critical infrastructure, the threat is even more significant.

A smart city has different technologies (IoT, AI, big data, cloud, mobile) in its core that are susceptible to cyberattacks. The IoT is maybe the most critical because it has a vast heterogeneous network of connected objects with humans in the loop interacting through apps. IoT ecosystems present several factors that could influence uncertainty as follows [6]:

- Error in the IoT design and implementation
- Physical variability, produced by actuators, sensors, and humans who are foreign to the system
- Presence of trusted agents and their roles
- Software-hardware used for data transfer
- Scalability
- User preferences for security and privacy
- Uncertainty in data-information-knowledge

3 Risk Analysis

The development of rigorous methods to evaluate cyber risk could minimize cyberattacks' effects and uncertainties on improving the smart city's cybersecurity posture and ensuring their long-term resilience.

Wu et al. [7] mention that the vulnerabilities are severe security threats that an attacker can misuse to obtain unapproved access to the system. Additionally, an attacker can strike additional hosts due to the interdependency among vulnerabilities. Moreover, Wu indicates that the attack success is the likelihood that (i) the vulnerabilities are favorably exploited and (ii) the vulnerability susceptibility depends on factors such as complexity, exploitability, and remediation level (see Eq. (1)).

$$P_i = \frac{Ac_i \times Av_i \times PR_i / RE_i}{m \backslash \left(AC_i \times Av_i \times PR_i / RE_i \right)_{i=1}} \tag{1}$$

where

- *Ac* denotes attack complexity.
- *Av* denotes attack vector; it reveals the context by which vulnerability exploitation is probable.
- *PR* denotes required privileges; it defines the level of privileges an attacker must hold.
- *RE* denotes the remediation level.

Broadly, risk analysis methods could be classified into two types:

1. Systematic risk, uncertainties generated by external factors that impact all. Generally, systematic risk is associated with "market risk"; it can analyze the economic risk in periods of economic weakness like recessions, wars, or fluctuations in currencies.
2. Unsystematic risk, specific uncertainty.

Other types of risk are the following: social risk, environmental risk, and operational risk.

3.1 Systematic Risk

According to Renn et al. [8], the four major components for dealing with systemic risk are complexity, uncertainty, ambiguity, and ripple effects beyond the source of risk.

The systemic crisis context considers two main phases: (i) the quiet phase and (ii) the crisis phase.

The inclusion of cognitive sciences (fuzzy engineering, Bayesian networks, convolutional networks) in cybersecurity could be considered to evaluate and predicted attacks, threats, and risk more adaptive and dynamically [9].

On the other hand, if the main objective of the attacker is to damage the control systems [10], therefore the attacker must follow the actions:

1. Infiltrate the field network.
2. Invalidate system functions.
3. Provoke incidents.

One typical modeling that is widely used to represent relationships or cause-effect is the Bayesian network.

3.2 Bayesian Network for Risk Analysis

Tundis et al. [28] mention that a growing interest in system risk analysis is the capability of showing the system under analysis interdependence with other systems and their interactions. Moreover, Tundis indicates that each systematic risk system's entity system generally belongs to a complex and heterogeneous network. Any network shock could be propagated in a non-uniform way, and it has the most significant speed depending on both exogenous and endogenous amplification factors. According to Tundis, modeling systemic risk is a complicated operation because need represents a risk in its dynamic and static characteristics. Tundis proposes developing a systematic risk analysis using Bayesian networks (BNs) based on goal diagrams. According to Tundis, the BN could describe the probabilistic relationships between faults/causes and failures/consequences. A probability function is used to associate each node with a particular set of values.

A BN is a probabilistic model that connects a set of random variables and their dependency relationships. These models use Bayesian inference, estimating the posterior probability of the unknown variables based on the known variables. BNs are a powerful form of machine learning to help decrease these models' false-positive rate.

The BN could establish models of attack knowledge and employ them to predict upcoming attacks and determine the risk [11, 12]. Likewise, BN can be used to evaluate the connectivity risk of protected core networking [13]. A risk assessment approach for telecommunication networks by utilizing the BN is introduced in [14] to examine the impact of attacks on the workflow. Zhang et al. [15] introduced a dynamic risk assessment using a fuzzy probability Bayesian network (FPBN) approach. Information risk factor analysis (FAIR) is one of the most traditional models for quantitative assessment of cybersecurity risks; Wang et al. [16] proposed a more flexible alternative approach (FAIR-BN), which implements the FAIR model using BNs. Zhu et al. [17] proposed a BN to analyze the spread of the attack over time and the consequences of the cyberattack on the industrial production process. They were evaluating dynamic cybersecurity risk quantitatively.

Judea Pearl first proposed the Bayesian network in 1985; these networks are also known as acyclic graph models or belief networks. It should be emphasized that BN as a risk management tool is one of the most important for the financial sector, because its significant contributions in the definition of probabilistic conditions of inference, with comprehensive management of variables, will be used to enhance the efficiency of the cybernetic attack detection systems.

Definition Let $G = (V,E)$ an acyclic directed graph, where V is a set of nodes and E is a set of edges. Then, let $X = (Xi)(i \in V)$, the random variable is represented by the node $i \in V$. The joint probability assigned to the node can be expressed as:

$$p(x) = i \in Vpx_i x_{pa(i)} \tag{2}$$

where $pa(i)$ is the node i parent. Moreover, for any random variable, their joint probability can be obtained by:

$$p(x_1,\ldots,x_K) = p(x_K \mid, x_1 \mid, \ldots \mid, x_{K-1}) \ldots p(x_2 \mid x_1) p(x_1). \tag{3}$$

Dynamic Bayesian networks (DBNs) are a temporal extension of BNs that allows to model dynamic processes. According to Onisko et al. [18], missing data signifies the most challenging task in developing a dynamic model. Nevertheless, there are several ways to handle this difficulty, one of which is to use an additional state to represent missing values. Onisko mentions that reasoning algorithms for BNs do not require complete information. Moreover, the posterior probability distribution over a variable in analysis can be derived given any subset of possible observations. Sequential or temporal information appears in many areas of engineering and science. On the one hand, the data analysis to predict future data may be of interest. On the other hand, analysis of the complete data sequence may also be necessary to identify patterns. This analysis can be carried out using dynamic Bayesian networks, a particular type of Bayesian networks specifically designed to model time series.

Ayele et al. [19] mention that DBNs are just Bayesian network applied for modeling temporal dependencies on time series structures. Simple BNs do not consider changes in time, and it cannot handle time-variant operating environments. DBNs are more useful for handling time-dependent risk scenarios.

The following steps are proposed to build DBNs:

1. Transforming indicator factors into a Markov chain process.
2. Define the discrete nodes' state.
3. Designate a marginal probability table for discrete root nodes and a conditional probability table for other discrete nodes.
4. Calculate the discretized conditional probability distributions of each continuous node.
5. Select the prior probability distribution for the system.
6. Construct the likelihood function, considering the system failure rate data.
7. Learning in a DBNs.
8. Computing the probabilistic inference (i.e., posterior distribution).

Some relevant aspects related to building DBNs which are mentioned by Ayele are the following:

1. A continuous variable (node) can take on a value between any other two values. Commonly, two approaches are used for managing continuous variables: static and dynamic discretization. The former needs to split the total range of the continuous variables into a finite number of intervals, while the latter provides fine-grained discretization in the regions that contribute notably to the density functions' structure.
2. The representation of a real-world problem by a DBN structure often necessitates the introduction of several nodes, and in such cases, conditional probabilities cannot be determined for all nodes precisely. This process is based on the expectation-maximization.

According to Frigault et al. [9], a model based on DBNs could be used to incorporate temporal factors; this graphical model is used for probabilistic inferences in dynamic domains that can permit users to monitor and update the system as time progresses and even predict other system behaviors [20]. Besides, they explain that in a standard DBN model, the system is represented as a BN sequence. Each BN represents a time interval of the DBNs corresponding to a given instant of time. Furthermore, the DBNs will have arcs between specific vertices of successive time sectors. In a DBN model, the Markovian property can be assumed to be satisfied, and the vertices can be classified as either observable or unobservable. The observable vertices value is known earlier during the analysis process, whereas that of unobservable vertices is not available but can be inferred.

According to Cabañas [21], a DBN is an extension of BNs in which random variables evolve over time. In the following image, you can see an example of DBNs with two variables, x_t and y_t, where the instant of time is denoted by t.

DBNs can contain two types of variables: observations represented by a square and hidden variables or state variables represented by a circle. In DBNs, the system's state only depends on the state in the previous instant y of the current observations (Markovian property). The DBNs in Fig. 2 models a hidden Markov model (HMM), which is the simplest type of DBNs: it contains a single discrete state variable and a single observation. However, DBNs are a more generic model. It may contain more variables, and the state variables may not be directly dependent on the observation. Given the topology of Fig. 2, the factorization of the DBNs at time T is:

$$P\left(X,Y\right) =_{t=1}^{T} p\left(x_t \mid x_{t-1}\right)_{t=0}^{T} p\left(y_t \mid x_t\right) p\left(x_0\right)$$

Therefore, to specify a DBNs, you need to define:

- Transition probability $p(x_t \mid x_{t-1})$.
- Probability of observation $p(y_t \mid x_t)$.
- Initial state probability $p(x_0)$.

The inference problem in DBNs consists of calculating $pX^T \mid Y^T$, where X^T denotes a finite set of observations and $Y^T = \{y_0, y_1, \ldots, y_T\}$ y X^T the set of corresponding hidden variables. Let $\alpha_t(x_t)$ be the joint probability of all observations y states up to time t:

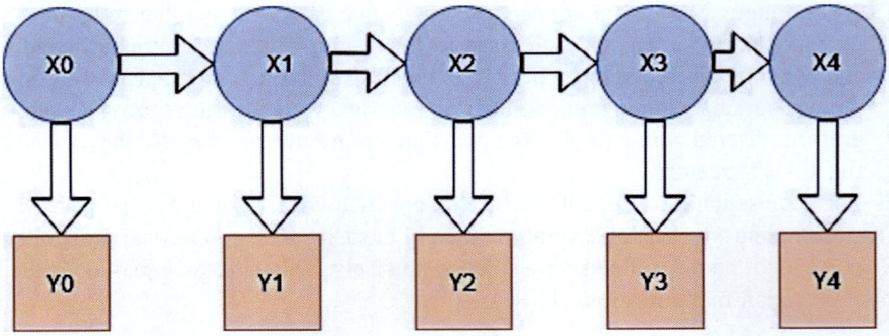

Fig. 2 HMM model DBNs [21]

$$\alpha_t\left(x_t\right) = p\left(y_t \mid x_t\right) \backslash p\left(x_i \mid x_{t-1}\right) \alpha_{t-1} x_t$$

with the initial condition of $\alpha_0(x_0) = p(x_0)$. If you want to calculate the most probable value of the hidden variable in the next instant, given the observations, you must apply Bayes' theorem:

$$p x_t \mid Y^t = \frac{), x_t p\left(x_{t+1} \mid x_t\right) \alpha_t\left(x_t\right)}{p\left(y_t\right)}.$$

3.3 *Digital Twin*

Cities are complex systems that connected economic, environmental, and social dynamics elements. According to Dembski et al. [22], complex systems include innovative and technological concepts, such as car-sharing and autonomous driving, decentralized smart energy grids, smart home, or digitization of administration tasks. A digital twin is associated with producing a parallel virtual version of the smart city that replicates humans' integration, infrastructure systems, and technology. According to Mohammadi and Taylor [19], connectivity and analytical capabilities enabled by the IoT are the base for the cognitive development of smart city digital twins. The IoT allows for the convergence of the physical and the virtual world and using machine learning algorithms focuses on the data provided by IoT solutions; digital twin tries modeling behaviors to detect anomalies and predict failures. Batty mentions that numerous studies define a digital twin as a "cyber-physical integration" for representing the real-world things with the same fidelity. On the other hand, the digital twin is considered the forefront of the Industry 4.0 revolution because it allows real-time decisions. The potential for digital twins within a smart city is associated with rapid developments in connectivity through the IoT. The digital twin is considered in Gartner's top ten strategic technology trends since 2017 [23].

According to the Smart Cities World Community [24], digital twins connect urban built environment spatial modeling, electrical and mechanical systems modeling based on deep learning informed training or mathematical descriptions, and real-time sensor data derived from IoT platform solutions. Among the advantages of modeling are cost savings, enhanced services for citizens, operational efficiencies, preventive maintenance, raised safety and security, and the inherent feasibility of automated generative design.

Digital twins can be employed for decision support, operator training, process control and monitoring, predictive maintenance, product development, real-time analytics, and behavior simulation [25]. Some sectors that considered the use of digital twins are the following: healthcare, maintenance, and urban sustainability.

Three main components constitute a digital twin (see Fig. 3):

- Physical environment that includes physical objects like human resources, cars, or buildings.
- Virtual environment that includes the virtualization model of the system or physical object.
- Connections of data between physical and virtual environment. Data is collected for sensors on physical objects, and then the data is processed using cognitive computational techniques; next, the virtualization model is built based on the behaviors or patterns detected on the data.

The main objective of a digital twin is to establish a decision context. There are four areas for decision-making in a digital twin [26]:

1. Knowledge representation means classifying and cataloging unstructured data using cognitive computing techniques.
2. Intelligent decisions, artificial intelligence (AI) can add context to these internal and external data sources and monitor processes.

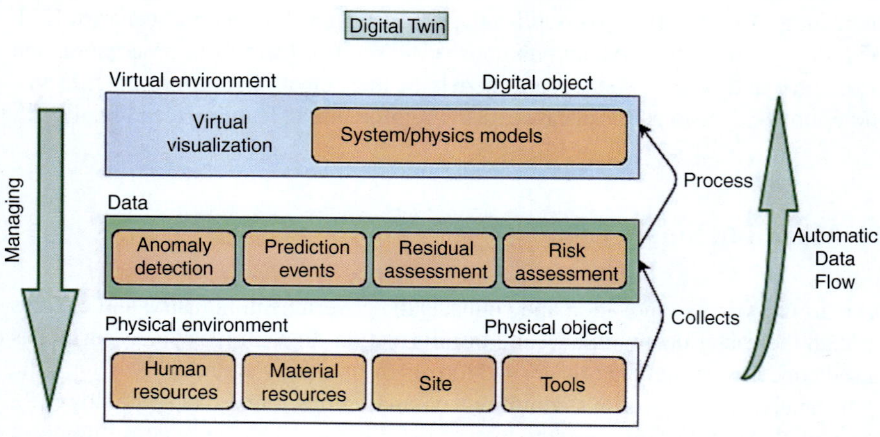

Fig. 3 Digital twin components

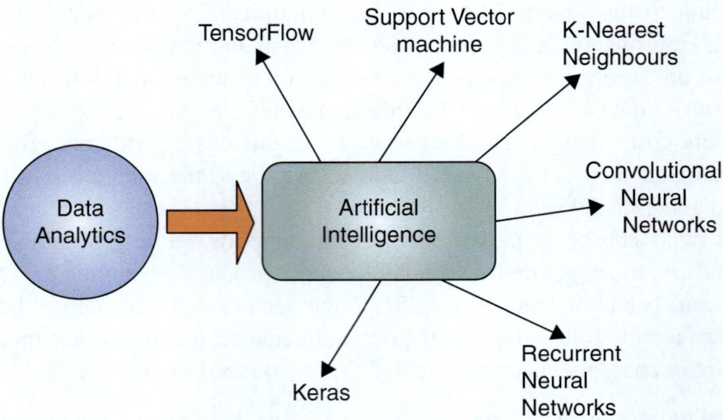

Fig. 4 Artificial intelligence applied to digital twin

3. Autonomous execution, AI identify values and thresholds for specific decision points based on past data for automated decision-making.
4. Enhanced assistance, multiple channels can allow users to interact with intelligent knowledge bases and drive value from the automated processes.

Industrial Internet of Things (IIoT) utilizes digital twins for implementation in the manufacturing industry. IoT systems can be controlled and optimized throughout their data lifecycle:

- Data from the manufacturing systems.
- Data from the Internet/users.
- Data from manufacturing.

Digital twin generates a virtual environment that can replicate an infinite number of scenarios. The simulated data is propagated and perpetuated through continuous embedding AI algorithms to establish the best decisions for specific scenarios [27]. With the introduction of AI, autonomous systems could learn from observation and experience and build the surrogate models of their environment to predict events and optimize decisions, for instance, using reinforcement learning (see Fig. 4).

4 Modeling Smart City

A smart city brings innovation and connects the government, industry, and citizens through the use of data with a wealth of information. In contrast, cybersecurity has raised concerns about data privacy and threats to smart city systems.

An analysis of the various definitions of smart cities found that technology is a constant element. For instance, the Institute of Electrical and Electronics Engineers (IEEE) suggests that a smart city support economy, mobility, environment, people, life, and governance through technology. TechTarget describes a smart city as a municipality that uses information and communication technologies to share information with the public, boost operational efficiency, and enhance the quality of government services and citizen welfare. Sivrikaya mentions that the smart city shows the following challenges [2]:

- Functionality: all information that is needed by smart city services has to be accessible.
- Heterogeneous Environment: the domain of a smart city consists of several subdomains. The data model needs to respond to a requester from a different domain.
- Dynamic Environment: the environment is permanently changing; this means that the data model used has to be dynamic.
- Huge amount of devices: there are thousands of devices or services in a city-scale environment.

Additionally, digital twins contain much information; however, it will be incomplete and imperfect. On the other hand, current risk models could be computationally heavy to run in real time during operation, or they do not capture the dynamics of risk for operational decisions due to lack of necessary detailing level.

It is essential to establish a dynamic risk model that can manage the lack of information and uncertainties under this context. Boje et al. [27] mentions that digital twin implementation could be represented in a semantic way through the acquired knowledge with the use of AI-enabled agents. On the other hand, Tundis based its process to model Bayesian network mapping rules from the SysML/UML of platform-independent models to risk analysis platform-dependent models. Information systems could be modeled for using UML diagrams [28]. This subsection defines one approach to modeling a smart city and determining its nodes and relationships. Dembski et al. [22] mention that cities are complex systems connected to economic, environmental, and social conditions and their changes. Additionally, they are also characterized by the perceptions and interests of citizens and stakeholders. A knowledge base is required with vocabularies and ontologies to manage the information diversity and overload [29]. Austin et al. [33] mention that a digital twin is a cyber representation of a physical system on the real time through monitoring and synchronization of data with events. According to Austin, social and natural domains generate difficulties in defining semantics and rules for the interaction on the smart city.

When faced with the challenge of representing a complex socio-technological system, OWL models' use is a critical step to ensure correct alignment among multiple domains such as actors, sensors, management workflows, Web resources, and BIM model data, among others. Petrova-Antonova proposes that digital city modeling follows the concept of a digital twin for providing data-driven decision-making. According to proposal of Petrova-Antonova, the city consists of *entities* and performs different *functions* summarized as follows:

1. *Entity* could be an *object* or an *actor*.
2. *Entities* have *state*.

3. *Actors* have *goals*.
4. *Goal* can be quantified using *indicators*.
5. *Actor* can perform *actions*.
6. *Functions* are used by a *group* of *actors*.
7. *Functions* can be grouped into *services*.
8. *Services* can be supported by *process*.
9. *Entities* participated in a *process*.
10. *Process* consisted of *events*.
11. *Actions* can change *state* and impacted *entity*.

Some instances of *object* are car, SCADA system, or traffic light, *actor* is driver or analyst, and *functions* are water and energy supply, waste collection, and traffic control. The *events* are recording according to the vocabulary of Recording and Incident Sharing (VERIS).

On the other hand, Austin proposes the following ontologies and rules for modeling the smart city.

- Urban building block level ontologies and rules: requirements, sensors, network, and control.
- Meta-domain ontologies and rules: temporal, spatial, units, and currency.

4.1 Experiment

For modeling a smart city, we follow the structure in [30]. A risk index is introduced to identify the vulnerability of a distribution system to be studied under cyberattack; this index is calculated as follows:

$$RI = V * C,$$

where V denotes the probability of the vulnerability used to initiate a successful attack and C indicates the outcomes produced by the attack.

Given that a dynamic network is used, according to Wang et al. [31] we need to consider the matrix of Table 1.

To build the Bayesian dynamic network of Fig. 5, the model of components of smart city shown in the Fig. 1 was taken as a reference.

The Bayesian server software helps to visualize the temporal nodes in all the times with which one works, as shown in Fig. 5. For each time, the network is the same, but the temporal node will have different probabilities depending on the time (see Fig. 6).

The following results have been obtained for each temporal node. These results indicate the probability with which the node that refers to one of the aspects of the city (SO, ECO, ENV) presents an attack or not knowing the information of all the parent nodes. The table results were obtained from Bayesian server software.

Table 1 Probability of cyberattacks on smart city

	$(t + 1)$ time				
(t) time	Very high	High	Middle	Low	Very low
Very high	0.4	0.2	0.2	0.1	0.1
High	0.2	0.4	0.2	0.1	0.1
Middle	0.1	0.2	0.4	0.1	0.1
Low	0.1	0.1	0.2	0.4	0.2
Very low	0.1	0.1	0.2	0.2	0.4

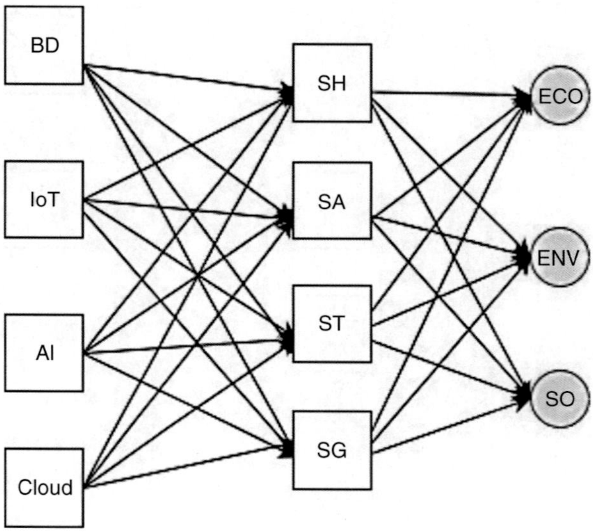

Fig. 5 BN of the smart city

In Table 2, we can see all the possible combinations of the parent nodes and the node's corresponding probabilities that refer to the city's social aspect regarding all the times considered. The resulting probabilities indicate that it is very likely that this aspect will be violated if attacks are carried out on the parent nodes because most of the probabilities of state 2 are greater than 50%.

On the other hand, in Table 3, we can observe all the possible combinations of the parent nodes and the ECO node's corresponding probabilities that refer to the city's economic aspect referring to all the times considered. The resulting probabilities indicate that the behavior is maintained over time; that is, if the node at time $t = 1$ has been violated in subsequent times, the probability that it will be violated again is similar to the probability presented at time $t = 1$ and in the same way if the node is not violated.

Finally, in Table 4, we can observe all the possible combinations of the parent nodes and the ENV node's corresponding probabilities that refer to the city's environmental aspect regarding all the times considered. The scenario is to change since if an attack co-occurs, the same thing will not necessarily happen later. From all the

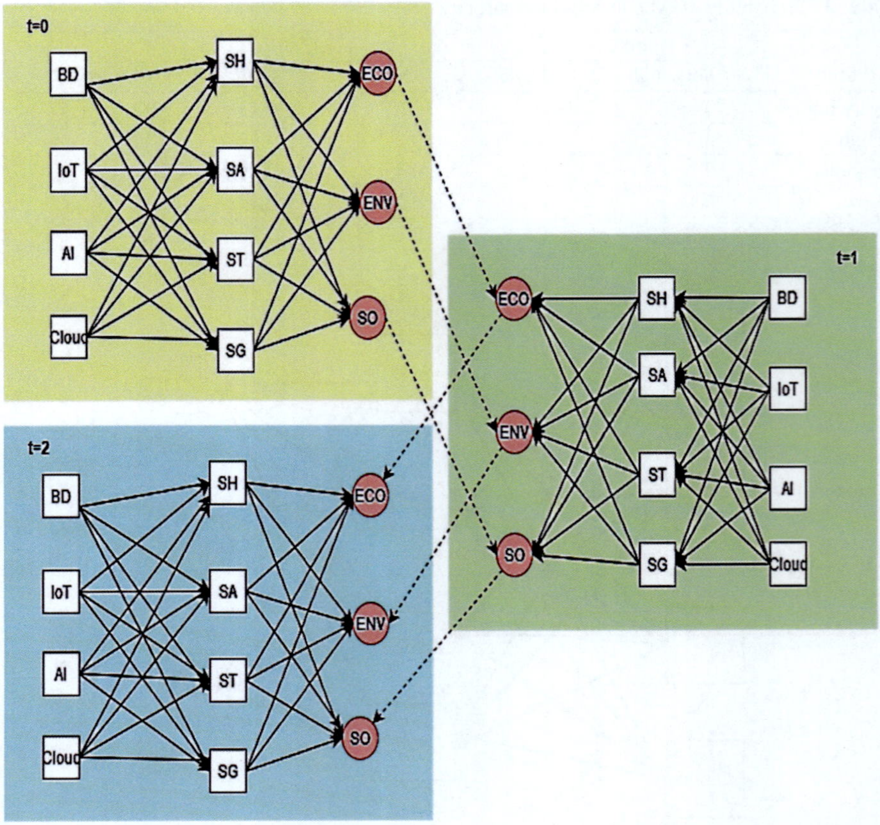

Fig. 6 DBN model of smart city

possible scenarios, it can be seen that the resulting information suggests that the node has a tendency to be attacked.

Additionally, we can observe possible scenarios that occur on the Internet, such as the following: From the model, the probability that the smart traffic node (ST) is attacked (state 2) or that it is not attacked (state 1) can be obtained by knowing information about all the parent nodes, which are the IoT, BD, AI, and cloud.

Table 5 shows all the possible scenarios that can happen in the smart traffic node concerning the attack or not of the parent nodes. It is logical that if the attacker decides not to take any action, the probability that there is no attack on the ST node should be greater than 0.5, which occurs. In the opposite case, if the attacker manages to damage all the parent nodes' systems, the probability of the existence of an attack on the ST node is high.

Another exciting result of this table is that if the attacker decides to damage only the IoT and AI systems, the probability of an attack on the ST node is 82%. Of all

Table 2 State of attacks technological components (columns BD, AI, Cloud, IoT), State of impact in social domain during time (columns 5 to 12)

BD	AI	Cloud	IoT	SO[t++1] = State1 F	SO[t++1] = State2 T	SO[t++2] = State1 F	SO[t++2] = State2 T	SO[t++3] = State1 F	SO[t++3] = State2 T	SO[t++4] = State1 F	SO[t++4] = State2 T
State1 false	State1 false	State1 false	State1 false	0.417	0.583	0.455	0.545	0.667	0.333	0.615	0.385
State1 false	State1 false	State1 false	State2 true	0.538	0.462	0.458	0.542	0.655	0.345	0.385	0.615
State1 false	State1 false	State2 true	State1 false	0.429	0.571	0.538	0.462	0.465	0.535	0.604	0.395
States false	State1 false	State2 true	State2 true	0.545	0.155	0.388	0.612	0.521	0.479	0.404	0.596
State1 false	State2 true	State1 false	State1 false	0.314	0.686	0.535	0.465	0.375	0.625	0.444	0.556
State1 false	State2 true	State1 false	State2 true	0.488	0.513	0.458	0.542	0.345	0.655	0.415	0.585
State1 false	State2 true	State2 true	State1 false	0.469	0.531	0.519	0.481	0.409	0.591	0.409	0.591
State1 false	State2 true	State2 true	State2 true	0.512	0.488	0.544	0.456	0.537	0.463	0.534	0.466
State2 true	State1 false	State1 false	State1 false	0.75	0.25	0.333	0.667	0.5	0.5	0.5	0.5
State2 true	State1 false	State1 false	State2 true	0.4	0.6	0.583	0.417	0.444	0.556	0.313	0.688
State2 true	State1 false	State2 true	State1 false	0.25	0.75	0.44	0.56	0.333	0.667	0.591	0.409
State2 true	State1 false	State2 true	State2 true	0.575	0.425	0.452	0.548	0.364	0.636	0.405	0.595
State2 true	State2 true	State1 false	State1 false	0.261	0.739	0.304	0.696	0.227	0.773	0.357	0.643

(continued)

Table 2 (continued)

BD	AI	Cloud	IoT	SO[t++1] = State1 F	SO[t++1] = State2 T	SO[t++2] = State1 F	SO[t++2] = State2 T	SO[t++3] = State1 F	SO[t++3] = State2 T	SO[t++4] = State1 F	SO[t++4] = State2 T
State2 true	State2 true	State1 false	State2 true	0.605	0.395	0.613	0.387	0.567	0.433	0.522	0.478
State2 true	State2 true	State2 true	State1 false	0.424	0.576	0.467	0.533	0.383	0.617	0.453	0.547
State2 true	State2 true	State2 true	State2 true	0.475	0.525	0.538	0.462	0.39	0.61	0.48	0.52

Table 3 State of attacks technological components (columns BD, AI, Cloud, IoT), State of impact in economic domain during time (columns 5 to 12)

BD	AI	Cloud	IoT	ECO[t++1] = State1	ECO[t++1] = State2	ECO[t++2] = State1	ECO[t++2] = State2	ECO[t++3] = State1	ECO[t++3] = State2	ECO[t++4] = State1	ECO[t++4] = State2
State1 false	State1 false	State1 false	State1 false	0.8	0.2	0.417	0.583	0.636	0.364	0.571	0.429
State1 false	State1 false	State1 false	State2 true	0.515	0.485	0.25	0.075	0.533	0.467	0.606	0.394
State1 false	State1 false	State2 true	State1 false	0.375	0.625	0.406	0.594	0.625	0.375	0.612	0.388
State1 false	State1 false	State2 true	State2 true	0.389	0.611	0.363	0.638	0.417	0.583	0.382	0.618
State1 false	State2 true	State1 false	State1 false	0.486	0.514	0.542	0.458	0.453	0.547	0.459	0.541
State1 false	State2 true	State1 false	State2 true	0.495	0.505	0.346	0.654	0.448	0.552	0.461	0.539
State1 false	State2 true	State2 true	State1 false	0.463	0.537	0.469	0.531	0.487	0.513	0.528	0.472
State1 false	State2 true	State2 true	State2 true	0.441	0.559	0.465	0.535	0.532	0.468	0.489	0.511
State2 true	State1 false	State1 false	State1 false	0.625	0.375	0.5	0.5	0.444	0.556	0.714	0.286
State2 true	State1 false	State1 false	State2 true	0.353	0.647	0.25	0.75	0.7	0.3	0.529	0.471
State2 true	State1 false	State2 true	State1 false	0.19	0.81	0.533	0.467	0.471	0.529	0.565	0.435
State2 true	State1 false	State2 true	State2 true	0.435	0.565	0.31	0.69	0.485	0.515	0.419	0.581
State2 true	State2 true	State1 false	State1 false	0.353	0.647	0.429	0.571	0.5	0.5	0.462	0.538

(continued)

Table 3 (continued)

BD	AI	Cloud	IoT	ECO[t++1] = State1	ECO[t++1] = State2	ECO[t++2] = State1	ECO[t++2] = State2	ECO[t++3] = State1	ECO[t++3] = State2	ECO[t++4] = State1	ECO[t++4] = State2
State2 true	State2 true	State1 false	State2 true	0.6	0.4	0.579	0.421	0.519	0.481	0.367	0.633
State2 true	State2 true	State2 true	State1 false	0.327	0.673	0.362	0.638	0.516	0.484	0.508	0.492
State2 true	State2 true	State2 true	State2 true	0.317	0.683	0.341	0.659	0.348	0.652	0.32	0.68

Table 4 State of attacks technological components (columns BD, AI, Cloud, IoT), State of impact in environmental domain during time (columns 5 to 12)

BD	AI	Cloud	IoT	ENV[t++1] = State1	ENV[t++1] = State2	ENV[t++2] = State1	ENV[t++2] = State2	ENV[t++3] = State1	ENV[t++3] = State2	ENV[t++4] = State1	ENV[t++4] = State2
State1 false	State1 false	State1 false	State1 false	0.471	0.529	0.533	0.467	0.667	0.333	0.6	0.4
State1 false	State1 false	State1 false	State2 true	0.611	0.389	0.063	0.031	0.8	0.2	0.645	0.355
State1 false	State1 false	State2 true	State1 false	0.659	0.341	0.661	0.333	0.633	0.367	0.675	0.325
State1 false	State1 false	State2 true	State2 true	0.388	0.612	0.505	0.494	0.43	0.57	0.452	0.548
State1 false	State2 true	State1 false	State1 false	0.486	0.514	0.462	0.338	0.553	0.447	0.581	0.419
State1 false	State2 true	State1 false	State2 true	0.644	0.356	0.507	0.493	0.527	0.473	0.675	0.325
State1 false	State2 true	State2 true	State1 false	0.463	0.537	0.5	0.5	0.518	0.482	0.045	0.55
State1 false	State2 true	State2 true	State2 true	0.556	0.444	0.598	0.402	0.542	0.458	0.596	0.404
State2 true	State1 false	State1 false	State1 false	0.4	0.6	0.333	0.667	0.4	0.6	0.5	0.5
State? True	State1 false	State1 false	State2 true	0.75	0.25	0.455	0.545	0.5	0.5	0.1	0
State2 true	State1 false	State2 true	State1 false	0.315	0.625	0.32	0.68	0.368	0.632	0.222	0.778
State2 true	State1 false	State2 true	State2 true	0.636	0.364	0.537	0.463	0.39	0.61	0.605	0.395
State2 true	State2 true	State1 false	State1 false	0.476	0.524	0.583	0.417	0.667	0.333	0.478	0.522

(continued)

Table 4 (continued)

BD	AI	Cloud	IoT	ENV[t++1] = State1	ENV[t++1] = State2	ENV[t++2] = State1	ENV[t++2] = State2	ENV[t++3] = State1	ENV[t++3] = State2	ENV[t++4] = State1	ENV[t++4] = State2
State2 true	State2 true	State1 false	State2 true	0.471	0.529	0.651	0.349	0.733	0.267	0.714	0.286
State2 true	State2 true	State2 true	State1 false	0.49	0.51	0.468	0.532	0.5	0.5	0.396	0.604
State2 true	State2 true	State2 true	State2 true	0.491	0.509	0.533	0.467	0.505	0.495	0.416	0.584

Table 5 Table of *P*(ST |IoT, BD, cloud, AI)

BD	IoT	Cloud	AI	ST = State1 false	ST = State2 true
State1 false	State1 false	State1 false	State1 false	0.843679144654573	0.156320855345427
State1 false	State1 false	State1 false	State2 true	0.581630990939282	0.418369009060718
State1 false	State1 false	State2 true	State1 false	0.404503680913654	0.595496319086346
State1 false	State1 false	State2 true	State2 true	0.926541832777806	0.0734581672221936
State1 false	State2 true	State1 false	State1 false	0.388417623255147	0.611582376744853
State1 false	State2 true	State1 false	State2 true	0.177589741034746	0.822410258965254
State1 false	State2 true	State2 true	State1 false	0.762891350464351	0.237108649535649
State1 false	State2 true	State2 true	State2 true	0.0146072597080552	0.985392740291945
State2 true	State1 false	State1 false	State1 false	0.540310390366177	0.459689609633823
State2 true	State1 false	State1 false	State2 true	0.89408127817979	0.10591872182021
State2 true	State1 false	State2 true	State1 false	0.638550973598681	0.361449026401319
State2 true	State1 false	State2 true	State2 true	0.516640393909009	0.483359606090991
State2 true	State2 true	State1 false	State1 false	0.516330120053403	0.483669879946597
State2 true	State2 true	State1 false	State2 true	0.8408S8418620032	0.159111581379968
State2 true	State2 true	State2 true	State1 false	0.671032592476116	0.328967407523884
State2 true	State2 true	State2 true	State2 true	0.064902111275846	0.935097888724154

the possible combinations, it can be seen that for the ST node to present an attack, the IoT nodes must have mainly been compromised.

Now consider another example in which it includes the leaf nodes (temporary nodes); we have the scenario where:

$$P\left(\mathrm{ECO}\left(t=1\right),|,\mathrm{ENV}\left(t=3\right),|,\mathrm{SO}\left(t=4\right),|,\mathrm{IoT},|,\mathrm{Cloud}\right)$$

In Table 6, it is observed that the probabilities of the possible events at time $t = 4$ in the social field, knowing that what happened in the environmental field at $t = 3$ and the economic one at $t = 1$, considering the parent nodes IoT and cloud, it is very unlikely that the city has presented an attack since the probabilities presented are very low in state 1 (no attack) and state 2 (attack).

Table 6 Probability of effects in Economic, Enviromental, Social domains due to attacks in IoT and Cloud infrastructures

IoT	Cloud	Economic ($t = 1$)	Environment ($t = 3$)	Social ($t = 4$) = state1 false	Social = state
State1 false	State1 false	State1 false	State1 false	0.11	0.13
State1 false	State1 false	State1 false	State2 true	0.15	0.11
State1 false	State1 false	State2 true	State1 false	0.1	0.14
State1 false	State1 false	State2 true	State2 true	0.15	0.12
State1 false	State2 true	State1 false	State1 false	0.16	0.087
State1 false	State2 true	State1 false	State2 true	0.2	0.083
State1 false	State2 true	State2 true	State1 false	0.11	0.13
State1 false	State2 true	State2 true	State2 true	0.14	0.091
State2 true	State1 false	State1 false	State1 false	0.11	0.09
State2 true	State1 false	State1 false	State2 true	0.14	0.07
State2 true	State1 false	State2 true	State1 false	0.13	0.17
State2 true	State1 false	State2 true	State2 true	0.15	0.14
State2 true	State2 true	State1 false	State1 false	0.13	0.12

Since the temporal nodes do not have connections between them, this network does not provide interesting information if you want to know information regarding several different temporal nodes at different times, for example, if we calculate $P(\text{ECO } (t = 1), \text{SO } (t = 2) \mid \text{BD, Cloud, AI, IoT})$. The model gives the results that if all the parent nodes are violated, and there has been a failure in the economic aspect in time 1, in time 2 in the social aspect, there is a 34% probability of an attack and 26% if there is no attack; note that the results are not significantly different compared to the previous result where information is obtained from the same temporal node. Following this logic, it is 25% probable that there is no failure in the economic aspect and 38% probable that there is a failure given that the environmental aspect and all the parent nodes of the network have been violated.

5 Conclusions

Cities are complex systems that connected economic, environmental, and social dynamics elements. The different infrastructures and services in the smart city's different domains may present vulnerabilities and expose the smart city to cyberattacks that expose the aspects of confidentiality, integrity, and smart city availability. Digital twins can be applied for developing decision support, real-time analytics, and behavior simulation in a smart city context that could support the modeling to detect cyberattacks [32–34].

Ecosystems, organizations, or informatics systems enhanced their cybersecurity posture based on reducing vulnerabilities. Attackers exploit vulnerabilities to perform cyberattacks. In the IoT environment, security should be considered a key fact since various devices communicate with multiple communication technologies. When heterogeneous devices interact, security vulnerabilities of each device are gathered, and new security vulnerabilities may happen.

The IoT is a crucial element of the evolution of a smart city. In the upcoming years, billions of devices will be raised and interconnected by the IoT, as endorsed by international consulting firms. It is worth noting that the IoT is vulnerable to cybersecurity attacks; this would strike the safety of the smart cities.

The modeling of the dynamic Bayesian network is interactive, and this allows a knowledge of the network with which one is working. Our model can be improved if a priori information given by experts is considered; Since the nodes are of type Boolean, state 1 represents no attack in the node, and state 2 represents the existence of an attack. The current network can be improved by considering more states in the temporary nodes.

The development of rigorous methods to minimize cyberattacks' effects and uncertainties could improve a smart city's cybersecurity posture. Incorporating risk management could improve the smart cities' capability to assure long-term resilience.

We could combine the proposal of risk-based layered security with machine learning or data mining techniques to evaluate, enhance, and measure the cybersecurity on IoT solutions adopted in smart cities.

The simulation scenario is based on simulated data from cyberattacks, we considered as future work to include dataset with a greater variety of cybersecurity attacks to evaluate the accuracy of the risk analysis model on smart cities. We consider important to validate the relationships between the different nodes to evaluate the possible entry and exit points of the attack surface.

References

1. United Nations. (2020). *Smart cities for sustainable development*. United Nations University. [online] Available at: https://unu.edu/projects/smartcitiesfor-sustainabledevelopment.html. Accessed 23 Sept 2020.

2. Sivrikaya, F., Ben-Sassi, N., Dang, X.-T., Gorur, O. C., & Kuster, C. (2019). Internet of smart city objects: A distributed framework for service discovery and composition. *IEEE Access*, 14434–14454. https://doi.org/10.1109/access.2019.2893340
3. Gordon, L., & McAleese, G. (2020). *Resilience and risk management in smart cities*. [online] Center for Infrastructure Protection & Homeland Security. Available at: https://cip.gmu.edu/2017/07/06/resilience-risk-management-smart-cities/. Accessed 17 Sept 2020.
4. Andrade, R. O., & Yoo, S. G. (2019). A comprehensive study of the use of LoRa in the development of smart cities. *Applied Sciences, 9*, 4753.
5. Kaur, M. J., Mishra, V. P., & Maheshwari, P. (2019). The convergence of digital twin, IoT, and machine learning: Transforming data into action. *Digital Twin Technologies and Smart Cities*, 3–17. https://doi.org/10.1007/978-3-030-18732-3_1
6. Alagar, V., & Wan, K. (2019). *Understanding and measuring risk due to uncertainties in IoT*. 2019 IEEE international conference on Smart Internet of Things (SmartIoT). https://doi.org/10.1109/smartiot.2019.00088
7. Wu, W., Kang, R., & Li, Z. (2015). *Risk assessment method for cybersecurity of cyber-physical systems based on inter-dependency of vulnerabilities*. 2015 IEEE international conference on industrial engineering and engineering management (IEEM). https://doi.org/10.1109/ieem.2015.7385921
8. Renn, O. (2020). New challenges for risk analysis: Systemic risks. *Journal of Risk Research*, 1–7. https://doi.org/10.1080/13669877.2020.1779787
9. Frigault, M., Wang, L., Singhal, A., & Jajodia, S. (2008). Measuring network security using dynamic Bayesian network, 23–30. https://doi.org/10.1145/1456362.1456368
10. Zhang, Q., Zhou, C., Xiong, N., Qin, Y., Li, X., & Huang, S. (2016, October). Multimodel-based incident prediction and risk assessment in dynamic cybersecurity protection for industrial control systems. *IEEE Transactions on Systems, Man, and Cybernetics: Systems, 46*(10), 1429–1444. https://doi.org/10.1109/TSMC.2015.2503399
11. Poolsappasit, N., Dewri, R., & Ray, I. (2012, January/February). Dynamic security risk management using Bayesian attack graphs. *IEEE Transactions on Dependable and Secure Computing, 9*(1), 61–74.
12. Xie, P., Li, J. H., Ou, X., Liu, P., & Levy, R. (2010, June). Using Bayesian networks for cyber security analysis. In *Proceeding of the IEEE/IFIP international conference of dependable systems and networks (DSN), Chicago, IL, USA*, pp. 211–220.
13. Wrona, K., & Hallingstad, G. (2010). Real-time automated risk assessment in protected core networking. *Telecommunication Systems, 45*(2–3), 205–214.
14. Szpyrka, M., Jasiul, B., Wrona, K., & Dziedzic, F. (2013). Telecommunication networks risk assessment with Bayesian networks. In *Computer information systems and industrial management* (LNCS 8104) (pp. 277–288). Springer.
15. Zhang, Q., Zhou, C., Tian, Y., Xiong, N., Qin, Y., & Hu, B. (2018). A fuzzy probability Bayesian network approach for dynamic cybersecurity risk assessment in industrial control systems. *IEEE Transactions on Industrial Informatics, 14*, 2497–2506.
16. Wang, J., Neil, M., & Fenton, N. (2020). A Bayesian network approach for cybersecurity risk assessment implementing and extending the FAIR model. *Computers and Security, 89*, 101659.
17. Zhu, Q., Qin, Y., Zhou, C., & Gao, W. (2018). Extended multilevel flow model-based dynamic risk assessment for cybersecurity protection in industrial production systems. *International Journal of Distributed Sensor Networks, 14*, 155014771877956.
18. Onisko, A., & Marshall Austin, R. (2015). Dynamic bayesian network for cervical cancer screening. *Lecture Notes in Computer Science*, 207–218. https://doi.org/10.1007/978-3-319-28007-3_1
19. Mohammadi, N., & Taylor, J. E. (2017). *Smart city digital twins*. 2017 IEEE symposium series on computational intelligence (SSCI). https://doi.org/10.1109/ssci.2017.8285439
20. Mihajlovic, V., & Petkovic, M. (2001). *Dynamic Bayesian networks: A state of the art*. University of Twente.
21. Cabañas, R. (2011). *Reconocimiento de gestos mediante Redes Bayesianas Dinámicas*. USLM.

22. Dembski, F., Wössner, U., Letzgus, M., Ruddat, M., & Yamu, C. (2020). Urban digital twins for smart cities and citizens: The case study of Herrenberg, Germany. *Sustainability, 12*(6), 2307. https://doi.org/10.3390/su12062307
23. Gartner. (2020). *Top 10 strategic technology trends for 2017: Digital twins*. [online] Available at: https://www.gartner.com/en/documents/3647717. Accessed 23 Sept 2020.
24. Smart Cities World. (2020). *The rise of digital twins in smart cities*. [online] Available at: https://www.smartcitiesworld.net/special-reports/special-reports/the-rise-of-digital-twins-in-smart-cities. Accessed 21 Sept 2020.
25. Fuller, A., Fan, Z., Day, C., & Barlow, C. (2020). Digital twin: Enabling technologies, challenges and open research. *IEEE Access*, 108952–108971. https://doi.org/10.1109/access.2020.2998358
26. Cronrath, C., Aderiani, A. R., & Lennartson, B. (2019). *Enhancing digital twins through reinforcement learning*. 2019 IEEE 15th international conference on automation science and engineering (CASE). https://doi.org/10.1109/coase.2019.8842888
27. Boje, C., Guerriero, A., Kubicki, S., & Rezgui, Y. (2020). Towards a semantic construction digital twin: Directions for future research. *Automation in Construction, 114*, 103179. https://doi.org/10.1016/j.autcon.2020.103179
28. Tundis, A., Garro, A., Gallo, T., Saccà, D., Citrigno, S., Graziano, S., & Mühlhauser, M. (2017). Systemic risk modeling and evaluation through simulation and bayesian networks, 1–10. https://doi.org/10.1145/3098954.3098993
29. Petrova-Antonova, D., & Ilieva, S. (2021). Digital twin modeling of smart cities. In T. Ahram, R. Taiar, K. Langlois, & A. Choplin (Eds.), *Human interaction, emerging technologies and future applications III* (pp. 384–390). Springer.
30. Dai, Q., Shi, L., & Ni, Y. (2018). *Risk assessment for cyber attacks in feeder automation system*. 2018 IEEE power & energy society general meeting (PESGM), Portland, OR, 2018, pp. 1–5. https://doi.org/10.1109/PESGM.2018.8586312
31. Wang, J., Fan, K., Mo, W., & Xu, D. (2016). *A method for information security risk assessment based on the dynamic Bayesian network*. 2016 International conference on networking and network applications (NaNA), Hakodate, pp. 279–283. https://doi.org/10.1109/NaNA.2016.50
32. Ahram, T., Taiar, R., Langlois, K., & Choplin, A. (Eds.). (2021). Human interaction, emerging technologies and future applications III. *Advances in Intelligent Systems and Computing*. https://doi.org/10.1007/978-3-030-55307-4
33. Austin, M., Delgoshaei, P., Coelho, M., & Heidarinejad, M. (2020). Architecting smart city digital twins: Combined semantic model and machine learning approach. *Journal of Management in Engineering, 36*(4), 04020026. https://doi.org/10.1061/(asce)me.1943-5479.0000774
34. Zhang, C., Xu, W., Liu, J., Liu, Z., Zhou, Z., & Pham, D. T. (2019). A reconfigurable modeling approach for digital twin-based manufacturing system. *Procedia CIRP, 83*, 118–125. https://doi.org/10.1016/j.procir.2019.03.141

Cognitive Internet of Things: Challenges and Solutions

Ali Mohammad Saghiri

1 Introduction

The Internet of Things (IoT) refers to an ecosystem including a set of devices, objects, and everything that has a unique identifier and the ability to communicate and transmit data over a network [1]. IoT-based systems generally include sensors, actuators, and control elements with limited resources in terms of computational, storage, and power. These systems will be the leading platform for deploying new businesses and play a crucial role in developing industries in the near future. IoT-based systems require a suitable design to organize their structure in a self-organized manner because of their distributed and dynamic nature. Therefore, many management approaches based on Artificial Intelligence (AI) and cognitive systems have been utilized in the IoT. As two results of these approaches, Cognitive Internet of Things (CIoT) and Artificial Intelligence of Things (AIoT) have been reported in [2, 3].

The rationale behind utilizing AI and cognitive systems theory is described as follows. From technical perspectives, the size of generated data by IoT devices is increasing, resources are distributed, and the characteristics of these systems are changing dynamically. Because of these characteristics, IoT systems require AI-based algorithms for management in a self-organized manner [4, 5]. On the other hand, from human perspective, human agents cannot solve challenges in many situations. Some of challenges that lead to the inefficiency of human agents are explained as follows. The reaction time of humans is not appropriate for controlling modern drones and vehicles. Humans are not able to make appropriate decisions considering big data generated by the IoT. The distributed and dynamic nature of the IoT leads to many managerial problems that are not solvable by humans

A. M. Saghiri (✉)
ICT Research Department, Niroo Research Institute, Tehran, Iran
e-mail: amsaghiri@nri.ac.ir

S. Pal et al. (eds.), *Artificial Intelligence-based Internet of Things Systems*,
Internet of Things, https://doi.org/10.1007/978-3-030-87059-1_13

manually. Therefore, human agents will be replaced by some intelligent systems such as cognitive systems in the near future, and then we will be faced with many cognitive systems that are combined with the IoT. Since cognitive systems are designated based on cognitive processes of the human brain and mind, utilizing these systems will have high priority in the development of CIoT because we can utilize human logic to make decisions with high accuracy and speed.

The main drawback of recently reported algorithms as management algorithms in CIoT is that most of them only focus on utilizing AI without considering fundamental issues of cognitive systems. In other words, many of recently reported algorithms for CIoT should be classified as applications of AI in the IoT. In the absence of human agents and deploying self-organized management mechanisms based on AI, we will be faced with an unknown situation that might be hurtful. This is because many problems of AI such as controllability and predictability are not solved properly [6–8]. Therefore, a management mechanism based on AI may inherit these problems implicitly. This means that utilizing the ultimate power of AI-based systems in designing CIoT may lead to a self-organized system that can hurt humans easily.

More specifically, challenges such as security, predictability, and complexity do not have a same meaning in all intelligent versions of the IoT [9]. Ignoring these challenges leads to missing the goals of cognitive systems in CIoT. In the literature, there is no study on challenges in designing CIoT considering associated technologies such as IoT, AI, and cognitive systems.

In this chapter, at first, AI, IoT, and cognitive systems are studied independently. Then, fusion between these technologies are analyzed. Finally, some of challenges in designing CIoT are highlighted with a special focus on cognitive systems. The rest of this chapter is organized as follows. In Sect. 2, associated technologies to CIoT and their challenges are explained. In Sect. 3, challenges and solutions in designing CIoT with a particular focus on cognitive systems are summarized. Section 4 is dedicated to conclusions.

2 Internet of Things, Artificial Intelligence, and Cognitive Systems: Definitions, Combinations, and Challenges

In this section, all possible combinations among AI, cognitive systems, and IoT are studied considering definitions and design challenges (Fig. 1). In the next three subsections, we give short descriptions for each field independently. Then, comparisons between each pair of technologies are given. In addition, some notes that highlight the importance of developing CIoT technology are given in the last part of this section.

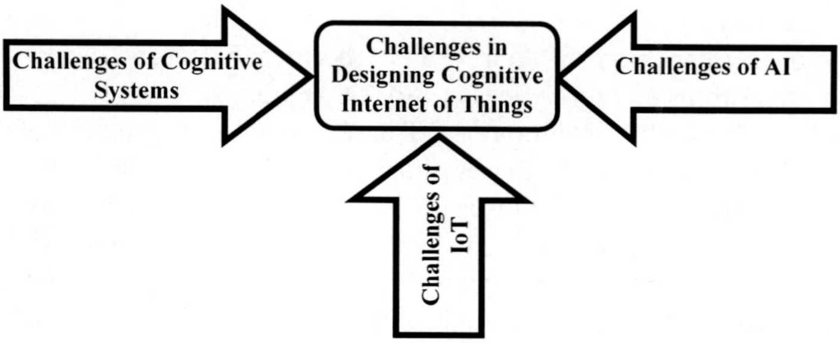

Fig. 1 Challenges in Designing Cognitive Internet of Things

2.1 Internet of Things

IoT ecosystem was introduced by Kevin Ashton in 1999 [4]. In the IoT, the "things" can include everything from a smart-board to a smart traffic control system. In these systems, each thing on the earth can be connected to other things. Thing-to-thing communication can be used to organize many scenarios in different fields, such as healthcare and educational systems. The integration of these systems with new computing technologies, including cloud, edge, and fog computing, can significantly benefit users, businesses, and markets. Some advantages of IoT-based systems are given below [10–13]:

- Efficiency in real-time decision-making for management and control can be obtained. This is because sensors gather a considerable amount of required information in the IoT ecosystems.
- Transparency can be obtained using information gathered by a wide range of sensors that monitor different parts of IoT-based systems in online fashion.
- Automation is an attractive characteristic that can be obtained via IoT-based systems. Utilizing IoT-based protocols, thing-to-thing communication can be done without requiring human involvement.
- Monitoring systems based on IoT sensors are very cheap and also efficient in terms of energy and cost.
- The integration among different IoT-based systems, including sensors, processors, and actuators, results in many benefits such as efficient data analyses and decision-making to resolve large-scale systems' complexity.
- IoT-based systems can be used to organize efficient mechanisms considering safety and security concerns for humans.

The above benefits lead to emerging smart governments, smart cities, smart healthcare, and smart manufacturing. The main challenges in designing IoT-based systems are given below [13], [14]:

- **Heterogeneity:** In IoT-based systems, the number of devices is increasing and the type of them are very different from each other. Therefore, designing management algorithms among devices is very challenging.
- **Scalability:** During the evolution of IoT-based systems, the number of sensors, actuators, and processers increases drastically. Therefore, in these systems, designing management algorithms in a scalable manner is vital. The primary versions of IoT-based systems focus on client-/server-based architectures, including cloud, fog, and edge computing. Recently, client-/server-based concept has been alternated with peer-to-peer communication and processing to handle distributed computations more efficiently.
- **Big data:** Since the number of devices is increasing, the variation of generated data by instruments is high, and the rate of data generation is excessive. Therefore, IoT-based systems lead to appearing big data with many managerial challenges.
- **Network management:** The network infrastructure for IoT-based systems is constructed over a wide range of devices that are distributed over various geographical positions. Therefore, network management algorithms should be equipped with policies to face with different environments (noisy and dynamic).
- **Privacy:** Privacy preserving in IoT-based systems is a difficult task due to the large-scale data analytics and heavy computations related to privacy concerns. This problem becomes more challenging after appearing machine-generated data because data owners are not clearly identified.
- **Security:** During the development and expanding IoT ecosystems, accessibility of devices increases day by day. As a result, designing secure algorithms in these systems is very difficult. In developing the Web of Things, this problem will be challenging because of increasing the availability of things.
- **Legal liability and responsibility:** When a thing makes a decision, and consequently many hurtful actions occur, the main question that arises is who is in charge of the results.
- **Sensors and actuators:** Designing inexpensive, accurate, and energy-efficient sensors and actuators belong to the main goals of designers of the IoT. Achieving these goals requires many high-tech methods and also researches that are not available in some situations.
- **Data analysis and decision-making:** Distributed, dynamic, and large-scale nature of the IoT leads to appearing many challenges in data analyzing and decision-making.
- **Interoperability and standardization:** The number of players in designing and utilizing IoT-based systems is increasing. In this domain, many platforms are closed-source, and other open-source platforms do not use standard architectures. It seems that defining some standards may be useful, but it will not easily happen because of the nature of software development in this field and weak interoperability.

2.2 *Artificial Intelligence*

Artificial intelligence does not mean the same concept to all scientists. In [15], Professor Russell introduced four definitions of artificial intelligence. According to his definitions, a wide range of devices can be considered as intelligent systems. Recently, a new classification for intelligent systems is introduced in the literature to provide a big picture of the future of intelligent systems. According to [9], intelligent systems are classified into three classes as explained below (Fig. 2):

- **Artificial Narrow Intelligence (ANI):** this type of intelligence refers to intelligent systems that can do specific tasks. For example, an agent with capabilities such as face recognition and games playing can be identified as artificial narrow intelligence. These agents are programmed to do tasks and cannot detect and formulate unknown tasks in a self-organized manner. We do not expect to see self-awareness in these agents. Some authors refer to this type of intelligence as weak artificial intelligence.
- **Artificial General Intelligence (AGI):** the concept of this type of intelligence does not refer to a unique thing in the mind of all leading scientists of artificial intelligence. Most researchers use artificial general intelligence for those agents that their intelligence is equivalent to human agents.
- **Artificial Superintelligence (ASI):** the author of [16] introduced three types of superintelligence: speed ASI, collective ASI, and quality ASI. Speed ASI refers to an agent faster than human, collective ASI refers to decision-making capabilities similar to a group of humans, and quality ASI refers to an agent that can do works that humans can't.

Recently, some changes have been made on the above classification. In [17, 18], the authors argue that human-level intelligence is different from AGI. They argue that human characteristics may put some assumptions and limitations in the computations of the machine. Therefore, human-level intelligence may not lead to general intelligence to solve a wide range of problems than human agents suffer from solving them. Using human-level intelligence to design AGI may lead to determining some implicit upper bounds in the machine and defeat the generalization

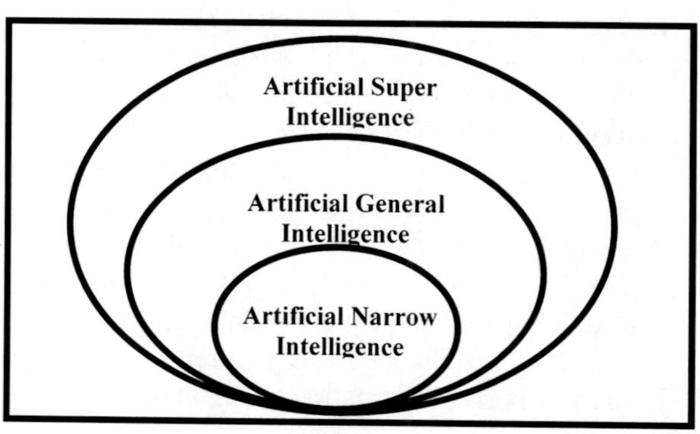

Fig. 2 Venn diagram for definitions of AI

capabilities. On the other hand, authors of some papers such as [18] argue that there is no difference between AGI and ASI when the definitions of AGI is not limited to human-level intelligence. According to [16, 18], if AGI-based agents' capabilities are further than human intelligence and there is no exact definition for ASI capabilities, then there is no need to differentiate AGI from ASI. All in all, definitions for AI is changing based on observations and expectations of humans. Some of the advantages of AI-based systems are given below:

- Estimation of values in management procedures with high-precision such as energy consumption and peak hours in power grid [19]
- Big data analytics with high precision [20]
- Increase the efficiency of management systems and asset management [19]

Although AI provides significant changes and improvements for digital business and facilitates smart services and digital transformation, it challenges present tremendous risks for individuals, organizations, and society. A list of challenges in designing AI-based systems is given below:

- **Energy Consumption, Global Warming, and Environmental Pollution:** One of the challenges of most machine learning algorithms is high energy consumption. In [21], environmental pollution and global warming are reported as other side effects of high computational power usage in AI-based systems.
- **Data Issues:** A category of AI invests in data-driven algorithms to construct machine learning models. It should be noted that, in some situations, there are some problems in data that lead to many difficulties in data-driven machine learning algorithms [22, 23]. For example, we may face with issues related to data incompleteness, data heterogeneity, data insufficiency, imbalanced data, untrusted data, and data uncertainty.
- **Security:** Security is a critical issue that has received much attention in recent years. An emerging field called adversarial machine learning was the first attempt to solve some security problems in data-driven machine learning methods [24].
- **Privacy:** Many of data-driven machine learning methods are fed by data of huge amount of users. During the execution of these methods, some different roles are determined, including data owner, data manipulator, and data visualizer. Many efforts should be done by researchers to solve the problems related to saving privacy considering different roles [25].
- **Fairness:** An algorithm is fair when its results are independent of some variables such as gender, ethnicity, and sexual orientation [26].
- **Safety:** In mission-critical, real-world environments, there is little tolerance for failure that can damage humans and devices. In these environments, existing approaches are not sufficient to support the safety of humans.
- **Beneficial:** A beneficial AI system is designated to behave in such a way that humans are satisfied with the results. Designing these systems will be required but extending their theory is still an ongoing process. In these systems, the agent is initially uncertain about the preferences of humans, and human behavior will be used to extract information about human preferences [27].

- **Predictability:** One of the most critical issues in designing intelligent systems is predictability [6–8]. Some problems, including ambiguity and paradox, may lead to an unpredictable AI-based system.
- **Explainable AI:** Explainable AI refers to AI methods such that the results of the solution obtained by them can be understood by human experts [28]. Since some machine learning algorithms such as deep learning invest in non-explainable symbols to do tasks, the results of these algorithms cannot be explainable.
- **Complexity:** The primary versions of intelligent systems invested in limited set of algorithms to do their jobs and therefore their complexity was limited to simple algorithms. This assumption is not correct any longer. The complexity of intelligent systems is increasing day by day.
- **Monopoly:** Many of AI-based solutions require substantial computational power. There are few companies (IBM, Amazon, and Microsoft) and countries which invest in AI and also high computational power devices. For example, few countries are pioneers in quantum computation that will be one of the enablers of artificial general intelligence and superintelligence. This capability may lead to the appearance of a monopoly in the scope of AI.
- **Responsibility:** AI-based systems such as self-driving drones will act autonomously in our world. In these systems, a challenging question is "who is liable when a self-driving system is involved in a crash or failure?" This problem has many dimensions. In [29], some interesting points related to responsibility issues are covered for a specific case study.
- **Controllability:** Eventually, AI-based systems surpass human's abilities in general intelligence and become superintelligent. In this situation, a superintelligence could become powerful and difficult to control for humans [30].
- **Reproducibility:** This feature refers to the ability of an agent to be recreated [31]. Existing AI-based methods lack this ability.
- **Continual Learning:** This ability received much attention in recent years. In supervised, unsupervised, and reinforcement learning methods, traditional algorithms only focus on fixed sets of data as training and testing sets. In real-world applications, we must consider a stream of data, and the learning process is ongoing. Continual learning methods enable agents to adapt to a continually changing environment [32].

2.3 Cognitive Systems

Cognitive computations can be interpreted with different perspectives in all types of intelligence, such as ANI, AGI, and ASI. Cognitive computing focuses on designing computations similar to computations that occur in the human brain. Many fields, including computational phycology and neurosciences, are used to organize cognitive systems. Some references such as [33] report a close relationship between cognitive systems and human-level intelligence. According to [34], this field enables computer programs with the abilities of knowing, thinking, and feeling as explained below:

- **Computer with the ability of knowing:** in this domain, several concepts such as fuzzy logic, the theory of causality, ontologies, denotational mathematics, object-attribute relation theory, concept algebra, process algebra, and inference algebra are used to design cognitive systems with the ability of knowing.
- **Computer with the ability of thinking:** in order to organize this capability in machines, two approaches are introduced in the literature. The first approach takes advantage of brain-like computer machinery, and the second approach focuses on determining causal relationships among concepts.
- **Computer with the ability of feeling:** emotional intelligence, sentiment analyses, and opinion mining are used to organize cognitive processes considering feeling capability.

All in all, cognitive computing requires many concepts from philosophy, psychology, linguistics, artificial intelligence, anthropology, and neuroscience. Some of the applications of cognitive systems are given below [35]:

- **Cognitive computing and robotics:** utilizing cognitive computing to drive robots was one of the goals of engineers in the last centuries. Recently, robots have been used to solve simple and routine tasks. Existing robots lack of cognitive abilities for analyzing smart cooperation among humans and robots but this style will be changed.
- **Emotional communication systems:** currently, communication between humans and robots suffer from many problems such as ambiguity and paradox. These problems become more challenging after digital transformation and arising some technologies such as virtual reality, augmented reality, and artificial life. The emotional communication based on cognitive systems facilitates communication between those entities that understand emotional symbols. This technology may affect the treatment of humans with some diseases such as autism.
- **Medical cognitive systems:** as an important application of cognitive systems, medical systems can be nominated, especially for managing chronic diseases. Because of the multidisciplinary origin of cognitive systems that focus on technologies such as machine learning, AI, and natural language processing, cognitive systems are going to detect modes and relations among diseases from data with high accuracy.

In [9], a list of challenges in designing cognitive systems is reported. This list is explained as below:

- **Problem Identification and Formulation:** these terms refer to two primary cognitive abilities in humans. Existing approaches in designing cognitive systems suffer from a lack of automated mechanisms to implement mentioned abilities.
- **Inspired Models:** a wide range of researches in cognitive systems focuses on inspired models considering the human brain and mind. These researches are not

mature enough to implement the cognitive abilities of humans, and therefore their abstractions may not lead to an appropriate cognitive system.

- **Connectionist and Symbolic Approaches:** recently, connectionist approach has received much attention to design learning algorithms because of extraordinary capabilities of deep neural networks. However, this approach does not invest in human-readable elements, and therefore it is not appropriate for implementing some cognitive processes. On the contrary, the symbolic approach focuses on some cognitive abilities and also human-readable elements; however, their performance in real-world problems is not appropriate.
- **Layered and Flat Cooperation methods:** cognitive systems may be organized as either layered or flat architectures. The architecture affects the growth of cognitive systems when cognitive processes try to improve their abilities.
- **Architecture Design:** for more than 10 years, designing cognitive architecture has received much attention. This problem is still open, and there is no generalized framework to cover all cognitive abilities [36]. This problem becomes more challenging in designing artificial general intelligence [37].
- **Evolution Strategy:** cognitive abilities of humans have been evolved over centuries based on several mechanisms, including genetic algorithms. Every cognitive system requires an internal or external mechanism to improve its cognitive abilities over time.
- **Programming Strategy:** selecting or developing a programming strategy to cover all cognitive abilities is a challenging problem.
- **Semantic and Communication:** the process of organizing communications among cognitive systems has attracted particular interest in recent years. To organize such process, many concepts such as ontologies and semantic web must be customized.
- **Morality:** designing morality at the core of the cognitive system is a challenging problem. Since the basic principles for morality are not fixed, programming this concept as computer code in the cognitive system is not easy.
- **Learning, Creativity, and Innovation:** analyzing the cognitive abilities of humans lead to appearing these concepts. Among these concepts, only learning abilities are developed in artificial intelligence, and other concepts are missed these days.
- **Machine Selection:** cognitive systems that manage the human brain are based on biochemical mechanisms. We may organize cognitive systems on other types of machines.
- **Explainable Systems:** humans are able to explain the reasons behind their decisions. This cognitive ability should be implemented in cognitive systems. Most existing AI-based agents are not explainable, and therefore they cannot give reasons for their decision-making.
- **Trust:** since cognitive systems will be used as self-organized mechanisms in a wide range of complex systems, designing algorithms for preserving trust will be vital.

- **Safety:** many cognitive processes may be emulated in machines but implementing those cognitive processes that their main goal is the safety of humans will be vital.
- **Predictability:** implementing some cognitive systems such as those inspired by humans may lead to designing unpredictable systems.
- **Security:** security in cognitive systems refers to those mechanisms that secure cognitive processes. The concepts behind securing cognitive processes may be different from traditional security mechanisms.
- **Energy Consumption:** executing a cognitive system may be energy-consuming because of utilizing many learning algorithms in different parts of cognitive processes that are usually executed in a parallel manner.
- **Computational Power:** The computational power required by learning-based algorithms is increasing. Therefore, utilizing cognitive systems that widely deploy learning algorithms leads to implementing algorithms that require high computational power.
- **Complexity:** identifying and organizing cognitive processes are complex problems. A dimension of complexity problem refers to the infancy knowledge of humans about the nature of cognitive abilities in humans.
- **Storage (Memory):** in all types of cognitive systems, memory plays an essential role. In some cases, more storage result in more cognitive abilities. Since the size of data gathered by these systems is increasing, utilizing data for analyzing and decision-making requires efficient algorithms.

2.4 Artificial Intelligence and Cognitive Systems

For more than 10 years, researchers have focused on identifying AI, cognitive systems, and relevant concepts. The relation between AI and cognitive systems is studied in the next two paragraphs [35]:

From cognitive systems perspectives: Many concepts of AI such as fuzzy logic, inference engines, reinforcement learning models, and ontologies are obtained through high-level cognitive analyses on humans. In this domain, some examples are given below [33, 34]:

- Fuzzy logic, ontologies, and theory of causality have a close relationship with knowing systems in humans.
- Reinforcement learning has a close relationship with repeated learning processes based on reward and penalties in humans and animals.
- Artificial neural networks are designated based on neuro-like interactions that come from neuroscience analyses.

From an artificial intelligence perspective: In many algorithms such as search mechanisms, function approximation, and statistical learning methods, big data analytics will be key enablers of cognitive systems in the next centuries. In addition, big data technology has received much attention for revolutionizing cognitive systems. In comparison to big data analyses that focus on solving challenges related to volume, velocity, variety, and veracity of data, cognitive computation focuses on organizing processing methods that mimic data processing in the human brain to do tasks. It should be noted that human brain memory is limited, but its image processing is efficient. Therefore, the goal of the learning systems designers is to organize cognitive processes that are similar to cognitive processes in the brain.

The progress in both mentioned perspectives is still ongoing. In addition to challenges of cognitive systems and AI that will be transferred to hybrid systems based on these technologies, the following challenges can be highlighted that should be considered [35].

- Efficient data collection (high speed and high quality)
- Efficient natural language processing and human-computer interaction with emotional cognition
- New processor design considering challenges of cognitive systems and AI

It is obvious that some of challenges such as security and safety are common challenges between AI and cognitive systems although their objectives are different from each other according to the context of their applications.

2.5 *Artificial Intelligence and the Internet of Things*

In the last decade, different combinations between AI and the IoT have been reported in the literature. For example, Intelligence of Things is studied in [14], Internet of Intelligent Things (IoIT) is explained in [5], and Artificial Intelligence of Things (AIoT) is used in [3]. In all of these examples, AI-based algorithms are used to bring more capabilities to the IoT such as intelligent sensing, intelligent connection management, and intelligent data processing. There is no fixed boundary among definitions of AI-based IoT systems. Overall, AI can increase the value of the IoT. On the other hand, IoT can promote the capabilities of AI. In the rest of this part, benefits and challenges in designing hybrid systems based on AI and IoT are studied.

All benefits from AI and IoT can be obtained in hybrid systems based on these technologies. Some benefits of these systems are given below [38]:

- From application perspective, AI can be used to design smart business, home, healthcare, and industries.
- Infrastructure of business intelligence will be organized on a combination of AI with the IoT. In these systems, intelligent decision-making and processing will be beneficial. In these systems, the IoT is in charge of collecting data and AI is in charge of processing the data in order to make sense of it.
- Personalized services such as personalized healthcare and medicine may be organized based on IoT and AI.
- AI plays a vital role in IoT to manage the flood of data and provide the analytics required to extract knowledge from data.

Challenges that arise during fusion of AI into the IoT are explained as below [14].

- **Complexity:** in IoT-based systems, many elements play different roles. These systems may be classified as complex systems. This means that, because of high complexity of IoT-based systems, deploying AI in them results in many problems. In addition, IoT limitations such as processing power, memory, and delay in real-time applications will lead many challenges to design AI-based algorithms.
- **Heterogeneity:** hybrid systems based on AI and IoT inherited different forms of heterogeneity from IoT. IoT systems suffer from challenges that come from wide heterogeneity of devices, platforms, operating systems, and services that exist and will be used to construct applications. Therefore, these challenges should be considered during adopting AI.
- **Security and privacy:** AIoT inherited these challenges from both AI and the IoT. In other words, these challenges are very serious in hybrid systems based on the IoT and AI. For example, AI-based mechanisms require data to learn a model for a security mechanism, but keeping the privacy of these data in some devices in IoT systems cannot be preserved because of hardware/software weaknesses in devices.
- **Standardization:** this challenge can be considered for AI, IoT, and their combinations. Since problems related to interoperability and heterogeneity are not solvable easily, determining standards for hybrid systems based on IoT and AI results in many problems.
- **Accuracy and speed:** in AI-based systems, data-driven machine learning algorithms require an adequate amount of data to increase their accuracy. But, in IoT-based systems, gathering more information results in consuming more time, and therefore in real-time applications of IoT, many difficulties arise during usage of AI-based methods.
- **Client-server and peer-to-peer:** popular architectures for IoT-based systems are based on edge, fog, and cloud computing concepts. However, there are many challenges while deploying these architectures in practice. For instance, the edge computing should be enhanced for optimizing the latency of transmitting massive data of the IoT in networks and achieving real-time responses with high-performance for analyzing the vast data using AI. All in all, client-server-based architectures would not be able to provide requirements for developing IoT systems in the future. Therefore, moving the IoT systems into a decentralized path may be the right decision. One of the famous decentralized architecture is peer-to-peer system. Peer-to-peer systems such as blockchain have received much attentions in recent years.
- **Legal aspects:** in hybrid systems based on AI and IoT, many entities play roles. Therefore, legal challenges and legal relationships should be studied and clarified. This challenge will be serious after appearing crashes and faults among things (such as drones and vehicles).
- **Artificial stupidity:** this problem has many dimensions in the IoT. In data-driven machine learning, sufficient data is required to train an appropriate learning model, and without data, the learning model has no trustable accuracy. Since

transferring vast amounts of data in IoT ecosystem is costly, the learning algorithms may be constructed based on inappropriate data. In other words, lack of data in the IoT leads to inappropriate learning model, which lead to error-prone decision-making mechanisms.

2.6 *Cognitive Systems and Internet of Things*

Cognitive systems are widely used in the IoT [39]. The relation between cognitive systems and the IoT is an interesting issue because of the following perspectives.

- **From IoT perspective:** The distributed, dynamic, and large-scale nature of IoT leads to high complexity in designing the management algorithms of these systems. Therefore, human thinking models as digitalized models are required for fast and accurate decision-making. Cognitive systems are applicable to not only the IoT but also other systems such as those illustrated in Fig. 3 [39–50].
- **From cognitive systems perspective:** The IoT refers to the huge amount of physical devices scattered around the world, all collecting and sharing data, that makes new services. These services provide valuable data for designing learning models and approaching to organize fully featured cognitive systems and also artificial general intelligence.

In order to merge cognitive systems with the IoT, several architectures are reported in the literature. According to [51–53], CIoT can be designated based on three-layered architecture explained as follows:

- **Requirement Layer:** in this layer, the goals and behavior of the system are determined using a language.
- **Cognitive Process Layer:** in this layer, cognitive processes are organized. Each cognitive process may be designated to manage several tasks.
- **Things Management Layer:** this layer manages IoT system using decisions made by cognitive processes.

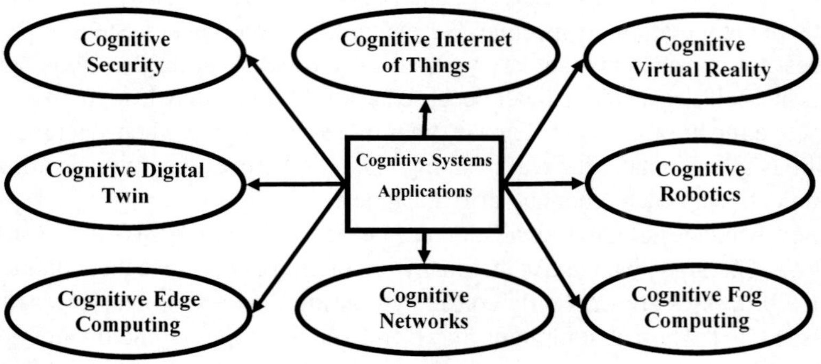

Fig. 3 Cognitive systems applications [39–50]

Hybrid systems based in IoT and cognitive systems bring numerous benefits which some of them are summarized as below [54]:

- In contrast to existing intelligent systems that only focus on structured data and simple learning mechanisms to solve problems, deploying cognitive systems may rely on multiple cognitive processes that invest in both structured and unstructured data.
- Similar to the human brain that utilizes information from multiple sources to do tasks, cognitive systems are able to implement the same functionality in the IoT.
- As an important feature of cognitive computing, big data gathered by the IoT ecosystem can be used for continually learning and reasoning based on cognitive processes similar to the learning process in humans.
- Cognitive computing will bring many abilities to process natural language to learn quickly from structured and unstructured data collected from various sources.

As it was previously mentioned, challenges in designing CIoT come from three sources: cognitive systems, AI, and IoT. Therefore, CIoT may inherit all of challenges from these sources. It should be noted that some of those challenges that are mentioned for AI and cognitive systems, such as trust, safety, security, and predictability, do not refer to exactly similar concepts and their meaning depends on the context. For example, to design a cognitive system, cognitive processes and the human mind are involved in the assumptions of computation, but in an AI-based system, there is no fixed set of assumptions. Some of challenges of CIoT will be discussed in Sect. 3.

2.6.1 Priority of Cognitive Internet of Things

An important issue that should be considered during the usage of AI in the IoT is that a wide range of AI-based systems does not consider the cognitive processes of humans. Therefore, they are not suitable to be used in CIoT. Because of this issue, more much efforts should be done on designing cognitive systems and their applications to engineering fields such as IoT and robotics, instead of utilizing AI without a cognitive computation perspective. For example, an intelligent system in the IoT that does not consider human safety may easily execute harmful actions against humans in a self-organized manner because harmful actions may lead to optimizing an objective function. This problem becomes more challenging when machines lack capabilities to understand the real meaning behind sentences that humans utilize to communicate with each other and also machines.

Some advantages of CIoT against other intelligent versions of IoT are explained as follows. The first advantage is that many important assumptions that conduct the cognitive processes are implicitly considered during the development of cognitive systems in CIoT. These assumptions may be missed in designing intelligent systems based on AI. The second advantage is that during the development of CIoT, many assumptions are injected into the knowledge of machines, which leads to facilitate the communication between humans and machines. The last advantage is that controllability and predictability of machines that are critical concerns in developing intelligent systems based on AI can be better managed through cognitive systems.

3 Challenges and Solutions in Designing Cognitive Internet of Things

In this section, some challenges in designing CIoT are explained in more details. The list of terms that present potential challenges is given below.

• Problem identification and formulation	• Morality, privacy, responsibility
• Energy consumption and environmental pollution	• Inspired models
• Explainable systems	• Predictability and control ability
• Layered and flat cooperation methods	• Security and trust
• Architecture design	• Semantic and communication
• Safety	• Machine selection

3.1 Problem Identification and Formulation

Thinking ability is considered as one of the main abilities in cognitive systems. CIoT inherited challenges of designing thinking machines from cognitive systems. Among numerous problems that should be solved for approaching an acceptable level of thinking ability, problem identification and formulation are considered as fundamental problems for autonomously organizing IoT. For example, CIoT systems should be able to automatically detect problems behind performance degradation and then deploy an appropriate mechanism to solve the problems. According to [15], detecting the problem (single-state, multiple-state, and contingency) and formulating it in a well-defined manner are the first steps of problem-solving procedure based on AI. According to [55], three approaches may be utilized for analyzing complex structures such as CIoT, as given below:

- Top-down (forward engineering)
- Bottom-up (backward engineering)
- Hybrid (forward engineering and backward engineering)

One approach that has attracted much attention in recent years is based on knowledge-based systems such as the Cyc project [56] and [57]. In the Cyc project, we can insert some facts about the problems into a knowledge base and then extract new problems using an inference engine. In [57], a rules-based configuration for problem detection is proposed.

All in all, in the field of CIoT, there is no general framework to identify and formulate problems in a self-organized manner. Most of the existing frameworks follow the bottom-up approach because the top-down approach for finding problems in different domains leads to challenging problems. For example, in the literature, a machine learning method is suggested for solving a problem by a scientist and other scientists try to modify that machine learning method for solving new versions of that problem. This process is usual in human societies. This means that few persons determine new problems, and others try to reuse problem-solving strategies that are not substantially novel. As a powerful strategy and based on bottom-up approach, genetic algorithms can be used to create a fully featured cognitive system. Genetic algorithms performed well in nature, and scientists argue that computer programs may apply this functionality as an evolutionary strategy.

3.2 Energy Consumption and Environmental Pollution

A wide range of intelligent systems invest in a high volume of data and also processing power [58]. In existing computers that mainly require high energy to support storage and processing power, executing intelligent algorithms leads to high energy consumption and sometimes high environmental pollution. In [21], the author studied these challenges in deep learning methods. In cognitive systems, these problems become more critical because many learning algorithms may be involved in the basic functionality of cognitive systems such as self-awareness. In CIoT that invest in architectures based on cloud, fog, and edge computing, each thing can execute learning algorithms with the aid of cloud computing or other elements such as fog and edge. However, high energy consumption problem will exist, and its position will be exchanged from things to other parts of CIoT ecosystem.

It should be noted that high-energy consumption problem is not only associated with learning elements of CIoT but also depends on the size of CIoT that is increasing because of the drastic growth of the size of IoT systems. On the other hand, IoT devices may use a wide range of radiofrequencies to communicate with each other, and the side effects of this phenomenon are not clear. In the next paragraph, some solutions reported in the literature are summarized.

In [59], green industrial IoT architecture is reported. In this architecture, an energy-efficient solution for industrial IoT (IIoT) is suggested. In addition, sleep/wake up-based scheduling mechanism is reported to define energy-efficient algorithms. In [60], wireless energy harvesting technique for the IoT is applied. In this solution, the energy harvesting technique is customized for smart cities. This technique refers to a promising solution for extending the lifetime of low-power devices. In [61], several algorithms for efficient energy management for the IoT in smart cities are suggested. Two novel solutions that may be considered in this domain are explained as follows:

- From a hardware standpoint, we may use energy harvesting techniques [62] (thermal, optical, and waves) to gather wasted energy from devices which are involved in learning processes.

- From a software standpoint, we can share files (models) in client/server or peer-to-peer fashion. These files can be the output of every learning process. With this approach, everything may use the shared model without paying its learning costs. Transfer learning theory may be used in this domain [63]. In addition, blockchain technology can be used to share knowledge (models) in fully distributed and protected manner. Considering this issue, an interesting discussion is given about the future of the Internet of Things (IoT), blockchain, and Internet of Minds (IoM) in [64].

3.3 Explainable Systems

A system is called explainable when the reasons behind its selected actions can be understood or interpreted by a human. In AI field, designing explainable system is known as a challenging problem [28, 65, 66]. CIoT inherited this challenge from AI. In CIoT ecosystems, machines will make many decisions, and therefore explainable systems will be useful. In the next paragraph, this problem is studied in the IoT and CIoT.

In [67], an explainable AI approach is used to the IoT in agriculture. In this approach, a fuzzy rule-based system that is interpretable is reported. In [68], explainable AI is applied to a healthcare framework for combating COVID-19-like pandemics. Another application of explainable AI in healthcare is reported in [69]. In this application, wearable devices were considered in cognitive processes and also some mechanisms are introduced to achieve some characteristics such as accountability. In [70], the authors invest in utilizing explainable AI to build cognitive cities based on the IoT. Since the consequences of decisions in industries are very important and industrial IoT is going to be very popular, some papers such as [71] suggests some mechanisms for utilizing explainable AI in industries.

In summary, since the penetration of CIoT is increasing along with vanishing of human roles in automation, cognitive systems will make many decisions at the core of CIoT. Therefore, finding reasons and interpretations behind each decision will be vital. In some applications such as healthcare systems, this problem will be challenging because the consequence of each decision in these systems can be challenging.

3.4 Layered and Flat Cooperation Methods

CIoT inherited layered and flat cooperation design challenges from IoT and cognitive systems. In what follows, we study these challenges from two perspectives.

From IoT design perspective: The following approaches have been used to organize IoT ecosystems:

- Layered cooperation among nodes of a network leads to the evolution of client-server architecture to cloud, fog, and edge ecosystems [72, 73].
- Flat cooperation among nodes of the network leads to the evolution of peer-to-peer systems to blockchain systems. These ecosystems may be more scalable than layered approaches [74].

From cognitive systems design perspective: the following approaches are used to handle complexities in designing cognitive systems:

- Layered cooperation among agents leads to organize multilevel cognitive systems to handle complexities in the environment of these systems. Deep learning can be classified as a result of this approach [75, 76].
- Flat cooperation leads to organize single layer cognitive systems. In order to handle more complexities, more layers are required [76].

In [76], layered architecture for cognitive Internet of vehicles that is an example of CIoT is reported. This paper focuses on security and then assumes an infrastructure based on cloud computing to execute learning algorithms. In [77], a layered architecture for the IoT is reported that considers the environmental complexities surrounding the devices. In [2], a cognitive management framework for the IoT for enabling autonomous mechanisms is reported in which a three-layered cognitive engine is utilized to bring self-healing capability.

3.5 Architecture Design

CIoT inherited several design challenges such as architecture design from IoT and cognitive systems. Therefore, the challenge of architecture design in CIoT can be analyzed from the following perspectives:

- **From IoT design perspectives:** there are numerous architectures such as those reported in [52, 54, 74] to organize the IoT ecosystem, and there is no standard to build a unique solution. It should be noted that peer-to-peer systems such as blockchain are used to change traditional architectures of IoT that invest in cloud computing and client-server concepts. This change results in appearing complex architectures of the IoT.
- **From cognitive systems design perspective:** according to a research reported in [36], more than 80 cognitive architectures are developed since 40 years ago. This report also gives several taxonomies of cognitive architectures. Among notable architectures, cognitive systems architectures ACT-R and Soar have influential positions [78, 79]. Selecting appropriate architecture considering problems of CIoT leads to a challenging task. Cognitive systems try to organize cognitive abilities related to perception, attention, action selection, memory, learning, reasoning, and metareasoning. The main problem that occurs during applying cognitive systems is that many limitations related to computations, memory, and interconnection patterns should be applied to organize a customized cognitive

system for CIoT. For example, memory that is considered as a simple element of cognitive systems may be converted to a complicated element of CIoT because we have no infinite size of memory, and increasing the size of information leads to increasing the access time. In addition to the mentioned challenges, many parameters with the origin of the infrastructure of IoT may affect the functionality of the cognitive systems.

3.6 Safety

CIoT applications will be vital in decision-making in a wide range of systems such as autonomous vehicles and drones. Because of the widespread CIoT applications, the cost of harmful and unsafe actions of them is very high. Some researchers focus on designing safe models considering the cost of harmful actions [80]. From another point of view, some papers such as those reported [81–83] try to judge the actions of an agent considering the safety of human. This problem is going to very challenging because of the vanishing role of humans and appearing fully self-organized systems. Unfortunately, this problem is not considered adequately in CIoT, and most of papers such as [84, 85] only focus on using CIoT and IoT to build secure systems such as secure homes and secure vehicular networks to bring safety for humans.

3.7 Morality, Privacy, and Responsibility

These terms refer to different dimensions of challenges in designing CIoT. The reason for discussing them in a group is that these concepts have many shared attributes among themselves. For example, keeping the privacy of data during the learning process can be classified under both privacy and morality [86]. From another perspective, defining morality in the cognitive systems of CIoT refers to novel concepts that are not related to privacy. Since the cognitive systems approaching to abilities of artificial general intelligence, designing moral cognitive systems will be an open issue.

Cognitive systems will be used as self-organized mechanisms in different fields. Since human decision-makers will be replaced by these systems, we must have appropriate mechanisms to find those entities that are responsible for each action done in the system. For example, who is responsible of an accident among autonomous vehicles? This problem becomes more challenging after rising some tools such as GANs [87]. These tools are able to generate samples for data. This capability may be used to generate fake texts, pictures, and videos by CIoT in a self-organized manner [88]. Humans may not able to control some of these abilities in machines. For example, a machine may be involved in a crime and then generate fake evidence for it. As two main challenges in this area, we may refer to two critical abilities studied in the next paragraph.

Deception and cheating belong to the human agents, but these characteristics may appear in a cognitive system. This is because cognitive systems are going to mimic the behavior of humans. For artificial narrow intelligence-based agent, we can't say that deception and cheating come from the intention of the machine, because these agents have no mind or consciousness. Deploying artificial general intelligence at the core of cognitive systems in CIoT may lead to these challenges. These challenges have not been considered yet in the literature of CIoT.

3.8 Inspired Models

Most cognitive systems such as Soar and ACT-R are inspired by the human brain [78, 79]. This is because the human brain is used to determine cognitive processes, and many studies are done on identifying and categorizing these processes based on neural and psychological sciences. Brain-inspired cognitive systems focus on either emulating the functions of different parts of the brain or input/output of total parts. For example, deep learning methods focus on emulating neural interactions in the brain, while the learning automata theory focuses on emulating human behavior [75, 89]. In addition, many cognitive processes are inspired by the collective behavior of humans in societies reported in the literature as game theory, social sciences, and swarm intelligence fields [90–92]. Implicitly, many nature-inspired and also bio-inspired algorithms are reported in the literature of CIoT. This is because many processes that manage our environment, from biological to ecological, can be seen as cognitive processes. In the next paragraph, some of these efforts are summarized.

In [93], a bio-inspired solution to distributed spectrum allocation problem in CIoT is reported. In this solution, a cognitive radio management algorithm is merged with a biological mechanism called reaction-diffusion to provide efficient spectrum allocation algorithm. In [94], conceptual descriptions of nature-inspired cognitive cities are analyzed. The authors of this work assumed cognitive city as a complex system that will be approaching to complex adaptive systems. They organized their model based on principles such as decentralized control and multi-directional networking. In [95], a cognitive model of task scheduling for IoT is suggested based on ant colony and genetic algorithms. In [96], cognitive packet networks for secure IoT are reported. The theory of cognitive packet networks relies on random neural networks that are brain-inspired learning techniques. In [97], a deep reinforcement learning technique is used to solve a task assignment problem in an intelligent version of the IoT.

Since there is no well-known framework that both communities of cognitive systems and the IoT agree on, cognitive systems experts may not accept some solutions that are published in the literature as cognitive solutions. As it was previously mentioned, we cannot consider every AI-based system as a cognitive system. In other words, more efforts should be made in this field.

3.9 *Predictability and Controllability*

These terms can be explained from two perspectives as given below:

- **From CIoT design perspective:** CIoT inherited predictability and controllability challenges from AI. In the literature, some papers such as [30] show that the controllability problem in AI-based solutions will not be solvable in many cases. In addition, according to well-known theorems such as Godel theorem and some challenges such as ambiguity and paradox come from the complexity field [6, 98], many problems cannot be formulated and solved based on mathematics. All of these challenges may be transferred to CIoT trough cognitive computation, and therefore CIoT systems will not be predictable and also controllable.
- **From CIoT applications perspective:** IoT systems are known as useful tools for online monitoring. AI-based tools may be used to utilize specific information to predict and control other systems. In this perspective, cognitive systems and machine learning algorithms are utilized to make decisions with high accuracy. For example, in [99], the authors use intelligent energy management of solar-powered sensor devices in a dynamic environment. They designated an algorithm that selects a model among a set of prediction models. In [100], the cognitive amplifier for the IoT is represented.

Overall, CIoT systems may be used to solve many problems such as controllability and predictability, although these problems exist in the core of CIoT.

3.10 *Security and Trust*

These challenges are multidimensional and very critical. We study them based on two perspectives, as explained in the following.

- **From CIoT design perspective:** since cognitive systems will be self-organized, they should be designated securely, because there will be no human agent to analyze attacks and then define corresponding defense mechanisms. Analyzing the security of cognitive systems and AI-based mechanisms is not easy, and few papers focus on this issue. For example, adversarial machine learning is a field that focuses on securing machine learning algorithms that invest in data-driven algorithms [24]. It seems that attacks to CIoT might be very complicated and also based on high-tech AI-based mechanisms. Therefore, security mechanisms for CIoT must be organized carefully. On the other hand, because of the increasing availability of things over the Internet, the reaction time to defense and recovery is very low. It seems that cognitive systems might be used to organize the attacks, and therefore we must design cognitive defense mechanisms [101]. During the design of CIoT, some issues are shared between security and trust. For example, most of AI-based security mechanism are based on trusted data. Therefore, attackers may use some features of data to hack security mechanisms.

- **From CIoT applications perspective:** many papers such as [44, 45] focus on using cognitive systems to design secure systems.

All in all, every software, including cognitive systems, may be hacked. Analyzing security challenges in cognitive systems is in the early stages.

3.11 Semantic and Communication

In the era of CIoT, challenges related to semantic and communication will be crucial. Most of systems such as computer networks, robotics, and the IoT will deploy cognitive systems. If all of these systems utilize standard language [102] and also ontology [103], cognitive systems of them will be able to increase their cognitive abilities through interactions among themselves. Some related papers are explained in the next paragraph.

In [104], authors focus on the role of object-oriented ontology in the IoT. According to this study, designers of IoT systems may reimagine data items, devices, and users, as significant as actors in a flat ontology. In [105], the authors focus on ontology to organize unified and formalized descriptions to solve the challenges of semantic heterogeneity in the IoT security domain. They reported four key sub-domains to cover a security situation in the IoT. These domains are context, attack, vulnerability, and network flow. In addition, user-defined rules can compensate for the limited description ability of ontology. Therefore, this model can enhance the reasoning ability of its ontology.

Many existing cognitive systems such as those reported in [2, 50] are designated to solve specific problems and do not utilize communication and knowledge sharing. So, the communication among cognitive systems embedded in devices around humans will not happen with existing approach. One of the techniques that can be used in CIoT to share knowledge among things is based on peer-to-peer file sharing systems and also blockchain [106].

3.12 Machine Selection

Organizing ecosystems for IoT and CIoT based on different technologies such as nano, molecular, and biological have attracted particular interests in recent years [107–109]. In these ecosystems, each thing may be equipped with sensors, actuators, and processing elements. In every technology, the type of cognition may be different because of the quality of processors and also interconnections among things. In the rest of this section, challenges and solutions of machine selection are summarized.

Nanotechnology, biotechnology, and quantum computation can be used to generate, process, and transmit data at different scale. These technologies will form the infrastructure of IoT systems according to the literature. In [107], the Internet of Nano Things (IoNT) is reported that refers to the intersection of three technologies: nanoscale devices, communication networks, and IoT. This technology focuses on the state-of-the-art methods in electromagnetic communications among nanoscale devices. In [110], the Internet of Multimedia Nano-Things (IoMNT) is introduced. IoMNT focuses on issues related to multimedia contents in IoNT. These systems have a high potential to manage a high volume of data generated at the nanoscale. On the other hand, in [111], the challenges in designing Internet of Multimedia Things (IoMT) are studied. It should be noted that, this technology is fused with nanotechnology to organize IoMNT. In [109], the potential of molecular computation is utilized to present the Internet of Molecular Things (IoMT). In [108], the Internet of Bio-Nano Things (IoBNT) is introduced. The rationale behind IoBNT is to merge concepts of synthetic biology and nanotechnology to define engineering principals for biological embedded computing devices. This technology enables applications such as intra-body sensing and actuation networks.

In summary, different ecosystems based on nanotechnology, biotechnology, and molecular mechanisms might be used to organize infrastructure for future generations of CIoT. In each ecosystem, many problems should be solved to support main functionalities of cognitive systems. Therefore, more much efforts should be done to organize CIoT based on the mentioned ecosystems.

4 Conclusions

In this chapter, challenges in designing CIoT have been summarized. The main contribution of this chapter was to study the challenges from three perspectives: AI, IoT, and cognitive systems. It seems that many of the recently reported algorithms that only use AI to solve problems of CIoT will be evolved into cognitive processes. Therefore, we may accept them as the first steps in organizing CIoT. Since the progress in this field is still ongoing, there are no fully featured and unified solutions for CIoT. The proposed study opens a new horizon to deploy the concepts of cognitive systems in CIoT.

References

1. Wortmann, F., & Flüchter, K. (2015). Internet of things. *Business & Information Systems Engineering, 57*, 221–224.
2. Foteinos, V., Kelaidonis, D., Poulios, G., Vlacheas, P., Stavroulaki, V., & Demestichas, P. (2013). Cognitive management for the internet of things: A framework for enabling autonomous applications. *IEEE Vehicular Technology Magazine, 8*, 90–99.
3. Navale, M. P., & Navale, R. G. (2020). Artificial Intelligence and Internet of Things (AIoT): Opportunities and Challenges. *CLIO An Annual Interdisciplinary Journal of History, 6*, 199–206.

4. Atzori, L., Iera, A., & Morabito, G. (2010). The internet of things: A survey. *Computer networks, 54*, 2787–2805.
5. Arsénio, A., Serra, H., Francisco, R., Nabais, F., Andrade, J., & Serrano, E. (2014). Internet of intelligent things: Bringing artificial intelligence into things and communication networks. In *Inter-cooperative collective intelligence: Techniques and applications* (pp. 1–37). Springer.
6. Dawson, J. (1996). *Logical dilemmas: The life and work of Kurt Gödel*. AK Peters/CRC Press.
7. Yampolskiy, R. V. (2019). Unpredictability of AI. *arXiv preprint arXiv:1905.13053*.
8. Hofstadter, D. R. (2007). *I am a strange loop*. Basic books.
9. Saghiri, A. M. (2020). A Survey on Challenges in Designing Cognitive Engines. In *2020 6th International Conference on Web Research (ICWR)* (pp. 165–171). IEEE.
10. Bekara, C. (2014). Security issues and challenges for the IoT-based smart grid. *Procedia Computer Science, 34*, 532–537.
11. Yun, M., & Yuxin, B. (2010). Research on the architecture and key technology of Internet of Things (IoT) applied on smart grid. In *2010 international conference on advances in energy engineering* (pp. 69–72). IEEE.
12. Soumyalatha, & Hegde, S. G. (2016). Study of IoT: understanding IoT architecture, applications, issues and challenges. In *1st International Conference on Innovations in Computing & Net-working (ICICN16), CSE, RRCE. International Journal of Advanced Networking & Applications*.
13. Firouzi, F., Farahani, B., Weinberger, M., DePace, G., & Aliee, F. S. (2020). IoT fundamentals: Definitions, architectures, challenges, and promises. In *Intelligent internet of things* (pp. 3–50). Springer.
14. Atlam, H. F., Walters, R. J., & Wills, G. B. (2018). Intelligence of things: opportunities & challenges. In *2018 3rd Cloudification of the Internet of Things (CIoT)* (pp. 1–6). IEEE.
15. Russell, S. J., & Norvig, P. (1994). *Artificial intelligence: A modern approach* (3rd ed.). Prentice Hall.
16. Boström, N. (2014). *Superintelligence: Paths, dangers, strategies*. Oxford University Press.
17. Chollet, F. (2019). On the measure of intelligence. *arXiv preprint arXiv:1911.01547*.
18. Yampolskiy, R. V. (2020). Human $$\backslash$neq $ AGI. *arXiv preprint arXiv:2007.07710*.
19. How does artificial intelligence help? What are some advantages and disadvantages? - Quora. 2020. https://www.quora.com/How-does-artificial-intelligence-help-What-are-some-advantages-and-disadvantages. Accessed June 30.
20. Jaseena, K. U., & David, J. M. (2014). Issues, challenges, and solutions: Big data mining. *CS & IT-CSCP, 4*, 131–140.
21. Strubell, E., Ganesh, A., & McCallum, A. (2019). Energy and policy considerations for deep learning in NLP. *arXiv preprint arXiv:1906.02243*.
22. Baig, M. I., Shuib, L., & Yadegaridehkordi, E. (2019). Big data tools: Advantages and disadvantages. *Journal of Soft Computing and Decision Support Systems, 6*, 14–20.
23. Sivarajah, U., Kamal, M. M., Irani, Z., & Weerakkody, V. (2017). Critical analysis of Big Data challenges and analytical methods. *Journal of Business Research, 70*, 263–286. Elsevier.
24. Huang, L., Joseph, A. D., Nelson, B., Rubinstein, B. I. P., & Tygar, J. D. (2011). Adversarial machine learning. In *Proceedings of the 4th ACM workshop on Security and artificial intelligence* (pp. 43–58). ACM.
25. Yang, Q., Yang, L., Chen, T., & Tong, Y. (2019). Federated machine learning: Concept and applications. *ACM Transactions on Intelligent Systems and Technology (TIST), 10*, 1–19. ACM New York.
26. Bellamy, R. K. E., Dey, K., Hind, M., Hoffman, S. C., Houde, S., Kannan, K., Lohia, P., Martino, J., Mehta, S., & Mojsilović, A. (2019). AI Fairness 360: An extensible toolkit for detecting and mitigating algorithmic bias. *IBM Journal of Research and Development, 63*, 4–1. IBM.
27. Russell, S., Dewey, D., & Tegmark, M. (2015). Research priorities for robust and beneficial artificial intelligence. *Ai Magazine, 36*, 105–114.

28. Došilović, F. K., Brčić, M., & Hlupić, N. (2018). Explainable artificial intelligence: A survey. In *2018 41st International convention on information and communication technology, electronics and microelectronics (MIPRO)* (pp. 0210–0215). IEEE.
29. Neri, E., Coppola, F., Miele, V., Bibbolino, C., & Grassi, R. (2020). *Artificial intelligence: Who is responsible for the diagnosis?* Springer.
30. Yampolskiy, R. V. (2020). On Controllability of AI. *arXiv preprint arXiv:2008.04071*.
31. Gundersen, O. E., & Kjensmo, S. (2018). State of the art: Reproducibility in artificial intelligence. *AAAI*, 1644–1651.
32. Shin, H., Lee, J. K., Kim, J., & Kim, J. (2017). Continual learning with deep generative replay. *Advances in Neural Information Processing Systems, 2017*, 2990–2999. USA.
33. Cassimatis, N. L. (2006). A cognitive substrate for achieving human-level intelligence. *AI Magazine, 27*, 45–45.
34. Gutierrez-Garcia, J. O., & López-Neri, E. (2015). Cognitive computing: a brief survey and open research challenges. In *2015 3rd international conference on applied computing and information technology/2nd international conference on computational science and intelligence* (pp. 328–333). IEEE.
35. Chen, M., Herrera, F., & Hwang, K. (2018). Cognitive computing: architecture, technologies and intelligent applications. *Ieee Access, 6*, 19774–19783. IEEE.
36. Kotseruba, I., & Tsotsos, J. K. (2020). 40 years of cognitive architectures: Core cognitive abilities and practical applications. *Artificial Intelligence Review, 53*, 17–94.
37. Lieto, A., Bhatt, M., Oltramari, A., & Vernon, D. (2018). The role of cognitive architectures in general artificial intelligence. *Cognitive Systems Research, 8*, 1–3.
38. Katare, G., Padihar, G., & Quereshi, Z. (2018). Challenges in the integration of artificial intelligence and Internet of things. *International Journal of System and Software Engineering, 6*, 10–15.
39. Ploennigs, J., Ba, A., & Barry, M. (2017). Materializing the promises of cognitive IoT: How cognitive buildings are shaping the way. *IEEE Internet of Things Journal, 5*, 2367–2374.
40. Beetz, M., Mösenlechner, L., & Tenorth, M. (2010). CRAM—A Cognitive Robot Abstract Machine for everyday manipulation in human environments. In *IEEE/RSJ international conference on intelligent robots and systems* (pp. 1012–1017). IEEE.
41. Amjad, A., Rabby, F., Sadia, S., Patwary, M., & Benkhelifa, E. (2017). Cognitive edge computing based resource allocation framework for Internet of Things. In *Second international conference on fog and mobile edge computing* (pp. 194–200). IEEE.
42. Prabavathy, S., Sundarakantham, K., & Mercy Shalinie, S. (2018). Design of cognitive fog computing for intrusion detection in Internet of Things. *Journal of Communications and Networks, 20*, 291–298.
43. Cordeschi, N., Amendola, D., & Baccarelli, E. (2014). Reliable adaptive resource management for cognitive cloud vehicular networks. *IEEE Transactions on Vehicular Technology, 64*, 2528–2537.
44. Greenstadt, R., & Beal, J. (2008). Cognitive security for personal devices. In *Proceedings of the 1st ACM workshop on Workshop on AISec* (pp. 27–30). ACM.
45. Kinsner, W. (2012). Towards cognitive security systems. In *11th international conference on cognitive informatics and cognitive computing* (pp. 539–539). IEEE.
46. Rahman, S. M. M. Cognitive Cyber-Physical System (C-CPS) for Human-Robot collaborative manufacturing. In *14th Annual Conference System of Systems Engineering (SoSE)*. IEEE.
47. Eng, K., Siekierka, E., Cameirao, M., Zimmerli, L., Pyk, P., Armin, D., Erol, F., Corina, S., Bassetti, C., & Kiper, D. (2007). Cognitive virtual-reality based stroke rehabilitation. In *World congress on medical physics and biomedical engineering 2006* (pp. 2839–2843). Springer.
48. Du, J., Zhu, Q., Shi, Y., Wang, Q., Lin, Y., & Zhao, D. (2020). Cognition digital twins for personalized information systems of smart cities: Proof of concept. *Journal of Management in Engineering, 36*, 04019052.

49. Saghiri, A. M., & Meybodi, M. R. (2016). An approach for designing cognitive engines in cognitive peer-to-peer networks. *Journal of Network and Computer Applications, 70*, 17–40. https://doi.org/10.1016/j.jnca.2016.05.012
50. Mahmoud, Q. H. (2007). *Cognitive networks*. Wiley Online Library.
51. HamlAbadi, K. G., Saghiri, A. M., Vahdati, M., TakhtFooladi, M. D., & Meybodi, M. R. (2017). A framework for cognitive recommender systems in the Internet of Things (IoT). In *2017 IEEE 4th international conference on Knowledge-Based Engineering and Innovation (KBEI)* (pp. 0971–0976). https://doi.org/10.1109/KBEI.2017.8324939.
52. Saghiri, A. M., Vahdati, M., Gholizadeh, K., Meybodi, M. R., Dehghan, M., & Rashidi, H. (2018). A framework for cognitive Internet of Things based on blockchain. In *2018 4th International Conference on Web Research (ICWR)* (pp. 138–143). https://doi.org/10.1109/ICWR.2018.8387250.
53. Vahdati, M., HamlAbadi, K. G., Saghiri, A. M., & Rashidi, H. (2018). A self-organized framework for insurance based on Internet of Things and blockchain. In *2018 IEEE 6th International Conference on Future Internet of Things and Cloud (FiCloud)* (pp. 169–175). https://doi.org/10.1109/FiCloud.2018.00032.
54. Park, J.-h., Salim, M. M., Jo, J. H., Sicato, J. C. S., Rathore, S., & Park, J. H. (2019). CIoT-Net: a scalable cognitive IoT based smart city network architecture. *Human-Centric Computing and Information Sciences, 9*, 29. SpringerOpen.
55. Bontempi, F., Gkoumas, K., & Arangio, S. (2008). Systemic approach for the maintenance of complex structural systems. *Structure and Infrastructure Engineering, 4*, 77–94. Taylor & Francis.
56. Lenat, D. B., Guha, R. V., Pittman, K., Pratt, D., & Shepherd, M. (1990). Cyc: toward programs with common sense. *Communications of the ACM, 33*, 30–49.
57. Helgren, M. J., Little, M. E., Bingham, P. E., Jr., Martin, R. G., & Treece, A. J. (2006). *Rules-based configuration problem detection*. Google Patents.
58. Sutton, R. S., & Barto, A. G. (1998). *Reinforcement learning: An introduction*. Cambridge University Press.
59. Wang, K., Wang, Y., Sun, Y., Guo, S., & Jinsong, W. (2016). Green industrial Internet of Things architecture: An energy-efficient perspective. *IEEE Communications Magazine, 54*, 48–54. IEEE.
60. Kamalinejad, P., Mahapatra, C., Sheng, Z., Mirabbasi, S., Leung, V. C. M., & Guan, Y. L. (2015). Wireless energy harvesting for the Internet of Things. *IEEE Communications Magazine, 53*, 102–108. IEEE.
61. Ejaz, W., Naeem, M., Shahid, A., Anpalagan, A., & Jo, M. (2017). Efficient energy management for the internet of things in smart cities. *IEEE Communications Magazine, 55*, 84–91. IEEE.
62. Priya, S., & Inman, D. J. (2009). *Energy harvesting technologies*. Springer.
63. Pan, S. J., & Yang, Q. (2010). A survey on transfer learning. *IEEE Transactions on Knowledge and Data Engineering, 22*, 1345–1359.
64. Wang, F.-Y., Yuan, Y., Zhang, J., Qin, R., & Smith, M. H. (2018). Blockchainized Internet of minds: A new opportunity for cyber–physical–social systems. *IEEE Transactions on Computational Social Systems, 5*, 897–906.
65. Gunning, D. (2017). Explainable artificial intelligence (xai). *Defense Advanced Research Projects Agency (DARPA), nd Web* 2.
66. Holzinger, A. (2018). From machine learning to explainable AI. In *2018 world symposium on Digital Intelligence for Systems and Machines (DISA)* (pp. 55–66). IEEE.
67. Tsakiridis, N. L., Diamantopoulos, T., Symeonidis, A. L., Theocharis, J. B., Iossifides, A., Chatzimisios, P., Pratos, G., & Kouvas, D. (2020). Versatile Internet of Things for Agriculture: An eXplainable AI Approach. In *IFIP international conference on artificial intelligence applications and innovations* (pp. 180–191). Springer.

68. Hossain, M., Shamim, G. M., & Guizani, N. (2020). Explainable AI and mass surveillance system-based healthcare framework to combat COVID-I9 like pandemics. *IEEE Network, 34*, 126–132. IEEE.
69. Pawar, U., O'Shea, D., Rea, S., & O'Reilly, R. (2020). Explainable AI in healthcare. In *2020 international conference on Cyber Situational Awareness, Data Analytics and Assessment (CyberSA)* (pp. 1–2). IEEE.
70. Alonso, J. M., & Mencar, C. (2017). Building Cognitive Cities with Explainable Artificial Intelligent Systems. In *CEx@ AI* IA*.
71. Gade, K., Geyik, S. C., Kenthapadi, K., Mithal, V., & Taly, A. (2019). Explainable AI in industry. In *Proceedings of the 25th ACM SIGKDD international conference on knowledge discovery & data mining* (pp. 3203–3204). ACM.
72. Rao, B. B. P., Saluia, P., Sharma, N., Mittal, A., & Sharma, S. V. (2012). Cloud computing for Internet of Things & sensing based applications. In *2012 sixth International Conference on Sensing Technology (ICST)* (pp. 374–380). IEEE.
73. Parwekar, P. (2011). From internet of things towards cloud of things. In *2011 2nd International Conference on Computer and Communication Technology (ICCCT-2011)* (pp. 329–333). IEEE.
74. Ghorbani, M., Meybodi, M. R., & Saghiri, A. M. (2019). An architecture for managing Internet of Things based on cognitive Peer-to-peer networks. In *2019 5th International Conference on Web Research (ICWR)* (pp. 111–116). IEEE.
75. Goodfellow, I., Bengio, Y., Courville, A., & Bengio, Y. (2016). *Deep learning* (Vol. 1). MIT Press Cambridge.
76. Cox, M. T. (2005). Metacognition in computation: A selected research review. *Artificial Intelligence, 169*, 104–141.
77. Darwish, D. (2015). Improved layered architecture for Internet of Things. *International Journal of Computing Academic Research (IJCAR), 4*, 214–223. Citeseer.
78. Anderson, J. R. (1996). ACT: A simple theory of complex cognition. *American Psychologist, 51*, 355–365.
79. Laird, J. E. (2012). *The Soar cognitive architecture*. MIT Press.
80. Varshney, K. R. (2016). Engineering safety in machine learning. In *2016 Information Theory and Applications Workshop (ITA)* (pp. 1–5). IEEE.
81. Gordon-Spears, D. F. (2002). Asimov's laws: Current progress. In *International workshop on formal approaches to agent-based systems* (pp. 257–259). Springer.
82. Haddadin, S. (2013). *Towards safe robots: Approaching Asimov's 1st law*. Springer.
83. Murphy, R., & Woods, D. D. (2009). Beyond Asimov: The three laws of responsible robotics. *IEEE Intelligent Systems, 24*, 14–20.
84. Chen, M., Tian, Y., Fortino, G., Zhang, J., & Humar, I. (2018). Cognitive internet of vehicles. *Computer Communications, 120*, 58–70. Elsevier.
85. Fraga-Lamas, P., Fernández-Caramés, T. M., Suárez-Albela, M., Castedo, L., & González-López, M. (2016). A review on internet of things for defense and public safety. *Sensors, 16*, 1644. Multidisciplinary Digital Publishing Institute.
86. Luccioni, A., & Bengio, Y. (2020). On the morality of artificial intelligence [Commentary]. *IEEE Technology and Society Magazine, 39*, 16–25.
87. Goodfellow, I., Pouget-Abadie, J., Mirza, M., Xu, B., Warde-Farley, D., Ozair, S., Courville, A., & Bengio, Y. (2014). Generative adversarial nets. *Advances in Neural Information Processing Systems*, 2672–2680.
88. Blitz, M. J. (2018). Lies, line drawing, and deep fake news. *Okla. L. Rev., 71*, 59.
89. Narendra, K. S., & Thathachar, M. A. L. (1989). *Learning automata: An introduction*. Prentice Hall.
90. Saghiri, A. M., & Meybodi, M. R. (2015). A self-adaptive algorithm for topology matching in unstructured Peer-to-Peer networks. *Journal of Network and Systems Management, 24*, 393–426. https://doi.org/10.1007/s10922-015-9353-9

91. Wang, B., Wu, Y., & Liu, K. J. (2010). Game theory for cognitive radio networks: An overview. *Computer Networks, 54*, 2537–2561.
92. Sanfey, A. G. (2007). Social decision-making: Insights from game theory and neuroscience. *Science, 318*, 598–602.
93. Li, J., Zhao, H., Hafid, A. S., Wei, J., Yin, H., & Ren, B. (2019). A bio-inspired solution to cluster-based distributed spectrum allocation in high-density cognitive Internet of Things. *IEEE Internet of Things Journal, 6*, 9294–9307. IEEE.
94. Machin, J., & Solanas, A. (2019). Conceptual description of nature-inspired cognitive cities: Properties and challenges. In *International work-conference on the interplay between natural and artificial computation* (pp. 212–222). Springer.
95. Basu, S., Karuppiah, M., Selvakumar, K., Li, K.-C., Islam, S. K. H., Hassan, M. M., & Bhuiyan, M. Z. A. (2018). An intelligent/cognitive model of task scheduling for IoT applications in cloud computing environment. *Future Generation Computer Systems, 88*, 254–261. Elsevier.
96. Nowak, M., Nowak, S., Domańska, J., & Czachórski, T. (2019). Cognitive packet networks for the secure internet of things. In *2019 Global IoT Summit (GIoTS)* (pp. 1–4). IEEE.
97. Zhao, R., Wang, X., Xia, J., & Fan, L. (2020). Deep reinforcement learning based mobile edge computing for intelligent Internet of Things. *Physical Communication, 43*, 101184. Elsevier.
98. Gödel, K. (2013). *Kurt Gödel: Collected works*. Oxford University Press on Demand.
99. Braten, A. E., & Kraemer, F. A. (2018). Towards cognitive iot: Autonomous prediction model selection for solar-powered nodes. In *2018 IEEE International Congress on Internet of Things (ICIOT)* (pp. 118–125). IEEE.
100. Huang, B., Bouguettaya, A., & Neiat, A. G. (2020). Cognitive amplifier for Internet of Things. *arXiv preprint arXiv:2005.06914*.
101. Ateniese, G., Felici, G., Mancini, L. V., Spognardi, A., Villani, A., & Vitali, D. (2013). Hacking smart machines with smarter ones: How to extract meaningful data from machine learning classifiers. *arXiv preprint arXiv:1306.4447*.
102. Chaib-draa, B., & Dignum, F. (2002). Trends in agent communication language. *Computational Intelligence, 18*, 89–101.
103. Maedche, A., & Staab, S. (2001). Ontology learning for the semantic web. *IEEE Intelligent Systems, 16*, 72–79.
104. Lindley, J., Coulton, P., & Cooper, R. (2017). Why the internet of things needs object orientated ontology. *The Design Journal, 20*, S2846–S2857. Taylor & Francis.
105. Xu, G., Cao, Y., Ren, Y., Li, X., & Feng, Z. (2017). Network security situation awareness based on semantic ontology and user-defined rules for Internet of Things. *IEEE Access, 5*, 21046–21056. IEEE.
106. Teslya, N., & Smirnov, A. (2018). Blockchain-based framework for ontology-oriented robots' coalition formation in cyberphysical systems. In *MATEC Web of Conferences* (Vol. 161, pp. 03–18). EDP Sciences.
107. Akyildiz, I. F., & Jornet, J. M. (2010). The internet of nano-things. *IEEE Wireless Communications, 17*, 58–63. IEEE.
108. Akyildiz, I. F., Pierobon, M., Balasubramaniam, S., & Koucheryavy, Y. (2015). The internet of bio-nano things. *IEEE Communications Magazine, 53*, 32–40. IEEE.
109. Kuscu, M., & Akan, O. B. (2015). The Internet of molecular things based on FRET. *IEEE Internet of Things Journal, 3*, 4–17. IEEE.
110. Jornet, J. M., & Akyildiz, I. F. (2012). The internet of multimedia nano-things. *Nano Communication Networks, 3*, 242–251. Elsevier.
111. Zikria, Y. B., Afzal, M. K., & Kim, S. W. (2020). Internet of Multimedia Things (IoMT): Opportunities, challenges and solutions. *Sensors, 20*, 23–34.

Part II
Applications

An AI Approach to Rebalance Bike-Sharing Systems with Adaptive User Incentive

Yubin Duan and Jie Wu

1 Introduction

Bike-sharing systems (BSSs) have been deployed in major cities across the world. The environmental and economic benefits brought by the BSS speed up the deployment of the BSS [7]. Riding bikes is more environment friendly and helps to reduce the CO_2 emission. The BSS also brings economic benefits to the public. Deploying a BSS in a local neighborhood could increase accessibility with local businesses. In addition, BSSs address the "last mile-first mail" issues in cities by decreasing journey times and increasing users' mobility. Driven by the idea of green commuting and the economic benefits, more and more people tend to use shared bikes for the daily commute. Many BSS operators also upgrade their bikes to utilize IoT devices such as GPS and alarm sensors. This makes maintaining the BSS easier, especially for bike rebalancing with AI systems.

Efficiently rebalancing the BSS is necessary and challenging with the expansion of the BSS. Without bike rebalancing, the asymmetric users' demands for bikes in temporal and spatial domains would cause *overflow* and *underflow* events as shown in Fig. 1. In docked BSSs, such as the *Citi bike* system in NYC, the overflow events occur at stations that are full of bikes. Users cannot return bikes to those overflow stations, which increases the detour distances of users. The underflow events occur at stations that have no bikes. Users cannot rent bikes at those underflow stations, and the BSS loses the potential profit. Although there are no concepts of station capacities for dockless BSSs, such as the *Mobike* in China, too many bikes clustered in a small region would also cause overflow events, since it would cause congestion in the city. The underflow events where there are no bikes in an area also make the BSS operators lose users. Those negative impacts caused by bike overflow and

Y. Duan (✉) · J. Wu
Temple University, Philadelphia, PA, USA
e-mail: yubin.duan@temple.edu; jiewu@temple.edu

S. Pal et al. (eds.), *Artificial Intelligence-based Internet of Things Systems*,
Internet of Things, https://doi.org/10.1007/978-3-030-87059-1_14

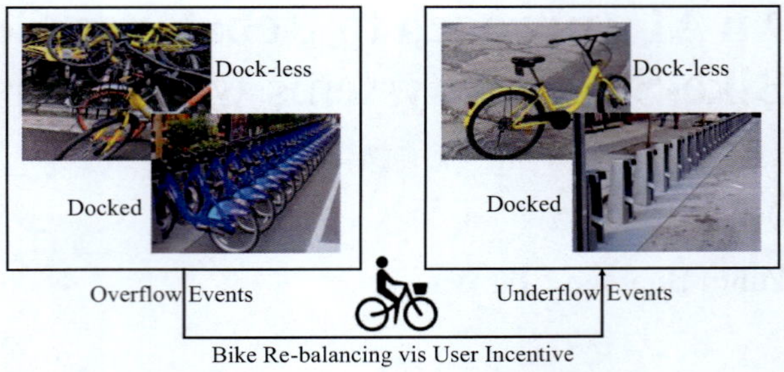

Fig. 1 Resolving underflow/overflow through bike rebalancing

underflow events motivate the BSS operators to rebalance their systems in a timely and cost-efficient manner.

Motivated by the quick development of AI algorithms, we investigate the bike rebalancing problem in this paper and attempts to optimize it with reinforcement learning. We follow the user incentive approach to rebalance the BSSs. Specifically, the BSS operator would provide a monetary incentive for users if they rent/return bikes at locations specified by the operator. We consider both *source incentive* and *destination incentive* for renting and returning bikes, respectively. Compared with truck-based rebalancing approaches, the user-based approach is more flexible and has been implemented in real-world BSSs, such as the *Bike Angle project*[1] in NYC. Our objective is to maximize the number of bike usages during a day. Maximizing the number of bike usages is critical since it could benefit both system operators and users. A larger number of bike usages means a better service level of the system. It could satisfy more user demands for bikes and enlarge the profit of BSS operators. Our constraint is the budget constraint that means the summation of incentives provided to users in a day is limited by a constant value. The budget constraint is essential for BSS operators since they need to make profits for long-term operation.

The problem we investigated is different from existing research. The state-of-the-art user-based bike rebalancing scheme [22] only considers the source incentives but ignores the power of destination incentives. Specifically, their scheme only considers encouraging users to rent bikes from nearby regions with a source incentive. We notice that destination incentives that let users return bikes at alternative locations can also help to rebalance the system. Adaptively combining those two incentives could bring extra benefits for bike balancing. Besides, we extend the problem scenario of which only considers the dockless BSSs. Both dockless and docked rebalancing problems are considered in this paper.

The benefits of the destination incentive are shown by the example in Fig. 2. In a dockless BSS, there are two users u_1 and u_2, and only one available bike located at R on the map. We assume u_2 arrives at the system right after u_1 reaches its destination u_1' . Firstly, we only consider source incentives. If a user cannot find bikes nearby his/her source location, the BSS would incentivize the user to enlarge its

[1] https://www.citibikenyc.com/bikeangels/

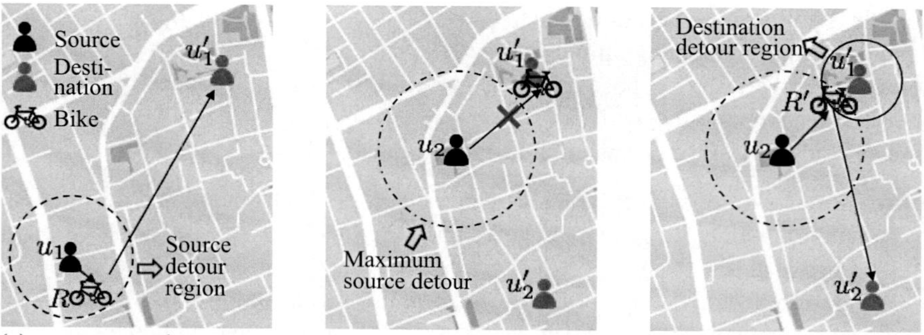

(a) u_1 accepts the source incentive

(b) u_2 cannot rent the bike w/o dest. incentive

(c) Dest. detour of u_1 helps u_2

Fig. 2 An illustration of the benefit of destination incentive

source detour region and pick up a bike, as shown in Fig. 2a. Note that users have maximum detour distances [24] since their mobility is limited by walking. Therefore, as illustrated in Fig. 2b, user u_2 cannot rent the bike, and the BSS operator loses the user. However, we notice that the detour distance of user u_1 is relatively small and does not exceed the detour distance limitation. u_1 could have another detour near its destination if an incentive is provided. If we also consider destination incentive, the user u_1 could return the bike at location R' instead of its destination, as shown in Fig. 2c. Then, the user u_2 could successfully rent the bikes with source incentives.

The example shows that the number of bike usages or the service level is increased from 1 to 2 by providing destination incentives along with source incentives.

Inspired by the motivation example, we propose a user-incentive-based rebalancing scheme that considers both source and destination incentives. However, it is not trivial to design such a scheme. The first challenge is the complex user dynamics.

[10] and [22] have shown the user dynamics in both temporal and spatial domains in docked and dockless BSSs, respectively. For dockless BSSs, another challenge is the extremely large number of bikes in a city. For example, Mobike plans to deploy hundreds of thousands of bikes in Guangzhou, China. Determining the incentive price for each bike is computationally complex. Another challenge for both dockless and docked BSSs is to adaptively adjust the source and destination incentive prices for each time slot in a day.

We propose to use reinforcement learning approaches to solve those challenges. Although the user dynamics are complex in both temporal and spatial domains, there exist usage patterns. Liu et al. [20] have shown that the user demands could be predicted by using machine learning techniques. The reinforcement learning agent could learn the pattern from its exploring experience. Besides, the reinforcement learning agent could adaptively adjust its pricing policy according to the reward function. Reinforcement learning algorithms fit our problem scenario. In addition, we could divide the city into multiple regions and let bikes in the same region have identical incentive prices. The problem scale could be reduced, as well as the complexity of training a reinforcement learning agent. Therefore, in this paper, we extend the reinforcement learning framework proposed by [22] that only considers source incentives and propose a hybrid incentive scheme that makes use of both source and destination incentives.

The contributions of our paper are summarized as follows:

- We propose to rebalance the BSS by offering users both source and destination incentives, which brings extra benefits compared with the source-incentive-only schemes.
- We analyze the advantage of the destination incentives and propose to combine the source and destination incentives by splitting the rebalancing budget.
- We adapt the state-of-the-art reinforcement learning framework for rebalancing dockless BSSs to determine the destination incentive prices.
- We further extend our scheme to docked BSSs by adding the capacities of each station into the state space of the reinforcement learning agent.
- We test the performance of our hybrid incentive schemes through experiments on a real-world dataset.

The remainder of the paper is organized as follows. Section 2 presents our problem statement, which contains the notations and system models. Section 3 presents our hybrid incentive scheme, which adaptively adjusts source and destination incentive strength. Section 4 discusses the extended scheme for docked BSS rebalancing. Section 5 illustrates the experiment that is conducted on a real-world dataset. Section 6 reviews the related works. Section 7 concludes the paper.

2 Problem Statement

2.1 Overview

In our model, we propose an adaptive approach for rebalancing dockless BSS. Given a limited budget, which is not sufficient enough to totally balance the BSSs, our approach adaptively allocates the budget to incentive users to conduct a detour at source and/or destination based on the underflow/overflow distribution across time and space. The objective is to maximize the overall service level of the system over a day. The service level is quantified by the number of satisfied users or the number of bike usages.

Specifically, the incentive used to encourage users to rent bikes at neighbor regions of their sources is denoted as a source incentive, while the incentive is called a destination incentive on the other side. For a source incentive, the BSS operator provides locations of available bikes to each user, along with incentive prices of bikes in neighbor regions. For a destination incentive, the operator suggests that users return bikes to neighbor regions of the user's destination. The price of source and destination incentive is determined by the incentive scheme. A reinforcement

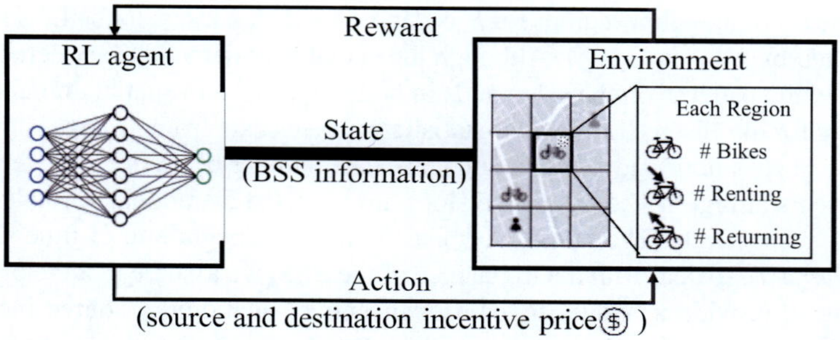

Fig. 3 An overview of the reinforcement learning framework

Table 1 Table of notations

Notations	Description
H	The region set and $H = \{h_i, .., h_n\}$ the slotted time set and $T = \{t_i, ..., t_m\}$ the user set and $U = \{u_i, ..., u_o\}$
T	Number of bikes in h_i at the beginning of t
U	Number of bike rented from h_i during timeslot t
$\varphi_i(t) D_i(t)$	Number of bike returned to h_i during timeslot t
$\Lambda_i(t)$	Neighbor regions of h
$N(h_i)$	u_k 's cost of moving from h_i to h_j with distance o
$c_k(i,j,\delta) T_{ij}(t)$	Number of users in h_i who rent bike from h_j and return to h_j during time t
$Pi + (t)/ pi - (t)\ B^+/B^-$	Source/destination incentive price of region h_i at t Budget provided by the BSS operator for the source/destination incentive

learning-based price scheme for source incentive has been studied in [22]. We propose to jointly consider source and destination incentives inspired by the benefits of destination incentive that we have observed. Users' choice of accepting incentives or not is simulated by the environment model. The performance of the rebalance is evaluated via the service level, which equals the number of satisfied users or the number of bike usages. The reinforcement learning framework of our hybrid incentive scheme is shown in Fig. 3 (Table 1).

2.2 *Incentive Scheme Model*

In the incentive scheme, we discretize time and space into time slots and square regions, respectively. The BSS operator provides differential source and/or destination incentive price for each region at each time slot. Specifically, each day is separated into m time slots in the time domain, denoted by $T = \{t_1, t_2, \ldots, t_m\}$. In the spatial domain, a city H is divided into n square regions, i.e., $H = \{h_1, h_2, \ldots, h_n\}$. The neighbors of a region h_i are defined as the four regions that are directly adjacent to

h_i, and the set of neighbor regions for h_i is denoted as $N(h_i)$. Users to the BSS system are denoted by $U = \{u_1, u_2, \ldots\}$. Although the actual user demands vary across time and space, the patterns on their demands in both temporal and spatial domains provide basis for our discretization. Our statistic on traces data from *Mobike* shows the existence of rush hour and demand hot spots. The number of users' rent events and return events at region h_i during time slot t are modeled as random variables $D_i(t)$ and $\Lambda_i(t)$, respectively. The number of bikes in h_i at the beginning of time slot t is denoted as $\varphi_i(t)$.To deal with the imbalance of the BSS, we assume that the provider is willing to provide a budget B for user incentive, including a source incentive budget B^+ and a destination incentive budget B^-. Our incentive scheme helps BSS operators to decide the differential price of source incentive $p_i^+(t)$ and destination incentive $p_i^-(t)$ for each region h_i at each time slot t. That is, if a user rents bikes at a neighborhood region h_i of his/her source region during time slot t, he/she can obtain an incentive $p_i^+(t)$. Each neighbor region may contain more than one bike, and the bikes in the same region have the same incentive price. Similarly, a destination incentive $p_i^-(t)$ is given to users who return bikes to h_i that are adjacent to users' destination region during time slot t. Different from the source incentive, we assume that each region only contains one potential return location, which is the center of the region. This simplification is to reduce the complexity of the model.

2.3 Environment Model

The environment mainly models user dynamics and provides feedback to the incentive scheme. Based on the source and destination incentive price vector generated by the scheme, the environment simulates each user's choice of accepting the incentive or not. We assume each user has a cost if he/she goes to alternative locations to rent bikes (with source detours) or return bikes (with destination detours). We follow the user cost model in [22, 24]. Specifically, a user u_k has an initial cost C for either source or destination detour. Besides, the cost is also relevant to the detour distance δ. Specifically, let $c_k(h_i, h_j, \delta)$ and $c_k'(h_i, h_j, \delta')$ denote the source and destination detour cost, respectively. h_i and h_j represent regions where u_k rents and returns a bike, respectively. δ and δ' are the corresponding source and destination detour distance. If the user u_k rents (or returns) a bike at a region, which is the neighbor of his/her source (or destination), his/her source detour cost $c_k(h_i, h_j, \delta) = C + \eta\delta^2$ (or destination detour cost $c_k'(h_i, h_j, \delta') = C + \eta\delta'^2$), where η is a constant coefficient. We assume users are not willing to rent or return bikes at regions further than neighbor regions, and that the cost of renting or returning bikes in these regions is infinity. If the user u_k rents (or returns) bikes in the same region as his/her source (or destination), there is no cost. Note that if a user detours at both source and destination, he/she will receive both source and destination incentives in one trip, which helps to resolve the overflow and underflow problem of the BSS.

Users make decisions on whether to accept source and/or destination incentive before they start riding. A user decides whether to accept the source incentive first, and then makes a decision regarding the destination. If a user in h_i requests a bike during time slot t, and he/she decides to rent a bike in h_j and to return it in h_k due to the incentives, then we increment $\tau_{ij}(t)$ by one to record this trace.

2.4 An Existing Pricing Scheme for Source Incentive

A pricing algorithm for source incentive is proposed by Pan et al. [22]. Their pricing scheme is based on a Markov decision process (MDP) and is optimized by using a reinforcement learning approach inspired by the hierarchical reinforcement learning and Deep Deterministic Policy Gradient algorithm [17]. The pricing algorithm is briefly stated in the section, and our adaptive incentive scheme is built upon it.

The MDP is used to model the interaction between the pricing scheme and the environment. Specifically, the MDP is a 5-tuple $(S, A, P, r, and\ \gamma)$, where S is the set of states $\{s_t\}$, A is the set of actions $\{a_t\}$, P describes the transition possibility between states under an action, r denotes the immediate reward, and γ is the discount factor. The weight of future rewards and the present reward is determined by the discount factor $\gamma \in [0, 1]$. $\gamma = 1$ represents that the future rewards share the same importance as the present reward, i.e., the overall reward is the additive sum of the reward from each time slot. The pricing scheme takes source incentive prices given to all regions as an action and the number of satisfied users as a reward. The MDP ends when the budget B is used up. The pricing scheme finds a policy π_θ, which maps states to actions, through optimizing the MDP based on reinforcement learning. The number of bikes rented from h_i and returned to h_j during time slot t is denoted by $\tau_{ij}(t)$.

Formally, their pricing scheme is defined as:

$$\max \quad \sum_{t=1}^{m}\sum_{i,j=1}^{n}\tau_{ij}(t)$$

$$\text{s.t.} \quad \sum_{t=1}^{m}\sum_{i=1}^{n}p_i^{+}(t) < B$$

$$\sum_{j=1}^{n}\tau_{ji}(t) - \sum_{j=1}^{n}\tau_{ji}(t) \le \varphi_i(t), \forall i, t$$

$$\varphi_i(t+1) = \varphi_i(t) + \sum_{j=1}^{n}\left(\tau_{ji}(t) - \tau_{ji}(t)\right) \forall i, t.$$

The objective is to maximize the number of valid traces over all regions and time slots. The first constraint is the incentive budget limitation. The second constraint means that the number of bikes in each region should not be less than zero at any region during any time slot. The last constraint represents the evolution of the number of bikes in each slot among different time slots.

The MDP is optimized by applying reinforcement learning algorithms. The reinforcement learning aims to train a parameter set θ in π_θ such that the overall rewards can be maximized by following a policy π_θ. Formally, the overall reward brought by the policy π_θ is:

$$J_{\pi_\theta} = E\left[\sum_{t=0}^{\infty} \gamma^k r\left(a_t, s_t\right) |, \pi_\theta |, s_0\right].$$

2.5 Problem Formulation

Based on the system and environment models, the BSS rebalancing problem is proposed. We aim to maximize the service level of a BSS in a one-day service circle. In each service circle, the BSS operator provides budget B^+ and B^- for source and destination incentives. Formally, our problem can be expressed as:

$$\begin{aligned} \max \quad & \sum_{t=1}^{m} \sum_{i,j=1}^{n} \tau_{ij}(t) \\ \text{s.t.} \quad & \sum_{t=1}^{m} \sum_{i=1}^{n} p_i^+(t) < B - B^- \\ & \sum_{t=1}^{m} \sum_{i=1}^{n} p_i^-(t) \le B^- \\ & \sum_{j=1}^{n} \tau_{ji}(t) - \sum_{j=1}^{n} \tau_{ji}(t) \le \varphi_i(t), \forall i, t \\ & \varphi_i(t+1) = \varphi_i(t) + \sum_{j=1}^{n} \left(\tau_{ji}(t) - \tau_{ji}(t)\right) \forall i, t \end{aligned}$$

Note that the difference with the existing price scheme can be found in the first two constraints, where we consider two kinds of incentives. The overall budget of source and destination incentives remains as B. The difference is that some part of the budget B^- is assigned to conduct the destination incentive:

3 Hybrid Incentive Scheme

3.1 Benefits of Destination Incentive

Although the source incentive is a straighforward way to increase the service level, it cannot fully unitize the power of user incentive. Besides source incentive, we use Fig. 4 to illustrate that the dockless BSS can also be balanced through the

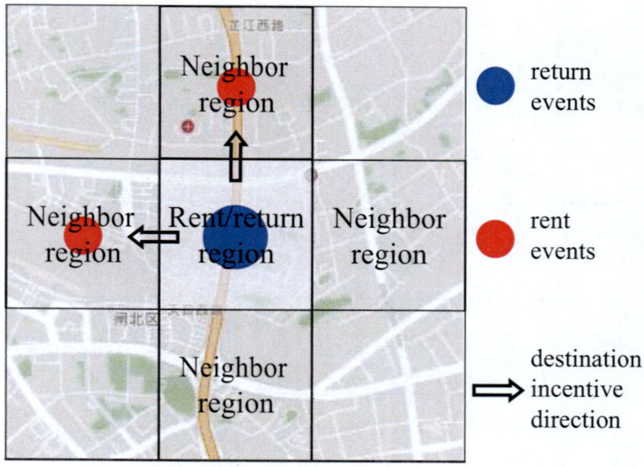

Fig. 4 An illustration of destination incentive

destination incentive. In the figure, the blue nodes represent areas whose return events are more frequent (i.e., the number of bikes on the area increases) in the given time window. The amount of extra return events for each region is shown in the figure, and each node's area is proportional to the extra value. The red nodes have the opposite meaning. By applying the destination incentive, i.e., incentivizing users to return bikes to neighbor regions, the imbalance usage can be greatly eased. A possible assignment for users is shown in the figure. The arrows indicate the destination incentive direction, and the number of users needed is shown along the arrows. However, the imbalanced demand cannot be totally satisfied since the number of extra return events could be different from the number of extra rent events. Although the spatial distribution cannot be fully balanced, the service level of the system can be improved because more users are able to rent bikes.

Algorithm 1 The hybrid incentive schema.

Input: The source and destination of user u_k
Output: Alternative bike b to rent and alternative location b' to return

$h_i, h_j) \leftarrow$ index of the region of location u_k and u'_k
$N(h_i), N(h_j)) \leftarrow$ neighboring regions of h_i and h_j
Incentive price set $I = (p_i^+(t), p_i^-(t)) \leftarrow$ the pricing scheme learned by the reinforcement learning agent
for all bikes in $N(h_i)$ and return locations in $N(h_j)$ **do**
    u_k calculates the net profit of source and destination detours, i.e., incentive prices minus detour costs
u_k chooses the bike b and return location b' with maximum net profit which is denoted as $p_{\max}$
if $p_{\max} < 0$ **then**
    u_k refuses incentives.
return b, b' as the user's alternative pick-up and drop-off locations.

The unique benefit of the destination incentive is illustrated through the toy example in Fig. 5. By applying the destination incentive, the service level can increase to 2. Only applying the source incentive cannot achieve such a service

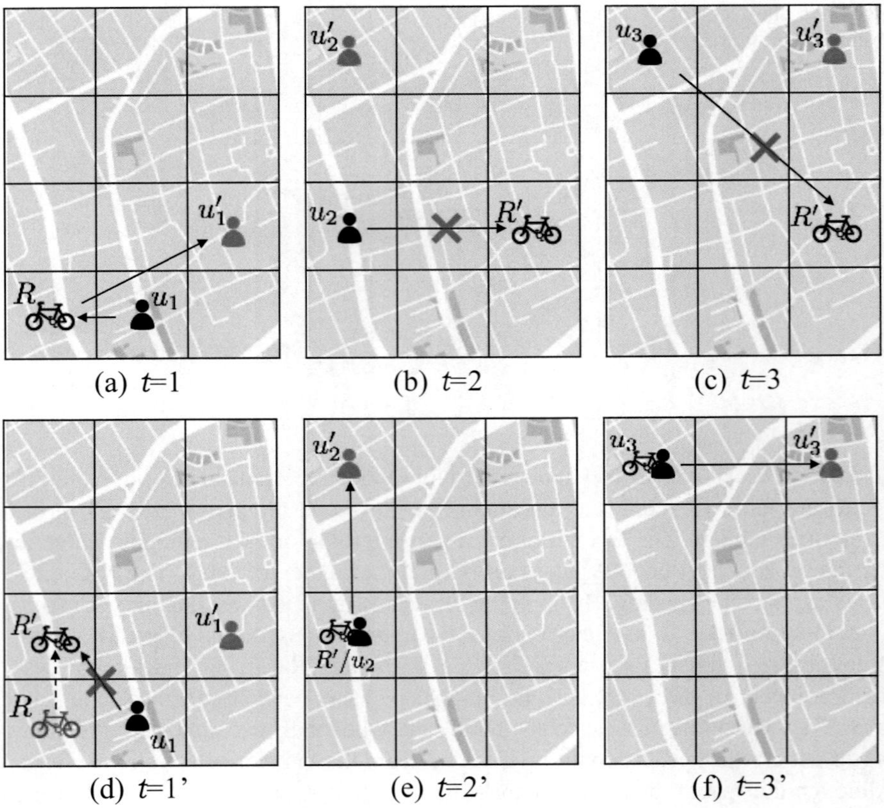

Fig. 5 An example showing the benefit of returning at alternative destinations

level. In the example, we consider that there are three users appearing in time slot 1, 2, and 3 at location u_1, u_2, and u_3 shown in the figure, respectively. Their corresponding destination is *u*1!, *u*2!, and *u*3!. There is only one bike in the map, and it is returned at location R at $t = 1$ without considering destination incentive. As shown in Fig. 5a–c, only user u_1 can successfully finish her trip, and users at u_2 and u_3 cannot rent a bike since the detour distance exceeds the limitation. In this case, the number of satisfied users is 1. However, if the destination incentive is allowed at $t =$ 1, the bike can be returned to location $R^!$ instead of R. Although it is too far for user u_1 to rent the bike, users u_2 and u_3 can finish their journeys without any detours. That is, the number of satisfied users increases to 2 with a well-designed alternative return location suggestion, and the users' total walk distance is the same as following the source or destination incentive scheme.

Although the destination incentive may have a better control on the trend of the bike flow, deciding the return location is complex due to the large size of the potential returning points. Especially for dockless BSSs, a bike could be returned to any location as along as it does not block others. The large solution space makes the

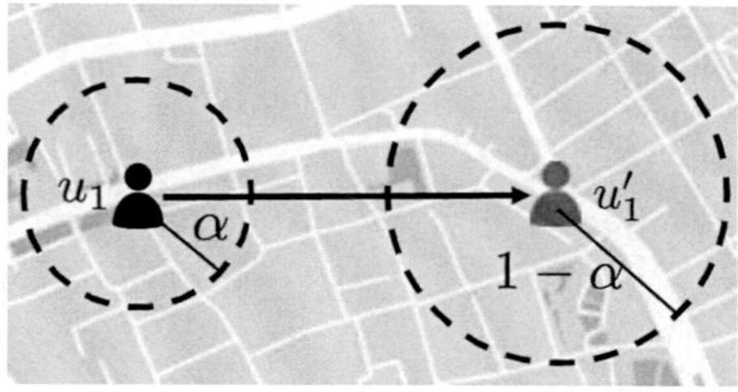

Fig. 6 Adaptively adjusting source and destination incentive

learning algorithm hard to converge. To cast off that complexity, we propose to sample the metric middle point of each region as the potential returning point. It means that if a user accepts the destination incentive, the system would suggest the metric middle point of the destination region as the returning point. This additional constraint not only reduces the action space for reinforcement learning algorithms but also reduces management difficulties for the system operator (Fig. 6).

The trade-off is that the efficiency of the destination incentive might be reduced. That is because choosing the middle point as returning point may cause the unnecessary detour, which may increase the detour distance of a user. If the pay-off brought by the destination detour is not more than detour distance wasted on a certain user dynamic, the destination detour is not useful anymore.

3.2 A Hybrid Incentive Scheme

Either the source incentive or destination incentive itself has its own shortage on certain user dynamics. Therefore, besides considering incentivizing users to just pick up or just drop off bikes in neighbor regions, we propose to combine these two kinds of incentives and build a hybrid incentive scheme. The intuition behind the hybrid incentive scheme is that the scheme could adaptively adjust the source and destination incentive based on different imbalance situations.

In the hybrid incentive scheme, the state and action space in the MDP is enlarged because of the destination detour budget B^- and price p^-. Specifically, a state vector s_t is constructed by $\sum_{h_i} \varphi_i(t)$, $\sum_{h_i} D_i(t-1)$, $\sum_{h_i} \Lambda_i(t-1)$, $B^+ - \sum_{h_i,t} p_i^+(t)$, $B^- - \sum_{h_i,t} p_i^-(t)$, and unserviced events in previous time slots. The first term represents the number of unused bikes over the city at the beginning of t. The total amount of bikes over the city is constant, but the number of unused bikes may vary over time due to the fluctuated usage of users. The $\sum_{h_i} D_i(t-1)$, $\sum_{h_i} \Lambda_i(t-1)$ represents the total number of rent and return events over the city, which captures the temporal bike usage information to the MDP. The $B^+ - \sum_{h_i,t} p_i^+(t)$ calculates the remaining budget for the source incentive, and $B^- - \sum_{h_i,t} p_i^-(t)$ calculates the remaining budget for the source and destination incentive, respectively. The MDP ends either when t

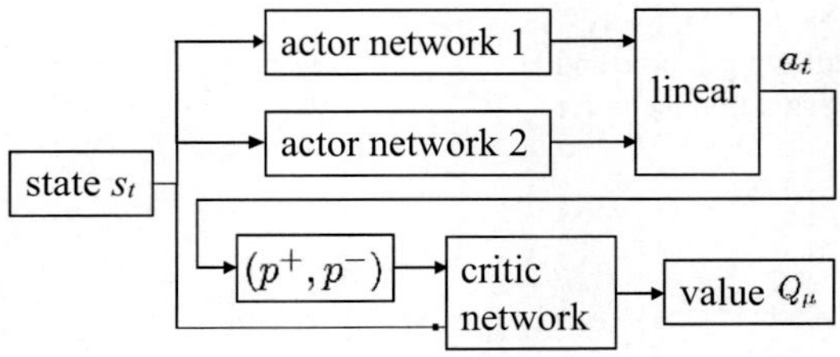

Fig. 7 The learning framework for hybrid incentive scheme

reaches the time slot upper bound, or when any of the remaining budget for source or destination incentive is empty.

An action vector a_t in A for time slot t is constructed by the source incentive price vector $\left(p_i^+(t), i = 1, \ldots, n\right)$ and source incentive price vector $\left(p_i^-(t), i = 1, \ldots, n\right)$. Although the transmission probability P is built on the enlarged state and action spaces, the state transmission still can be simulated via our environment model. The reward r of the hybrid incentive scheme is constructed by rewards from source incentive $r^+(s_t, p^+)$ and rewards from the destination incentive $r^-(s_t, p^-)$.To adapt the modification to the MDP, we extend the actor-critic framework in [22]. The size of the actor network is enlarged as shown in Fig. 7. The actor network 1 is used to learn the source incentive prices $p^+(t)$, and the actor network 2 is used to learn the destination incentive prices $p^-(t)$. As for the critic network, the sub-Q-value of each region h_i at step t is evaluated based on $\left(p_i^+(t), p_i^-(t)\right)$ instead of just considering $p_i^+(t)$, and the estimation of the Q-value changes correspondingly.

3.3 *Adaptively Adjust Source and Destination Incentive*

Besides adjusting the learning framework, we also propose two different ways to adjust the ratio of source and destination incentive price. One way is to adjust the strength of source and destination incentive by controlling the ratio of source and destination incentive budget. It is achieved by splitting the total budget into source and destination incentive budgets based on a ratio ρ.

Definition 1 (Budget Division) Assume the total budget available is B, and the budget division ratio is ρ. Then the budget appointed to source incentive is ρ, and the remaining $(1 - \rho)B$ is used for the destination incentive.

Under this scheme, the remaining budget of source and destination incentive in the initial state s_0 becomes:

$$B^+ = \rho B, B^- = (1 - \rho) B$$

The overall reward during a day under policy n_0 becomes:

$$J_{\pi_\theta} = E\left[\sum_{k=0}^{\infty}\gamma^k\left(r^+\left(a_k,s_k\right)+r^-\left(a_k,s_k\right)\right)|,\pi_\theta|,s_0\right]$$

The other way is to adjust the ratio between detour distances of source and destination incentive. It is achieved by adding the maximum detour constraint to users in the environment model. Let l denote the maximum detour distance that a user can accept, including source and destination detour. The value of l can be extracted from a user survey when applying the scheme in the real world. We split l into two parts l^s and l^d, which correspond to the maximum source detour and maximum destination detour, respectively. Let a denote the adjust parameter between l^s and l^d.

Definition 2 (Detour Distance Division) Given the maximum detour distance l of each user and parameter a, the maximum detour under source incentive is $l^s = al$, and the detour under destination incentive is $l^d = (1 - a)l$, as shown in Fig. 6.

To keep consistent with the environment model, we assume the user rejects the detour either if his/her detour distance exceeds the limitation or if he/she cannot gain profit from the detour. Through this setting, we try to limit each region's source and destination incentive among all time slots.

Formally, based on a, we attempt to limit the source and destination incentive as:

$$p_i^+\left(t\right) < C + \eta\alpha^2 l^2 \text{ and } p_i^-\left(t\right) < C + \eta\left(\left(1-\alpha\right)l\right)^2 \ \forall t \in T$$

The budget division strictly imposes restrictions on budgets of source and destination incentives, while the detour distance division restricts the source and destination incentive price on estimation. Either kind of incentive is adaptive among regions, and the sum of incentive prices cannot exceed the corresponding budget. The budget division is applied to the initial state of the MDP. The detour distance division is applied to the environment, the incentive greater than the limitation cannot bring benefits to the scheme.

3.4 *Properties*

In addition to the adaptive adjustment, the hybrid incentive scheme can help to break a long detour distance of one user into two short detour distances of different users. More clearly, we use the example in Fig. 8 to show the benefits brought by applying source and destination incentives at the same time. We consider there are two users arriving at time slots 0 and 1, respectively. Their origins are u_1 and u_2, and destinations are u_1' and u_2', respectively. There is one bike located at R at the beginning time slot $t = 0$. The solid lines show users' paths if only the source incentive is allowed. By following the solid lines, user u_1 rents the bike from R and returns it to u_1'. Then, user u_2, who comes after u_1 returns the bike, has to rent the bike at u_1' with

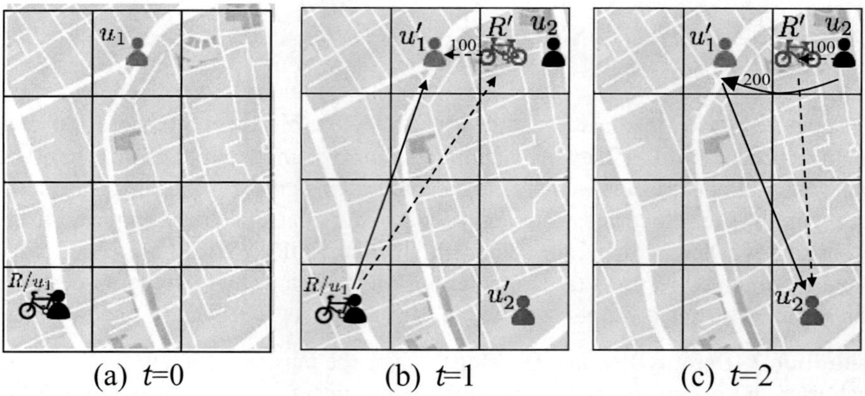

Fig. 8 Illustration of combining source and destination incentive

a 200 m detour, while the detour distance of user u_1 is 0. The 200 m detour of u_2 may exceed his/her maximum detour limitation. In contrast, the dashed lines show paths when both the source and destination incentives are allowed. User u_1 could return the bike to R' with a 100 m destination detour. In this case, user u_2 could rent the bike from R' with a 100 m source detour. The relatively long detour of user u_2 is equally shared by u_1 and u_2 in our hybrid incentive scheme.

According to the survey [24], the growth rate of each user's detour cost is proportional to the square of his/her detour distance. Our hybrid system provides a probability that let users with relatively shorter detour distances help share the long detour distance. It mitigates the burden for users with relatively longer detour distances and helps to attract more users to accept the incentive. It may help to incentive more users to become involved in the rebalancing and increase the number of satisfied users to the system. With more potential users joining rebalancing, the system is easier to choose proper users for rebalancing.

4 Hybrid Incentive in Docked BSS

In this section, we consider the docked BSS in which users must rent or return bikes at bike stations deployed by the BSS operator. In the docked BSS scenario, our objective is also maximizing the service level of the BSS in the daily service period.

In the docked BSS rebalancing, let H denote the set of stations rather than regions, and let h_i denote the station i. The docked BSS suffers more from the imbalance bike distribution. Specifically, it would cause *overflow* and *underflow* stations. The overflow stations are the stations that are full of bikes. Users cannot return bikes to overflow stations. The underflow stations have no bikes, and users cannot rent bikes. We also follow the user incentive approach to resolve the overflow and underflow issues. Specifically, at each time slot t, the BSS operator would assign source and destination incentives for each station h_i. The source incentive price is denoted as $p_i^+(t)$, which is the amount of monetary incentive for a user who rents bikes from the station. We provide source incentives to resolve the overflow issues. Similarly, the destination incentive price $p_i^-(t)$ represents the incentive for a user who returns bikes to h_i. It is used to resolve the underflow issues of a station. The source and destination incentive prices would not be provided at the same station. That is, if there is a source incentive at station h_i at time t, the destination incentive price $p_i^-(t) = 0$.

The challenge of rebalancing a docked BSS is the capacity limitation of each bike station. Let c_H denote the vector of station capacities of stations in H. Specifically, even if a station is located in a popular area and has a large number of bike renting demands, the station cannot hold more bikes than its capacity. The destination incentive might be infeasible for these fully occupied stations. The reinforcement learning agent needs to know the current number of bikes in each bike station at each time slot, along with the capacity of the station.

Besides the capacity limitations, there is no concept of neighborhood region in the docked BSS rebalancing. For the dockless BSS rebalancing scenario, we use neighborhood regions to reduce the complexity for the reinforcement learning agent. The incentive price for each region is affected by its four neighborhood regions rather than all regions in the map. For the docked scenario, we also need to reduce the state space for the agent. In particular, there are usually hundreds of stations in a city. When determining the incentive price for a station, taking the states of all stations into consideration would also lead to a large solution space. Therefore, similar to using neighborhood regions, we only consider states of k *nearby stations* when determining the incentive price for each station in a docked BSS. The k nearby stations of station h_i are defined as the first k nearest stations to h_i. Same as neighborhood regions, the k nearest stations of h_i are determined and could be hard coded into the reinforcement learning agent.

Considering the differences between docked and dockless BSSs, we extend our reinforcement framework to learn the source and destination incentive price for docked BSSs by enlarging the state space (adding station capacities as features) and replacing neighborhood regions with k nearby stations. The state and actor vectors of the reinforcement learning agent are updated to:

$$s_t' = (s_t, c_H), a_t' = (b(t) \in \{1, -1\}, p_i(t), i = 1, \ldots, n).$$

Specifically, the state vector of the reinforcement learning becomes $s_t' = (s_t, c_H)$ where s_t is the state vector used for dockless BSS rebalancing and c_H is the vector of capacities of stations in H. To make sure the source and destination incentive prices at a station would not be positive simultaneously, we update the action vector to $a_t' = (b(t) \in \{1, -1\}, p_i(t) \in R^+, i = 1, \ldots, n)$ where $b(t)$ is used to specify whether it is the source incentive or the destination incentive and $p_i(t)$ represents the incentive price for station h_i.

The environment model is also modified to fit the docked BSS rebalancing scenario. The detour cost model used for dockless BSSs cannot be used for docked BSSs since there are no neighborhood regions in docked BSSs. As a result, we update the cost model for docked BSSs and remove the limitations caused by neighborhood regions. Instead, we set maximum detour distances for users based on their original trip lengths. Assume a user plans to rent bikes from the station h_i and return to h_i'. The alternative pick-up station is h_j. Then, in docked BSSs, the source detour cost of the user is:

$$c\left(h_i, h_j\right) = \begin{cases} C + \eta \cdot dis\left(h_i, h_j\right)^2 & dis\left(h_i, h_j\right) < \kappa \cdot dis\left(h_i, h_i'\right). \\ +\infty & \text{otherwise.} \end{cases}$$

where the parameter κ is used to adjust the maximum detour distances. For example, when we are setting $\kappa = 1$, we assume that a user would not be willing to take a detour whose length is longer than the user's original journey. The cost model for destination detour is updated in the same way.

By updating the state and action spaces of the reinforcement learning agent and the environment model, we extend our hybrid incentive scheme for dockless BSSs to docked BSSs. The performance of the extended scheme is tested on real-world datasets, and the results are illustrated in Sect. 5.

5 Experiment

5.1 Dataset

We use the data published by *Mobike* to construct the dockless dataset and use the trip history data published by NYC to construct the docked dataset. The NYC dataset contains more than 1.5 million trip records for 328 bike stations. The Mobike dataset contains more than 100 k trip records of Shanghai. The record of each trip includes trip duration (in seconds), trip start (end) time and date, start (end) latitude and longitude, etc. The summary of the Mobike dataset is shown in Table 2. We first visualize the imbalanced bike distributions in those datasets.

The spatial imbalance of the Mobike dataset is shown in Fig. 9. To illustrate the spatial imbalance, we plot the usage of region in Fig. 7 for AM rush hours (7:00–9:00 AM) and PM rush hours (6:00 PM–8:00 PM). The blue nodes represent stations

Table 2 The Mobike dataset

Data source	Mobike traces (Shanghai)	
Timespan	Aug. 1 2016 to Sep. 1 2016	
Weekdays (weekends)	24 (8) days	
Bike data	# Bikes 79,063	# Trips 102,361

(a) Pick-up events in 8 a.m. - 9 a.m. (b) Pick-up events in 6 p.m. - 7 p.m.

Fig. 9 The temporal and spatial imbalanced distribution in the dockless BSS dataset

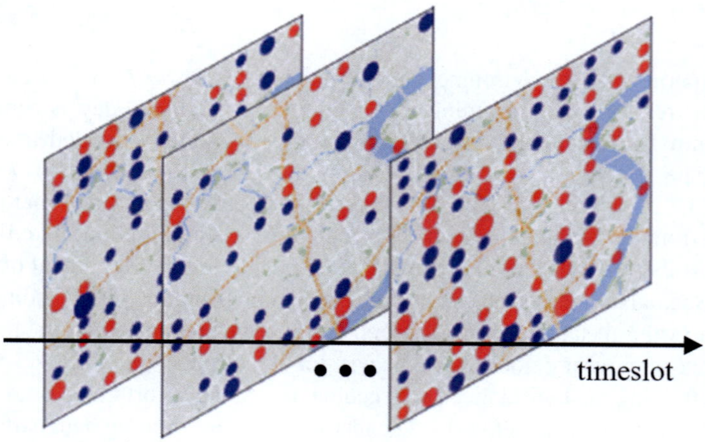

Fig. 10 Rebalancing among multiple time slots

whose return events are more frequent (i.e., the number of bikes on stations increase) in the given time window. The node's diameter is proportional to demands of the corresponding station. The red nodes have the opposite meaning. Figure 10 shows the variation of spatial imbalance over multiple time slots.

In addition, we investigate the bike imbalance of the NYC dataset. Figure 11a shows the statistics on a bike user's trip duration. Figure 11b illustrates the statistics of the temporal usage distribution of trips on 08/01/2016 (Monday). It shows that more than 55% of trips are shorter than 10 min. Figure 11b further shows that the demands of bikes are not even during a day. There also exist morning and evening peak hours.

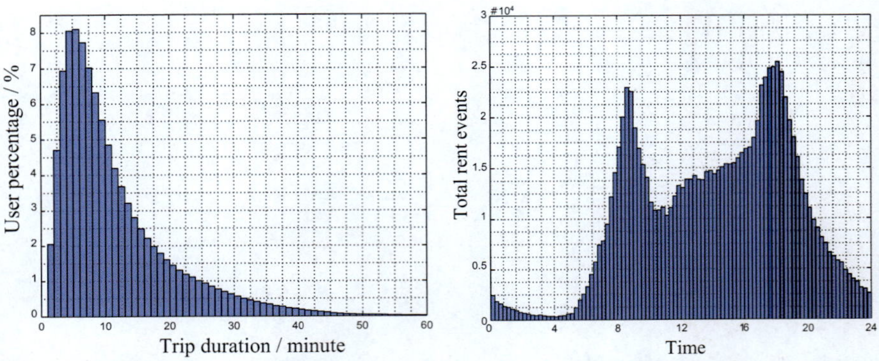

(a) Duration distribution (b) Rent event distribution

Fig. 11 Statistics of the NYC dataset

5.2 *Experiment Setup*

In our experiment, the environment model is built on *OpenAI Gym*, a toolkit for comparing reinforcement learning algorithms. Specifically, a day is temporally divided into 24 time slots, and the Shanghai city is spatially divided into 20–40 regions. The effective area of the city is bounded by [30.841°N, 31.477°N] and [120.486°E 121.971°E]. Users' time requests, locations, and destinations are extracted from the Mobike trace data. Through the statistics of unique bike ID, there is a total of 79, 063 bikes used in the dataset. Considering the retirement of broken bikes, the actual number of bikes may be less than that amount. Users' riding speed is chosen as the mean speed of all users, and the walking speed is assigned as 5 km/h. The cost of user detour obeys the cost model introduced in Sect. 2.3.

When training the hybrid incentive scheme, the Adam algorithm is used to optimize both actor and critic networks. The learning rates for training both parts are set as 10–4. In each step, to explore more action space, a Gaussian noise is added to each action generated from the actor network. Although [17] proposed to add Uhlenbeck-Ornstein noise to actions, the Gaussian noise is used for simplicity. The discount factor γ in the MDP is chosen as 0.99.

In the first set of experiments, we compare the performance of our algorithm with others under different budgets. The budget is varied from 1000 to 2000, and the performance is quantified by the decreased unserviced ratio defined in [22]. The number of unserviced users increases by one if the user cannot find a bike to ride, or he/she is not satisfied with any source incentive offered by the system. Let N_1 denote the number of unsatisfied users with no incentive, and let N_2 denote the number of unsatisfied users with incentive. Then, the corresponding unserviced ratio is defined as $(N_1 - N_2)/N_1$.

The second set of experiments focuses on the number of satisfied users. The number of satisfied users is proportional to the income of the BSS. We assume the BSS operator charges 1 for each user who rents the bike. Therefore, the profit of the operator can be calculated by subtracting the budget spent for incentive from the overall income of a day.

We also test the influence of the initial bike amount. If the initial bikes are sufficient enough, then each user can find a bike without a detour and the maximum service level is achieved. However, the number of bikes is limited in each region. Therefore, wisely spending the budget to achieve a better service level is important. The last set of experiments focuses on the rebalance performance across multiple days. Our first comparison algorithm is the source-incentive-only scheme [22], which is denoted as HRP. The second comparison algorithm is the DBP-UCB [24], which is one of the state-of-the-art bike rebalancing approaches based on user incentive. A randomized incentive scheme is used as a baseline.

Then, we tested our docked BSS rebalancing scheme. We compare our learning-based scheme with a fixed incentive scheme used by the operator of the NYC Citi bike. Specifically, the Citi bike launched the "Bike Angels" project and gave users fixed points if they would rent/return bikes at specific locations. Each point was worth about \$0.1. During the experiment, we denote this rebalancing scheme as *Fixed*. Besides, we use the *Random* scheme, which assigns incentive prices to stations randomly, as the baseline.

5.3 Results

We illustrate our experiment results on decreased unserviced ration in Fig. 12a. From the figure, we can conclude that the performance of our hybrid approach achieves better performance than other approaches. Comparing with the HRL that just considers the source incentive, we can conclude that adaptively allocating incentive on source detour, and destination detour can bring additional benefits on the service level. It is reasonable since the source and destination incentives are

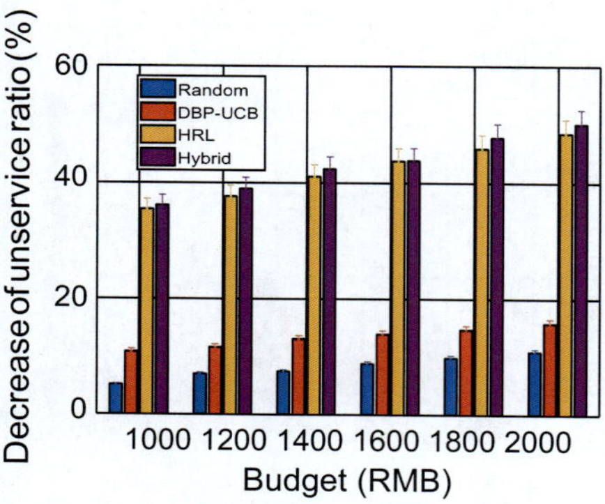

(a) Comparison on DUR.

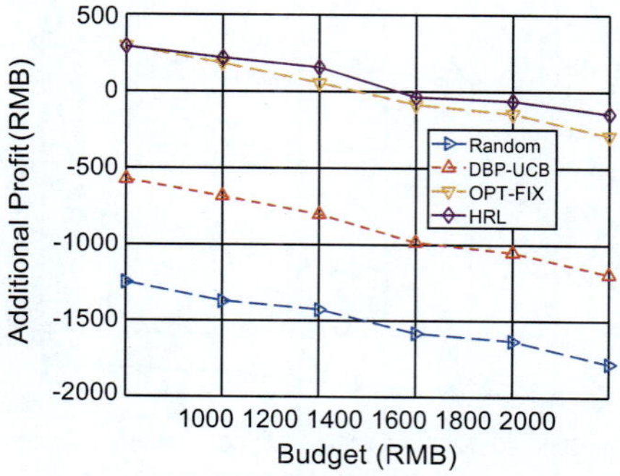

(b) Comparison on profit.

Fig. 12 Comparison on DUR and additional profit with of varying budget

included in the action spaces of the hybrid incentive scheme. By comparing HRL and DBP-UCB, we can conclude that the reinforcement learning can greatly improve the service level since it considers further reward when choosing the action for each state. The performance trend of all approaches shows that more user requests can be satisfied with a higher budget, even for the randomized policy.

The additional profit brought by the incentive is illustrated in Fig. 12b. As stated in [22], the HRL can bring additional benefits to the BSS operator when the budget is not too large. The hybrid incentive scheme can also gain profits from the incentive, which is arguably one of the most important features to BSS operators. However, as the budget increases, the profit decreases. It illustrates that the number of satisfied users increases more slowly with the increasing budget. That is to say, the BSS operator may not find it worthwhile to totally rebalance. The totally rebalanced system means that all user requests can be satisfied. The DBP-UCB and randomized scheme can bring additional profits to the system with a budget less than 1000 within the Mobike dataset.

Figure 13a shows the influence of the initial bike amount. With more initial bikes placed in the city, incentive schemes are more likely to achieve better performance. The increasing ratio in the figure is not as sufficient, and the reason for this could be that the initial bikes are uniformly distributed among the city, and adding bikes to regions with nearly no user requests may waste bikes. If the distribution of initial bikes could fit user requests, the increased initial bike amounts may greatly improve the service level.

Figure 13b shows the rebalance performance over multiple days. We count the decreased unservice events since it is additive. The difference between the HRL and the hybrid incentive increases as the number of days increases. It shows that the hybrid incentive scheme keeps a better distribution than the HRL. As we have shown in the previous section, the destination incentive, to a certain extent, is eager to place bikes in regions with more requests. These bikes are more likely to be used

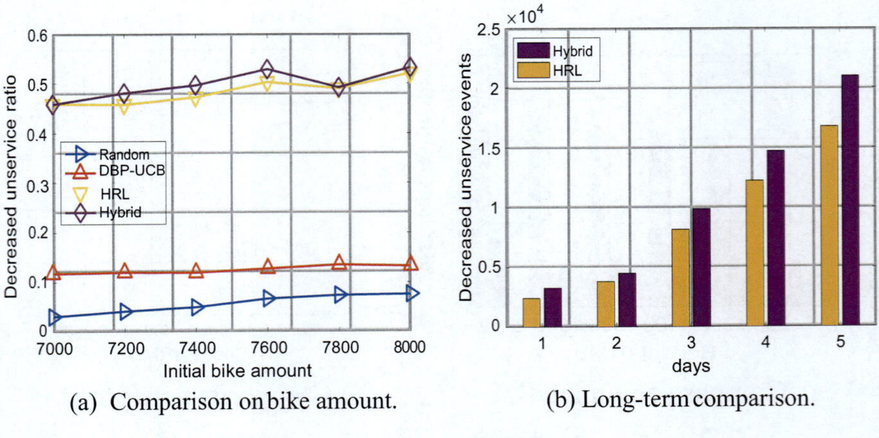

(a) Comparison on bike amount. (b) Long-term comparison.

Fig. 13 Cumulative density function and long-term performance comparison

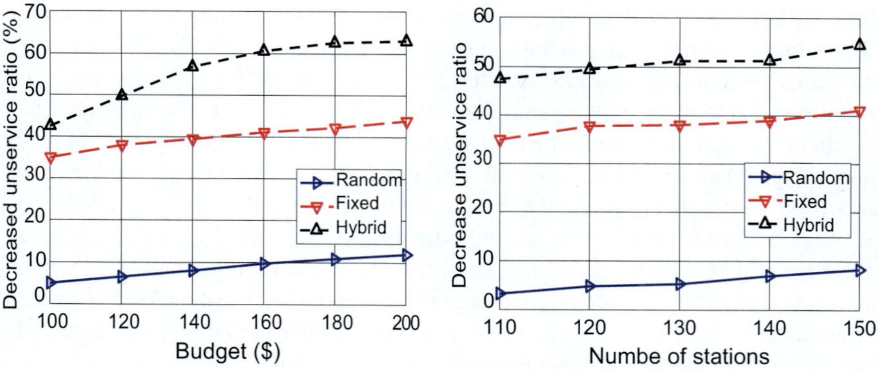

(a) Varying the budget amount. (b) Varying the number of stations.

Fig. 14 Performance comparison on docked BSS dataset

when the number of time slots increases. It may explain that the advantage of the hybrid scheme is more obvious with a larger number of time slots.

The experiment results on the docked BSS dataset—the NYC dataset is shown in Fig. 14. Figure 14a shows the performances of different schemes under different incentive budgets. Not surprisingly, all schemes could achieve a higher service level when a larger budget is provided. Among them, our hybrid incentive scheme makes better use of the budget and has the highest service level. It shows that the reinforcement learning agent could learn a better way of allocating incentives among bike stations, compared with a fixed incentive scheme. Figure 14b shows the experiment results under different numbers of stations. The budget per station is fixed during the experiment. This result shows that our hybrid incentive scheme is robust when the scale of the docked BSS is expanded. The learning agent is still capable to allocate incentives wisely to keep the service level at a high level, and it outperforms the fixed incentive scheme by about 30.7%–35.3%.

6 Related Work

With the booming of BSSs, more and more researchers are devoting their efforts to related issues including user demand prediction [5, 16, 21, 26], bike rebalance strategy [23, 19, 24, 28, 14, 11], station location optimization [18, 4], bike lane planning [1], and suggestion of user's journeys [27, 29, 6]. We focus on the studies that have been conducted on rebalance strategy designing issues, which are closely related with our work.

Before designing efficient bike rebalancing scheme, accurately predicting user demands for BSSs is critical. The existing demand predication methods could be group as station level and cluster level prediction approaches. The station level prediction is designed to predict the number of rent/return events at each bike station in docked BSSs, such as [13, 15]. However, it ignores the potential depends among bike stations and may not generate inaccurate demand predictions [16]. To overcome this, Li et al. [16] proposes to cluster similar stations. Specifically, they proposed to use transition patterns and station locations to cluster bike stations first. Instead of predicting demand of each bike station, they predict the demand of each cluster. Chen et al. [5] further improve the prediction accuracy by considering more features such as traffic and social event. By adding those factors into consideration, the prediction accuracy is improved accordingly. Du et al. [8] adapt a convolutional neural network (CNN) for the demand prediction. Their algorithm could find virtual stations by using density peak–based clustering. Then, they use the CNN to predict the demand of each virtual station. By utilizing demand prediction, our approach aims to optimize the worker assignment during rebalancing.

Rebalancing strategies designed for docked BSSs have been widely studied. Typically, there are two major approaches, which are the truck-based and the user-based approach. The truck-based approach such as [3, 12] means the BSS operator hires a fleet of trucks to transport bikes from overflow stations to underflow stations. Chemla [3] proposed a single-vehicle rebalancing problem, where each station could be used as a buffer to temporarily store some bikes. Their rebalancing algorithm is based on brand and bound, which can be used for small size BSS systems. When the number of stations exceeds 100, the time cost is significant. However, in real-world BSS systems, the number of stations in a large city could easily exceed 100. Liu et al. [19] proposed a method that first clusters bike stations according to geographic information and station status and then assigns a truck to each cluster. They model the bike rebalancing as an integer programming problem and use integer programming solvers to optimize route for trucks used for bike rebalancing. The number of bikes that need to be moved within a cluster is usually small. Therefore, the integer programming solvers could solve the routing problem efficiently. Different from those works, we follow the user-based approach, which is more flexible and cost-efficient. Specifically, we recruit a group of workers to rebalance the system instead of hiring a fleet of trucks.

The dockless BSSs becomes more and more popular in major cities. The advantage of the dockless BSSs is that user could return bikes at their destination instead of finding a return station. However, it also brings challenges for rebalancing. The scale of the rebalancing problem is enlarged since each bike's location should be considered. The rebalancing scheme designed for docked BSSs cannot be directly applied because of the enlarged solution space. New rebalancing approaches are required. For dockless BSS rebalancing, existing researches usually follow a user-based approach. As for the user-based approach like [25, 24], the BSS operator gives incentive to users and suggests them to rent or return bikes at certain stations. User-based approaches expect that the BSS can achieve self-balance. They improve the overall service level by controlling users' dynamics through incentive. The user-based approach is more flexible. Unlike truck-based approaches, which could only apply a few rounds of rebalancing in a day, the user-based rebalancing lasts continuously, only if the budget is sufficient.

Designing the pricing mechanism is the key problem in these approaches since it directly affect the user dynamics. Waserhole [25] presented a dynamic pricing mechanism that incentivizes users to redistribute bikes by providing alternative rental prices. Singla et al. [24] proposed a pricing mechanism to incentivize users via crowdsourcing. For dockless BSSs, besides the source incentive scheme based on reinforcement learning proposed by Pan et al. [22], there are many other approaches. Liu et al. [21] propose a demand prediction method. Their inference model combines convolutional neural network and factor analysis techniques. Based on an precise demand prediction, some docked rebalanced scheme may be extended to the dockless scenario. Caggiani et al. [2] proposed a dynamic bike rebalance method including a prediction scheme of the number and position of bikes and a relocation decision system. Our hybrid scheme is an end-to-end system, and the incentive price can be given without demand prediction. Besides rebalancing, BSS operators usually set an electric fence [30] to restrict the mobility range of bikes, which helps to solve the imbalanced bike distribution. By applying the electric fence policy, users need to return bikes into designated zones. Adaptively adjust the capacity of each electric fence can avoid the case where too many users return bikes to a certain station. Through the adjustment of the capacity, the imbalanced issue of the system could be solved to some extent.

7 Conclusion

We investigate the bike rebalancing problem for both dockless and docked BSSs in this paper. We illustrate that the imbalanced bike distribution in BSSs might cause bike overflow and underflow events. Those events may bring congestion to the city or decrease the service level of BSSs, and rebalancing BSSs in a timely manner is necessary. We follow the user-based approach for rebalancing and propose to adaptively provide both source and destination incentives to users with the objective of maximizing the service level. We adapt a reinforcement learning framework in [22] to overcome the complex user dynamics for dockless BSSs. In addition, we extend the learning framework to the docked BSS rebalancing problem. The capacity of each station is added to the state space of the reinforcement learning agent, and the environment model is also updated to fit the docked BSS scenario. We use real-world trace data from Mobike and NYC Citi bike to test our dockless and docked rebalancing scheme, respectively. Experiment results show that providing both source and destination incentives could achieve a higher service level compared with the state-of-the-art source-incentive-only scheme. Our extended scheme outperforms the fixed incentive scheme, which is currently implemented by the City bike in NYC. Besides bike-sharing, electric-car and e-scooter sharing become more and more popular nowadays. Those systems also have rebalancing issues. The scales of those systems are different from BSS. Although we can directly apply our rebalancing scheme to those systems, designing a new rebalancing system that makes full use of IoT sensors on those devices would be a more efficient solution.

Acknowledgement This research was supported in part by the National Science Foundation grants CNS 1824440, CNS 1828363, CNS 1757533, CNS 1618398, CNS 1651947, and CNS 1564128. This paper is an extended version of the conference paper [9] published in IEEE MDM 2019.

References

1. Bao, J., He, T., Ruan, S., Li, Y., & Zheng, Y. (2017). Planning bike lanes based on sharing-bikes' trajectories. In *Proceedings of the 23rd ACM SIGKDD international conference on knowledge discovery and data mining* (pp. 1377–1386).
2. Caggiani, L., Camporeale, R., Ottomanelli, M., & Szeto, W. Y. (2018). A modeling framework for the dynamic management of free-floating bike-sharing systems. *Transportation Research Part C: Emerging Technologies, 87*, 159–182.
3. Chemla, D., Meunier, F., & Calvo, R. W. (2013). Bike sharing systems: Solving the static rebalancing problem. *Discrete Optimization, 10*(2), 120–146.
4. Chen, L., Zhang, D., Pan, G., Ma, X., Yang, D., Kushlev, K., Zhang, W., & Li, S. (2015). Bike sharing station placement leveraging heterogeneous urban open data. In *Proceedings of the 2015 ACM International Joint Conference on Pervasive and Ubiquitous Computing* (pp. 571–575).
5. Chen, L., Zhang, D., Wang, L., Yang, D., Ma, X., Li, S., Wu, Z., Pan, G., Nguyen, T. M. T., & Jakubowicz, J. (2016). Dynamic cluster-based over-demand prediction in bike sharing systems. In *Proceedings of the 2016 ACM International Joint Conference on Pervasive and Ubiquitous Computing* (pp. 841–852).
6. Cheng, P., Xu, C., Lebreton, P., Yang, Z., & Chen, J. (2019). Terp: Time-event-dependent route planning in stochastic multimodal transportation networks with bike sharing system. *IEEE Internet of Things Journal, 6*(3), 4991–5000.
7. DeMaio, P. (2009). Bike-sharing: History, impacts, models of provision, and future. *Journal of Public Transportation, 12*(4), 3.
8. Du, B., Hu, X., Sun, L., Liu, J., Qiao, Y., & Lv, W. (2020). Traffic demand prediction based on dynamic transition convolutional neural network. *IEEE Transactions on Intelligent Transportation Systems*.
9. Duan, Y., & Wu, J. (2019). Optimizing rebalance scheme for dock-less bike sharing systems with adaptive user incentive. In *2019 20th IEEE International Conference on Mobile Data Management (MDM)* (pp. 176–181). IEEE.
10. Duan, Y., & Wu, J. (2019). Optimizing the crowdsourcing-based bike station rebalancing scheme. In *2019 IEEE 39th International Conference on Distributed Computing Systems (ICDCS)* (pp. 1559–1568). IEEE.
11. Duan, Y., & Wu, J. (2020). Spatial-temporal inventory rebalancing for bike sharing systems with worker recruitment. *IEEE Transactions on Mobile Computing*, 1–1.
12. Duan, Y., Wu, J., & Zheng, H. (2018). A greedy approach for vehicle routing when rebalancing bike sharing systems. In *2018 IEEE Global Communications Conference (GLOBECOM)* (pp. 1–7). IEEE.
13. Froehlich, J., Neumann, J., Oliver, N., et al. (2009). Sensing and predicting the pulse of the city through shared bicycling. *Proceedings of IJCAI, 9*, 1420–1426.
14. He, S., & Shin, K. G. (2020). Dynamic flow distribution prediction for urban dockless e-scooter sharing reconfiguration. *Proceedings of The Web Conference, 2020*, 133–143.
15. Kaltenbrunner, A., Meza, R., Grivolla, J., Codina, J., & Banchs, R. (2010). Urban cycles and mobility patterns: Exploring and predicting trends in a bicycle-based public transport system. *Pervasive and Mobile Computing, 6*(4), 455–466.

16. Li, Y., Zheng, Y., Zhang, H., & Chen, L. (2015). Traffic prediction in a bike-sharing system. In *Proceedings of the 23rd SIGSPATIAL International Conference on Advances in Geographic Information Systems* (pp. 1–10).
17. Lillicrap, T. P., et al. (2015). Continuous control with deep reinforcement learning. *arXiv preprint arXiv, 1509.02971.*
18. Liu, J., Li, Q., Qu, M., Chen, W., Yang, J., Xiong, H., Zhong, H., & Fu, Y. (2015). Station site optimization in bike sharing systems. In *2015 IEEE International Conference on Data Mining* (pp. 883–888). IEEE.
19. Liu, J., Sun, L., Chen, W., & Xiong, H. (2016). Rebalancing bike sharing systems: A multi-source data smart optimization. In *Proceedings of the 22nd ACM SIGKDD International Conference on Knowledge Discovery and Data Mining* (pp. 1005–1014).
20. Liu, J., Sun, L., Li, Q., Ming, J., Liu, Y., & Xiong, H. (2017). Functional zone based hierarchical demand prediction for bike system expansion. In *Proceedings of the 23rd ACM SIGKDD International Conference on Knowledge Discovery and Data Mining* (pp. 957–966).
21. Liu, Z., Shen, Y., & Zhu, Y. (2018). Inferring dockless shared bike distribution in new cities. In *Proceedings of the eleventh ACM international conference on web search and data mining* (pp. 378–386).
22. Pan, L., Cai, Q., Fang, Z., Tang, P., Huang, L.: Rebalancing dockless bike sharing systems. arXiv preprint arXiv:1802.04592 (2018).
23. Singla, A., & Krause, A. (2013). Truthful incentives in crowdsourcing tasks using regret minimization mechanisms. In *Proceedings of the 22nd international conference on World Wide Web* (pp. 1167–1178).
24. Singla, A., Santoni, M., Bartók, G., Mukerji, P., Meenen, M., & Krause, A. (2015). Incentivizing users for balancing bike sharing systems. *AAAI, 1*, 723–729.
25. Waserhole, A., & Jost, V. (2016). Pricing in vehicle sharing systems: Optimization in queuing networks with product forms. *EURO Journal on Transportation and Logistics, 5*(3), 293–320.
26. Yang, Y., Heppenstall, A., Turner, A., & Comber, A. (2020). Using graph structural information about flows to enhance short-term demand prediction in bike-sharing systems. *Computers, Environ ment and Urban Systems, 83*, 101521.
27. Yoon, J. W., Pinelli, F., & Calabrese, F. (2012). Cityride: a predictive bike sharing journey advisor. In *2012 IEEE 13th International Conference on Mobile Data Management* (pp. 306–311). IEEE.
28. Yoshida, A., Yatsushiro, Y., Hata, N., Higurashi, T., Tateiwa, N., Wakamatsu, T., Tanaka, A., Nagamatsu, K., & Fujisawa, K. (2019). Practical end-to-end repositioning algorithm for managing bike-sharing system. In *2019 IEEE International Conference on Big Data (Big Data)* (pp. 1251–1258). IEEE.
29. Zhang, J., Lu, P., Li, Z., & Gan, J. (2018). Distributed trip selection game for public bike system with crowdsourcing. In *IEEE INFOCOM 2018-IEEE Conference on Computer Communications* (pp. 2717–2725). IEEE.
30. Zhang, Y., Lin, D., & Mi, Z. (2019). Electric fence planning for dockless bike-sharing services. *Journal of Cleaner Production, 206*, 383–393.

IoT-Driven Bayesian Learning: A Case Study of Reducing Road Accidents of Commercial Vehicles on Highways

Wilson Nwankwo, Charles Oluwaseun Adetunji, and Akinola S. Olayinka

1 Introduction

Internet of Things (IoT) belongs to the pervasive computing paradigm. Its goal is enabling communication and relay of data between different devices across the Internet [7]. IoT is an innovation in which the data from the participating devices or entities are stored in the cloud where it could be easily accessed, processed, and managed. Typical participating devices include sensors and actuators, which are easily integrable into more complex systems and used for gathering data and transmission to specified nodes on the Internet. The advancement of cloud computing has helped promote the relevance of IoT as cloud systems now provide requisite mechanisms for storing data, analysis, data gathering, and data visualization. According to Chamandeep [13], the characteristics of cloud include on-demand service provision, resource pooling and elasticity, and so on. In general, IoT offers highly developed connectivity of devices, systems, and services that are somewhat ahead of machine-to-machine communications, covering a variety of protocols, domains, and applications. IoT variants have applicability to nearly all fields of automation permitting advanced applications like smart home, smart traffic management, smart grid, etc. Other prominent applications include heart monitoring implants, biochip transponders on animals, automobiles with built-in sensors, and

W. Nwankwo
Informatics and CyberPhysical Systems Laboratory Department of Computer Science, Edo State University Uzairue, Edo State, Nigeria

C. O. Adetunji (✉)
Applied Microbiology, Biotechnology and Nanotechnology Laboratory, Department of Microbiology, Edo State University Uzairue, Edo State, Nigeria

A. S. Olayinka
Computational Science Research Unit, Department of Physics, Edo State University Uzairue, Edo State, Nigeria

S. Pal et al. (eds.), *Artificial Intelligence-based Internet of Things Systems*, Internet of Things, https://doi.org/10.1007/978-3-030-87059-1_15

field operation equipment that aid firefighters during search and/or rescue operations. Typical applications include thermostat systems and washer that utilize Wi-Fi for remote monitoring [4]. IoT is bringing the human civilization closer to direct communication between machines [62]. One of such areas that has attracted interests in the application of IoT is smart traffic control and vehicle management [58] including accident prevention and control. Road accidents can be fatal, resulting in the death of motorists and other road users. Some causes of road accident identified by Muthusamy et al. [50], George et al. [22], and Oluwaseyi and Gbadamosi include bad roads, stress of drivers, age of the vehicle, carelessness, using mobile phone while driving, ignoring the red signal in traffic signals, over speeding, poor weather conditions, dangerous bends, abandoned vehicles, animals not under control, obstruction on the road, driving while drunk, and insensitivity and irresponsibility on the part of state authorities. Awareness and knowledge of causes of these accidents may help motorists to avoid them.

The effect of these accidents on the society has been discussed under economic and social perspectives.

Economically, road accidents are computed at a cost of about 3% of a country's GDP globally [76]. WHO [76] also estimates that 93% of global road fatalities are associated with countries within the low and middle-income category, respectively. Research has shown that such accidents are a major economic challenge in developing countries, especially in Africa [39, 75, 77]. Economic costs of fatalities range from 2% to 5% of GDP in many countries [51]. These costs do provide requisite basis for local transport safety authorities to plan and undertake improvement projects including hazard profiling of locations, periodic road audits, and other preventive steps that would help mitigate these accidents.

Socially, fatal traffic incidents lead to profound sorrow, death, and losses in developmental resources. Disabilities, loss of properties, sustainable property, safety, basic freedoms, and human right violations may result [23]. Researchers such as Abdelfatah et al. [2], Shahid et al. [67], Rusli et al. [64], and Usman et al. [73] have lamented the physical, social, economic, and psychological impact that result whenever there is a road accident. Presently, road fatalities is a global health concern. Nak and Mauricio [51] report that some 1.3 million people die yearly from such fatalities and up to 50 million injured globally. With the rising concerns, new approaches need be explored toward proffering solutions. This chapter aims at investigating road accident mitigation using an IOT-driven Bayesian system. The goal of this chapter is to discuss a system that could offer a forward adjustment on road status conditions through progressive updates and computation of risks associated with specific road segments using Bayesian learning. The risk is expressed in terms of criticality, i.e., the outlook of a road accident occurring as a function of likely road condition, fatality, and injury rates. The proposed system not only would present on real time the risk status information but also would provide audible alerts to the hearing of drivers and passengers in a vehicle plying that road at all times.

2 IoT Strategies and Implementations

In this section, we provide brief but detailed background on IoT, machine learning concepts, useful sensor technologies in road and vehicle management, traffic models, and finally some relevant proposed and implemented solutions.

The IoT concept was initiated in 1999 by Kevin Ashton [25]. According to Hamid et al. [25], the concept had changed as technology evolves in the last decade, as its default goal was to evolve a digital communication device that does not require human intervention. With advances in Internet technologies, device connectivity is emerging at an unparalleled scale and pace [7]. With the ongoing success in the establishment of connectivity across buildings, household devices, vehicles, embedded with hardware devices, software, sensors, and with resultant nonintrusive data exchange [40, 66], IoT is poised to emerge as the greatest contributor to big datasets with nearly unlimited use across various smart applications [31].

Quite a number of industries like agriculture, transportation, and mining are implementing IoT to enhance the overall efficiency of their processes and improve their control mechanism. The IoT concept is conceived as an extended machine-to-machine (M2M) communication. Remote systems such as sensors are connected to servers with little or no human involvement, whereas IoT advances M2M connectivity by integrating web/mobile applications and cloud systems [20]. For instance, the adoption of IoT in traffic management could have several benefits including: remote monitoring of vehicles to improve fleet and road usage efficiency, safety, reduce accidents, and vehicle usage, to provide quick response service to customers, and to enable smart interaction between drivers and the environment [21]. A typical IoT system is an embedded system with capability to support several concurrent device connections to engender a large dataset to be transmitted, processed, and stored in the cloud. Javadi et al. [31] stated four components of a typical IoT platform:

(a) Sensors and other hardware devices: fundamental components that collect data from the environment
(b) Communication network (Wi-Fi or cellular technologies – 3G, 4G, and 5G)
(c) Big data
(d) The cloud where the data are stored and processed

2.1 *Machine Learning (ML)*

ML produces and automates prescriptive, predictive, and numerical models and algorithms directed at process optimization and performance enhancement in different aspects of the human. ML uses advancements from empirical methods, operations research, and statistics. It finds hidden knowledge in data without delay in looking for it [17]. Hamid et al. [25] categorized ML into supervised, reinforcement, and unsupervised learning. Unsupervised learning (UL) is a group of techniques, which are driven by input data alone. A major UL algorithm is clustering. In clustering, data points are grouped against a set of data values or domain points [20]. Supervised learning (SL) involves the use of a model to make prediction as a result of knowledge acquired from training dataset, which include both the input and expected output. There are two major SL approaches: regression and classification. In ML, regression is useful when predicting event outcomes based on variables characteristics in domain data obtained from the dataset, whereas classification is used to identify a group to which a new instance belongs and for predicting the target class for each category of data [40].

Reinforcement learning (RL) involves an agent that learns via a participatory interaction with the environment through trial-and-error response from its own knowledge, experiences, and actions. RL attempts to determine a decent mapping that describes observations undertaken for actions addressing situations for the decision-maker collaborating with an environment. RL is an authoritative method to speed up preliminary learning with outstanding results [35]. RL has gain popularity in the field of computational science, computational neuroscience, computer science, applied mathematics, control and automation engineering, and mechatronics. Recent advances in RL over the years have led to its application in addressing multi-agent problems.

SL finds usefulness in the following domains:

(a) *Bioinformatics* – where biological information of people, such as fingerprints, iris texture, and earlobe, are manipulated to support intelligent reasoning like authentication and authorization [63].
(b) *Speech Recognition* – here, the algorithm learns the human voice and is able to recognize it. Some applications are found in virtual assistants, e.g., Google Assistant and Siri.
(c) *Spam Detection.*
(d) *Object-Recognition and Vision* [56].
(e) Control

However, SL has its downsides and these include:

(a) Overfitting [43], hence complementary approaches should be used to reduce this effect on the developed model.
(b) Relatively high computation time [44].
(c) Reduction in accuracy due to incoherent data.
(d) Preprocessing requirements.
(e) Precision and usefulness are dependent on correctness of the dataset and the algorithm arising therefrom.

Similarly, UL is ideal in the following scenarios:

(a) Automatic splitting of dataset into groups based on similarities [20].
(b) Detection of anomalies in datasets. This is useful in spotting fraudulent connections.

(c) Association mining: This is useful in identifying associated dataset that occur together in a dataset.

Like SL, UL has its weaknesses too. These include:

(a) Difficulty in getting precise details on data sorting and data output used in UL.
(b) Unknown and unlabeled input data leads to less accuracy of the results.
(c) Spectral classes hardly agree with informational classes.
(d) More time is spent on interpreting and labeling the classes that correspond to the classification.
(e) Spectral properties of classes change with time, making it impossible to have the identical class information while moving from one image to another [35].

In SL, classification is a very vital ML technique that finds practical applications to many problems such as: speech recognition, handwriting recognition, biometric identification, object-recognition for vision, spam detection, bioinformatics, document classification, etc. [43]. Classification incorporate many vital algorithms. These have been categorized by Lakshmana and Ragupathy [36] to include: decision trees (DT), nearest neighbor (N-N), support vector machines (SVM), boosted trees (BT), random forest (RF), Naive Bayes (NB) classifier, and neural networks (NN).

Bayes Theorem and Bayesian Learning

Bayes' theorem explains the probability of an event, founded on previous knowledge of circumstances, which might be connected to the event under consideration. For instance, if the danger of coming up with health problems is directly proportional to age, Bayes's theorem would enable the danger of an individual of a known age to be assessed more precisely than simply presupposing that the individual is distinctive of the population as a whole [12]. Bayes's theorem is useful in Bayesian inference application. When used, the probabilities involved in Bayes' theorem may lead to a different understanding. With Bayesian probability interpretation, the theorem conveys how a degree of belief, expressed as a probability, should rationally change to give evidence of the availability of related data ([36]; [46]; [78]). Bayesian learning is founded on the concept of Bayes' theorem. Bayesian classification may be used to establish causal relationships. Thus, it can enable the understanding of a problem domain including the prediction of the intervention consequences. Two major forms are identified: the general Bayes classification and the Naïve Bayes (NB) classification (a more prominent classifier these days). In the traditional Bayes classification (Bayes nets or Bayesian belief networks), the user determines the conditionally dependent and independent variables, respectively. NB is hinged on the primordial assumption of independence among predictors. That is, NB classifiers presupposes that a particular feature present in a class exhibits no relationship with another feature present or that all of these properties possess independent contribution to the probability. NB classifier is easy to build, scales well with large datasets, and can outperform other sophisticated classification algorithms *[5]*. The NB classifier uses prior, posterior, and class conditional probabilities for its computation *[5]*. The strengths of the NB algorithm as shown in Francois-Lavet et al. [20] and

Kusy and Zajdel [35] include: small training data, simple computing, ease of implementation, time efficiency, big data handling, compatibility with incomplete data or missing values, and insensitivity to irrelevant features and noise.

2.2 *Transportation and Sensor Technologies*

For decades, sensor technologies have emerged ubiquitous attracting a lot of interests. They find wide utilization in health care, agriculture and forestry, and vehicle and marine monitoring. Sensor technologies are embedded in devices for traffic control, entertainment, and safety [29, 71]. Lately, sensors for tire pressure and rear-view visibility are mandatory in the design of vehicles, and smart transportation components aimed at providing services that improve on the convenience and comfort of motorists and reduce road hazards and traffic congestion [74]. Vehicle manufacturers sometimes install various sensors for data gathering and monitoring of the vehicle's performance and activity, which in turn provide higher effectiveness and help for drivers [33].

Classification of Sensors
Sensors may be categorized according to:

(a) Location in a vehicle, e.g., powertrain, chassis, and body.
(b) Functionality, e.g., diagnostics sensors, convenience sensors, environment monitoring sensors, driving sensors, motion detection, night vision sensor, safety sensors, and traffic monitoring sensors [16].
(c) Mechanism of operation, e.g., proximity, ultrasonic, electromagnetic, and optical.
(d) Nature of material used for manufacturing of the sensor.

In intelligent transport systems (ITS), identifying the sensors that address problems such as traffic bottleneck and parking hitches, longer travelling times, increase in gas emissions levels, and the menace of accidents is very relevant toward improving transport infrastructure and commuters' experience [16].

Modern vehicles incorporate an array of sensors that capture data for use by the vehicle control unit in regulating specific functionalities in the vehicle. These sensors include engine sensors, interior and exterior survey sensors, vehicle dynamics sensors, etc. A typical application is tire-pressure monitoring, which uses auditory, light, or vibration warning to alert the driver once the tire air pressure is low [15].

Electromagnetic, Ultrasonic, and Proximity Sensors
Electromagnetic, ultrasonic, and proximity sensors are employed in parking assistance and reverse warning applications. Proximity sensors could detect closer objects; however, their accuracy is impaired by temperature and humidity. Ultrasonic sensors use sonar to identify distance of vehicle from an object and alert the driver once the vehicle approaches beyond a set distance. Electromagnetic sensors alert

the driver when an object is in the electromagnetic field created around the front and back bumpers.

Gyroscopes and Accelerometers

These sensors are used to discover a vehicle's spatial and kinetic parameters, e.g., vehicle position, orientation, and velocity [33]. They are used with Global Positioning Systems (GPS) to advance accuracy [74]. Radar and speed sensors are used in applications that provide advice and decision support to drivers regarding potential danger when changing lanes or veering out of the appropriate lane. The driver is generally warned by a vibration in the seat or steering wheel or by the use of alarm [29].

Light Detection and Ranging (LIDAR)

LIDAR is incorporated in autonomous vehicles and enables self-driving cars to maintain a 360-degree view and depth. LIDAR sensors use beams of laser light and measure the period taken by the emitted light to hit the sensor [33].

Inflatable Road Tube Sensors

These are also called pneumatic/air-filled road tube sensors. The sensor-enabled tube is placed across the road traffic lanes to enable data collection about vehicles. When a vehicle's tire moves over the inflatable tube, the sensor sends a burst of air pressure with a resultant electrical signal to a processor [16]. Vehicles count is done as well as classification based on the data captured by the sensor.

The Inductive Loop Detector (ILD)

This sensor finds common deployment in traffic management. It is used to merge traffic flow, vehicle's occupancy, length, and speed. It has a long-coiled loop of wire that could be placed into or under the surface of the road. When a vehicle passes over the sensor, it produces an electrical signal, which is transmitted to a processor [41].

Magnetic Sensors

These detect vehicles whenever a change occurs in the earth's magnetic field. They are used to sense occupancy, flow, speed, and vehicle length [60].

Piezoelectric sensors detect vehicles. When a vehicle traverses the sensor, a change in the sensor's voltage occurs and is transmitted to a processor. These sensors could be used on four lanes.

Video Image Processor System

This comprises computer, two or more video cameras, and intelligent software for image processing. The video cameras capture video images of traffic scenes that depict flow, volume, and occupancy. However, their performance is easily impaired by poor weather conditions [41].

Radar Sensors

These transmit low-energy microwave radiations reflected from objects or entities within the detection zone. Different radar systems exist. A prominent variant is the Doppler system that uses frequency shifts in tracking vehicles and computing vehicle speed. The frequency-modulated continuous wave variant from radars radiates continuous transmission power to quantity flow volume, speed, and presence. Generally, radar sensors exhibit good accuracy, ease of deployment, support for several detection zones, and continuous operation irrespective of time; however, their major setback is high susceptibility to electromagnetic interference [54].

Infrared Sensors
These sensors convert reflected energy into electrical signals. The converted signals are transmitted to the processor where they are used for computations. Two categories of infrared sensors are recognized: passive infrared (detects vehicles using emissions or reflections from vehicle presence, flow volume, and occupancy) and active infrared (uses light emitting or laser diodes to capture data on speed, flow volume, classification, and vehicle presence including traffic density) [37].

Ultrasonic Sensors
These sensors also capture data from traffic flows and vehicle speed. They compute the distance between vehicles depending on the elapsed time of a sound wave transmitted at frequencies within 25–50 KHz and reflected to the sensor by an object. The received energy is translated to electrical signals then transmitted to processing unit (PU). However, their performance is constrained by environmental issues. They detect sound energy generated by a vehicle in motion within a coverage area. Acoustic sensors employ replacing magnetic induction loops to calculate traffic volume, occupancy, and average speed of vehicles [41].

Road Condition Sensors
This category utilizes combined laser and infrared technologies to examine road conditions prior to human interventions through traffic safety and road maintenance campaigns. It should be noted that these sensors are prone to regular intermittent maintenance requirements, otherwise their performance deteriorates [8].

Radio-Frequency Identification (RFID) Sensors
These are advanced sensors that uses radio frequencies to identify moving vehicles. The RFID sensors are very convenient to use and easily capture data from these vehicles automatically and exhibit wider range of integration [19].

2.3 Traffic Models

2.3.1 Collision Avoidance Model (CAM)

The CAM was introduced in 1959 through the work of Kometani and Sasaki [32]. They had attempted to describe the relationship in terms of distance between vehicles. The CAM stipulates a safe distance that a driver following a vehicle should maintain to prevent any collision with an oncoming vehicle. A collision avoidance system is seen as an automobile safety system designed to either prevent or mitigate a collision impact. The model would ensure that the speed and distance between the two vehicles are kept under check. The goal is to afford the driver a precautionary measure always and to engage appropriate reaction to avoid possible collision with the vehicle behind or ahead of it. The CAM is prescriptive [57] and identify useful technologies such as camera, radars, and GPS sensors [38].

2.3.2 The Action Point Model (APM)

The APM is a psychophysical model used to explain the car-following behavior in terms of thresholds and action points. The APM contemplates the inability of drivers in tracking arbitrarily small changes in stimuli, such as relative speed, thus, a driver's driving behavior is expected to be adjusted only if a certain threshold is attained. The commands defined by these thresholds show in what circumstances the driver would respond [59].

2.3.3 Intelligent Driver (ID) Model

Considered a microscopic model, the ID model is a nonmathematical model that provides a reference when assessing an individual's traffic behavior in relation to an oncoming vehicle. IDM is easy to adapt during tasks involving adaptive cruise control system [24].

2.3.4 Latent Class (LC) Model

The LC model is a framework that separates longitudinal driving behavior into different driving characteristics, e.g., car following, free flowing, and emergency braking [47]. The driver's decision qualities in each system is related to the strength of different stimuli around the immediate environment. These decisions are probabilistic, for instance, the possibilities in a car-following system may be expressed in three states: increased acceleration (A), deceleration (D), and neither accelerate nor decelerate. These three states are dependent on the stimuli from the leading vehicle(s). Similarly, in a free-flow system, a driver's behavior may assume same states as in car-following state; however, the stimulus differs and may be under the perfect control of the driver. The emergency system exhibits one major state, i.e., deceleration in contemplation of a possible collision [47]. The driver may be in one of three discrete states: deceleration, acceleration, and neither accelerate nor decelerate.

2.4 Review of Relevant Developments

Maria [42] presented a use case and system components that could be used to evaluate and report accident-prone areas to elicit increased response rates from the authorities mandated to oversee traffic and road incidents. The system defines usability components and harps on portability and use of the tool on different devices such as tablets, mobile phones, and desktops. User-friendliness is emphasized to promote further accessibility and utility to the global community of motorized. The system also captures information on optimized routes to nearby support station. Occurrence of accidents can be reported easily and get alerts, while the support station would carry out the review of the reported incidence, determine the geographical location and coordinates of the accident area, and trigger necessary action.

Babitha et al. [11] implemented an automated system for road safety surveillance using a Hybrid CNN-LSTM approach. In the study, drowsiness and underage driving were identified as two major contributory factors to accident. The study proposed a face image descriptor-based combination using convolution neural network and the ResNet50 architecture to forecast age. The other aspect of the system utilizes a persistent neural network and the LSTM architecture to sense when a driver is drowsy and provide useful alerts that prevents sleepiness when driving. They developed face recognition algorithms using image processing to enable to predict ages of drivers, thereby preventing underage drivers. In addition to differentiating drivers of different ages, the system could detect fatigue in drivers. The system's accuracy in prediction was rated at 96%.

Truong [72] adopted a systematic approach for the analysis of self-organization possibilities of flows and situational modeling of heterogeneous transport systems using fuzzy logic. The study was aimed at predicting and preventing transport accidents. The author investigated and formalized fuzzy-multiple and fuzzy-logical problems for prediction of accidents and also considered the task of developing an expert system that would help assess and classify risks. The result yielded a structured approach with tools especially for transport tasks that are poorly formalized and in conditions where little or no awareness in respect of the transport system or changes in the transportation environment.

Maria et al. [43] developed a system architecture with activity diagrams for system for emergency response and intended to improve the communication and coordination processes of emergency response units, thus offering increased community awareness in respect of the various factors leading to traffic incidents. The system used a system architecture to aid in analyzing traffic accident profiles of Legazpi City associated with recklessness leading to homicide, physical injury, and damage to property, respectively. The study attempted to integrate spatial analysis to aid in localizing accidents. It incorporates user-friendly mobile application through which traffic accident data could be submitted. Through the system, users could also report the incidents and their locations to the nearby emergency response units.

Maria et al. [44] utilized three algorithms jointly: NB, DT, and k-nearest neighbor (KNN), respectively, to perform a heart disease classification and prediction. Naive Bayes yielded 86% accuracy, which is adjudged higher than that recorded in previous works where other algorithms are used.

Ayman [10] proposed a cost-effective and simple novel vehicle speed identification electronic system. His method was based on the image processing using vehicle plate numbers captured using camera, as inputs. The experiment is intended to provide a validation and reliability assessment of the approach proposed. The results obtained demonstrated that the proposed system could offer some efficiency in speed detection at lesser cost. Kulwant et al. [34] employed image processing principles in the design of an IoT-based pothole detection system. The system was an embedded component and placed in a vehicle to continuously scan the road surface while in motion. It identifies potholes, bumps, etc. and alerts the driver on time to enable him avoid the bumps or pothole. Their emphasis was on detection tasks especially bad spots on the road. Image processing algorithms were applied to process data captured from ultrasonic sensor in the vehicle. The device uses GPS module to pothole positions. The data on potholes may be relayed to the local data store via a GPRS or Bluetooth module. The locally stored data may be transferred to the cloud via Wi-Fi or a compatible higher network technology connected to the system. The cloud data are accessible to the authorities including road maintenance agencies and the general public. This would help drivers to avoid roads that are rated unsafe or adopt safe driving techniques should they decide to ply such roads.

Swetha et al. [70] proposed a system that comprises an ultrasonic sensor, microcontroller, and GPS, respectively. The sensors capture the distance between the vehicle and the pothole. The captured data are relayed to the microcontroller. Depending on the computed distance, the driver is alerted through voice commands. The GPS captures the geographical coordinates of the potholes, humps, and other bad spots. The captured data are relayed to the *ThingSpeak* cloud for analysis and maintenance authorities to take necessary action. Lakshmana and Ragupathy [36] designed and implemented a hybrid probabilistic stock sentiment prediction model based on the real-time market data and integrated real-time stock data. Captured dataset include news on stocks and stock momentum to forecast investors' disposition toward stock buying or selling. They also implemented a novel stock sentiment score and data conversion procedures to normalize the market trend for intraday data. They used an enhanced Bayesian network method on their clean data to predict and test the stock trend on the clean data. Their results from the experiment showed that their anticipated system might be more efficient in terms of its accuracy and its error rate when compared with other models for classification.

Christy et al. [14] proposed a technology to automatically control a decelerating system by proactively sensing the accident-prone zones and averting accidents. In their experiment, they deployed Arduino microcontroller, L293d motor driver, and ultrasonic sensors while adopting machine learning approach with RFID protocols for the control of the decelerating system. The machine learning enabled system, which is reinforced by a secured RFID, was mounted to a vehicle's dashboard in order to effectively control the decelerating system of the vehicle. Pattern of driving of the driver was also monitored under several circumstances in order to minimize traffic incidents. This was achieved with the aid of Car Trips data log and acelerolinear_terra dataset. A mobile app was developed based on the machine learning model created from the dataset collected. The mobile app can help the driver in controlling the decelerating system and averting accidents and rash driving when fitted into the windshield of the car.

Aaron et al. [1] integrated vision sensors and NN for traffic control system. The captured data are processed and utilized for traffic flow optimization. The NN was trained through simulation and optimized to intercept the data from each stoplight.

Hussein et al. [28] studied the influence of technology essentials on the achievement of big data analytics and IoT-enabled transportation system implementations. The data utilized for the study was gathered through systematic survey of available literature and in-depth interview. Their findings showed that technology has significant role to play in any implementation of successful big data and IoT-enabled system for transportation, which should not be undermined.

Supriyono et al. [69] developed an Arduino-based (nano-microcontroller) manipulator for a motorcycle engine controller (EC). Their goal was to generate more power and torque compared with the available standard variant. The researchers deployed an oxygen (O_2) sensor (narrow band variant) to detect O_2 in the exhaust manifold and maintain air–fuel ratio (AFR) at a level comparable with the approved stoichiometric level. The manipulator was positioned between the O_2 sensor and the EC. They utilized two sets of source codes for the experiment. The two code sets were for combustion maintenance under high O_2 and low O_2 conditions, respectively. They evaluated the power and torque produced by the test engine using dynamometer and emission tests, respectively. The experiment recorded a success in that the manipulators showed increased power and torque output in the engine. Statistically, the manipulator when applied to the EC maintains AFR at a ratio of 14:2:1.

Aditya [3] designed pothole detection and identification system using a microcontroller. The author's focus was on the development of a prototype that comprises a microcontroller, ultra-violet (UV) sensor, and a WI-FI Modem that processes the data. The system alerts the two-wheeler driver using an interface by developing a software application in android. Amitha et al. [6] introduced a low cost economic solution for a continuous road monitoring system to detect the potholes and to inform the concerned authority about the pothole. The real-time pothole detection and road monitoring system used ultrasonic sensor and accelerometer to measure the depth of the pothole and jerking, respectively. The GPS module captures the pothole location, which is subsequently stored in the cloud database (DB). The DB provides the authorities, experts, researchers, and vehicle operator with valuable data. The public could access information in the cloud through a web server. It is believed that this solution would enable precautionary measures to be taken to prevent and/or control future incidences.

Pathan and Khan [62] designed a detection and alert system. The system detects potholes and also alerts the driver when the computed value from the GPS exceeds the preset threshold value. The detection component comprise two ultrasonic sensors. The GPS is used for localization of potholes. The alert system comprises a buzzer and vibrator (attached to the steering). The detection algorithm computes and localizes bumps and potholes as the vehicle moves.

Jackelou et al. [30] designed a road accident case status prediction system that integrates a modified C4.5 algorithm. The DT ensemble approach was used to analyze traffic incidents. A total of 25 variables were used, and the dataset comprises

799 records of road accidents. Their experiment showed that alcohol intake by motorists was the commonest cause of road accidents.

Partha et al. [61] proposed a cost-effective fever detection prototype. The system integrates a NB classifier into NodeMCU microcontroller and a web server. The users (patients) are required to submit their responses (Yes or No format) on eight symptoms (cough, sore throat, fatigue, headache, chills, fever, sneezing, and muscular pain) through a website. The NB classifier would then predict whether or not the fever is connected to common cold or flu. Following data submission, the patient undergoes checks by a doctor. The computed values following diagnosis were independently classified under common cold and flu patients. In a voluntary experiment involving 22 patients, the computed statistics, p (0.068 > 0.05) and p (0.089 > 0.05), were not considered significant in the classification.

Marimuthu et al. [45] designed a system that automatically detects potholes and bumps. The system alerts vehicle drivers to drive safely. A mobile application was used to provide timely alerts on potholes and bumps while ultrasonic sensors are used to identify the potholes, bumps, and bad sectors including the measurement their height and depth, respectively. The system uses a GPS receiver to compute the location coordinates of bumps, potholes, bad sectors, etc. on the road. The data captured include bump height, geographic location, and pothole depth. The data are stored in a database. The system generated audio alerts to the driver.

Shubham et al. [68] designed pothole detection system for monitoring road using IoT. The authors propose a non-noisy technique that utilizes sensors present on cell phones. Accelerometer, GPS sensor readings for traffic, and street conditions location were used. The system was able to identify potholes and had two vital functionalities: distinction of potholes, which is done through a multisensor subsystem comprising of accelerometer and gyrator, and driver advising. Data generated in the process are stored on a cloud base, which could be retrieved and used by other road users. Knowledge about the location and position of the pothole would enable appropriate concerned authorities to take relevant action.

Hossain and George [27] proposed a system for monitoring the eye closure ratio of the driver and detects the drowsiness of the driver by using Pi camera. If the eye closure ratio is less than the standard ratio, the driver gets an alert from a buzzer. The study focuses on building an alert system to mitigate drowsiness of a driver. There is no speed control mechanism of the system.

Nanda et al. [52] conceived a system that could detect accidents via accelerometers and vibration sensors. Accident location is computed through a GPS and relayed through the GSM module. Through the system, nearby authorities and/or hospitals could be informed by way of text messages. The prosed system would also integrate drowsiness (by monitoring eye blink) and alcohol detection components, respectively. This system is limited in that it would mandate a driver to wear glasses at all times to be connected to the system, and this is considered quite uneasy for drivers especially those who are not used to wearing glasses.

Wadhahi et al. [74] proposed a system with accident prevention and detection mechanisms. It uses IR sensors to detect accidents. Predefined GSM numbers are provided through which text alerts containing location address may be sent through the GSM module. Warning signals are also conveyed to the driver once the distance between vehicles is beyond a certain threshold. However, the system neither incorporated a speed-limiting mechanism nor any intelligence component such as prediction model.

Navod and Bindu [53] designed and implemented a system for vehicle tracking, control, and status reporting. In the system, vehicle door, parking lights, and side mirrors were monitored and controlled by a mobile phone.

Sarasvathi et al. [66] proposed an alert system either for an accident or theft by providing the accident location using microcontrollers, embedded sensors, and cloud services. However, no mechanism was incorporated to prevent accidents.

Mohd et al. [49] proposed an integrated smartphone-oriented system that enables a driver with a smartphone to view the road condition while driving. The goal of the proposed device was to ensure safety while driving. They designed a mobile app (Android) through which data from a vehicle are relayed to a proximal IoT-Fog server for further processing. The intelligent component of the system was the k-means clustering model. With the algorithm, location of bad road sectors and accident-prone zones could be computed. The driver could use his smartphone's Google map for display. The data generated are stored in the cloud for remote access by other users. The proposed system was evaluated using statistical and experimental simulation. The simulated result shows that the system could offer better performance when compared with existing approaches.

The Bayesian network theory has been applied to risk and causation analysis for road accident analysis [78]. The case study was Adelaide Central Business District (CBD) in South Australia. The implemented model uses Netica platform and integrates a K2 algorithm with expert knowledge into the network structure, with Expectation–maximization algorithm as parameter learning algorithm.

Das et al. [18] anticipated an accident monitoring system, which tracks and monitor accident location. The system offers a device to reduce catastrophes by monitoring driver's eye blinking, an indicator for drowsiness, obstacles on the road, and the drunken state of the driver. Accident and the location of the vehicle are sensed by this system.

Anusha et al. [9] implemented a system using LPC2148. The system integrates a database, GPS, and GSM modules. The engine temperature and alcohol consumption were also detected by the prototype. The detected values could be viewed from a website, thus providing safety travelers in the vehicle.

Mayuresh [48] proposed a system which adopts open-source platform to collect data and track vehicle location in real time. The system monitors the rate of fuel consumption, temperature of the vehicle engine, and the speed of the vehicle at various time. Communication module deployed in the system has capabilities for GPS, GPRS, and GSM, thereby ensuring smooth data transmission. The captured data are stored in the web server database for further processing and necessary action.

Harum [26] suggested a framework based on Raspberry Pi. It is connected to a 3G/4G dongle. The component is designed to receive signals from a mobile tower. It may be attached to a vehicle from where it could transmit data to a web server. The vehicle location could be accessed from remote stations in real time.

Data mining approach was deployed by Sachin and Durga [65] in identifying specific locations where accidents occur regularly in order to isolate contributory factors to the accident occurrences through the analysis of resulting dataset. Their approach involved segmenting of the location of incidence into k groups with the aid of k-means clustering algorithm depending on incidence frequency counts. Following the segmentation, they applied association rule mining to discover how distinct attributes in incidence dataset relate to established location characteristics. The extent of damage during an accident can be simulated with the aid of various algorithms like neural networks (NN) trained with hybrid learning, decision tree (DT), support vector machines (SVM), and synchronized mixed models of NNs and DT. The result of Sachin and Durga [65] measured through experiment favored hybrid method of NNs and DT, which offered superior accuracy than a single algorithm used alone.

Aishwarya et al. [4] developed an accident-prevention system that incorporates IoT for tracking incidences during night trips. They integrated a component for monitoring eye blink of drivers. The system alerts drivers once a threshold indicating drowsiness is reached. The embedded component explores some ideal psychological states, which a driver may exhibit while driving. In addition to blinking, eye and head movements which are indicators during an initial phase of sleep are also utilized. Two devices are employed: an infrared sensor to monitor the blinking rate of drivers and an accelerometer for detecting head movement. The data from both devices are used to estimate the physiological sleep state. Analysis and computations ensured that a normal blink rate is disregarded as a factor in computing the system output. In an extreme state of sleep, the infrared sensor would receive an abnormal blinking rate, and this would trigger an alarm, which subsequently awakens the driver. The sensors used are IoT-enabled allowing the transmission of captured or sensed data across the network for further uses such as emergency response. Netica was used to compute posterior probabilities and descriptive and inferential investigation. The study demonstrated that Bayesian networks might be successfully utilized in solving complex issues like road accident analysis and prediction by establishing variables and the various interrelationships they exhibit in a contemporary accident domain. The model was applied to predict incidence probabilities on certain road conditions, identify reasons, and generate state(s) that result to accidents. Consequently, their results provide hypothetical backing for urban road administrators to do critical analysis of contributory issues against road accidents cases, thereby improving safety measures on urban road system and minimizing road accidents.

3 Case Study

We adopted a case study approach to reflect specificity. The case study is a popular trunk "A" road, the Benin-Auchi Highway. This highway is a major segment of the A2 highway, one of the major highways in Nigeria. It begins from Warri to Benin

City to Lokoja to Abuja to Kaduna to Zaria to Kano and terminates at Kongolam, Niger Republic border.

The data on accidents over a 40-month period (first quarter of 2017 to the first quarter of 2020) are used for analysis and prediction. The data originate from the office of the federal road safety in Benin City, Edo State Nigeria. The road is segmented into 12 sections (locations) using marked similarities such as height above sea level, width of road, traffic density, and distance from a reference point (Benin city). The traffic density per road segment is computed as the total number of vehicles within a segment of the road in an hour. The location where accidents occurred, road condition, and the number of lives lost (male and female) are extracted from the source data and used to map each segment of the road to a degree of vulnerability to accidents. The degree of vulnerability is rated either low or high. This vulnerability is contextually regarded as the risk value and is represented with red (high) and low (green) in the prototype implementation [55].

The risk associated with a vehicle plying the road at each point of the road constitutes the target variable that would be predicted using the Bayesian system.

3.1 Materials

The materials used in this study are: Breadboard and jumper wires, ESP8266-12E (see Table 1) development board, sim900 GSM, GPS (uBlox Neo 6 M), 16 × 2 I2C LCD, Buzzer, Relay, Arduino IDE, server-grade PC (HP elite 2.8Ghz core i7 PC @16GB RAM), and USB cables. For the prototyping, the ThingSpeak open IOT prototyping platform is used. However, a road safety coordinating base station incorporating a 170 m radio tower is recommended around the Ekpoma axis of the highway where a central analytics server system would reside for live deployment.

Table 2 shows the interfacing between the NodeMCU with the GPS module. Uploading of data from the development system to the board is through RXD0 and TXD0 and is reserved.

Table 1 ESP8266-12 Microcontroller features

Processor	L106 32-bit ESP-12E @ 80-160 Hz
Firmware	NodeMCU
Data pins	16 GPIO for interfacing with sensors, switches, LEDs, etc.
ADC channel	1(10-bit) accessible through A0
Communication	UART, SPI, I2C, SPI
RAM	4 MB
SPI pins	4 (SCK, CS, MISO, MOSI) for SPI communication
I2C pins	Available
UART pins	2.
Power supply	3.3 V (max)

Table 2 Component interfacing with ESP8266-12

NEO-6 M GPS	ESP8266-12	16 × 2 I2C LCD	GSM Sim900A	Buzzer
VCC (2.7–5 V)	3.3 V		Uses separate power input 4.7–5 V (5 V adapter is needed)	
RX(receive)	D1(GPIO5)	SCL		
TX(transmit)	D2(GPIO4)	SDA		
GND	GND	GND	GND	SG1
	VIN	VCC		
	D3		TX	
	D4		RX	
	D0			SG2

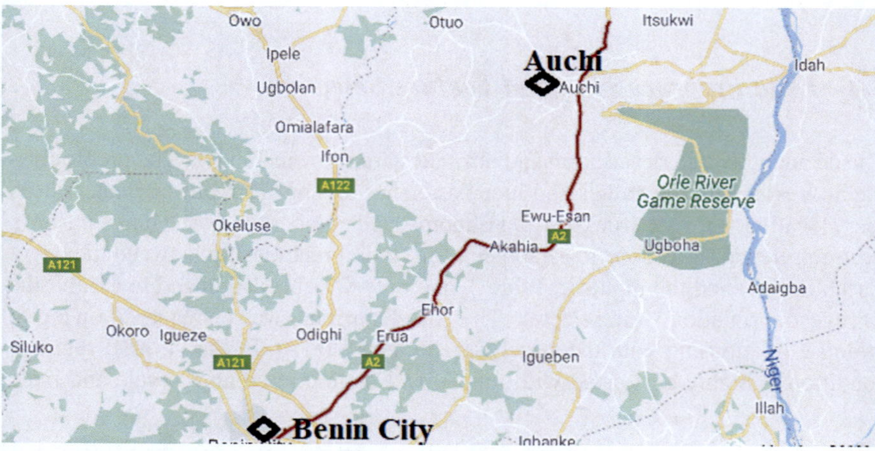

Fig. 1 Map showing Benin-Auchi Highway

3.2 *Description of the Case Study*

The Benin-Auchi highway in Edo State is popular for its linkages. Figure 1 shows the road. This highway is a major segment of the A2 highway, one of the major highways in Nigeria. It is connected as follows: Warri → Benin City → Lokoja → Abuja → Kaduna → Zaria → Kano → Kongolam, Niger Republic border. The road's socioeconomic relevance is judged by its connection of the nation's south to its northern counterpart. The road is roughly 110 km in length and spans through the following major landmark local councils in Edo State: Ikpoba-Okha, Uhunmwode, Esan West, Esan Central, and Etsako West, respectively, and covering the following 12 landmarks: Eyean axis of Benin City, Idokpa, Idumwunha, Urhokuosa, Ehor, Iruekpen, Ekpoma, Ewu, Irrua, Agbede, Aviele, and Auchi. It is one of the busiest highways in Nigeria with some peculiarities, i.e., of the nature of vehicles that ply it. Commercial vehicles, heavy trucks, and trailers conveying petroleum products (from South) and livestock (from the North) are common sights at all hours of the day.

3.3 Design Approach

The system is divided into two components: the client device and the server. The client device resides in the vehicle and communicates with the server through the WIFI and/or GSM and GPS. It also incorporates a register for storing manifest data and an audio device. The data is transmitted to the server (at a coordinating location) where a radio tower (mast) is available to provide a long-distance connectivity along the highway. An LTE-compliant 5-20GHZ omni-radio resides on the tower and connects through a gateway to a web/analytics server, and all communications from the recipient devices are sent to the server through the radio tower. Figure 2 shows the logical model of the design. Figure 3 shows the schematic of the client device.

3.4 Data Preprocessing and Feature Selection

The accident dataset description and attributes are show in Table 3. The preprocessing addresses missing values and inconsistencies arising from out-of-range values (e.g., location, between Iruekpen and Ekpoma, and number of injured male, unsure). Feature selection was done to establish a correlation relationship between the target variable (risk) and the inputs in Table 3, and only variables adjudged to exhibit the strongest correlation were selected. The "date" attribute was disregarded owing to its loose relationship with the target variable (see Fig. 4). In Fig. 4, road segment condition is highly correlated with female and male injured and female and male

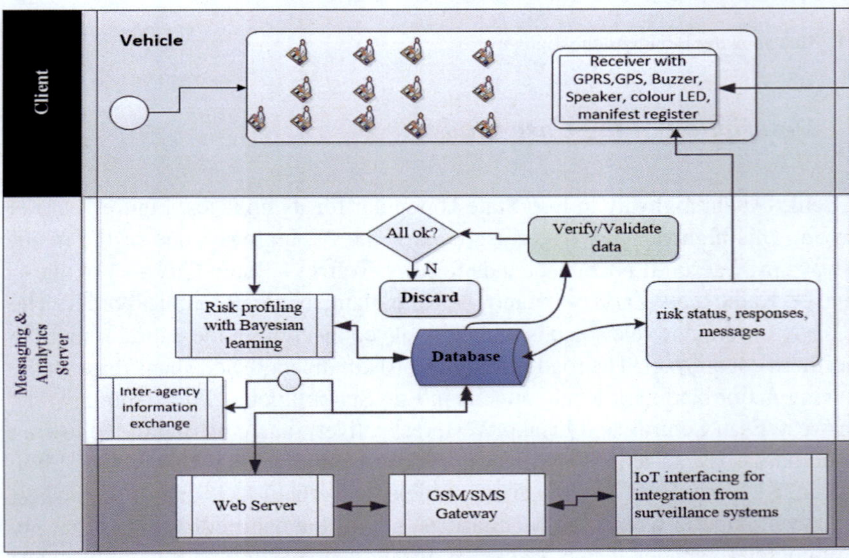

Fig. 2 Logical model of the proposed system

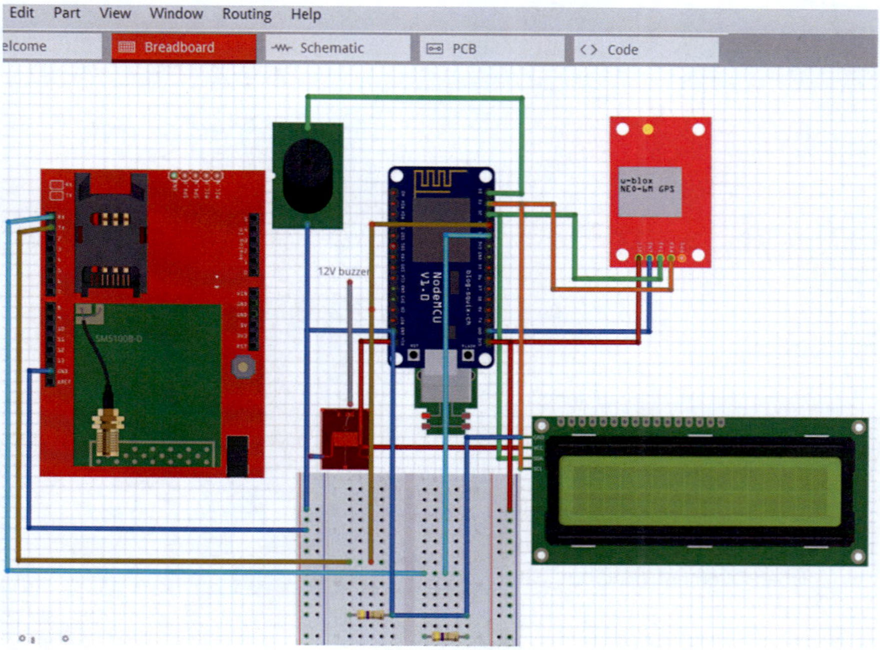

Fig. 3 Schematic of the client device

Table 3 Description of selected variables in the cybercrime dataset

S/N	Label	Description
1	Risk	*Target variable*
2	Date	Predictor representing day of accident e.g., Sunday
3	Number_motorists	Number of motorists in a commercial vehicle
4	Male	Number of males
5	Female	Number of females
5	Injured_male	Injured males
6	Injured_female	Injured females
7	Male_fatality	Number of male fatalities
8	Female_fatality	Number of female fatalities
9	Road_segment_condition	Road status condition
10	Location	Location on the road at any time

fatalities. The above plot can be helpful in feature selection. It can help to prevent multicollinearity in linear models. However, the attributes, injured_male, injured_female, male_fatality, and female_fatality, are not considered feasible attributes that could be easily computed in a nonchaotic system (free moving vehicle under no accident), hence not be used as inputs for the actual Bayesian prediction. The other attributes such as number_of_ motrists, male and female could be preset into the device from the passenger manifest at the commencement of a trip.

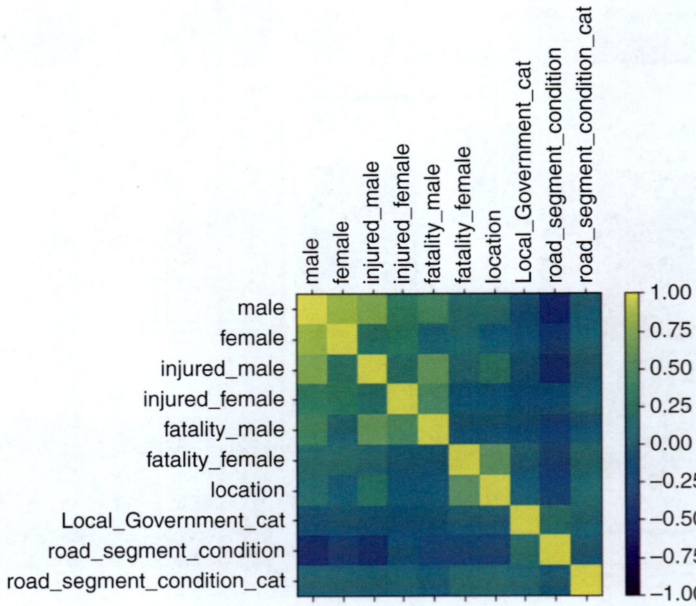

Fig. 4 Correlation plot between feature variables

3.5 Bayesian Model Building and Testing

The Naïve Bayes algorithm is considered appropriate for this problem in that the various predictors appear naturally independent; for instance, there is no overt relationship between the males and females in commercial bus. The same appears true for injured males and females and fatality rates. The primary Bayesian rule therefore is to predict a risk of accident: $P(R|X)$, given the feature domain X (number_motorists = N, male = M, female = F, road_segment_condition = S, location = T).

Mathematically,

$$P(R \mid X) = \frac{P(X \mid R) P(R)}{P(R)} \quad (1)$$

The algorithm implemented to compute $P(R|X)$ is:

- Divide the dataset into the training set (80%) and test set (20%).
- Determine attribute probabilities conditional on the class value.
- Compute joint conditional probability for the attributes using the product rule.
- Neglect attribute with missing values.
- Where an attribute value does occur regularly with the class value, insert a probability of zero (0).

- Use Bayes rule to calculate the conditional probabilities for the class variable.
- Compare the probabilities.
- Compute the mean and standard deviation of the set.
- Return class with the highest probability.
- Apply results to a testing dataset.

4 Results

4.1 Bayesian Model

The result of the Bayesian model building using Django and Python are summarized in Table 4 while Table 5 shows the confusion matrix from the analysis. The NB model performance showed produced an accuracy of 98.1395% with 211 instances classified accurately and 1.8605% (i.e., 4 instances) of incorrect classification with 0.9559 learning capacity (Kappa statistic). The mean absolute error (MAE), RMSE, relative absolute error, and root relative square error (RRSE) are: 0.2339%, 0.2535%, 54.4335%, and 54.7352%, respectively. Figure 5 shows the scatter plot of the risk profile against the number of injured motorists based on gender. Figure 6 presents more information on the relationship that exists among accident victims in respect of injured victims and fatalities. It appears that females are more predisposed to higher risks during accidents.

4.2 The Remote Server and Its Operation

The proposed implementation server (not the ThingSpeak cloud platform) would perform multiple functions, and as shown in Fig. 2, it would incorporate the following: database server (for storing data generated within the system and that received

Table 4 Detailed accuracy by class

Class	TP rate	FP rate	Precision	Recall	F-measure	MCC	ROC area	PRC area
HIGH	1.000	0.060	0.974	1.000	0.987	0.957	1.000	1.000
LOW	0.940	0.000	1.000	0.940	0.969	0.957	1.000	0.999
Weighted avg.	1.000	0.981	0.041	0.982	0.981	0.981	0.957	1.000

TP true positive, *FP* false positive, *MCC* Matthews correlation coefficient

Table 5 Confusion matrix

$N = 215$	Predicted low	Predicted high
Actual high	0	148
Actual low	63	4

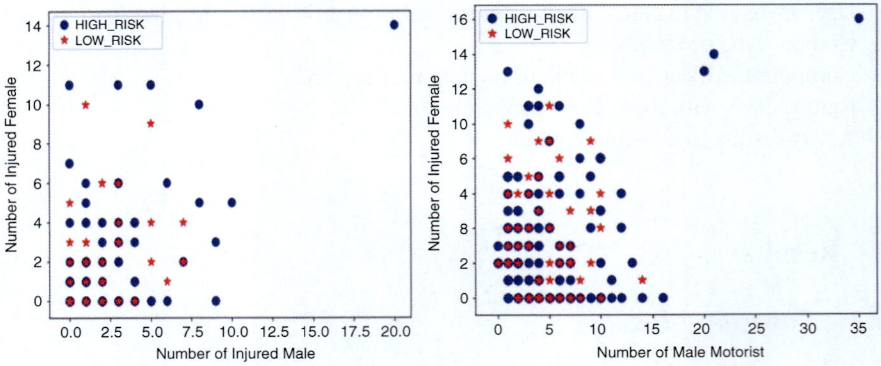

Fig. 5 Scatter plot of risk profile against number of injured motorists based on gender

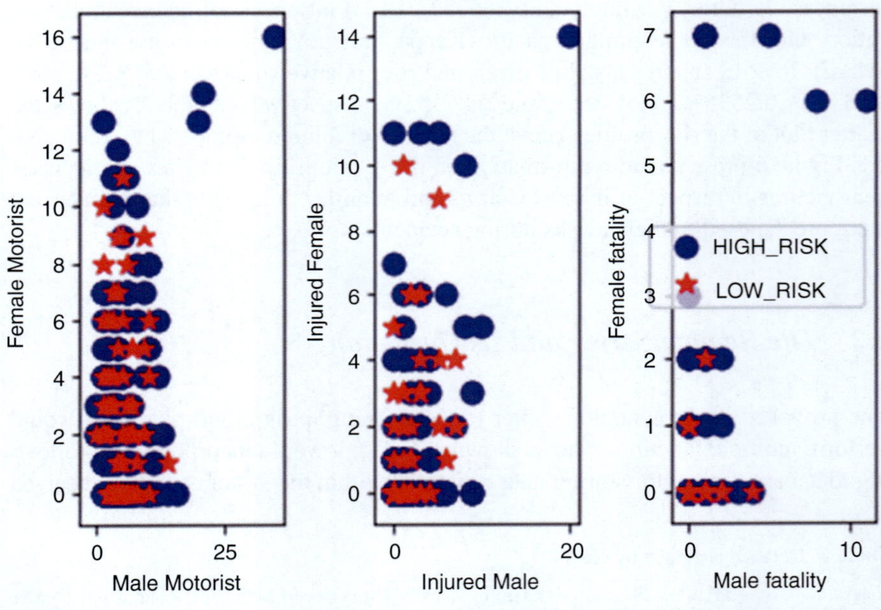

Fig. 6 Scatter plot of risk profile as function of accident victims split into injured victims and fatalities

from the client through the web and the SMS gateway), web server (to provide public access through the web), analytics server (to analyze inputs from remote vehicles housing the client using the developed model), messaging and polling (to relay text and voice messages to vehicles plying the highway in response to receiving the geographical coordinates of the vehicle), and API to enable extensive connection with other surveillance systems.

Operationally, road status conditions would be updated at the server end, and this may be through human intervention (i.e., following road maintenance of particular sections, the status is updated, which would elicit a forward adjustment in the model) or through the integration of ultrasonic or laser sensors in the client device. The server stores data on the 12 road segments (stated above) in a database table with the following schema:

$$F(\text{Location}) = \text{Schema}(\text{location\_name}, \text{begin\_cordinate\_x1}, \text{begin\_cordinate\_y1}, \quad \text{end\_cordinate\_x2}, \text{end\_cordinate\_y2}, \text{road\_status})$$

When a location coordinate is received from the client, the server checks the compares it with the stored references and translates it to the equivalent location identifier, which is subsequently added to the existing variable list (male, female, number_motorists, etc.) used for prediction. The server also retrieves the updated value of the road segment corresponding to the location coordinate received. Following execution of the prediction model using the inputs, the server polls the client device through the sms gateway and returns an encoded voice message that details the risk profile of the location including warnings on the likelihood of accidents and fatalities should the driver fail to adhere to safety rules. The server only polls the client device when the prediction involves a high-risk level. This would reduce the stress on the server.

5 Conclusion

Various developments in the IoT as applied to the domain of road safety and accident prevention campaigns have been discussed. It is stressed that while profound theorization, propositions, prototyping, and integration have been made; machine learning and AI are reforming these developments. This adopts a case study approach and proposes an integration of a Bayesian model into a real-time and centralized road safety infrastructure. The requirements and procedures are captured and prototyping explained. In conclusion, IoT could be extensively applied to critical infrastructure (at national, regional, and local council levels) including road traffic sensitization and safety campaigns. The incorporation of MLMs into existing and future accident control and prevention infrastructure has significant prospects and could reduce losses connected with road accidents.

References

1. Aaron, D. M. A., Francis, X. A., Janos, L. T., Raymund, M., & Francisco, A. M. (2019). Sensor-based traffic control network with neural network. *International Journal of Advanced Trends in Computer Science and Engineering, 8*(4). https://doi.org/10.30534/ijatcse/2019/01842019

2. Abdelfatah, A., Al-Zaffin, M. S., & Hijazi, W. (2015). Trends and causes of traffic accidents in Dubai. *Journal of Civil Engineering and Architecture, 9*(2).
3. Aditya, S. J. (2019). Detection and identification of potholes using microcontroller. *Iconic Research and Engineering Journals, 3*, 257. https://irejournals.com/formatedpaper/1701420.pdf
4. Aishwarya, S. R., Ashish, R. C., Prasanth, M. A., & Savitha, S. C. (2015). An IoT based accident prevention & tracking system for night drivers. *International Journal of Innovative Research in Computer and Communication Engineering, 3*(4) http://www.rroij.com/open-access/an-iot-based-accident-prevention-tracking-system-for-night-drivers.pdf
5. Aji, P. W., Ahmad, C. K., Della, M. P. M., Risky, P. A., Sandika, M. P., Sulton, A. K., & Youngga, R. N. (2019). Naïve Bayes classifier for journal quartile classification. *International Journal of Recent Contributions from Engineering, Science & IT, 7*(2). https://doi.org/10.3991/ijes.v7i2.10659
6. Amitha, T., Shiji, J. T. D., Sinu, V., Sooraj, P., & Indrasena, N. V. (2019). Real time pothole detection and road monitoring system. *International Journal of Innovative Research in Science, Engineering and Technology, 8*(6) www.ijirset.com
7. Ammar, M., Giovanni, R., & Bruno, C. (2018). Internet of Things: A survey on the security of IoT frameworks. *Journal of Information Security and Applications, 38*, 8–27.
8. Amr, S. E., Jin, L., Aboelmagd, N., Hossam, S. H., & Nizar, Z. (2018). Towards a practical crowdsensing system for road surface conditions monitoring. *IEEE Internet of Things Journal, 5*(6), 4672–4685. https://doi.org/10.1109/JIOT.2018.2807408
9. Anusha, A., Vikrant, K., Amol, K., & Tushar, A. (2017, July). *Vehicle tracking and monitoring system to enhance the safety and security driving using IoT*. 2017 international conference on recent trends in electrical, electronics and computing technologies (ICRTEECT).
10. Ayman, T. H. (2020). Design of electronic system for speed detection of vehicles. *International Journal of Advanced Trends in Computer Science and Engineering, 9*(1.2). https://doi.org/10.30534/ijatcse/2020/2791.22020
11. Babitha, D., Mohammed, I., Subrata, C., Ramya, G., & Kolla, B. P. (2020). *International Journal of Advanced Trends in Computer Science and Engineering*. https://doi.org/10.30534/ijatcse/2020/132922020
12. Berrar, D. (2018). Bayes' theorem and naive Bayes classifier. *ResearchGate*. https://www.researchgate.net/publication/324933572. https://doi.org/10.1016/B978-0-12-809633-8.20473-1
13. Chamandeep, K. (2020). The cloud computing and Internet of Things (IoT). *International Journal of Scientific Research in Science, Engineering and Technology*, 19–22.
14. Christy, A., Vaithyasubramanian, S., Viji, A. M., & Naveen, R. J. (2019). Artificial intelligence based automatic decelerating vehicle control system to avoid misfortunes. *International Journal of Advanced Trends in Computer Science and Engineering, 8*(6). https://doi.org/10.30534/ijatcse/2019/75862019
15. Chung, J., & Kim, H. (2020). An automobile environment detection system based on deep neural network and its implementation using IoT-enabled in-vehicle air quality sensors. *Sustainability, 12*(6). https://doi.org/10.3390/su12062475
16. Coffin, D., Oliver, S., & VerWey, J. (2019, November). *Building vehicle autonomy: Sensors, semiconductors, software and U.S. competitiveness*. Office of Industries Working Paper ID-063, November 2019. Available at SSRN: https://ssrn.com/abstract=3492778 or https://doi.org/10.2139/ssrn.3492778.
17. Crawford, B., Khayyam, H., Milani, A. S., & Jazar, R. (2019). Big data modelling approaches for engineering applications. In R. N. Jazar (Ed.), *Nonlinear approaches in engineering applications*. Springer.
18. Das, T., Sinmon, B., & Char, L. (2017). *Vehicle accident prevent cum location and monitoring system*. 2017 8th Annual Industrial Automation and Electromechanical Engineering Conference (IEMECON).

19. Doolan, R., & Muntean, G. M. (2017). EcoTrec—A novel VANET-based approach to reducing vehicle emissions. *IEEE Transactions on Intelligent Transportation Systems, 18*, 608–620. https://doi.org/10.1109/TITS.2016.2585925
20. Francois-Lavet, V., Henderson, P., Islam, R., Bellemare, M. G., & Pineau, J. (2018). An introduction to deep reinforcement learning. *Foundations and Trends® in Machine Learning, 11*(3–4), 219–354.
21. Gadam, S. (2018). *Artificial intelligence and autonomous vehicles*. Retrieved from https://medium.com/datadriveninvestor/artificial-intelligence-and-autonomous-vehiclesae877feb6cd2
22. George, Y., Athanasios, T., & George, P. (2017). Investigation of road accident severity per vehicle type. *Transportation Research Procedia, 25*, 2076–2083.
23. Gianfranco, F., Soddu, S., & Fadda, P. (2017). An accident prediction model for urban road networks. *Journal of Transportation Safety and Security, 10*(4), 387–405. https://doi.org/10.1080/19439962.2016.1268659
24. Haifei, Y., Changjiang, Z., Yi, Z., & Zhong, W. (2020). Model with the action point paradigm to enhance the performance of autonomous driving digital object identifier. *IEEE Access*. https://doi.org/10.1109/ACCESS.2020.2999648
25. Hamid, K., Bahman, J., Mahdi, J., & Reza, N. J. (2020). Artificial intelligence and internet of things for autonomous vehicles. https://www.researchgate.net/publication/335021813. https://doi.org/10.1007/978-3-030-18963-1_2
26. Harum, K. (2016). Vehicle detection and tracking system IoT based. *International Research Journal of Engineering and Technology (IRJET), 5*(8), 1237–1241.
27. Hossain, M. Y., & George, F. P., (2018). IOT based real-time drowsy driving detection system for the prevention of road accidents. *International Conference on Intelligent Informatics and Biomedical Sciences (ICIIBMS), Bangkok, 20*(18), 190–195.
28. Hussein, W. N., Kamarudin, L. M., Hussain, H. N., Hamzah, M. R., & Jadaa, K. J. (2019). Technology elements that influence the implementation success for big data analytics and IoT-oriented transportation system. *International Journal of Advanced Trends in Computer Science and Engineering, 8*(5). https://doi.org/10.30534/ijatcse/2019/74852019
29. Imteaj, B. (2015). *Smart vehicle accident detection and alarming system using a smartphone*. 2015 international conference on computer and information engineering (ICCIE).
30. Jackelou, S. M., Ariel, M. S., & Ruji, P. M. (2019). Road traffic accident case status prediction integrating a modified C4.5 algorithm. *International Journal of Advanced Trends in Computer Science and Engineering, 8*(5). https://doi.org/10.30534/ijatcse/2019/114852019
31. Javadi, B., Calheiros, R. N., Matawie, K. M., Ginige, A., & Cook, A. (2018). Smart nutrition monitoring system using heterogeneous internet of things platform. In G. Fortino, A. B. M. S. Ali, M. Pathan, A. Guerrieri, & G. & Fatta D. (Eds.), *Internet and distributed computing systems* (pp. 63–74). Springer.
32. Kometani, E., & Sasaki, T. (1959). Dynamic behaviour of traffic with a non-linear spacing speed relationship. In *Proceedings of the symposium on theory of traffic flow* (pp. 105–119). Elsevier.
33. Kong, W., Lin, W., Babiloni, F., Hu, S., & Borghini, G. (2015). Investigating driver fatigue versus alertness using the granger causality network. *Sensors, 15*, 19181–19198. https://doi.org/10.3390/s150819181
34. Kulwant, S., Sristy, H., Chandra, M., Sushanth, G., & Sanjith, G. (2020). Iot based real time potholes detection system using image processing techniques. *International Journal of Scientific & Technology Research, 9*(2) http://www.ijstr.org/final-print/feb2020/Iot-Based-Real-Time-Potholes-Detection-System-Using-Image-Processing-Techniques.pdf
35. Kusy, M., & Zajdel, R. (2015). Application of reinforcement learning algorithms for the adaptive computation of the smoothing parameter for probabilistic neural network. *IEEE Transactions on Neural Networks and Learning Systems, 26*, 2163–2175.
36. Lakshmana, P. M., & Ragupathy, R. (2019). A new sentiment score based improved Bayesian networks for real-time intraday stock trend classification. *International Journal*

of Advanced Trends in Computer Science and Engineering, 8(4). https://doi.org/10.30534/ijatcse/2019/10842019
37. Lanatà, A., Valenza, G., Greco, A., Gentili, C., Bartolozzi, R., Bucchi, F., Frendo, F., & Scilingo, E. P. (2015). How the autonomic nervous system and driving style change with incremental stressing conditions during simulated driving. *IEEE Transactions on Intelligent Transportation Systems, 16*, 1505–1517. https://doi.org/10.1109/TITS.2014.2365681
38. Lim, H. S. M., & Taeihagh, A. (2019). Algorithmic decision-making in AVs: Understanding ethical and technical concerns for smart cities. *Sustainability, 11*(20), 5791. https://doi.org/10.3390/su11205791
39. Liyanage, T. D., & Rengarasu, T. M. (2015). Traffic accident analysis and development of accident prediction model. In *Annual session of IESL* (pp. 105–109). The Institute of Engineers.
40. Mahdavinejad, K., & Mohammad, S. (2018). Machine learning for Internet of Things data analysis: A survey. *Digital Communications and Networks, 4*(3), 161–175.
41. Makandar, D. A., & Mulimani, D. (2018). VIP methods for sports video – An analysis. *Asian Journal For Convergence In Technology (AJCT), 3*(3) Retrieved from http://www.asianssr.org/index.php/ajct/article/view/249
42. Maria, C. A. A. (2020). Integrating spatial data analysis for road traffic incident response system. *International Journal of Advanced Trends in Computer Science and Engineering, 9*(1.2). https://doi.org/10.30534/ijatcse/2020/3291.22020
43. Maria, C. A. A., Rosemarie, T. B., Jocelyn, O. T., & Daniel, E. M., Jr. (2020a). Analysis and system architecture design for road traffic incidents. *International Journal of Advanced Trends in Computer Science and Engineering, 9*(12). https://doi.org/10.30534/ijatcse/2020/3391.22020
44. Maria, T. V., Eric, M., Riyanto, J., & Tuga, M. (2020b). Heart disease prediction system using data mining classification techniques: Naïve Bayes, KNN, and decision tree. *International Journal of Advanced Trends in Computer Science and Engineering, 9*(3). https://doi.org/10.30534/ijatcse/2020/82932020
45. Marimuthu, B., Solaiyappan, S. & Gowthami, N. (2018). Automatic detection of potholes and humps for controlling the vehicle speed. *International Journal of Engineering Research in Computer Science and Engineering (IJERCSE), 5*(3). https://www.semanticscholar.org/
46. Markus, D., Matthias, S., Bryan, T. A., & Soto, B. G. D. (2015). A Bayesian network model to predict accidents on Swiss highways. *Infrastructure Asset Management, 2*(4). https://doi.org/10.1680/jinam.15.00008
47. Masami, Y., Nobuhiro, U., & Toshiyuki, N. (2015). Latent class analysis for driving behavior on merging section. *Transportation Research Procedia, 6*, 259–271, 2352–1465 ©2015. www.sciencedirect.com. ScienceDirect International Symposium of Transport Simulation.
48. Mayuresh, D. (2017, February). *Internet of Things based vehicle monitoring system*. Fourteenth international conference on wireless and optical communications networks (WOCN) IEEE.
49. Mohd, A. A. M., Jinat, A. P., & Amit, K. D. (2017). *An intelligent smartphone based approach using IoT for ensuring safe driving*. International conference on electrical engineering and computer science (ICECOS) 2017. https://doi.org/10.1109/ICECOS.2017.8167137.
50. Muthusamy, A. P., Rajendran, M., Ramesh, K., & Sivaprakash, P. (2015). A review on road traffic accident and related factors. *International Journal of Applied Engineering Research, 10*(11).
51. Nak, M. S., & Mauricio, R. (2015). *Road crashes have more impact on poverty than you probably thought* [online]. Available from: https://blogs.worldbank.org/transport/road-crashes-have-more-impact-poverty-you-probably-thought. Accessed 20 Aug 2020.
52. Nanda, S., Joshi, H., & Khairnar, S. (2018). *An IOT based smart system for accident prevention and detection*. Fourth International conference on computing communication control and automation (ICCUBEA), Pune, India, pp. 1–6.
53. Navod, F., & Bindu Tushara, D. (2018). Vehicle monitoring, controlling and tracking system by using android application. *International Journal of Technical Research and Applications, 4*(1). https://doi.org/10.13140/RG.2.2.10243.40480
54. Niraj, P. B., & Geetha Priya, M. (2016). Radar and its applications. *International Journal of Computer Technology and Applications, 9*(28), 1–9. https://www.researchgate.net/pub-

lication/317427496_RADAR_and_its_Applications/link/593a6c57aca272bcd1ea1c40/download
55. Nwankwo, W., & Olayinka, A. S. (2019). Implementing a risk management and X-ray cargo scanning document management prototype. *International Journal of Scientific and Technology Research, 8*(9), 93–105.
56. Nwankwo, W., & Ukhurebor, K. E. (2020a). Data centres: A prescriptive model for green and eco-friendly environment in the cement industry in Nigeria. *International Journal of Scientific and Technology Research, 9*(5), 239–244.
57. Nwankwo, W., & Ukhurebor, K. E. (2020b). Web forum and social media: A model for automatic removal of fake media using multilayered neural networks. *International Journal of Scientific and Technology Research, 9*(1), 4371–4377.
58. Nwankwo, W., Olayinka, A. S., & Ukhurebor, K. E. (2019). The urban traffic congestion problem in Benin City and the search for an ICT-improved solution. *International Journal of Science and Technology, 8*(12), 65–72.
59. Olstam J., & Tapani, A. (2004). Comparison of car-following models. In *Transport forum*. VTI, 2(4).
60. Omar, H., Zhuang, W., & Li, L. (2012). VeMAC: A TDMA based MAC protocol for reliable broadcast in VANETs. *IEEE Transactions on Mobile Computing, 2012*(12), 1724–1736. https://doi.org/10.1109/TMC.2012.142
61. Partha, P. R., Dinesh, D., & Debashis, D. (2018). Design of a low cost and novel Naive Bayes classifier based NodeMCU web server for fever type detection. *Biomedical Research, 29*(12), 2500–2503. www.biomedres.info
62. Pathan, A., & Khan, A. K. (2018). IoT based Pothole Detection & Alert System. *International Journal for Innovative Research in Multidisciplinary Field, 4*(4) https://www.ijirmf.com/wp-content/uploads/201804021.pdf
63. Rajeswari, R. P., Juliet, K., & Aradhana, N. (2017). Text classification for student data set using Naive Bayes classifier and KNN classifier. *International Journal of Computer Trends and Technology, 43*(1), 8–12. https://doi.org/10.14445/22312803/ijctt-v43p103
64. Rusli, R., Haque, M. M., King, M., & Voon, W. S. (2015) A comparison of road traffic crashes along mountainous and non-mountainous roads in Sabah, Malaysia. In: *The 2015 Australasian road safety conference*.
65. Sachin, K., & Durga, T. (2016). A data mining approach to characterize road accident locations. *Journal of Modern Transportation, 24*(1), 62–72.
66. Sarasvathi, N., Fong, J., & Zu, X. (2018). Study and implementation of internet of things (IoT) based vehicle safety alert and tracking system. *INTI Journal, 1*(10), 1–11. ISSN e2600-7920.
67. Shahid, S., Minhans, A., Che Puan, O., Hasan, S. A., & Ismail, T. (2015). Spatial and temporal pattern of road accidents and casualties in peninsular Malaysia. *Jurnal Teknologi, 76*(14).
68. Shubham, I., Pragati, A., Krushna, K., & Manjushri, M. (2018, December). Pothole detection system for monitoring road using IoT. *International Journal of Advance Engineering and Research Development, 5*(12), 212581938.
69. Supriyono, S., Iman, M., & Waluyo, A. S. (2019). The development of engine control module manipulator module based on Arduino to increase power and torque of motorcycle engine. *International Journal of Advanced Trends in Computer Science and Engineering, 8*(6). https://doi.org/10.30534/ijatcse/2019/76862019
70. Swetha, K., Punithgowda, K. M., Lalithesh, D., Deepak, S. B., & Shivuprasad, U. J. (2020). Automatic detection and notification of potholes and humps on roads using IoT. *International Research Journal of Engineering and Technology (IRJET), 7*(6). www.irjet.net
71. Syafrudin, M., Alfian, G., Fitriyani, N., & Rhee, J. (2018). Performance analysis of IoT-based sensor, big data processing, and machine learning model for real-time monitoring system in automotive manufacturing. *Sensors, 18*, 2946.
72. Truong, T. T. (2020). Predicting and preventing transport accidents with fuzzy relationships. *International Journal of Advanced Trends in Computer Science and Engineering, 9*(2). https://doi.org/10.30534/ijatcse/2020/89922020

73. Usman, T., Fu, L., & Miranda-Moreno, L. F. (2016). Injury severity analysis: Comparison of multilevel logistic regression models and effects of collision data aggregation. *Journal of Modern Transportation, 24*(1).
74. Wadhahi, N. T. S. A., Hussain, S. M., Yosof, K. M., Hussain S. A., & Singh, A. V. (2018). *Accidents detection and prevention system to reduce traffic hazards using IR sensors*. 2018 7th international conference on reliability, infocom technologies and optimization (trends and future directions) (ICRITO), Noida, India, pp. 737–741.
75. Wang, C., Xu, C., Xia, J., & Qian, Z. (2017). Modeling faults among e-bike-related fatal crashes in China. *Traffic Injury Prevention, 18*(2).
76. WHO. (2020). *Road traffic injuries* [online]. Available from: https://www.who.int/news-room/fact-sheets/detail/roadtraffic-injuries. Accessed 29 Oct 2020.
77. World Bank. (2018). *Road deaths and injuries hold back economic growth in developing countries* [online]. Available from: https://www.worldbank.org/en/news/press-release/2018/01/09/road-deaths-and-injuries-hold-backeconomic-growth-in-developing-countries. Accessed 29 Oct 2018.
78. Xin, Z., & Wen, L. Y. (2017). A Bayesian network approach to causation analysis of road accidents using netica. *Hindawi Journal of Advanced Transportation*. https://doi.org/10.1155/2017/2525481

On the Integration of AI and IoT Systems: A Case Study of Airport Smart Parking

Vinh Bui, Alireza Alaei, and Minh Bui

1 Introduction

Recent advances in Internet of Things (IoT) and artificial intelligence (AI) technologies have revolutionised many areas of our lives, including infrastructure and transportation. For example, organisations responsible for managing cities have started using a large number of cameras to capture the flow of traffic across cities every day. This resulted in a huge amount of data, which if can be processed efficiently, will allow the efficient use of the city transportation infrastructure, quickly disperse and avoid traffic jams and mitigate/prevent accidents and crimes. In recent years, AI technologies have played an important role in processing this type of data, for example, counting cars and people crossing intersections [16, 52], identifying vehicle licence plates [15, 50] and searching for parking space [5, 38].

While IoT devices are characterised by the ability to collect and transfer data over a network without requiring human-to-human or human-to-computer interactions, AI brings life to these devices by enabling them to learn and take actions based on the data they have already collected from various sources. The combination of AI and IoT technologies allows researchers and industry experts to create intelligent/smart machines, which can perform human-like tasks, or even mimic human behaviours. These two technologies, therefore, usher in the concept of "smart devices" starting from smartphones and smart cameras to smart houses and smart cities. The concept of smart parking has also emerged in this context.

V. Bui (✉) · A. Alaei
Faculty of Science and Engineering, Southern Cross University, Gold Coast, QLD, Australia
e-mail: vinh.bui@scu.edu.au; ali.alaei@scu.edu.au

M. Bui
University of Queensland, Brisbane, QLD, Australia
e-mail: q.m.bui@uqconnect.edu.au

S. Pal et al. (eds.), *Artificial Intelligence-based Internet of Things Systems*,
Internet of Things, https://doi.org/10.1007/978-3-030-87059-1_16

In the area of smart parking, researchers and engineers have been actively working on solving a wide range of parking problems, from providing customers with information about parking slots' availability in real time to maximising the utilisation of parking facilities and developing smart parking systems using IoT and AI [3]. Although the technology for addressing individual technical challenges in this area is currently available, the integration of these technologies under a unified system to address the parking problem in general and airport parking in particular still presents some substantial challenges. For instance, it is necessary to address the trade-off between the high demand in computational power of AI algorithms and the low computational capability of IoT devices, the high communication bandwidth requirement of imagery data and the low power consumption of IoT devices and the interoperability and inconsistent technology ecosystem issues caused by variations in hardware and software of IoT devices [39]. In addition, data privacy and security [25], energy consumption [17] and environmental sustainability [46] are other issues to be addressed.

This chapter, therefore, discusses the system integration issues through a case study of airport smart parking. It demonstrates how existing AI, IoT and Cloud computing technologies can be integrated into a unified system to solve the airport parking problem. In particular, a system that integrates IoT parking sensors, the IBM IoT management platform and OpenALPR library was developed to tackle technical challenges, such as the automatic detection of plate number, model and colour of the vehicle in each parking slot in both indoor and outdoor parking environments. This functionality enables the system to provide managers and users with real-time analytics about each parking area. It will also discuss important issues related to the system design, including the system architecture, hardware and software, sensing and network technologies and detection accuracy and errors, just to name some of them. Although the proposed system was developed for the airport parking problem, the discussion around it is relevant to problems in other domains from the system design and integration perspective.

The rest of the chapter is organised as follow. Section 2 reviews the related works. Section 3 introduces the airport parking problem and specifies requirements for an airport smart parking management system. In Sect. 4, the design solution with the focus on the system architecture and design considerations for the airport smart parking problem is presented. Section 5 presents some experimental results and discusses lessons learnt from a practical system deployment. Finally, concluding remarks and directions for future works are provided in Sect. 6.

2 Related Work

IoT is a paradigm where myriads of interconnected objects with sensing and actuating devices provide the ability to collect and share data through the Internet to enable innovative applications [23]. Artificial intelligence, especially machine learning (ML) techniques, can be used to gain knowledge from data generated by

IoT devices. Thus, the combination of IoT and ML enables the development of valuable services for the society in various domains and especially in infrastructure and transportation [3, 63, 27].

The idea of smart parking was initially introduced to solve the problem of parking management in megacities, in which IoT and ML algorithms were used to, for example, track parking slot's availability, maximise utilisation of parking facilities and minimise customers' searching time for available parking spaces [3].

Tracking the parking slot's availability is the foundation of any smart parking management system. This problem has been addressed in both the literature and in practice using a wide range of sensing techniques, which can be classified into two groups: techniques using occupancy sensors and techniques based on computer vision. In the first group, small, sturdy, low power and low-cost occupancy sensors with different sensing technologies, including magnetic, ultrasonic, infrared, and loop sensors, were used to detect occupancy of each parking space [42, 43]. The advantage of this group of occupancy tracking techniques is the simplicity, reliability and accuracy of occupancy detection algorithms. On the other hand, scalability is a major issue with this group, as a huge number of sensors are required to be deployed and maintained for any tasks. In addition, these sensors do not provide any other information apart from vehicle occupancy, hence limiting their applications.

The second group of occupancy tracking techniques relies on advances in computer vision technology to process data collected from vision-based sensors, such as CCTV cameras, to detect the availability of parking spaces. ML algorithms, especially deep learning, are the core component of these occupancy tracking techniques [4, 28, 59, 62]. The main advantage of this group of techniques is the scalability and deployability, as cameras are relatively easy to deploy and a single camera can cover a large parking area. In addition, vision-based sensors provide richer data compared with other types of sensors. For instance, the video stream collected from CCTV cameras can be processed by ML algorithms to obtain vehicle plate numbers [6, 40], makes and models [10, 11, 53]. This provides managers with greater support for their management and decision making. However, the computer vision-based techniques are sensitive to weather and operating environments as the image quality of vision-based sensors, e.g., CCTV cameras, is greatly affected by lighting condition, dust and humidity. Moreover, vision-based sensors often generate a huge volume of data, which causes higher costs of data transmission and storage. ML algorithms also require high computational power, which may not be available in IoT devices. Furthermore, machine/deep learning techniques often need annotated data for training and verification, which may not be available in some smart parking applications.

Beside tracking parking slot's availability, ML algorithms were used to maximise utilisation of parking facilities, minimise customers' searching time for available parking space [19, 35, 48] and create efficient smart parking pricing systems [32, 45, 51]. For example, Aydin et al. [9] proposed a smart parking system that can assist users in finding an available parking space and to minimise the time spent in locating the nearest available car park using the genetic algorithm. Shoeibi and Shoeibi [49] introduced an automated valet parking based on hybrid robotic valets and Deep Q-Learning algorithm to optimise the usage of parking space. More recently, Tekouabou et al. [57] proposed an ensemble-based model to optimise the prediction of the availability of parking spaces in smart parking. Moreover, Saharan et al. [45] introduced a machine learning-based approach to predict the occupancy of parking slots, which in turn deduced occupancy-driven prices for arriving vehicles.

Research works have also been carried out in developing architectures for smart parking applications. A smart parking architecture based on cloud computing was initially a common idea to utilise the advantage of cloud computing for storing and processing big data [29, 44]. However, as vision-based sensors are increasingly used in smart parking, the data transmission process from sensors to cloud services becomes the bottleneck of the cloud-based architectures. In addition, this type of architectures is not suitable for real-time data processing due to communication latency. More recently, fog computing [14, 37, 56] and edge computing architectures [13, 33, 47] were proposed to address the limitation of the architectures based on cloud computing. The main idea behind the proposal of these new architectures is to process data as close to the location where the data were obtained as possible to reduce the time and amount of data transmission. However, not all data processing algorithms, especially ML algorithms, can run at the fog level and especially, at the edge devices due to the limitation of the device computational power [36]. Therefore, the selection of suitable architectures for smart parking should take into account problem-specific requirements as no architecture is one size fits all.

Despite the significant amount of works that have been carried out in the domain of smart parking design and development, there is a lack of research work addressing specials needs and characteristics of the airport parking problem [12]. Although few works have been done on studying the integration of AI and IoT, several design factors, such as system architectures, sensing and network technologies, hardware and software platforms, security and other application-specific requirements, could influence the performance and design of smart parking systems in practice [3]. Moreover, weather conditions and operating environments are critical factors to be considered in any smart parking system design. This chapter, therefore, is written to discuss this knowledge gap and provide readers with a practical solution. In the next section, the airport smart parking problem is thoroughly discussed to better understand its requirements.

3 Airport Parking Problem and Its Requirement Analysis

Parking income is a major contributor to the operating income of many airports around the world occupying around 40% of the non-aeronautical revenue or 25% of the commercial revenue [22, 61]. For example, in Australia, Sydney Airport recorded an operating profit of $97 million from car parking operations in 2016–2017, and this represented an operating profit margin of 71.9% of revenue [7]. However, the service quality in many airport car parks remains an issue [7]. A study of the Melbourne airport suggested that the airport parking is among those services that caused concerns for passengers and should urgently be addressed by airport managements [26]. Moreover, airport parking businesses have been looking for technological solutions to improve the efficiency and security of the airport parking facilities and to reduce the management cost to compete with the rise of the ride-hailing services offered by the transportation network companies, such as Uber, Lyft, Didi-Chuxing, Ola and Grab [60].

Smarter parking management systems have been an active research incentive of researchers in recent years, often done through the integration of IoT technology. Research in this domain primarily focuses on solving various technical challenges, such as finding the closest available parking space [31, 44], and achieving optimum usage out of parking facilities [49, 55]. These objectives were all achieved whilst taking into considerations the advantages of IoT technology. Additionally, works were also carried out in the development of various architectures and frameworks for smart parking [8, 24]. From the literature, it can be noted that numerous system architectures and frameworks were designed to address the issues related to smart parking in general, yet the specific issues regarding airport smart parking have not been adequately addressed. Although airport parking shares many similarities with other parking problems, it has different objective functionalities and constraints.

Currently, many parking management systems are operational in big airports and shopping complexes. However, most existing systems suffer from the following limitations:

- Customers are only able to see available parking spots in a small local area and will often compete for these spots. There is a lack of information about available spots in neighbouring areas and a mechanism to quickly reserve a free parking spot.
- Current systems are not designed with the ability to obtain detailed information about the current vehicle parked in each slot. Thus, customers are not able to ascertain the current status of their vehicles within the parking facilities. In the context of airport parking, it is also important to assist customers in retrieving their vehicles in a quick and timely manner upon their return trips. At the current stage, none of the existing airport parking management systems has this capability.
- Current systems are mostly designed to work within an indoor environment, while many (airport) parking facilities are of outdoor.

From an airport management perspective, product differentiation is the key to winning any competition. Real-time information about the vehicle, environment temperature and humidity in each parking slot would help to develop efficient systems for real-time vehicle monitoring, parking guidance, signage, way finding and real-time booking. The availability of such information would also help to quickly identify threats and hazards to the security of the parking facilities. Finally, real-time information of the parking facilities would allow solving parking optimisation problems, such as minimising terminal travel time for customers.

Considering the limitations of the existing airport parking systems, a set of functional and non-functional requirements for an airport smart parking management system were proposed. In particular, the smart parking system should have the following functional requirements:

- Real-time information about the parking slot occupancy and vehicle details including make, model, colour, plate number, parking duration and status.
- Turn-by-turn assistance for customers navigating inside the parking facility to their reserved parking slots or their vehicles on their return.
- Online and real-time booking, reservation and payment.
- Function to allocate available parking slots to customers in real-time subject to optimisation objectives, such as minimum walk time to terminals.

Concerning the non-functional requirements of an airport parking management system, it is important that airport parking is operational 24/7 in both indoor and outdoor environments. The system should also be scalable, as the parking capacity of a regional airport like the Gold Coast airport in Australia is already around 5000 parking slots. Besides scalability, interoperability and security are other important features since the system stores customers' data and interacts with other information systems, such as the payment, booking and flight information systems. Another important requirement of airport parking system is the simplicity of system deployment. The system should not require significant modification to the existing parking infrastructure since most airports already have operational parking facilities. Finally, the system should be cost-efficient.

4 System Design

This section discusses the design considerations, constraints and architectures as well as the selection of sensing technologies, IoT platforms and vehicle identification framework for the proposed airport smart parking system.

4.1 Design Considerations

4.1.1 Energy Consumption

IoT devices are characterised by low-energy hardware architecture with limited computational power. Many IoT devices are designed to operate on batteries. Typical IoT hardware devices, such as Raspberry Pi 3 Model B, are based on a Quad Core ARM Cortex-A53 1.2 GHz 64-bit CPU running on a Broadcom BCM2837 System on a Chip (SoC) with a Broadcom Video Core IV Graphical Processing Unit (GPU). The ARM processor works at frequencies ranging from 700 MHz to 1.2 GHz. When the system is idle, it consumes around 1.3 W, but the consumption increases when it performs heavy computation tasks, such as vehicle identification.

Table 1 The required computational power of common DNNs [2]

DNN architecture	Input size	Param size	FLOPs
MobileNet	224 × 224	16 MB	579 MFLOPs
AlexNet	227 × 227	233 MB	727 MFLOPs
CaffeNet	224 × 224	233 MB	724 MFLOPs
VGG-F	224 × 224	232 MB	727 MFLOPs
SqueezeNet	224 × 224	5 MB	837 MFLOPs
GoogleNet	224 × 224	51 MB	2000 MFLOPs
ResNet-18	224 × 224	45 MB	2000 MFLOPs
SENet	224 × 224	440 MB	21,000 MFLOPs

AI algorithms, on the other hand, often require significant computational power, which is proportional to energy consumption [18]. Computational power is often measured in floating-point operations per second or FLOPs. Table 1 shows the amount of computation required by common deep neural networks (DNNs) for a single pass in a typical image recognition task. As indicated in Table 1, even a small deep neural network like MobileNet would need 579 MFLOPs while larger neural networks like GoogleNet would require around 2GFLOPs of computational power.

The trade-off between energy consumption and computational power indicates that running AI algorithms directly on IoT devices will cause the device battery to drain quickly. To give a better estimation in terms of energy consumption and running time, common AI algorithms and frameworks were tested in a simple image classification task on Raspberry Pi 3 platform operating on a 10,000 mAh battery. Figure 1 shows the average energy consumption of the tested algorithms and frameworks. As shown in Fig. 1, OpenCV consumed less energy among the tested frameworks, and it is more suitable for IoT integration. Tensorflow also performed relatively well, especially on the Squeeznet architecture.

The integration of AI algorithms on IoT platforms, therefore, should take into account the requirement of computational power and energy consumption when choosing the algorithms, ML frameworks and the location of their deployment.

4.1.2 Network Technology

Network technology is featured with various characteristics, such as communication range, data rate, energy consumption and cost. Unfortunately, there is always a trade-off between these characteristics. Long communication range and high data rate also mean high energy consumption and high cost. IoT devices often operate on low-power and low-energy hardware, which also means IoT applications have to communicate in a lower data rate or a shorter distance. Figure 2 shows the classification of network technologies by their data rate, energy consumption, communication range and cost.

As shown in Fig. 2, low-power wide-area network (LPWAN) technologies designed specifically for IoT, such as NB-IoT. Sigfox and LoRa operate at a very

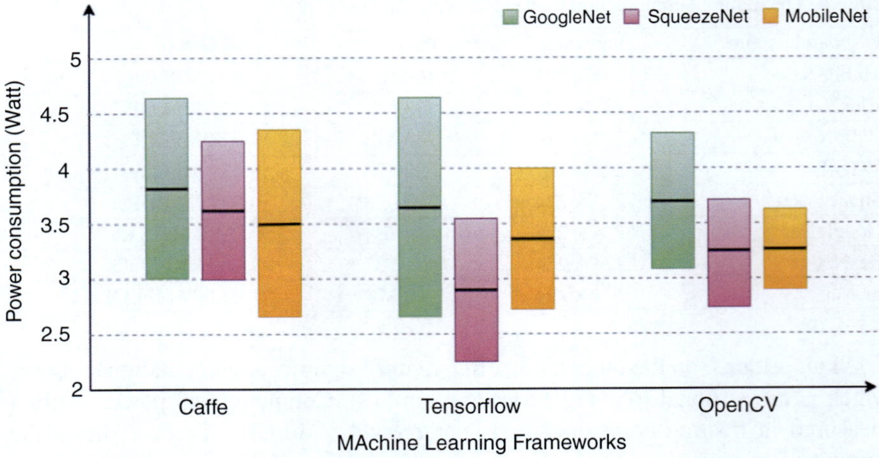

Fig. 1 The average energy consumption of common machine learning frameworks on Raspberry Pi 3 platform

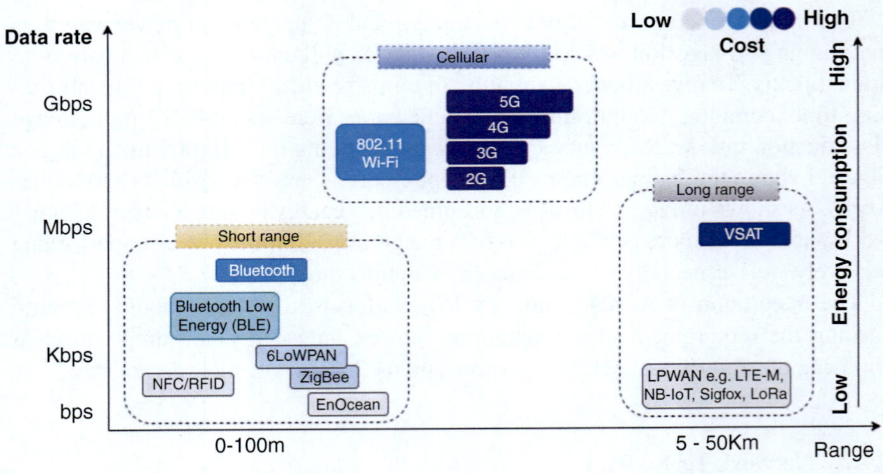

Fig. 2 Classification of network technologies by data rate, communication range, energy consumption and cost

low data rate (50–100 Kbps) and are only suitable for IoT applications that send and receive a very small amount of data, i.e., tens or hundreds of bytes per day. The advantages of these technologies are the long communication distance (5–50 km), extremely low-energy consumption and low cost that make them suitable for sensor networks.

On the other hand, short-range technologies like NFC, BLE and Bluetooth operate at a much higher data rate, i.e., up to 1Mbps, but in a much shorter distance, i.e., less than 100 m. Among the short-range technologies, BLE has the best data rate to energy ratio with the energy consumption from 10 to 30 mA while it operates at 1–3 Mbps. An important advantage of these technologies is the cost as they operate without the need for supporting infrastructure.

Finally, cellular network technologies have a good balance between energy consumption, communication range and data rate. However, the cost of deploying and operating these technologies is high due to the need for supporting infrastructure.

In summary, as IoT applications are generally data-centric, the selection of network technologies for an IoT platform should be based on the amount and characteristics of data that need to be exchanged between devices in the platform while considering other factors, such as communication range, level of energy consumption and cost.

4.1.3 Data Volume, Processing Time and Location

While many IoT devices, such as smart smoke detectors, only send and receive a very small amount of data every day, there are IoT devices, such as smart cameras, that need to send a large amount of data over networks. AI algorithms are often used to process such data to reduce the amount of data that are sent over networks. For example, instead of sending a photo of a vehicle for identification, AI algorithms can be used in smart cameras to recognise the vehicles make, model and plate number. This information will be sent over networks, enabling the use of low-energy and narrow-band network technologies, e.g., LoRa and NB-IoT.

For IoT applications that need to process data in real-time, the processing needs to occur near the site of data generation, i.e., the edge. However, most AI algorithms demand a huge amount of processing power, which is not always available in edge devices. Although the computation capability of edge devices has increased tremendously during the past decade, it is still challenging to perform sophisticated AI algorithms in these resource-constrained environments. Therefore, consideration should be taken on the selection of an AI algorithms and data processing location for delay sensitive IoT applications like the airport smart parking.

4.1.4 Weather and Environment

Other critical factors, which must be considered, in the design of any IoT applications are weather conditions and operating environments. Variations in the lighting condition, temperature and humidity greatly affect the camera photo quality, stability and accuracy of IoT sensors. As a result, the performance (accuracy and processing time) of AI algorithms when dealing with such data is also affected. Therefore, it is important to train AI algorithms with data collected from IoT sensors in different weather conditions and operating environments.

4.2 Design Architectures

4.2.1 Design Architecture for Generic IoT Applications

IoT applications are characterised by heterogeneous technologies, where the integration of technologies often suffers from interoperability problems [21]. The layered architecture design is often used to ensure the interoperability of different technologies.

Figure 3 illustrates the five-layer architecture commonly used by IoT applications [20, 27]. The five-layer architecture generally contains sensor–actuator, edge, fog, cloud, and application layers. In addition, the security layer is common to all five layers. The functionalities of each layer are provided in the following:

- The sensor–actuator layer, which is composed of sensor and actuator nodes, is responsible for environmental data acquisition and manipulation. Sensor nodes acquire data by converting physical, chemical or biological variables into readable digital signals or numerical properties. Actuator nodes manipulate the physical environment by turning control signals into some kind of physical actions, e.g., triggering an alarm if a high temperature is detected. In this layer, an important requirement is IoT sensor nodes that should be of low cost with very low power consumption and robust to noise, so that they can be deployed in a large scale, in unprotected environments and without a frequent battery change or with energy harvesting facilities.
- The edge layer provides a low-level interface to the sensor nodes. It is responsible for receiving data from sensor nodes and performing simple data processing, e.g., sampling, scaling, and compressing data, before sending the compressed data to the upper fog layer. This intermediate data processing step is necessary to significantly reduce the amount of data transmitted over communication networks in a large-scale deployment. The edge layer also provides a standardised software interface to configure and manage sensor nodes from the upper layers. AI algorithms can be deployed at the edge layer for processing delay-sensitive data or performing data compression tasks. However, it is common that only trained computational models are used at the edge layer due to the limited computational power in edge devices.

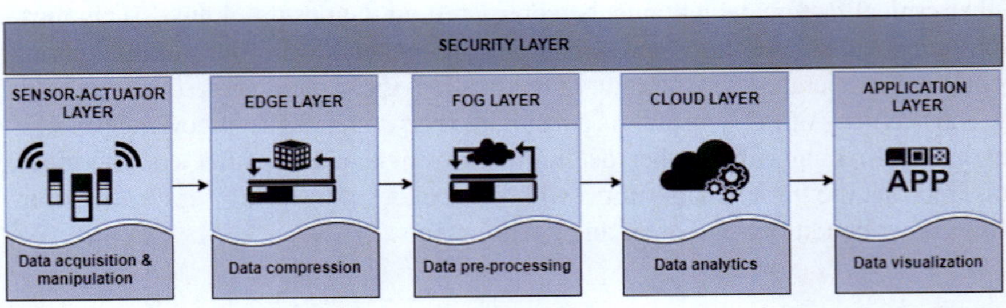

Fig. 3 The five-layer architecture of generic IoT applications

- The fog layer acts as a local hub for collecting data from local edge devices in a large-scale deployment. It is responsible for data preprocessing (e.g. filtering) and providing gateway, which connects IoT to the Internet. This layer is also responsible for implementing local data policy. AI algorithms can be deployed at the fog layer for data analysis tasks that are subject to the local data policy.
- The cloud layer is responsible for performing data analytics. It consists of cloud servers that provide different services, including global data storage, big data analysis, and IoT device management. Depending on the application, specific cloud services can additionally be implemented. This layer also provides mechanisms, such as graphical user interface (GUI) and application programming interface (API) for external applications to access IoT data, perform data analytics and manage IoT devices. AI algorithms are deployed at the cloud layer for complex data analysis tasks.
- The application layer is responsible for data visualisation and system-user interaction. It consists of desktop, mobile and web applications that consume services provided by the cloud layer to manage IoT devices and perform data analytics as requested by end-users and finally to visualise the results in user-friendly formats.
- The security layer is liable for providing security to all layers by implementing various security mechanisms, such as encryption, authentication, authorisation and auditing.

4.2.2 Proposed Architecture for the Airport Smart Parking

The specific requirements of the airport smart parking problems suggest the need for modification to the abovementioned five-layer architecture. In particular, due to the airport security requirement, several data analytics tasks need to be done locally. Therefore, the fog and cloud layers were combined to create a layer called the information aggregation layer to increase the data manageability. Furthermore, thanks to the availability of low-cost hardware platforms, such as Raspberry Pi Zero, it is more economical to combine the sensor and the edge layers in a single hardware platform to perform the functionalities of both layers. As a result, a three-layer architecture composed of the information sensing layer, the information aggregation layer and the application layer was proposed to address the airport smart parking problems. The proposed architecture is depicted in Fig. 4 and the details of each layer are discussed in the following subsections.

- The information sensing layer.

Data required for identifying the vehicle in each parking slot are collected in this layer. In addition, data of the surrounding environment, such as the humidity, temperature and carbon dioxide level can also be obtained. From the sensing technology perspective, there are several options to design this layer. Different types of sensors, including ultrasonic, magnetic, infrared, laser and CCTV, can be used to detect parking slot occupancy. Table 2 compares the common sensing technologies used for parking occupancy detection. As presented in Table 2, CCTV sensors are

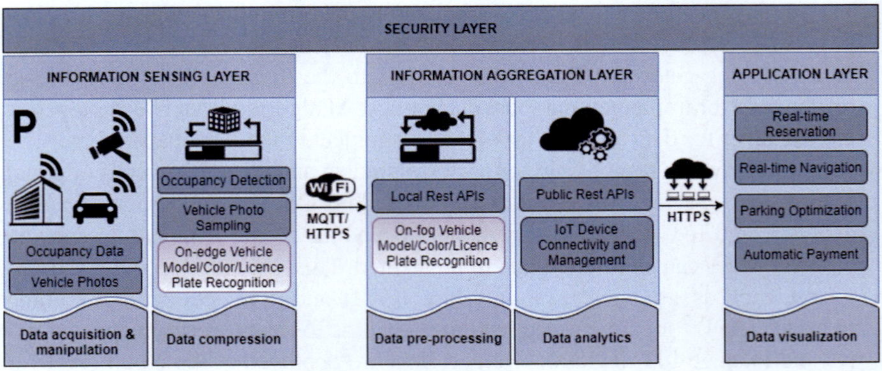

Fig. 4 The three-layer architecture of the airport smart parking system

Table 2 Comparison of sensing technologies used for smart parking occupancy detection [3, 43]

Sensor type	Intrusive deployment	Average power consumption	Speed detection	Multi-lane detection	Weather sensitive	Accuracy	Cost
Passive IR	–	74.35 mW	–	–	Yes	**	$$
Active IR	Yes	106.03 mW	Yes	Yes	Yes	**	$$
Ultrasonic	–	89.21 mW	–	–	–	***	$
LDR	–	103.11 mW	–	–	Yes	**	$
Magnetometer	Yes	82.54 mW	Yes	–	–	****	$
RFID	–	0.1–2 W	–	–	–	**	$
CCTV	–	1.0–10 W	Yes	Yes	Yes	****	$$$$
Inductive loop	Yes	0.7–1.0 W	Yes	–	Yes	****	$$
Microwave	–	1.0 W	Yes	Yes	–	***	$$

More stars indicate higher accuracy and cost

among the most accurate sensors for occupancy detection in the literature of IoT. However, they have a higher cost and energy consumption compared with other sensors. Meanwhile, ultrasonic sensors are one of the best occupancy detection sensors offering a good balance between the cost, accuracy and deployability.

In addition to the accuracy and cost, an important requirement for selecting a parking sensor is its ability to reliably operate in both indoor and outdoor environments, during the day and at night and in different weather conditions. The industry variant of the HC-SR04 ultrasonic sensor was selected to use in the proposed smart parking system for this reason. The sensor provides 2–400 cm noncontact measurement function with the ranging accuracies of up to 3 mm. Apart from occupancy detection, the information sensing layer is also responsible for taking photos of the vehicle in each parking slot. A 5MP auto-IR camera was chosen for the task of the photo collection. The built-in auto-IR function is required to allow the camera to take photos at both day and night.

To manage the sensors, the Raspberry Pi 3 platform was selected, although the Raspberry Pi Zero-W was sufficient for the task. Compared with other IoT hardware platforms, the Pi 3 platform provides better support for computer vision and machine learning at a relatively low cost. A pair of HC-SR04 ultrasonic sensors were integrated into the platform to increase the system redundancy and measurement accuracy. The integration of the ultrasonic sensors and the IR camera in a single hardware platform allow the development of algorithms to control the camera activation and ensure that vehicle photos are always taken at the right distance, increasing vehicle recognition accuracy and reduced energy consumption. To get a better idea about the hardware platform developed for detecting vehicles in the information sensing layer, a prototype of the system is shown in Fig. 5.

Another important function of the information sensing layer is to perform data compression, especially for imagery data, i.e., vehicle photos in this case. A possible solution to this problem is to perform vehicle identification in the information sensing layer to avoid sending vehicle photos to the information aggregation layer for the identification task. However, the limit in the computational power of the Raspberry Pi 3 platform suggests that further testing is necessary to confirm the solution feasibility. Some experimental results and discussions in regard to the proposed solution are provided in Sect. 5.

- The information aggregation layer.

This layer is responsible for collecting and aggregating data of each parking slot received from the edge devices, performing data processing and storage and undertaking data analytics. It also provides mechanisms to access data and manage edge devices subject to data management policies. Currently, there are several competitive cloud-based IoT platforms, including AWS IoT, Google Cloud IoT and the IBM Watson IoT Platform, which can be used in the information aggregation layer. All of these platforms support the various functionalities, including IoT device connectivity and management, data processing and data storage, required in the information aggregation layer. However, each platform supports different levels of data analytics, data security and edge computing solutions, which make them suitable for

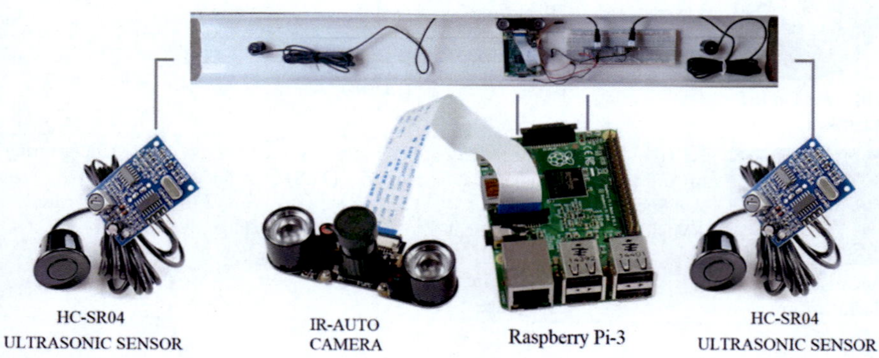

Fig. 5 The hardware prototype of the IoT smart parking device

different IoT applications. The key features of the common IoT platforms are summarised in Table 3.

Considering different aspects of the IoT platforms presented in Table 3, the IBM IoT platform was chosen for our proposed airport smart parking system. Our choice

Table 3 Key features of common IoT platforms

IoT platforms	Communication protocols	Key offering and its main functions	Security	Edge computing solutions	Top-3 use cases
Amazon AWS IoT Core	HTTP MQTT WebSockets	Connectivity Authentication Rules engine Development environment	Transport layer security (TLS) Authentication (X. 509)	*FreeRTOS* edge operating system *IoT GreenGrass* edge computing platform	Smart city Connected home Agriculture
The Cisco IoT Operations Platform	MQTT	Mobile connectivity eSIM as a service Machine learning to improve security	Transport layer security (TLS)	*Cisco iOX* edge development platform *Cisco Edge* intelligence	Smart city Manufacturing Connected vehicles
Google Cloud IoT Platform	HTTP MQTT	Connectivity Device management	Transport layer security (TLS/SSL)	*Edge TPU* chip enabling deployment AI at the edge	Smart parking Energy Transportation and logistics
IBM Watson IoT Platform	HTTP MQTT	Connectivity Device management Real-time analytics Blockchain	Transport layer security (TLS) Authentication (SSO) Identity management (LDAP)	*IBM Edge Application Manager* platform	Smart buildings Manufacturing Agriculture
Microsoft Azure IoT Platform	HTTP MQTT WebSockets AMQP over	Connectivity Authentication Device management Device monitoring IoT Edge	Transport layer security (TLS)	*IoT Edge* as an integral part of IoT Hub	Manufacturing Retail Healthcare

of the platform is also supported by the literature, as the IBM IoT platform was selected for the airport smart parking application in the system introduced by Ullah et al. [58]. In addition, the IBM IoT platform supports the integration with the IBM blockchain security technology, which has the capability of working with IoT devices, to track payment histories and automatically fulfil invoices and payments requirements [1]. Furthermore, the IBM IoT platform can also be deployed in private clouds, which is an airport security requirement.

- The application layer.

Several applications can be developed for airport parking management by using the information provided by the aggregation layer. To improve the customer experience, a web app or mobile app can be developed to allow customers to access real-time information about the parking facilities, provide booking and reservations facility and make payments for their booking. The application can also remind customers where their vehicles were parked and provide instruction to help customers find their vehicles in the parking facilities. Figure 6 illustrates an example of the information provided by the airport smart parking application.

To improve the security and the manageability of the parking facilities, a visualisation of the real-time parking information can be fed into the airport security and management team. Data about the parking environment, e.g., temperature, humidity and carbon dioxide level can further be used for early detection of fire hazard in the car park. Parking data and the analytics performed in the information aggregation layer are made available to the application layer via a set of RESTful API. Two sets of API, the local and global APIs, are implemented in the proposed framework to enhance system security. The local API provides functions to access all available features of the data, while the global API only provides access to some analytics of the data, such as the availability of parking facilities as shown in Fig. 6.

- The network connectivity.

The proposed platform relies on the IEEE 802.11n/ac technology for network connectivity, as most airports already have good Wi-Fi coverage. Other IoT network technologies, such as NB-IoT, Sigfox, and Lora, are inferior to Wi-Fi in this situation as the main strength of these technologies lies in their low-energy consumption rather than the data transfer rate, which in this case is insufficient. The MQTT

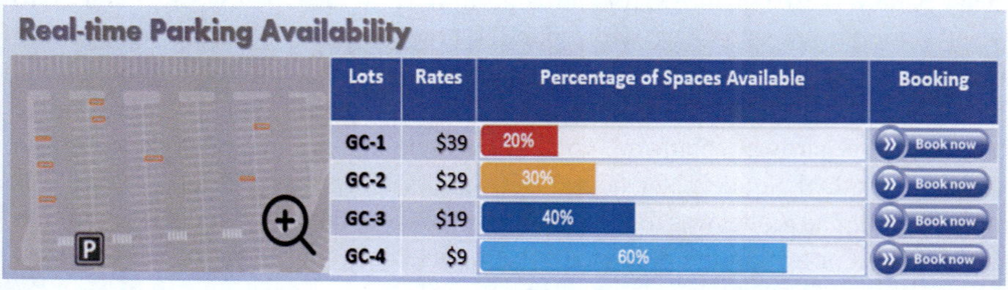

Fig. 6 Example of information provided by the airport smart parking application

protocol was used in the proposed smart parking system for exchanging data between the information sensing and the aggregation layer of the proposed system.

- The security layer.

In this layer, the connection between the sensor nodes and devices in the aggregation layer through the MQTT protocol is encrypted by using the standard Transport Layer Security (TLS). In addition, the IBM IoT platform also provides a set of security services, including device authentication, certificate provisioning and verification, secure booting and firewalls to enhance the proposed platform security.

4.3 Vehicle Identification Framework

An important functionality of the proposed smart parking system is the ability to perform automatic vehicle identification, including the recognition of the plate number, model and colour of the vehicle parked in each parking slot. The existing literature suggests several frameworks, both commercial and open-source that can be used to fulfil this function [6, 30]. Traditional pattern matching using local binary pattern (LBP) and histogram of oriented gradient (HOG) features and machine learning techniques such as k-nearest neighbours (KNN), support vector machine (SVM) and more modern deep learning have been used in these frameworks for the identification task [30]. Overall, these frameworks were reported to achieve good performance (i.e. >95% accuracy) in their testing environments [6]. To select a vehicle identification framework for this airport smart parking project, besides accuracy, additional selection criteria including computational complexity, processing time, robustness and cost were also considered.

To give a better idea about performance, processing time and robustness of different frameworks, Table 4 compares existing character recognition methods, which plays an important part in the vehicle identification frameworks. Considering the results provided in Table 4 and available vehicle identification frameworks in the literature [41], OpenALPR was selected to fulfil the task in the proposed framework. OpenALPR is an automatic licence-plate recognition library distributed with a commercial licence and an open-source version [41]. OpenALPR was developed on two important computer vision libraries, OpenCV and Tesseract OCR. The local binary pattern (LBP) feature extraction algorithm from the OpenAPLR was used to detect vehicle number plates, while the Tesseract's optical character recognition algorithm was used to recognise the plate numbers from the detected number plates. OpenALPR is capable of processing 5–7 frames, i.e., photos, per second using the Raspberry Pi 3 hardware platform. To detect vehicle' make and model, the approach based on residual SqueezeNet neural network proposed by Lee et al. [34] was implemented. The model was trained using the Vehicle Make and Model Recognition dataset [54]. For testing the proposed system, data collected from the sensors used in the proposed framework were considered.

Table 4 Relative comparison of the existing character recognition methods [6]

Methods	Convenience	Inconvenience	Accuracy (%)
Pattern matching features	More competent for recognising non-broken, fixed size, single font characters. Simple and straight forward technique	Higher processing time for processing all the pixels, not robust to the scale and rotation of the characters, noise, multi-font, broken characters, etc.	95.6–95.7
Extracted features	Faster recognition, capable of extracting salient features, robust to different noises	Recognition performance might get decreased by non-robust features, requires time for extracting features	98.3–98.6
Machine Learning e.g. ANN	Relatively simple implementation, high efficiency in the case of the huge volume of data	Additional processing time for training the network, processing complexity	96.9–98.5
Statistical classifiers	Capable of learning the differences and the uniformities of the multiple characters	Relatively complex, high processing time	95.7–99.5

5 Experiment Results and Discussion

Several experiments were conducted to evaluate the AI-IoT integration process, using the developed IoT smart parking device in a practical deployment scenario, with the focus on the level of energy consumption, the processing time and the detection accuracy at different times of the day and in various weather and lighting conditions. Details of each experiment are presented in the following subsections.

5.1 Energy Consumption

The first experiment was conducted to evaluate the level of energy consumption of the developed smart parking device. The device was tested in two AI integration modes: on edge and on fog. In the first case, OpenALPR framework was deployed directly on the smart parking device to perform vehicle identification task. Meanwhile, in the latter case, the device needed to make a Rest API call to a local network server that runs OpenALPR as a vehicle identification service. The device was powered by a standard 10,000 mAh power bank in both day and night time. The maximum temperature at the day time was 30 °C and the minimum temperature at the night time was 7 °C. The energy consumption of the device was tested in two operation regimes: normal and exhaustive. In the normal regime, the device performed one vehicle identification process every hour, while in the exhaustive regime, the device repeatedly performed the vehicle identification process. An experiment was also conducted on the use of energy harvesting technology to power the device. This is important for the deployment of the device in outdoor parking facilities, which may not have access to the main power supply. In this experiment, a 12,000

Table 5 The device energy consumption measured through battery life

Mode	Regime				
	Battery/power bank	Normal		Exhaustive	
		Min	Max	Min	Max
On edge	10,000 mAh	9 h 17 m	9 h 19 m	7 h 21 m	7 h 23 m
On fog	10,000 mAh	8 h 12 m	8 h 18 m	6 h 17 m	6 h 20 m
On fog	Solar 12,000 mAh	24 h+	24 h+	24 h+	24 h+

mAh power bank with a solar charger was used to power the device. The level of energy consumption was measured through battery life. The results obtained at different scenarios are presented in Table 5.

The results reported in Table 5 clearly indicate that the device consumed less energy in the on-edge AI integration mode. This implies that energy spending on Wi-Fi communication was significant, even greater, in comparison with the energy spending on running AI algorithms for the vehicle make, model, colour and plate number recognition. In addition, when the device was powered using a 12,000 mAh power bank with the solar charger, it could work for more than 24 h in both the normal and exhaustive regimes. This indicates that the proposed device can operate continuously using energy harvesting technologies, such as the solar-powered battery. This finding suggests the developed device suitable for outdoor deployment.

5.2 *Detection Accuracy*

Weather and lighting conditions could greatly affect the performance of the smart parking device camera. Therefore, the second experiment was conducted to evaluate the impacts of weather and lighting condition on the detection accuracy of the proposed smart parking device. Six vehicles from different manufactures, including Audi, Ford, Holden, Kia, Mazda and Toyota, were used in this experiment. The vehicles were registered in three different states (Victoria, New South Wales and Queensland) of Australia. The experiment was carried out during the day and at night and in sunny, cloudy, and rainy weather. Each vehicle was tested in the front and reverse parking directions. For each direction, three photos were taken by the parking device. In total, 180 vehicle images (i.e. six vehicles x two directions x three photos per direction x five conditions) were taken by the device. This means the vehicle identification algorithm was tested 180 times. The results of the experiments are summarised in Table 6. Sample vehicle photos taken by the device as part of the experiment are shown in Fig. 7.

From the results presented in Table 6, it can be noted that the lighting and weather conditions have no significant impact on the plate number recognition algorithm as reflected through a small variation in the algorithm detection accuracy. The algorithm achieved as high as 98.9% accuracy in a good, daytime condition and 96.7% at the night-time of rainy days.

Table 6 The accuracy of the smart parking device in different operating conditions

	Condition				
	Day			Night	
Vehicle	Rainy	Cloudy	Sunny	Rainy	Clear
Plate number	97.8%	98.9%	98.3%	96.7%	97.8%
Make	99%	99%	99%	48.9%	50.1%
Model	57.1%	57.1%	57.1%	42.8%	42.8%
Colour	100%	100%	100%	N/A	N/A

Fig. 7 Sample photos captured by the device in different conditions

In contrast, both algorithms for the recognition/identification of the vehicle makes and models were greatly affected by the variations of lighting conditions. The accuracy of the vehicle-make recognition algorithm reduced from 99% during the day to around 50% at night. Meanwhile, these figures went down from 57.1% to 42.8% for the vehicle model identification algorithm. It is worth mentioning that the vehicle colour recognition was 100% accurate in the daytime. In the night time, the camera was switched to the black-and-white mode, and the colour information was not available. Hence, the algorithm was not evaluated in this scenario.

Further investigation into the identification accuracies suggests that the errors in the plate number recognition were mainly due to the camera exposure, which was saturated by either the direct sunlight or the front light beam as seen in Fig. 7 (top-right) photo. The same issue was observed in the vehicle make recognition at night. Often, important features, such as the manufacture logo, were not visible in the front parking direction due to the impact of the direct light beams from the vehicle headlights. As a result, the algorithm failed to recognise the vehicles' make from the front parking direction. Reverse parking did not suffer from this issue, as there was no strong light beam at the vehicle rear. This explained the decrease of the vehicle make identification accuracy from 99% during the day to around 50% at night. The vehicle model recognition was the least accurate with only around 50% accuracy. This is due to the SqueezeNet model used for vehicle model identification that was trained using the Vehicle Make and Model Recognition dataset. The dataset did not contain vehicle images taken at the night time and vehicle models after 2016. In addition, there was not a sufficient number of training samples in the dataset for some vehicle makes, such as Holden and Kia.

5.3 Processing Time

The third experiment was conducted to compare the processing time of the on-edge and on-fog deployment modes of the proposed vehicle identification algorithms. We were also interested in evaluating the impact of the operating environment, e.g., the temperature and weather conditions on the processing time. Table 7 shows the average processing time of the two deployment modes in different time and weather conditions. As it is evident from the results provided in Table 7, in overall, the on-edge deployment mode has a notable smaller processing time, i.e. 3.62–3.77 s, compared with 4.53–4.72 s of the on-fog deployment mode. This is understandable if we take into account the time required to upload a 5 Mpx (2592 × 1944) photo over a 100Mbps Wi-Fi connection, which was approximately 1.2 s. There was not a notable difference in the processing time between day and night. However, the processing time can be significantly reduced by reducing the size of the captured photos. In particular, the processing time was reduced by approximately 40% when the image size was reduced by 20%. It is worth noting that when the photo resolution was decreased by a factor of two in both dimensions, the recognition accuracy was decreased by 16% on average.

5.4 Discussion

The results of the conducted experiments confirmed the deployability and advantages of the on-edge integration solution in the proposed smart parking design. The on-edge deployment of the AI algorithms for vehicle identification provided a better processing time and consumes less energy compared with the on-fog deployment. The level of energy consumption of the developed smart parking device was small enough to allow the device to be powered using an energy harvesting source, e.g., solar batteries. There was a notable impact when the proposed system was operated at different environmental and weather conditions. Training datasets also played a significant role in the success of AI algorithms integrating on IoT devices,

Table 7 Processing time in second (s) under different conditions

		Daytime			Night-time	
Condition node	Photo size	Sunny (30° C)	Cloudy (25° C)	Rainy (15° C)	Clear (10° C)	Rainy (7° C)
On edge	2592 × 1944	3.77 s ± 0.19 s	3.73 s ± 0.12 s	3.74 s ± 0.16 s	3.62 s ± 0.18 s	3.62 s ± 0.19 s
On fog	2592 × 1944	4.72 s ± 0.39 s	4.68 s ± 0.35 s	4.69 s ± 0.37 s	4.53 s ± 0.33 s	4.54 s ± 0.31 s
On edge	2048 × 1536	1.97 s ± 0.15 s	1.94 s ± 0.11 s	1.95 s ± 0.14 s	1.88 s ± 0.16 s	1.89 s ± 0.18 s
On fog	2048 × 1536	2.74 s ± 0.22 s	2.71 s ± 0.21 s	2.72 s ± 0.24 s	2.68 s ± 0.23 s	2.69 s ± 0.25 s

especially if AI algorithms were deployed for computer vision tasks. The experiments have shown that the accuracy of such algorithms can be reduced as much as 50% in certain conditions.

6 Conclusion and Future Avenue

In summary, the process of AI–IoT integration was discussed through a case study of smart airport parking in this chapter. A number of issues related to the integration of AI and IoT technologies from the system analysis and design perspectives, including requirement analysis, identification of design constraints, selection of system architectures and hardware and software technologies are addressed to handle the problems of the airport smart parking.

A prototype of the smart parking device was also built using a Raspberry Pi 3 platform integrating with industry-standard parking sensors, auto-IR camera, OpenALPR and the IBM IoT cloud platform to study the feasibility of the proposed design architecture. Moreover, the impact of operating environments and weather conditions on the performance of the smart parking device were evaluated. A series of experiments were conducted to evaluate the energy consumption, recognition and identification accuracy and processing time of the smart parking device at different times of the day and in different weather conditions. The experiment results indicated that the developed smart parking system is capable of providing real-time information about the parking facilities. It is also easy to deploy and cost-efficient. The developed smart parking device can detect vehicle plate numbers and vehicle colours with high accuracy. The accuracies of the vehicle make and model recognition algorithms have, however, been greatly impacted by the training dataset and lighting conditions.

Future works will be carried out on improving the vehicle make and model recognition algorithms by addressing the camera overexposure problem, enriching vehicle make and model datasets and improving the robustness of the identification algorithms. From the AI–IoT integration perspective, it is important to continue addressing the trade-off between high demand for the computational power of AI algorithms and the low-energy, low-computational power nature of IoT hardware platforms. This can be done by proposing innovative architectures for AI deployment, improving the efficiency of AI algorithms and developing low-energy but high-performance IoT hardware platforms.

References

1. Ahmed, S., Rahman, M. S., & Rahaman, M. S. (2019). A blockchain-based architecture for integrated smart parking systems. *Proceedings of the 2019 IEEE international conference on pervasive computing and communications workshops*, March 11–15; Kyoto, Japan, pp. 177–182.
2. Albanie. (2019). *Estimates of memory consumption and FLOP counts for various convolutional neural networks.* https://github.com/albanie/convnet-burden
3. Al-Turjman, F., & Malekloo, A. (2019). Smart parking in IoT-enabled cities: A survey. *Sustainable Cities and Society, 49*, 101608.
4. Amato, G., Carrara, F., Falchi, F., Gennaro, C., & Vairo, C. (2016). Car parking occupancy detection using smart camera networks and deep learning. *Proceedings of the 2016 IEEE symposium on computers and communication*, June 27–30, Messina, Italy, pp. 1212–1217.
5. Amato, G., Carrara, F., Falchi, F., Gennaro, C., Meghini, C., & Vairo, C. (2017). Deep learning for decentralized parking lot occupancy detection. *Expert Systems with Applications, 72*, 327–334.
6. Arafat, M. Y., Khairuddin, A. S. M., Khairuddin, U., & Paramesran, R. (2019). Systematic review on vehicular licence plate recognition framework in intelligent transport systems. *IET Intelligent Transport Systems, 13*(5), 745–755.
7. Australian Competition and Consumer Commission (ACCC). (2018). https://www.accc.gov.au/media-release/airport-profits-continue-to-grow#:%7E:text=The%20four%20airports%20earned%20a,terms%20from%20the%20previous%20year
8. Awaisi, K. S., Abbas, A., Zareei, M., Khattak, H. A., Khan, M. U. S., Ali, M., Din, I. U., & Shah, S. (2019). Towards a fog enabled efficient car parking architecture. *IEEE Access, 7*, 159100–159111.
9. Aydin, I., Karakose, M., & Karakose, E. (2017). A navigation and reservation based smart parking platform using genetic optimization for smart cities. *Proceedings of the 2017 5th international Istanbul smart grid and cities congress and fair*, April 19–21, Istanbul, Turkey, pp. 120–124.
10. Baran, R., Glowacz, A., & Matiolanski, A. (2015). The efficient real-and non-real-time make and model recognition of cars. *Multimedia Tools and Applications, 74*(12), 4269–4288.
11. Boonsim, N., & Prakoonwit, S. (2017). Car make and model recognition under limited lighting conditions at night. *Pattern Analysis and Applications, 20*(4), 1195–1207.
12. Bui, V., & Bui, M. (2019). A truly smart airport parking solution. *Proceedings of the 2019 IEEE Asia-Pacific conference on computer science and data engineering*, December 9–11, Melbourne, Australia, pp. 1–4.
13. Bura, H., Lin, N., Kumar, N., Malekar, S., Nagaraj, S., & Liu, K. (2018). An edge based smart parking solution using camera networks and deep learning. *Proceedings of the 2018 IEEE international conference on cognitive computing*, July 2–7, San Francisco, CA, USA, pp. 17–24.
14. Celaya-Echarri, M., Froiz-Míguez, I., Azpilicueta, L., Fraga-Lamas, P., Lopez-Iturri, P., Falcone, F., & Fernández-Caramés, T. M. (2020). Building decentralized fog computing-based smart parking systems: From deterministic propagation modeling to practical deployment. *IEEE Access, 8*, 117666–117688.
15. Chen, R. C. (2019). Automatic license plate recognition via sliding-window darknet-YOLO deep learning. *Image and Vision Computing, 87*, 47–56.
16. Chen, J., Su, W., & Wang, Z. (2020). Crowd counting with crowd attention convolutional neural network. *Neurocomputing, 382*, 210–220.
17. Ding, X., & Wu, J. (2019). Study on energy consumption optimization scheduling for Internet of Things. *IEEE Access, 7*, 70574–70583.
18. García-Martín, E., Rodrigues, C. F., Riley, G., & Grahn, H. (2019). Estimation of energy consumption in machine learning. *Journal of Parallel and Distributed Computing, 134*, 75–88.

19. Geng, Y., & Cassandras, C. G. (2013). New "smart parking" system based on resource allocation and reservations. *IEEE Transactions on Intelligent Transportation Systems, 14*(3), 1129–1139.
20. Gia, T. N., Qingqing, L., Queralta, J. P., Zou, Z., Tenhunen, H., & Westerlund, T. (2019). Edge AI in smart farming IoT: CNNs at the edge and fog computing with LoRa. *Proceedings of the 2019 IEEE AFRICON*, September 25–27, Ghana, pp. 1–6.
21. Gonzalez-Usach, R., Yacchirema, D., Julian, M., & Palau, C. E. (2019). Interoperability in IoT. In G. Kaur & P. Tomar (Eds.), *Handbook of research on big data and the IoT* (pp. 149–173). IGI Global.
22. Gu, H. (2019). Airport revenue diversification. *Journal of Management Science & Engineering Research, 2*(1), 25–28.
23. Gubbi, J., Buyya, R., Marusic, S., & Palaniswami, M. (2013). Internet of Things (IoT): A vision, architectural elements, and future directions. *Future Generation Computer Systems, 29*(7), 1645–1660.
24. Gupta, R., Budhiraja, N., Mago, S., & Mathur, S. (2020). An IoT-based smart parking framework for smart cities. In S. Neha, C. Balas, & V. Emilia (Eds.), *Data management, analytics and innovation* (pp. 19–32). Springer.
25. HaddadPajouh, H., Dehghantanha, A., Parizi, R. M., Aledhari, M., & Karimipour, H. (2019). A survey on internet of things security: Requirements, challenges, and solutions. *Internet of Things, 14*, 100129.
26. Jiang, H., & Zhang, Y. (2016). An assessment of passenger experience at Melbourne Airport. *Journal of Air Transport Management, 54*, 88–92.
27. Kassab, W. A., & Darabkh, K. A. (2020). *A–Z survey of Internet of Things: Architectures, protocols, applications, recent advances, future directions and recommendations* (Vol. 163, p. 102663). Journal of Network and Computer Applications.
28. Ke, R., Zhuang, Y., Pu, Z., & Wang, Y. (2020). A smart, efficient, and reliable parking surveillance system with edge artificial intelligence on IoT devices. *IEEE Transactions on Intelligent Transportation Systems, 22*, 1–13.
29. Khanna, A., & Anand, R. (2016). IoT based smart parking system. In *2016 international conference on internet of things and applications*, January 22–24, Pune, India, pp. 266–270.
30. Kiran, V. K., Parida, P., & Dash, S. (2019). Vehicle detection and classification: a review. *Proceedings of the 2019 international conference on innovations in bio-inspired computing and applications*, December 16–18, Odisha, India, pp. 45–56.
31. Kizilkaya, B., Caglar, M., Al-Turjman, F., & Ever, E. (2019). Binary search tree based hierarchical placement algorithm for IoT based smart parking applications. *Internet of Things, 5*, 71–83.
32. Kotb, A. O., Shen, Y. C., Zhu, X., & Huang, Y. (2016). iParker—A new smart car-parking system based on dynamic resource allocation and pricing. *IEEE Transactions on Intelligent Transportation Systems, 17*(9), 2637–2647.
33. Lee, C., Park, S., Yang, T., & Lee, S. H. (2019). Smart parking with fine-grained localization and user status sensing based on edge computing. *Proceedings of the 2019 IEEE 90th vehicular technology conference*, September 22–25, Hawaii, USA, pp. 1–5.
34. Lee, H. J., Ullah, I., Wan, W., Gao, Y., & Fang, Z. (2019). Real-time vehicle make and model recognition with the residual SqueezeNet architecture. *Sensors, 19*(5), 982–988.
35. Lin, T., Rivano, H., & Le Mouël, F. (2017). A survey of smart parking solutions. *IEEE Transactions on Intelligent Transportation Systems, 18*(12), 3229–3253.
36. Mahdavinejad, M. S., Rezvan, M., Barekatain, M., Adibi, P., Barnaghi, P., & Sheth, A. P. (2018). Machine learning for Internet of Things data analysis: A survey. *Digital Communications and Networks, 4*(3), 161–175.
37. Naranjo, P. G. V., Pooranian, Z., Shojafar, M., Conti, M., & Buyya, R. (2019). FOCAN: A fog-supported smart city network architecture for management of applications in the Internet of Everything environments. *Journal of Parallel and Distributed Computing, 132*, 274–283.

38. Ng, C. K., Cheong, S. N., & Foo, Y. L. (2020). Low latency deep learning based parking occupancy detection by exploiting structural similarity. In R. Alfred, Y. Lim, H. Haviluddin, & K. O. Chin (Eds.), *Computational science and technology* (pp. 247–256). Springer.
39. Noura, M., Atiquzzaman, M., & Gaedke, M. (2019). Interoperability in Internet of Things: Taxonomies and open challenges. *Mobile Networks and Applications, 24*(3), 796–809.
40. Omar, N., Sengur, A., & Al-Ali, S. G. S. (2020). Cascaded deep learning-based efficient approach for license plate detection and recognition. *Expert Systems with Applications, 149*, 113280.
41. Open Automatic License Place Recognition (OpenALPR). (2015). https://github.com/openalpr/openalpr
42. Paidi, V., Fleyeh, H., Håkansson, J., & Nyberg, R. G. (2018). Smart parking sensors, technologies and applications for open parking lots: A review. *IET Intelligent Transport Systems, 12*(8), 735–741.
43. Perković, T., Šolić, P., Zargariasl, H., Čoko, D., & Rodrigues, J. J. (2020). Smart parking sensors: State of the art and performance evaluation. *Journal of Cleaner Production, 262*, 121181.
44. Pham, T. N., Tsai, M. F., Nguyen, D. B., Dow, C. R., & Deng, D. J. (2015). A cloud-based smart-parking system based on Internet-of-Things technologies. *IEEE Access, 3*, 1581–1591.
45. Saharan, S., Kumar, N., & Bawa, S. (2020). An efficient smart parking pricing system for smart city environment: A machine-learning based approach. *Future Generation Computer Systems, 106*, 622–640.
46. Salam, A. (2020). Internet of Things for sustainability: Perspectives in privacy, cybersecurity, and future trends. In S. Abdul (Ed.), *Internet of Things for sustainable community development* (pp. 299–327). Springer.
47. Sarker, V. K., Gia, T. N., Ben Dhaou, I., & Westerlund, T. (2020). Smart parking system with dynamic pricing, edge-cloud computing and LoRa. *Sensors, 20*(17), 4669.
48. Shin, J. H., & Jun, H. B. (2014). A study on smart parking guidance algorithm. *Transportation Research Part C: Emerging Technologies, 44*, 299–317.
49. Shoeibi, N., & Shoeibi, N. (2019). Future of smart parking: Automated valet parking using deep Q-learning. *Proceedings of the 2019 international symposium on distributed computing and artificial intelligence*, June 26–28, Avila, Spain, pp. 177–182.
50. Silva, S. M., & Jung, C. R. (2020). Real-time license plate detection and recognition using deep convolutional neural networks. *Journal of Visual Communication and Image Representation, 71*, 102773.
51. Simhon, E., Liao, C., & Starobinski, D. (2017). Smart parking pricing: A machine learning approach. *Proceedings of the 2017 IEEE conference on computer communications workshops*, May 1–4, Atlanta, GA, USA, pp. 641–646.
52. Song, H., Liang, H., Li, H., Dai, Z., & Yun, X. (2019). Vision-based vehicle detection and counting system using deep learning in highway scenes. *European Transport Research Review, 11*(1), 51.
53. Tafazzoli, F., & Frigui, H. (2016). Vehicle make and model recognition using local features and logo detection. *Proceedings of the 2016 international symposium on signal, image, video and communications*, November 21–23, Tunis, Tunisia, pp. 353–358.
54. Tafazzoli, F., Frigui, H., & Nishiyama, K. (2017). A large and diverse dataset for improved vehicle make and model recognition. *Proceedings of the IEEE conference on computer vision and pattern recognition workshops*, July 21–26, Honolulu, HA, USA, pp. 1–8.
55. Tandon, R., & Gupta, P. K. (2019). Optimizing smart parking system by using fog computing. *Proceedings of the international conference on advances in computing and data sciences*, April 24–25, Valletta, Malta, pp. 724–737.
56. Tang, C., Wei, X., Zhu, C., Chen, W., & Rodrigues, J. J. (2018). Towards smart parking based on fog computing. *IEEE Access, 6*, 70172–70185.
57. Tekouabou, S. C. K., Cherif, W., & Silkan, H. (2020). Improving parking availability prediction in smart cities with IoT and ensemble-based model. *Journal of King Saud University-Computer and Information Sciences*. https://doi.org/10.1016/j.jksuci.2020.01.008

58. Ullah, M., Nardelli, P. H., Wolff, A., & Smolander, K. (2020). Twenty-one key factors to choose an IoT platform: Theoretical framework and its applications. *arXiv* preprint arXiv:2004.04924.
59. Vítek, S., & Melničuk, P. (2018). A distributed wireless camera system for the management of parking spaces. *Sensors, 18*(1), 69.
60. Wadud, Z. (2020). An examination of the effects of ride-hailing services on airport parking demand. *Journal of Air Transport Management, 84*, 101783.
61. Warrender, A. (2019). Airport commercial revenues in the time of the digital shopper. *Journal of Airport Management, 13*(4), 303–321.
62. Wu, Q., Huang, C., Wang, S. Y., Chiu, W. C., & Chen, T. (2007). Robust parking space detection considering inter-space correlation. In *2007 IEEE international conference on multimedia and expo*, July 2–6, Beijing, China, pp. 659–662.
63. Zantalis, F., Koulouras, G., Karabetsos, S., & Kandris, D. (2019). A review of machine learning and IoT in smart transportation. *Future Internet, 11*(4), 94.

Vision-Based End-to-End Deep Learning for Autonomous Driving in Next-Generation IoT Systems

Dapeng Guo, Melody Moh, and Teng-Sheng Moh

1 Introduction

The Internet of Things (IoT) systems have grown and become an essential part of everyday lives. Originated from simple and interconnected sensors, the ubiquitous presence of IoT is now indispensable for smart cities, smart factories, etc. Traditional IoT devices are usually equipped with very limited processing power since they are designed to perform simple identification, sensing, actuation, and networking tasks. The more complex data processing and decision making are usually done at the remote server.

This IoT model however inevitability introduces latency between sensing and actuation, which makes it less suitable for latency-sensitive applications, such as autonomous things that are expected to interact with the physical environment autonomously in real time [1]. Furthermore, sensor information exchange incurs heavy network traffic, which hinders the system scaling up in the current network infrastructure [2].

The next-generation IoT devices are expected to deliver intelligent knowledge and adaptive controls beyond simple sensor information processing. They will behave as autonomous or semiautonomous devices that both generate and consume information, evolving into the Internet of Autonomous Things (IoAT) [3]. The autonomous vehicle is an example of IoAT; it is an evidence of the emergence of the next-generation IoT, or IoAT, era.

Autonomous driving has received increasing attention in recent years. With the rapid development of artificial intelligent (AI) and IoT, autonomous driving evolves from a concept that exists in science fiction to concrete IoAT systems that are beneficial to society in numerous aspects.

D. Guo · M. Moh (✉) · T.-S. Moh
San Jose State University, San Jose, CA, USA
e-mail: dapeng.guo@sjsu.edu; melody.moh@sjsu.edu; teng.moh@sjsu.edu

S. Pal et al. (eds.), *Artificial Intelligence-based Internet of Things Systems*, Internet of Things, https://doi.org/10.1007/978-3-030-87059-1_17

First and foremost, autonomous driving improves road traffic safety. According to the National Highway Traffic Safety Administration (NHTSA), 94% of serious crashes are caused by human factors. Autonomous driving removes human factors from road traffic, which reduces road accidents to prevent injuries and save lives [4].

Second, autonomous driving reduces economic loss due to road traffic accidents. NHTSA suggests that billions of dollars have been lost each year because of road traffic accidents, in the form of lost workplace productivity, loss of life, and decreased quality of life. Since autonomous driving improves road traffic safety, it can greatly reduce such economic loss.

Third, autonomous driving reduces traffic congestion. Humans do not drive perfectly. For instance, in dense traffic, a small driver mistake can get amplified to cause traffic congestion. Autonomous driving can provide smoother vehicle control and help the vehicles drive at an optimized speed, which can ease the traffic congestion in the current road infrastructure. And eventually, with the help of IoAT networks, autonomous driving vehicles will be able to share their driving information with the network to enable global traffic optimization for better routing and facilitate coordination between vehicles to further improve safety and reduce congestion.

Finally, autonomous driving enables new applications for vehicles, which brings convenience to daily life. Fully autonomous driving requires no human intervention. This makes self-parking possible that passengers can be picked up and dropped off without worrying about finding a parking spot. Also, it allows the vehicle to be shared by multiple family members as the vehicle can be remotely summoned. Furthermore, it provides people who are not fit to drive, especially, the elder people, and visually impaired people a way for mobility.

The early attempt at autonomous driving dates to the 1930s, where an electric vehicle was designed to trace the electromagnetic fields generated by devices embedded under the road surface. This vehicle indicates a concrete effort toward autonomous driving. However, it did not generate the intelligence needed for the vehicle to interact with a dynamic physical environment. Its feasibility is also in doubt as it relies on a whole new road infrastructure.

In 1986, Ernst Dickmanns' VaMoRs Mercedes van pioneered in using computer vision approaches for autonomous driving. Multiple generations of autonomous driving systems had been tested in this vehicle, which not only demonstrated the capability of computer vision methods but also provided valuable experience in building a small yet powerful autonomous driving system. In 1994, the experience acquired from the 5-ton VaMoRs van was transferred into a VaMoRs-P sedan. The VaMoRs-P is also a computer vision-based autonomous driving system that drives like a human who takes both vision and inertial information into account [7]. Furthermore, a 4-D approach is used in the VaMoRs-P, which utilizes spatial-temporal information. The VaMoRs-P is not simply state of the art at its time; its design ideas have found their way into today's deep learning (DL)–based autonomous driving systems, such as using multiple cameras for visual perception, 3-D object detection, and object tracking. Most importantly, it explores temporal information in the input sequence, which has been underutilized in many modern systems.

In 2004, the Defense Advanced Research Projects Agency (DARPA) held the first DARPA Grand Challenge where 15 teams with their autonomous driving vehicles competed to finish an off-road route [5]. Although no team finished the whole route in 2004, this contest and the contests in the following years are an early push in accelerating the development of modern autonomous driving technology. With decades of advancement in sensor technologies, computing hardware, and deep learning methods, significant progress has been made toward autonomous driving on public roads, especially the ones utilizing deep learning.

Although we are on the edge of entering the era of autonomous driving, numerous vehicle manufactures, and autonomous driving solution providers have revised their roadmap and postponed the deployment of their fully autonomous driving systems. The roadblocks exist in both ethical and technological aspects. Ethically, there are moral dilemmas such as the trolly problem, which is not only difficult for the machine to solve but also hard for humans to reason with. If we cannot create a set of human acceptable moral standards for the machine, the fully autonomous driving vehicles might never be accepted by the public. Technologically, the current autonomous driving vehicles do not meet the criteria of being low cost, low latency, high accuracy, and high reliability.

Currently, we are still at a stage that human attendance and attention is needed when using autonomous driving functions in case the vehicle makes a mistake or needs to fall back to human control. Furthermore, current autonomous vehicles are generally designed to work individually without collaboration within the IoAT network. Each autonomous vehicle will make a reactive decision that is the most suitable for itself based on the information it directly precepts without taking any global information into account, which makes autonomous driving less effective. With IoAT networks, autonomous vehicles can make proactive decisions based on both the direct precepted and remotely provided information from the network to provide safer and smoother vehicle control and better navigation.

The rest of the chapter will be focusing on using vision-based end-to-end deep learning technologies to achieve autonomous driving. We will briefly discuss the ethics of autonomous driving, and then describe the autonomous driving paradigm and deep learning methodologies for autonomous driving. We finally propose a vision-based end-to-end deep learning model that is compatible with the next generation IoT system, accepts navigational command for controllable routing, utilizes temporal information for better vehicle control, and incorporates transfer learning for better generalization. We then compare and analyze its feasibility, strengths, and weakness against existing models.

2 Ethics of Autonomous Driving in IoAT

Technology is a double-edged sword; autonomous driving is no exception, which distributes both well-being and harm [6]. When under situations where harm cannot be avoided, how to distribute the harm becomes a moral dilemma that both the developer of autonomous driving systems and the policymakers are struggling with. Since the moral principle varies in different regions, religions, genders, ages, and even social status, it is difficult to find a set of unified moral standards. Although these scenarios are rare, we need to consider the difference and commonality in ethical preference to design a socially acceptable autonomous driving system.

The Moral Machine Experiment is a research carried out by MIT trying to expose such ethical preferences in both the global level and in the different subpopulation levels to guide in designing such a socially acceptable autonomous driving system. To achieve this goal, the experiment is designed into the form of a website that can be accessed globally by different groups of people. There are nine attributes to be considered, namely, intervention, relation to autonomous vehicles, gender, fitness, social status, law, age, number of characters, and species. The website presents the users with life-threatening moral dilemmas in scenarios created with the nine attributes where the autonomous vehicle needs to make the less evil decision and lets the users judge how the vehicle should react.

With 39.61 million decisions collected from 233 countries, there are multiple levels of findings through the experiment [6]. The first level of findings concerns the preference on a global level. The result shows, in general, that people prefer to save humans over other animals, more characters over fewer, younger over older, lawful over unlawful, and higher social status over lower [6]. The result at the global level points the autonomous driving system designers in a general direction of how the public prefers the autonomous driving systems to be. The second level of finding is about preference based on the individual characteristics of respondents. It finds there is no significant variation in preference between subpopulations, although the tendency to the same preference might be different [6]. The third level of findings suggests in the same culture clusters, such as countries sharing the same cultural origin, people have a closer tendency for the preference. And finally, the preference is highly related to the country the users come from as there are differences in the law, economy, human rights, and health system.

However, other than the preference of the public, we also need to take other factors, such as local regulations, into consideration. For instance, in 2017, the German Ethics Commission on Automated and Connected Driving provided a guideline for autonomous vehicles. Rule number 7 suggests, when facing a dilemma, protecting human lives has a higher priority over other animals, which is consistent with the result of the Moral Machine. However, rule number 9 states that it is prohibited to make decisions based on personal attributes, such as gender and age. This rule conflicts with the preference of the public. Since the variations exist, the designer will need to take both the general preference and local regulations into account.

We find it necessary to consider the ethical preferences and local regulations during the future design phase; however, improving the performance and reliability of the current autonomous vehicle under normal driving circumstances remains a challenge and should be prioritized in research. For the rest of the chapter, we will focus on the technical aspects of the autonomous vehicle.

3 Autonomous Driving Paradigms in IoAT

There are two major paradigms in deep learning-based autonomous driving, namely, the modular pipeline paradigm and the end-to-end learning paradigm. Although the performance of models in both paradigms has been greatly improved with the help of deep learning, each paradigm still has its strength and weakness.

3.1 Modular Pipeline Paradigm

The modular pipeline paradigm follows the divide-and-conquer principle, which has been widely used in robotic fields. Such a model divides the complex autonomous driving problem into clearly defined stages and subproblems that are easier to solve. Then, each subproblem is solved and optimized separately. Finally, the related subproblems are connected through pipelines with the assumption that the optimal output from the previous stage will be the optimal input for the next stage [7]. It is the most widely adopted approach since the early attempts on autonomous driving, including the VaMoRs-P mentioned above.

Since each module is only responsible for a well-defined subproblem, the modules can be distributed to developers with different expertise and tackled independently. Because the subproblems are well defined, it is easier to understand and explain the inner working and behaviors of the system. This approach also allows each module to be easily modified to meet specific requirements without affecting the performance of unrelated modules. However, human-defined subproblems and features may not be optimal for the overall system. Developing such a system often requires expert knowledge in the field of each subproblem, and the system may get very complex.

A generalized modular pipeline model is shown in Fig. 1. Four major modules are residing in four stages in this model. In the first stage, the perception module performs tasks such as objection detection, objection tracking, and semantic segmentation on inputs from a fusion of sensors including LiDARs, cameras, and ultrasonic radars. In the next stage, the localization module performs its tasks with input from GNSS, IMU, HD Maps, and the output of the perception module. The planning module is in the next stage, which consumes the outputs from both the perception module and localization module for tasks such as trajectory planning. Finally, in the last stage, the control module will turn the output from localization and planning modules into vehicle control signals of steer, accelerate, and brake.

Multiple works fall into the modular pipeline paradigm [8–13]. However, due to the high complexity of modular pipeline models, much of the works reviewed only focus on improving upon a specific subproblem, rather than providing a complete pipeline to drive a vehicle.

Nevertheless, [8] proposed a computer vision-based direct perception model, which falls in the modular pipelined paradigm but with part of it being like the

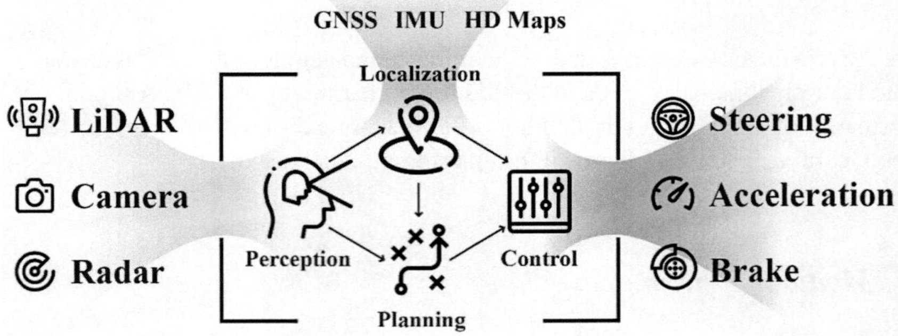

Fig. 1 A modular pipeline model

end-to-end learning paradigm. Instead of recognizing and tracking all driving-related objects to create a representation of its surrounding environment and make decisions based on all the information, [8] argues that a subset of this information is sufficient for the driving task and computing all the information increases the complexity. Therefore, a direct perception approach is proposed where meaningful indicators such as the angle of the car, distance to lane mark, and distance to adjacent cars are generated from the input image using convolutional neural networks (CNN) following end-to-end learning fashion. Then, based on the indicators, a closed-form controller is used to drive the vehicle. The model is mainly tested on recognizing and predicting the indicators, as well as generating steer control accordingly. The results indicate the model is relatively capable in evaluations with the KITTI dataset and the simulator. However, the major flaw is that the indicators this model predicts are handcrafted; in situations where the handcrafted indicators do not exist in the input, the proposed approach cannot work properly.

Both camera and LiDAR have been used as the main perceptual source for autonomous driving. Compared with cameras, LiDAR is considered expensive, which makes the autonomous vehicle less affordable. However, LiDAR-based 3-D object detection has higher accuracy compared with image-based 3-D object detection. The performance gap has been a roadblock that hinders the development of low-cost imaged-based autonomous driving. Generally, the performance gap between the two approaches is considered caused by the error of depth estimation grows quadratically with distance in the image-based approach while it only grows linearly with the LiDAR-based approach.

However, [9] suggests differently that the representation of data is a major cause of the performance gap. By converting the image-based depth estimation, which is usually represented as an additional channel of the image into a 3-D point cloud pseudo-LiDAR representation, the result of pseudo-LiDAR data is considered very similar to the ground truth LiDAR data. Furthermore, with the pseudo-LiDAR method, any 3-D object detection method designed for point cloud can be applied to image data, which provides more flexibility. The test results confirm the pseudo-LiDAR approach combined with point cloud-based 3-D detection methods outperforms the baseline state-of-the-art image-based 3-D detection methods, especially in the bird-eye-view format. However, LiDAR-based methods still outperform pseudo-LiDAR-based methods, especially in the far distance cases.

The work [10] proposed a flexible pipeline that applies any existing 2-D object detection method to 3-D object detection tasks. The state-of-the-art 2-D object detection methods have achieved very high accuracy. This proposed pipeline enables us to transfer the success in 2-D object detection into 3-D object detection. The pipeline is a LiDAR and camera fusion approach where one branch of the pipeline performs 2-D object detection on an image to produce a 2-D bounding box. Then, the 2-D bounding box is projected into the 3-D point cloud to select a subset of the points. Since a 2-D bounding box in 3-D space can contain points from different distances and different objects, the model proposes multiple 3-D bounding boxes to be fitted on three generalized vehicle models. The proposal with the highest fitting score will be further fine-tuned by another CNN to form the final bounding box. This pipeline ranks second among the 3-D object detection algorithms in the KITTI benchmark at its time. Although the performance is good on the KITTI dataset, there are many more types of vehicles and objects related to driving in the real world. The generalized vehicle model is handcrafted using a 3-D CAD dataset; this implies that to make the model able to detect other objects in the real world, we will need to manually create 3-D CAD models for many more objects. This will make it difficult to apply this pipeline in a more general term.

In [11–13], the research focuses on improving the speed of point cloud-based object detection. Both [12, 13] use bird-eye-view (BEV) representation of the point cloud data to better explore the depth information. In [11], it proposes to run a fully convolutional network on a 2-D projection of the frontal-view point cloud with dilated convolution to increase the receptive field. However, in [12], the work suggests a different approach that using dilated convolution will cause checkerboard artifacts in a higher-level feature map, which will decrease performance. Instead, [12] increases the number of convolutional layers and combines the residual of different convolution layers to reduce the detail loss. The work [13] suggests, directly performing 3-D convolution on dense data, is slow. Only certain parts of the data need to be processed, such as we only care about objects on the road. Therefore, the work proposed to divide the input into voxels and use a computation mask to make the model focus on objects on the road, which reduces the amount of computation needed.

3.2 *End-to-End Learning Paradigm*

On the other hand, models in the end-to-end paradigm use one single deep neural network that handles the complete self-driving task without any intermediate step. Such models directly map the inputs from sensors to vehicle control signals, without the need to understand any human-defined steps, such as objection detection,

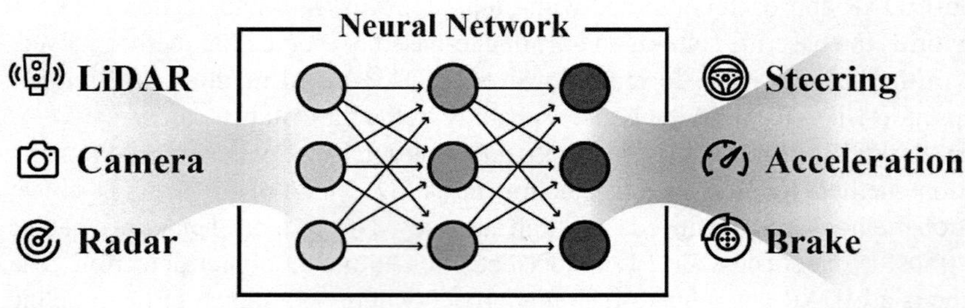

Fig. 2 An end-to-end learning model

path planning, etc. The assumption is that, with the appropriate deep learning model and training method, end-to-end learning will learn the internal features that are most suitable and optimize all the processing steps simultaneously, which will eventually lead to better performance. With the advancement of deep learning, end-to-end learning models can be much simpler in structure compared with modular pipeline models while maintaining comparable performance. Moreover, end-to-end learning models are also very flexible; the structures can be easily modified for improvement and adaption for new features.

A generalized end-to-end learning model is shown in Fig. 2. The model uses LiDARs, cameras, and ultrasonic radars as the main perceptual sensors. The inputs are directly mapped to vehicle control signals using a deep neural network.

Ever since Nvidia demonstrated that their image-based end-to-end deep learning autonomous vehicle can successfully drive itself on public roads in long-distance, the end-to-end learning paradigm has received an unprecedented amount of attention. Compared with modular pipeline models, end-to-end models are usually much simpler, and it is easier for small research groups to work on. Furthermore, most end-to-end models are paired with cameras rather than LiDARs; this makes such systems more affordable and easier to implement. Here, we are focusing on models using end-to-end deep learning technologies for computer vision in autonomous vehicles [14–25]. Even with simpler structures and lower sensors cost, the models surveyed achieve performance improvement in different ways.

Before we dive into the models, we need to understand another important component of the end-to-end approach, that is how the models are trained. Most end-to-end learning models in this section are trained with imitation learning, including our proposed models. Imitation learning is a type of supervised learning where the model focuses on imitating expert demonstrations. This training method is very intuitive and effective that, since humans can drive, if the model can mimic human actions, it will also be able to drive. In terms of autonomous driving, the expert demonstrations used as training data are easy to collect and label. We can collect the training data by recording the observations from the cameras, the corresponding control signals from the vehicle, and optionally, the navigational intentions from the driver. By syncing these data, the observations are implicitly labeled. Another way to train the end-to-end model is by using reinforcement learning. However, the action space for autonomous driving is continuous and large; it is inefficient to train the models using this way as it takes too long to converge [25].

With the recent advancement in computing hardware and labeled dataset, CNN has been widely adopted in pattern recognition tasks. In 2016, Nvidia published the research [14] to demonstrate that by using imitation learning, CNN can also be extended to autonomously steering a vehicle based on RGB images efficiently and effectively. In 2017, Nvidia published follow-up research [15] providing more details and internal feature analysis of their previous model. The visualization of the activations in the intermediate layer shows that the model not only implicitly learns to recognize driving-related objects that are included in training data, such as traffic signs, road lanes, vehicles, and unmarked road boundaries, but also recognizes driving-related objects that are not included in the training data, such as a construction vehicle exiting a construction site. Research [14] and [15] help shape the recent research trend in autonomous driving as we see an increasing number of researchers are utilizing end-to-end approach trained with imitation learning.

Although the following works improve end-to-end autonomous driving in different ways, oftentimes, they are improved upon each other and share much in common. For instance, the core structures are very similar as they all use variations of CNN as the main perceptual network and utilize imitation learning as part of the training process [14–25]. Furthermore, in [14] and [15], the authors proposed to collect training images by placing three front-facing cameras: left, middle, and right on top of the vehicle. The images captured by the left and right cameras are used to augment the training set. By offsetting the steering angles, such data teaches the car how it should recover itself when driving off-center. This method is also widely adopted in other research works [14, 15, 17–21, 23–25] and is critical in improving online performance [19]. Early research works mostly focus on autonomously lane keeping; the models proposed do not accept external navigational commands, where the vehicles will be roaming on the street and make random turns at intersections [14–16, 22]. In [18], the author suggests that autonomous vehicles need to accept external navigation commands and make meaningful turns at the intersections to get to a destination. Therefore, a high-level command is used to indicate the navigation intention, which is either integrated as the input to the model or as a switch to select a corresponding action branch [17–21, 23–25].

In an end-to-end learning model, we cannot precisely divide the model into different functional modules. However, by feature analysis, we know certain parts of the network are mainly learning image-related features and we refer to this part of the model as the perceptual module [14, 15]. And we also refer to the parts of the network that consumes the output of the perceptual module and mainly be used to generate control signals or other types of predictions as the prediction module. The performance of the model is decided by if the perceptual module can correctly and precisely capture the driving-related features, and if the prediction module is capable of reason with the features captured. In [14] and [15], the author proposed a perceptual module consisting of six convolutional layers, and a prediction module with three fully connected layers. The variation of this structure is widely used in the following works.

Testing is an important step in delivering reliable autonomous driving vehicles. Before an autonomous driving system can be road tested, the system is usually required to be tested in a simulated environment. By letting the model interact with the simulator, which creates a realistic imitation of the real world, we not only can evaluate how well the model performs, but also avoid the risk of running an immature system on the road. In [14], a rather primitive simulation method is used. The simulator takes prerecorded video clips and feeds the first frame into the end-to-end model, then transforms the following frame of the video and the ground truth control signals according to the predicted control signals to simulate vehicle movements. Since the image synthesized is simply shifting and rotating the image, there is no simulation of the physics such as object collision, inertia, or friction; therefore, it cannot accurately simulate the vehicle driving dynamic. Furthermore, we cannot customize the simulation configuration to change parameters such as weather, lighting, or traffic, as the simulation is limited to prerecorded videos only.

Autonomous driving powered by deep learning also requires a large amount of training data. How to collect large amounts of high-quality training data with high variety and how to test the system in a realistic yet safe environment becomes a problem every autonomous driving project needs to deal with. CARLA is an open-source driving simulation platform specifically designed for the development, training, and validation of autonomous driving systems [17]. It supports customizable environment and road layouts where users can test both urban and highway driving. It provides a flexible set of sensors including RGB images, LiDAR point clouds, semantic segmentation cameras, and depth sensors to enable more types of autonomous driving systems. It also allows the users to dynamically change the weather and lighting conditions, which helps the model better generalize in the dynamic environment. CARLA simulator has been successfully utilized in work [17–21, 23–25] to train and validate autonomous driving models with both imitation learning and reinforcement learning.

In [17], the work not only introduces the CARLA simulator and its functions, more importantly, but also presented a baseline performance comparison of three image-based models, namely, the modular pipeline (MP), imitation learning (IL), and reinforcement learning (RL) models. This research briefly introduces each of the three models where the MP model uses a semantic segmentation network based on ResNet pretrained on ImageNet and a waypoint planner. The IL model uses a CNN model like [14, 15] with the addition of a speed measurement module concatenated to the output of the image network. There are four control signal prediction branches selectively activated by navigational high-level commands (HLC). The RL uses A3C style training methods. The result suggests that when comparing the IL with MP, the performance of the two approaches is similar in most testing conditions. However, the IL performs better in lane-keeping, especially in a new testing environment, while the MP performs better in avoiding collision with obstacles. When comparing IL with RL, IL largely outperforms RL in most of the cases even if the RL is trained for 12 days. The work suggests this is due to urban driving with continuous action space is much more difficult than problems previously solved by RL; thus, the model training does not converge well.

The research [18] is a follow-up study of [17], focusing on the imitation learning method used in [17], which is referred to as conditional imitation learning (CIL). This paper also focuses on incorporating the navigational command into the model so the vehicle will follow a specific route in driving. Following navigational commands is an important feature in real life as we do want the vehicle to be able to get from a specific point A to point B rather than randomly cruising in the city. Two network structures are being compared in this work, namely, the command input model and the command conditional model.

There are three input modules in the command input model, for instance, the image perceptual module, the measurement module, and the command module. In both networks, the image perceptual modules are the same, which is a CNN utilizing dropout and batch normalization to reduce overfitting. The measurement modules used in the two models are also the same. It is a fully connected network that takes the speed of the vehicle and concatenates the output with the flattened output of the image module.

In the command conditional model, the control module consists of four control branches of fully connected networks, one for each of the HLC. The high-level navigational command is used as a switch to choose, which control branch to train or to make predictions on. While in the command input model, the HLC is fed as an input in the form of one-hot encoding vector into a fully connected command module, and the output is concatenated with the output of the image module and measurement module. Then a single control prediction module is used to produce a control signal.

To improve generalization, image data augmentation is used for both networks where a subset of image transformations including changes in contrast, brightness, tone, and the addition of Gaussian blur, Gaussian noise, salt-and-pepper noise, and region dropout are performed with randomly sampled magnitude. The two models are trained with imitation learning on human driving data collected in the CARLA simulator. Since human driving is mostly driving in the centerline, the authors introduce motion drifts during the data collection process and only records the recovery process of the human driver to make the dataset more balanced and helps the model to learn how to recover from mistakes. After the two models are trained with static images, they are tested in an unseen map during the simulation. The average distance per infraction and the success rate of finishing the route is recorded. The result suggests, the two models utilizing HLC proposed in this work perform much better than models that do not use the HLC or only use the direction of destination as input. The result also suggests the command conditional model follows the HLC better and achieves a higher route success rate while the command input model makes less driving mistakes and performs better in distance per infraction metric [18].

Evaluation of an end-to-end autonomous driving system trained by imitation learning is tricky. Since imitation learning can be categorized as supervised learning, the most natural way to evaluate such a model is by using its validation loss, which in most of the time; its mean squared error (MSE). However, [19] suggests, a model with excellent offline evaluation results may still perform poorly in an online simulation. The insight is that MSE has a very weak correlation with online performance results; thus, it is not a good metric to evaluate a model. Through the experiment, there are two methods found to improve offline evaluation effectiveness. The first is to carefully select the evaluation data used, where the data should contain examples of recovery from multiple kinds of mistakes. The other takeaway is that we need to choose the evaluation metric wisely; for example, the mean absolute error (MAE) has a stronger correlation to online performance; it might be a is choice when compared with MSE.

4 Proposed Model for Autonomous Driving in IoAT

In this work, we propose a vision-based end-to-end autonomous driving model that utilizes long short-term memory (LSTM) network to explore temporal information for better vehicle control, accepts HLC for meaningful navigation and incorporates speed regularization to better learn speed related features. The proposed model, namely, the temporal conditional imitation learning model is a low-cost system that focuses on improving the accuracy and reliability, while maintaining low latency and enabling global optimization in the IoAT network.

The structure of the temporal conditional imitation learning model is shown in Fig. 3. The model is based on the CNN-LSTM network structure, which is often used in sequence prediction tasks such as video sentiment classification, but with multiple modifications and improvements.

First, in terms of input handling, the proposed model is similar to the CIL command input model where three input modules are used, and the only perceptual sensor is the RGB camera [18]. Instead of a shallow CNN used in the CIL models,

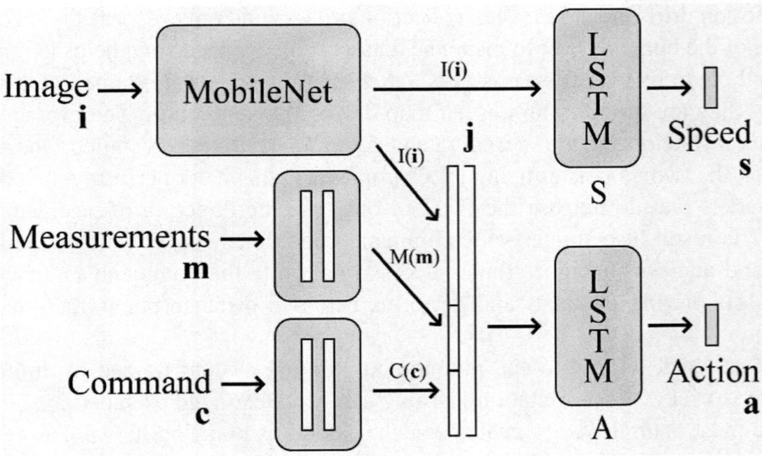

Fig. 3 Temporal conditional imitation learning model

the image perceptual module uses MobileNet, which is significantly deeper and more complex to better learn the image features. Furthermore, the MobileNet is pretrained on ImageNet, which allows us to take advantage of transfer learning to further improve the generalization ability of the image perceptual module.

A more complex and deeper network, however, does not imply that we will sacrifice the model's inference speed. By using depth-wise convolution, MobileNet is specifically designed to run on low computation power devices efficiently; therefore, it is suitable for IoAT applications where relative limited computing power is equipped. In this way, we run a more capable model on the vehicle itself to improve the reliability, without the need for extremely high-performance computing hardware, or offload the work to the cloud, which will add a significant amount of latency.

The measurement module and the command module are kept the same as the CIL command input model, where two fully connected layers are used for each module and with the HLC input being one-hot encoded. The HLC is organized into four types, each corresponding to a navigational option: following the lane, turn left at the intersection, turn right at the intersection, and go straight. The HLC is a way for humans and other systems to communicate the navigational intention to the autonomous driving system.

In our model, the HLC can be generated from both inside the vehicle or the cloud. From inside the vehicle, we have a local path planner when given a destination, it generates a route and the turn-by-turn HLCs to guide the vehicle to the destination. From outside the vehicle, we have a global traffic optimizer on the cloud, which analyzes all vehicle locations in the IoAT network and the traffic states, it can optimize the vehicle's route by sending turn-by-turn HLCs and alter the route in real time. Unlike observation inputs, which need to be processed dozens of times every second, HLCs can be updated in much longer intervals; thus, the network delay will not affect the vehicle performance as long as the HLC arrives in time before the turns.

The outputs of three input modules are concatenated, then fed into the action prediction module. In addition to the improvements in the image perceptual module, we also replace the fully connected action prediction module with an LSTM network, which allows us to explore the temporal information that exists in input sequences [23]. As we stated, the action from a previous time step has an impact on the next time step, and there is information that can only be retrieved from input in a sequence of time steps, such as the sense of relative movement and sense of speed. By exploring temporal information, the model will be able to understand the dynamic environment better and make a better decision.

To further improve vehicle performance in the dynamic world, we incorporate speed prediction regularization [24]. We jointly trained an LSTM speed prediction module connected to the image perceptual module, which forces the image perceptual module to learn speed-related features, thus the overall model will not overly rely on the speed measurement from the input, and it will learn a better sense of speed, which is extremely useful in avoiding accidents. For our proposed model, it is trained with imitation learning and mean absolute error (MAE) as the loss function.

5 Analysis of Deep Learning Models for Autonomous Driving

In this section, we provide a qualitative analysis and comparison of several main models that have been proposed for autonomous driving, including the one we proposed model, as presented in Sect. 4. Detailed experiments, including dataset and training, and performance results have been reported in another publication [26].

5.1 The Autonomous Driving Paradigm

In [9–13], the works follow the modular pipeline paradigm where each work only solves a subproblem of the whole driving task. And in our case, models in [9–13] mostly focus on achieving efficient object detection by using deep learning models. Research [8] falls in between the modular pipeline paradigm and the end-to-end learning paradigm. The deep learning part of the model can be considered as end-to-end learning, as it uses one single deep learning network like the one in [14, 15] to predict driving-related affordance. Then, a closed-form controller is used to control the vehicle according to the affordance predicted, which makes the overall model modular. For [14–25], the models follow the end-to-end learning paradigm; the autonomous driving system is represented by one single model that directly maps the perceptual and measurement inputs to vehicle control signals. For instance, in [14–16], a simple CNN network is used, and the only input is an RGB image. In [17–22, 24, 25], the models add more input branches on top of [14–16] to incorporate measurement of speed, depth, and HLC.

Our proposed model also follows the end-to-end learning autonomous driving paradigm. In this work, we improved upon [17, 18] where a CNN-LSTM network is used to map image, speed, and HLC inputs directly into vehicle control signals. Compared with modular pipeline models, the complexity of the proposed end-to-end approach is more manageable, thus more suitable for individual research. However, it is difficult to understand what the model has learned; thus, it might be helpful to visualize the activations of the inner layers of the model and analyze what kind of features have been learned.

5.2 The Perception Source

Multiple types of sensors have been used as the perception source in the previous works, including LiDAR, RGB camera, and depth camera. LiDAR uses laser light to illuminate the target and measures the reflection. By calculating the time difference, LiDAR can estimate the distance between the sensor and the target points, therefore, create a 3-D representation of the environment it surveyed. The LiDAR data is commonly represented by a point cloud, and it is often used in models that require an accurate 3-D representation of the world. For instance, in [10–13], LiDAR is used for 3-D object detection where driving-related objects are marked in bounding boxes. However, autonomous driving requires real-time 3-D objection detection on the high-resolution point cloud, which requires many computing resources onboard the vehicle which is expensive and hard to achieve. This explains why [10–13] focus on both accuracy and efficiency when performing 3-D objection detection.

The other common perception sensor is the RGB camera which has been used as the main perception source in all end-to-end learning models surveyed [14–25]. The camera technology has been developed and tested for decades, and the cameras feature high resolution, high frame rate images that are closer to how humans see the world and are massively available on the market. Compared to LiDAR and multisensor fusion models, monocular camera-based models have a simpler hardware structure, which results in lower implementation costs for both development and deployment.

The model proposed in this work is computer vision-based where the primary perception source is a center-mounted RGB camera. Since we are imitating human driving behaviors, we would like the model to perceive the world as close as to the way a human does. By using RGB cameras, we can keep sensor cost and complexity low, which makes the model more feasible to deploy in the real world. Furthermore, processing 2-D RGB images require less computing power compared with 3-D point cloud data, which makes the model more suitable in IoAT applications that have relatively limited computing power compared to the cloud.

However, the RGB cameras are easily affected by weather and lighting conditions compared with LiDAR. It requires large amounts of image augmentation in the training process to help the model generalize. There is also information that is difficult to be extracted from monocular RGB images, such as depth information. Therefore, research [20] uses a depth camera as a supplement to improve performance. Finally, the proposed model and other end-to-end models surveyed use only one center-mounted camera as the perceptual source during driving. Since the field of view is fixed, the blind spot may limit the models' ability to avoid crashes during turns.

5.3 The Training Method

Imitation learning is the most popular training method used in the end-to-end learning paradigm, where the models try to imitate expert demonstrations. The models in [14–24] are all trained with this method. To help the model better generalize with imitation learning, we need to provide a training dataset with as many varieties of driving scenarios as possible, including different weather, lighting, road type, and vehicles. Besides, we also need the dataset to be balanced on driving actions including demonstrations for the four HLC types and recovery from mistakes. In [25], the work combines imitation learning with reinforcement learning to further improve the driving performance and reduce training time for reinforcement learning. The work uses imitation learning to pretrained the deep neural network to achieve autonomous driving, then uses the trained weight as the initial weight in reinforcement learning to fine-tune the model, especially for the driving scenario not included in the imitation learning training data.

The proposed models are trained with imitation learning, where the models try to imitate expert demonstrations recorded in the CARLA simulator. In our case, the expert demonstrations are recorded from both the human driving and a built-in autopilot function that drives the vehicle in the center of the lane. The human data is mainly used to show the model how to recover from a mistake, while the autopilot data is used to teach all other normal driving situations.

In our work, we choose to use imitation for its simplicity and efficiency. Compared with reinforcement learning, we only need to collect driving footage, then simply trained the deep learning model in a supervised way. And since the action space for autonomous driving is continuous and large, it may take a very long time in training, and it is not guaranteed to converge well. However, the performance of models trained by imitation learning is limited by the training data quality. In the best case, the model can be as good as the expert demonstrations. If there are cases not covered in the training data, the model may perform poorly.

5.4 Controllable Routing

Accepting navigational command is a key part to achieve autonomous driving. Although we intuitively describe end-to-end autonomous driving is a mapping from the observation of camera images and measurements to control signals that drive the vehicle, this mapping does not hold if we want to build a model that will meaningfully navigate the road and drive from a specific starting point to a specific endpoint on map. It not only requires a model to generate one set of control signals to operate the vehicle safely, more importantly, for the same observation, but also requires the model to generate different sets of control signals corresponding to different navigational intentions. Early works like [14–16] do not take navigation command into account, and the model will make random turns at intersections. To help guide the vehicle meaningfully navigate the road, [18] proposed to use HLC to communicate the navigational intention. In [18], the author proposed two ways to incorporate HLC. The first one is a branched network using HLC as a switch, where corresponding prediction branches in the network are activated by HLC to guide the vehicle. The results indicate such a branched network performs the best in their test. This branched approach is adopted in most of the following research [19–21, 24, 25]. The other way is to use the HLC as an input, which is adopted in [23].

From our test, we find the input approach, which uses the HLC as an input to the model results in a more table trained model. Our proposed model takes HLC as an input in the form of one hot vector and use the same prediction branch for all HLCs.

In this input approach, the prediction branch is trained with a more variety of driving data, which helps it generalize better to unfamiliar scenarios. By incorporating HLC, it not only allows the human or local planner to communicate the navigational intention to the vehicle, but also allows the traffic optimization servers in IoAT networks to alter the route of the vehicle to avoid or reduce traffic.

5.5 *Exploring Temporal Information*

In [8], the author claims that end-to-end models that map input from a single time step to the output cannot understand the driving task well as the action in the previous time step has an impact on the action for the next time step. Although the perceptual inputs for the current time step may be similar in different scenarios, the action for the next time step can be very different; thus, it is hard to make a correspondent decision without knowing the previous states. For instance, when following a vehicle, we need to know the relative movement between the agent vehicle and the other vehicle to better control the speed. In [23], this problem is partially addressed that a CNN-LSTM structure is used, which will generate control signals based on five previous states.

Our proposed model also utilizes the LSTM network to explore the temporal information, which only exists in a sequence of observations. Our model uses five previous states to make one prediction. Temporal information is critical to improving the reliability of the models as it helps the model to learn the sense of speed and relative movement. The proposed model takes this idea one step forward by incorporating speed regularization, which forces the perceptual module to learn features about speed. Both features help the model to make a better judgment in a dynamic environment, especially avoiding crashing.

However, using LSTM also incurs more computation in both the training stage and testing stage. For instance, we use five previous observations, the training time is approximately five times that when using one observation. The increased computation also affects inference speed; slightly more powerful computing hardware is needed. Furthermore, since the five observations are sampled from a buffer with an equal time interval, we need to make sure the sampling time interval remains unchanged in testing.

5.6 *Image Perception Network*

In [14–23, 25], the image perception networks are mainly variations of CNN trained from scratch. And the depth of such image perception networks is usually shallow, and the size is small. Although it is easy and fast to train such networks, it is also easy to reach their limitation, especially they do not generalize well to unfamiliar environments. To improve the performance of an image-based end-to-end autonomous driving model, a more capable image perception network is needed. In [24], a ResNet34 pretrained on ImageNet data is used as the perception network; together with speed regularization, the models outperform most models that use self-trained shallow CNN in CARLA benchmark at its time.

Our proposed model uses MobileNet pretrained on ImageNet as the main perceptual network. MobileNet is a lightweight and efficient variation of CNN, which has about four million parameters compared with 22 million parameters in ResNet34. MobileNet is designed to run on low-powered devices such as mobile devices with little impact on accuracy. Since we are combining a deeper and more capable image perception network with LSTM, we strive to keep the computation requirement of the image perception low. The high efficacy and accuracy of MobileNet make it a suitable choice for our autonomous driving model.

5.7 Flexibility

As we have seen in a series of research in end-to-end autonomous driving systems, a core structure [14] has been evolved into multiple forms to incorporate HLC [18, 21], depth information [20], temporal information [22, 23], and pretrained image network [24]. Our proposed model follows the same core structure and already incorporates HLC, temporal information, and a pre-trained image network. It has the flexibility and can be easily modified to accept new features, such as depth information from the sensor, traffic information from the IoAT network, and even vehicle-to-vehicle information in the future. With more information we get from IoAT networks, the model will be able to make proactive decisions based on inputs that cannot be directly perceived by the onboard sensors, which will further improve the reliability of the system.

6 Conclusion and Future Research Directions

This chapter puts together several missing pieces when autonomous driving models were previously developed. We first provide an overview of research in autonomous driving, specifically on the use of end-to-end, computer vision-based deep learning methods. The ethics of autonomous driving has been briefly presented, followed by major autonomous driving paradigms and end-to-end deep learning methods. Besides, we have proposed a vision-based, end-to-end deep learning model that is high reliable, high accuracy, low latency, and low cost, suitable to be deployed in IoAT networks. The model accepts external HLC from a local planner or a remote server, which takes advantage of the global traffic optimization provided by IoAT network. Detailed analysis and comparison of major models of deep learning for autonomous driving have been described. Comparing with the previous works, the proposed model uses a deeper and yet more efficient image perceptual network. The perceptual network is pretrained on ImageNet, and we take advantage of transfer learning to convert the previously learned knowledge to help the model better capture the driving-related features. It explores temporal information in a sequence of inputs to capture the temporal dependency in the dynamic environment, together with speed regularization, the model can learn a sense of relative movement and speed, which helps the model make a better judgment in a dynamic environment.

Many improvements may be applied to the existing research in end-to-end learning for autonomous driving. Firstly, a model trained by imitation learning can only be as good as its expert demonstration. This prevents us from further improving the model performance, even if the model is still capable of learning. In the future, we can adopt the idea of using transfer learning to transfer the knowledge learned from imitation learning to reinforcement learning [25]. By using the trained network weights from imitation learning as the initial weights, there would be a great chance to reduce the time for the model to converge in reinforcement learning; therefore, the performance can be future improved or fine-tuned.

Also, we can incorporate depth information into the proposed model, by either using the depth as an input or jointly train a depth prediction module, to allow the perceptual network to pay attention to depth-related features. Furthermore, there are other ways to explore temporal information, such as using ConvLSTM or 3DCNN networks. Additionally, the model can take advantage of more information provided by the IoAT network. For instance, it might be difficult to recognize traffic light states in certain weather or may be due to the camera's view being block by other vehicles. If the traffic light states are provided by the IoAT network, we would be able to make correct decisions to improve the reliability even when we cannot directly perceive such information.

References

1. Lu, H., Liu, Q., Tian, D., Li, Y., Kim, H., & Serikawa, S. (2019). The cognitive internet of vehicles for autonomous driving. *IEEE Network, 33*(3), 65–73. https://doi.org/10.1109/mnet.2019.1800339
2. Marosi, A. C., Lovas, R., Kisari, A., & Simonyi, E. (2018). A novel IoT platform for the era of connected cars. 2018 IEEE international conference on future IoT technologies (Future IoT). https://doi.org/10.1109/fiot.2018.8325597
3. Keeley, T. O. M. (2016, April 30). IoT to IoAT: Internet of Autonomous Things devices provides solutions. https://www.controleng.com/articles/iot-to-ioat-internet-of-autonomous-things-devices-provides-solutions
4. NHTSA. (2020, June 15). Automated Vehicles for Safety. NHTSA. http://www.nhtsa.gov/technology-innovation/automated-vehicles-safety
5. Behringer, R., Sundareswaran, S., Daily, R., Bevly, D., Gregory, B., Elsley, R., … Guthmiller, W. (2004). The DARPA grand challenge – development of an autonomous vehicle. *IEEE Intelligent Vehicles Symposium*. https://doi.org/10.1109/ivs.2004.1336386
6. Awad, E., Dsouza, S., Kim, R., Schulz, J., Henrich, J., Shariff, A., … Rahwan, I. (2018). The moral machine experiment. *Nature, 563*(7729), 59–64. https://doi.org/10.1038/s41586-018-0637-6

7. Janai, J., Güney, F., Behl, A., & Geiger, A. (2020). Computer vision for autonomous vehicles: Problems, datasets and State of the Art. *Foundations and Trends in Computer Graphics and Vision, 12*(1–3), 1–308. https://doi.org/10.1561/0600000079
8. Chen, C., Seff, A., Kornhauser, A., & Xiao, J. (2015). DeepDriving: Learning affordance for direct perception in autonomous driving. *2015 IEEE international conference on computer vision (ICCV)*. https://doi.org/10.1109/iccv.2015.312
9. Wang, Y., Chao, W.-L., Garg, D., Hariharan, B., Campbell, M., & Weinberger, K. Q. (2019). Pseudo-LiDAR from visual depth estimation: Bridging the gap in 3D object detection for autonomous driving. *2019 IEEE/CVF conference on computer vision and pattern recognition (CVPR)*. https://doi.org/10.1109/cvpr.2019.00864
10. Du, X., Ang, M. H., Karaman, S., & Rus, D. (2018). A general pipeline for 3D detection of vehicles. 2018 IEEE international conference on robotics and automation (ICRA). https://doi.org/10.1109/icra.2018.8461232
11. Minemura, K., Liau, H., Monrroy, A., & Kato, S. (2018). LMNet: Real-time multiclass object detection on CPU using 3D LiDAR. 2018 3rd Asia-Pacific conference on intelligent robot systems (ACIRS). https://doi.org/10.1109/acirs.2018.8467245
12. Yang, B., Luo, W., & Urtasun, R. (2018). PIXOR: Real-time 3D object detection from point clouds. 2018 IEEE/CVF conference on computer vision and pattern recognition. https://doi.org/10.1109/cvpr.2018.00798
13. Ren, M., Pokrovsky, A., Yang, B., & Urtasun, R. (2018). SBNet: Sparse blocks network for fast inference. 2018 IEEE/CVF conference on computer vision and pattern recognition. https://doi.org/10.1109/cvpr.2018.00908
14. Bojarski, M., Del Testa, D., Dworakowski, D., Firner, B., Flepp, B., Goyal, P., … Zieba, K. (2016, April 25). End to end learning for self-driving cars. *Preprint submitted to arXiv, Nvidia*. https://arxiv.org/abs/1604.07316
15. Bojarski, M., Yeres, P., Choromanska, A., Choromanski, K., Firner, B., Jackel, L., & Muller, U. (2017, April 25). *Explaining how a deep neural network trained with end-to-end learning steers a car*. Preprint submitted to arXiv. https://arxiv.org/abs/1704.07911.
16. Chen, Z., & Huang, X. (2017). *End-to-end learning for lane keeping of self-driving cars*. 2017 IEEE intelligent vehicles symposium (IV). https://doi.org/10.1109/ivs.2017.7995975.
17. Dosovitskiy, A., Ros, G., Codevilla, F., Lopez, A., & Koltun, V. (2017, November 10). *CARLA: An open urban driving simulator*. Proceedings of the 1st Annual Conference on Robot Learning 1–16.
18. Codevilla, F., Muller, M., Lopez, A., Koltun, V., & Dosovitskiy, A. (2018). End-to-end driving via conditional imitation learning. 2018 IEEE International Conference on Robotics and Automation (ICRA). https://doi.org/10.1109/icra.2018.8460487
19. Codevilla, F., López, A. M., Koltun, V., & Dosovitskiy, A. (2018). *On offline evaluation of vision-based driving models* (Computer Vision – ECCV 2018 Lecture Notes in Computer Science) (pp. 246–262). https://doi.org/10.1007/978-3-030-01267-0_15
20. Xiao, Y., Codevilla, F., Gurram, A., Urfalioglu, O., & Lopez, A. M. (2020). Multimodal end-to-end autonomous driving. *IEEE Transactions on Intelligent Transportation Systems*, 1–11. https://doi.org/10.1109/tits.2020.3013234
21. Wang, Q., Chen, L., Tian, B., Tian, W., Li, L., & Cao, D. (2019). End-to-end autonomous driving: An angle branched network approach. *IEEE Transactions on Vehicular Technology, 68*(12), 11599–11610. https://doi.org/10.1109/tvt.2019.2921918
22. Chi, L., & Mu, Y. (2017). Learning end-to-end autonomous steering model from spatial and temporal visual cues. Proceedings of the workshop on visual analysis in smart and connected communities – VSCC '17. https://doi.org/10.1145/3132734.3132737
23. Haavaldsen, H., Aasbø, M., & Lindseth, F. (2019). Autonomous vehicle control: End-to-end learning in simulated urban environments. Communications in Computer and Information Science Nordic Artificial Intelligence Research and development, 40–51. https://doi.org/10.1007/978-3-030-35664-4_4

24. Codevilla, F., Santana, E., Lopez, A., & Gaidon, A. (2019). Exploring the limitations of behavior cloning for autonomous driving. 2019 IEEE/CVF international conference on computer vision (ICCV). https://doi.org/10.1109/iccv.2019.00942
25. Liang, X., Wang, T., Yang, L., & Xing, E. (2018). CIRL: Controllable imitative reinforcement learning for vision-based self-driving. Computer vision – ECCV 2018 lecture notes in computer science, 604–620. https://doi.org/10.1007/978-3-030-01234-2_36
26. Guo, D., Moh, M., & Moh, T. (2020). *Vision-based autonomous driving for Smart City: A case for end-to-end learning utilizing temporal information*. 5th international conference on smart computing and communication (SmartCom 2020).

A Study on the Application of Bayesian Learning and Decision Trees IoT-Enabled System in Postharvest Storage

Akinola S. Olayinka, Charles Oluwaseun Adetunji, Wilson Nwankwo, Olaniyan T. Olugbemi, and Tosin C. Olayinka

1 Introduction

Agriculture as a major player in national economies globally is threatened by serious challenges at the various production stages especially in postharvest preservation, translocation of harvested produce, and storage [8]. As the world population increases drastically toward the 9 billion mark in the year 2050, several international agencies and authorities have expressed fears on the possible threats of starvation, hunger, and poor living and health standards [39]. It has been documented that majority of agricultural produces in the developing countries are prone to postharvest losses [4–7]. Postharvest losses are a function of several factors. Some of the striking factors are: inadequate postharvest storage techniques, poor handling,

A. S. Olayinka
Computational Science Research Unit, Department of Physics, Edo State University Uzairue, Iyamho-Uzairue, Edo State, Nigeria

C. O. Adetunji (✉)
Applied Microbiology, Biotechnology and Nanotechnology Laboratory, Department of Microbiology, Edo State University Uzairue, Iyamho-Uzairue, Edo State, Nigeria

W. Nwankwo
Informatics and CyberPhysical Systems Laboratory Department of Computer Science, Edo State University Uzairue, Iyamho-Uzairue, Edo State, Nigeria

O. T. Olugbemi
Laboratory for Reproductive Biology and Developmental Programming, Department of Physiology, Edo State University Uzairue, Iyamho-Uzairue, Edo State, Nigeria
e-mail: olaniyan.olugbemi@edouniversity.edu.ng

T. C. Olayinka
Department of Mathematical & Physical Sciences, Samuel Adegboyega University, Ogwa, Edo State, Nigeria

S. Pal et al. (eds.), *Artificial Intelligence-based Internet of Things Systems*, Internet of Things, https://doi.org/10.1007/978-3-030-87059-1_18

the incidence of pests and diseases, and lack of basic amenities (road network, portable water, and lack of electricity) [1–8].

Consequently, postharvest challenges have been identified as a major threat to global food supply and a potential predictor of starvation and hunger especially in Africa with Nigeria as a major focal point, and according to the UN [50] apart from India and China, Nigeria's population is projected to be the third largest in the world by the year 2050 with the three nations accounting for 35% of the world's urban population growth from 2018 to 2050. This places Nigeria and jurisdictions with geometric birthrates and poor agricultural facilities in dare risk.

With the increasing pressure on local agricultural systems and the inadequate processing and postharvest storage infrastructure, which might be linked to the consistent and almost uncontrollable birthrates leading to contention, new measures, techniques, procedures, and innovative means for agricultural production are advocated to forestall near reality of global food shortage.

In response to the global awakening, modern agricultural systems are evolving, and emphasis is being made toward automated and smart agriculture also termed precision agriculture. Precision agriculture (PA) entails the deployment of information technology and allied devices in agricultural processes, procedures, and products (PPPs) in order to ensure optimum product yield, maximize profit, conserve the environment, and guarantee sustainability. PA has been adjudged as one of the best solutions that reduce losses arising from conventional agricultural operations, which is a major setback in agriculture. Through technologically enhanced observation, measurement and objective analysis could be undertaken to guide agro processes and procedure management for optimum performance.

PA is dependent on data. The data derived from interdisciplinary fields have been found useful and could drive innovations in agricultural productivity. These data could augment the historical data on agricultural processes over the years and data on ongoing operations. For instance, data on agricultural activities are easily available through semiautomated means, and in recent times, sensors, which could permit capturing of biological, agricultural, and environmental (an interface of weather conditions dynamic soil, crop) characteristics vis-à-vis the performance of agricultural machinery, have been found very useful in agricultural technology.

AI and machine learning (ML) have been identified as evolving and sustainable agricultural modernization technology alongside high-performance computing, big data, and IoT technologies that could be integrated into smart agricultural operations at different levels. The application of ML have been documented in several fields, which entails climatology [56, 57], economic sciences, bioinformatics [28, 35] agriculture, food security, biochemistry, robotics, medicine, aquaculture, and meteorology.

Therefore, ML and AI could play a significant role in the discovery and development of useful models and systems that could help in mitigating the aforementioned challenges. Hence, this chapter intends to provide a detailed discussion on the application of Bayesian learning and decision tree IoT-enabled system in postharvest storage of agricultural produce especially fruits.

2 Technology and Postharvest Management of Vegetables and Fruits

The natural sources of essential vitamins such as vitamins A and C in the daily meal of people across different jurisdictions are vegetables, root crops, and fruits. The nutritional elements from these vital food sources may be unintentionally reduced, impaired, or destroyed soon after the harvest. Experience has shown that many perishable food crops such as vegetables and fruits lose their organoleptic (color, taste, smell, texture, etc.) and physicochemical characteristics (e.g., enzyme actions, essential nutrients, etc.) on exposure over a period to the air, sunlight, microorganisms, water, etc.

Postharvest operations are directed toward the preservation of harvested food crops. Common strategies may include traditional preservation methods such as heating, drying, freezing, fermentation, chemical application (salting, oiling, vinegar juice, etc.), etc. Advances in agricultural technology has also led to the invention and use of hybrid preservation approaches such as evaporative cooling system [51]; evaporative technology variants such as, zero energy brick cooler, evaporative charcoal cooler, and pot-in-pot cooler [37]; subzero temperature preservation[33]; edible coatings including lipids, hydrocolloids, and composite coatings [53]; canning; chemical inhibitors such as ethylene scavengers [52]; ultrasonic treatment to eliminate pesticide remains, deactivation of enzymes (ascorbic acid oxidase, lipoxygenase, polyphenol oxidase, and pectic enzyme peroxidase), sterilization, stabilization of polysaccharides in cell walls, and elimination of putrefying microbes on harvested fruits and vegetables [27]; green technologies such as high-pressure processing, UV-radiation, cold plasma, acid electrolyzed water, and gamma irradiation [44]; hydrostatic and biological agents [38]; and nanocomposites ([31]) and other nanodevices[29].

Each method of preservation may have its peculiarities, which is connected to some physical and/or biochemical factors. For instance, a hybrid vegetable preservation that employs drying the vegetable may be treated with a gas such as sulfur (IV) oxide (SO_2) prior to drying to curtail enzymatic browning [22, 47]; vitamin C breakdown, and disinfection or elimination of susceptible putrefying microbes [11, 12, 29]. Note that the two major factors in the aforesaid drying process are the SO_2 level and temperature. In some cases, the pre-drying application of SO2 may be regarded as a prestorage action. In any case, these factors may be detected, measured, and consequent upon the measurement, it is possible to optimize the relevant factors that would enable reduction of losses due to damages and losses posed by extreme or unwanted physicochemical and/or biological elements [15, 22, 48]. This section discusses majority of the relevant factors. The perspective of the discussion is tilted toward relevance, detectability, and measurability. The essence is to appreciate these factors in physical terms since the relevance of IoT in this context is directed toward detecting the relevant and changing parameters (e.g., relative humidity, temperature, and CO_2 level) in a postharvest storage system, transmitting the changes in real time as live data to the remote or proximal and online data analysis server computer, which utilizes the data to effect some intelligent prediction that results in a feedback. The feedback is relayed back to the automated preservation system for optimization of parameters or for the execution of a definite control action other than optimization such as routine examination of the stored produce.

2.1 Environmental (Abiotic) Factors Affecting Fruit and Vegetable Preservation

Generally, agricultural activities including storage of harvested produce and the environment are interconnected and influence each other. Changes in the environmental conditions affect agricultural food storage in several ways. These environmental parameters are discussed briefly.

2.1.1 Temperature

Like agricultural production, storage of agricultural produce is dependent on optimum temperature and relative humidity. Changes in temperature affects biochemical activities in food crops, and any departure causes a decline in production output and food storage systems, hence optimum temperature is often canvassed [39]. Fruit preservation and storage are affected by the environmental elements such as temperature, relative humidity, chemicals such as ethylene, and ambient light [36]. Mahajan et al. [36] had identified temperature as the ultimate factor in the preservation of vegetables and fruits. However, this does not undermine the relevance of biotic factors such as putrefying microbes and cellular physiochemical reactions involving enzymes within the cells of the harvested produce [12, 44]. It is important to stress that even the enzymatic reactions brought about by normal cell functions, chemicals, and those engendered by biotic agents all require temperature as an important catalyst in such processes hence are bound to regress in the absence of optimum temperature. Vegetable and fruit storage are highly influenced by water losses. Temperature is a major controlling factor for water losses in postharvest vegetables and fruits [34]. According to Liu [32], ethylene synthesis is directly proportional to the temperature level; hence, postharvest fruits and/or vegetables should be maintained at low temperature, low O_2 level, high humidity, high CO_2 level, low ethylene level, and sterile environments.

High temperatures cause increased respiration rates, water loss [17], and deterioration in fresh produce [23]. The water loss rate, however, varies from one fruit to another owing to surface-area-to-volume ratio, surface structure, size and number of lenticels and stomata, and the cuticular thickness and composition [34]. Application of cold stress to vegetables and fruits restricts further growth and development, as well as distribution of biochemical matter [46].

2.1.2 Relative Humidity

Like temperature, relative humidity is a factor that may either promote the freshness of vegetables and fruits or impair it. The resultant effect of relative humidity on the freshness of a fruit or vegetable varies from fruit to fruit and from vegetable to vegetable [18]. According to the Food and Agricultural Organization, the main causes of losses in fresh agriculture produce including vegetables and fruits are high temperature, suboptimal relative humidity, and injury to harvested produce [20].

2.1.3 pH

Another interesting environmental factor is the pH of the environment. The reduction of pH impacts the growth of putrefying microbes. Many of these microbes are incapacitated and killed by low or extreme high pH values of the storage environment. Majority of the microbes associated with fruit spoilage grow well at near neutral pH.

Following the harvest of vegetables and fruits, some enzymes continue their intracellular activities. The continued catalysis of biochemical processes within the harvested produce lead to ripening and subsequent decay. Oxidative enzymes would continue to drive cellular respiration, i.e., oxygen-driven metabolism of glucose to produce energy. If this cellular respiration is not prevented, the shelf life of these fresh produce are shortened, and spoilage sets in. Cellular respiration and enzymatic catalysis are all driven by an optimum pH. Distorting this pH impairs cellular activities, thus enhancing the stability of stored harvested produce.

Application of some gases and acid solutions for the preservation of fruits and vegetables has been documented [55]. The most relevant gases are briefly discussed.

2.1.4 Nitrogen Oxide (NO)

First among the gases for preservative use is nitrogen oxide (NO). This gas is often derived from a donor compound such as sodium nitroprusside (SNP). Exogenous NO is noted for the following:

(a) Inhibition of ethylene biosynthesis [24].
(b) Enhancement of antioxidant system [9, 43].
(c) Induced defense system.
(d) Activation of the C-repeat binding factors (CBFs) pathway [46].
(e) Regulation of energy and sugar metabolisms [55].

2.1.5 Chlorine Dioxide (ClO_2)

It has been noted that controlled release of chlorine dioxide (ClO_2) gas in a vegetable and fruit storage system is effective as a sanitizer for the storage environment and also for the control of pathogenic and putrefying microbes that are responsible for the spoilage of most fruits and vegetables [49]. The researchers also did not undermine the negative effects the excessive use of such chemicals could have on the consumers of the vegetables hence due diligence must be done during the use of these chemicals to reduce any detrimental effect their use may have on the end consumers.

2.1.6 Carbon (IV) Oxide (CO_2)

CO_2 is a major player in the life cycle of majority of edible crops. Photosynthesis attests to the relevance of CO_2 in crop growth and development. Postharvest storage is influenced by the amount of CO_2 in the storage environment. CO_2 at higher concentrations lowers the metabolic rate of the produce, thus decreasing the senescence process. For desirable results, the CO_2 pressure and volume must be carefully controlled in which case an appropriate threshold is set to and maintained throughout the storage cycle. At appropriate levels, CO_2 has been found to:

(a) Retard growth of microbes and fungi [19].
(b) Reduce synthetic reactions in fruits [19].
(c) Influence some organic acid metabolism.
(d) Act as an ethylene scavenger and inhibitor [52].
(e) Reduce discoloration of fruits and vegetables.
(f) Inhibition of enzymatic actions in the produce [12].
(g) Increased retention of firmness [14].

2.1.7 Oxygen (O_2)

O_2 is notable for respiration in plants. Like CO_2, a postharvest fruit and/or vegetable storage system should operate optimally under optimized O_2 limits. The implication is that O_2 levels need be routinely modified as high O_2 concentrations generally favors growth in majority of microbes known to cause vegetable and fruit spoilage. Some results documented against modified O_2 concentrations by various researchers include:

(a) Elongation of the fruit/vegetable storage life [55].
(b) Inhibition of ethylene gas that promotes fruit ripening [52].
(c) Decreased rate of respiration in the stored produce [23].
(d) Inhibits ethylene gas production.

(e) Decreased degradation of soluble pectin.
(f) Decrease oxidation rates.

2.1.8 Arachidonic Acid

Exogenous arachidonic acid has been demonstrated as a good preservative for cherry tomatoes at 20°C. According to a study, arachidonic acid decreased the decay and loss of weight in the tomatoes; reduced the accumulation of malondialdehyde (MDA); increased the polyphenol oxidase (PPO) activity, catalase, and peroxidase (POD); delayed membrane permeability; sustained the cell membrane integrity; and decreased the total soluble solids and titratable acidity loss during later phases of storage [54]. They noted that using a 2.5 mg L^{-1} of the compound provides an optimum concentration to maintain the quality of the cherry tomatoes. Unlike other chemicals used for preservation, it was stressed that arachidonic acid is purely a green preservative and does not exhibit any detrimental aftereffects on the consumers of the stored produce.

2.1.9 Y-Aminobutyric Acid (GABA)

Gamma-aminobutyric acid (GABA) in combination with NO has shown to be effective in the preservation of cherry fruits [9, 43]. According to the studies, 5 mM GABA mixed with 500μM of sodium nitroprusside (donor of NO) and used to treat cherry fruits exhibited greater firmness. The result suggests that the chemicals when applied and the fruits subsequently stored at 4 °C reduced the enzymatic degradation of the cell wall hence the marked reduction in the browning of cherry fruits (through decreased H_2O_2 accumulation). The study presupposes that the chemicals induced increased activity of phenylalanine ammonia lyase enzyme while decreasing the enzymatic action of the polyphenol oxidase enzyme, which resulted to increased phenols, anthocyanins accumulation, flavonoids, and superior 2,2-Diphenyl-1-picrylhydrazy (DPPH) scavenging capacity [9]. Research has shown that the regulation of these biochemical activities and the modulation of the chemical agents introduced for preservation is done by temperature. The implication is that temperature remains a common denominator in every vegetable and fruit preservation program or facility. Prior to eventual storage, freshly harvested fruits would require precooling using a mixture of cold water, sodium hypochlorite, thiabendazole [13], or other safe disinfectants. Other prestorage activities include sorting of the fruits (to separate mechanically injured or unwholesome ones from the whole fruits) and grading to segment the fruits into various classes depending on firmness determined using a texture analyzer.

2.2 IoT as a Machinery for Relaying Data from Postharvest Storage Sites

IoT is a pervasive computing paradigm that potentially changes how computers and humans communicate, compute, and coordinate with each other. IoT systems largely involve the integration of sensing devices such as sensors, actuators, and processors that interact with each other to serve a significant purpose [40]. The IoT technology has developed in several domains including animals, hospitals, smart cities, airplane, computers, plants, offices, cars, etc. Consequently, the number of IoT devices is growing every day, as they provide comfort in human life and work and may provide much better results than humans [16, 45]. In order to deliver the functionality of the IoT effectively, the following elements, which include identifying, sensing, communication, computing, services, and semantics, are needed. However, there is no single consensus on architecture for IoT, which is agreed universally.

IoT, as a pervasive computing platform, offers unlimited support for real-time control and monitoring of experiments, production, and similar events that are subject to eventual fluctuations that may result to critical and undesirable conditions. It has been shown that controlled environments produce better results compared with uncontrolled environments [30]. It is noteworthy that there are three critical intrinsic parameters in an IoT-enabled postharvest storage instrumentation. These are temperature, relative humidity, and pH of the environment [23]. These intrinsic factors are critical to the operation of other parameters such as gases, enzymatic actions, and microbes. For instance, a fruit or vegetable storage chamber or environment should constantly be subjected to the optimal temperature, relative humidity, environmental oxygen, etc. otherwise the fresh produce would deteriorate. Similarly, the pH of the environment would also affect the stability of the stored produce in a similar way as temperature. Putrefying microbes thrives well at different pH as some are acidophiles (requiring acidic media), some neutrophiles (need relatively neutral pH), while some are alkaliphiles (growing best at pH 8-10.5), hence the identification of the major microbe(s) that are responsible for causing the putrefaction of a specific freshly harvested vegetable or fruit is very important and should be made prior to the selection of an appropriate pH for storage of the produce. Consequently, a monitoring and control solution should be in place in any of these instrumentations. As site monitoring may be highly inconvenient in high-risk environments, the use of IoT would simplify the entire process. Thus, pH, humidity, temperature, O_2, and CO_2 sensors may be used in the storage facility, and such IoT-enabled devices could relay continuous measurements to remote analytics centers where the transmitted data are analyzed and feedback sent immediately to the storage plant for requisite control action. This would greatly aid the optimal performance of the vegetable or fruit storage plant.

2.3 IoT Requirements for Postharvest Vegetable and Fruit Storage Instrumentation

The IEC 30141 standard categorizes the IoT platform into six domains [26] namely: user, sensing and controlling, physical entity (PE), management and operations, application/service, and communication and access domains, respectively. The user domain is at the top of the hierarchy, whereas the PE sits at the bottom and include the environment incorporating the terminal devices to which the sensing and controlling domain relates with.

Simply, an IoT implementation is a conglomerate of devices that interoperate to deliver a solution. The IoT is an emerging technology that enable the interconnection of buildings, household devices, vehicles, manufacturing plants, storage facilities, etc. using embedded devices such as sensors, software, and network connectivity, providing a platform for the data collection and exchange without human intervention.

A typical IoT platform can be seen as integrated system, which is able to support millions of concurrent device connections to generate a large volume of data to be transported and processed by cloud systems. Four vital components in a typical IoT platform have been identified: sensors and hardware devices, communication network (Wi-Fi; cellular technologies, 3G, 4G, 5G; Li-Fi; etc.), data, and the cloud (where the data are stored, processed, and accessed). IoT applications on cloud platforms can provide feedbacks and decisions to the cyberphysical systems.

An IoT implementation would require a careful analysis of the project for which IoT is a critical component. Requirements engineering includes specification of the data requirements, sensing technologies, network connectivity requirements, remote processing site (analysis server), and hardware specifications [42]. The data requirements revolve around the data items to be captured by the sensors including the format and types, the source, endpoint, necessary control feedbacks from the endpoint servers, and the nature of data transmitted back to the control devices such as dehumidifiers, heaters, O_2/CO_2 monitors, etc. The sensing technologies include the various embedded sensors such as pH sensors, gas sensors (oxygen/CO_2), humidity sensors and temperature sensors, chemical, and biosensors, respectively.

The choice of sensing technologies is based on features ranging from sensitivity, durability, versatility, and ease of integration. The network requirements are very critical to the success of the IoT implementation. The factors that are considered prior to network device selection include: distance, climatic factors, location (indoor or outdoor), market availability, integrability, and compatibility with the micro integration platform such as the microcontroller board. The remote processing site may be cloud-based and may require medium to high-end servers with appreciable computational power including software, speed, and storage resources. Later in this section, a prototype implementation of an IoT solution using cost-effective hardware would be presented.

2.4 Supervised Learning

Supervised learning (SL) develops and uses models to make predictions. The model is often developed using existing data. Two SL approaches exist: regression—an approach to find the correlation between variables and classification—is used to identify which set a new observation of data belongs, i.e., making a forecast target class for an observed instance or occurrence. The SL algorithms discussed in this chapter are the classification algorithms. Classification algorithms are grouped into: Bayes classifier, nearest neighbor, support vector machines, boosted trees, decision trees, random forest, and neural networks. The Bayesian classification and decision trees would be discussed.

Bayes' theorem explains the probability of an event, based on prior knowledge of conditions that might be related to the event under consideration. For example, if the risk of developing health problems is certain to increase with age, Bayes' theorem would enable the risk of an individual of a known age to be assessed more accurately than simply presupposing that the individual is typical of the population as a whole. One of the many areas of applications of Bayes' theorem is Bayesian inference. When used, the probabilities involved in Bayes' theorem may lead to a different understanding. Bayesian learning is based on the Bayes' theorem. Bayesian classification may be used to establish causal relationships. Thus, it can enable the understanding of a problem domain including the prediction of the intervention consequences. Two major forms are identified: the general Bayes classification and the naïve Bayes classification (a more prominent classifier these days). In the traditional Bayes classification (Bayes nets or Bayesian belief networks), the user determines the conditionally dependent and independent variables, respectively. However, the Naïve Bayes is guided by the assumption of independence among predictors. That is to say that, Naive Bayes classifiers (NBC) assumes that the presence of a particular feature in a class is not related to the presence of any other feature or that all of these properties possess independent contribution to the probability. This family of classifiers is relatively easy to build, particularly useful for very large datasets, and it is highly scalable. Along with simplicity, Naive Bayes is known to outperform even highly sophisticated classification methods *[10]*. The NBC is constructed around three components: posterior, prior, and conditional probabilities. The strengths of the NBC, according to Francois-Lavet et al. [21] include:

(a) Small training data.
(b) Simple computing.
(c) Easy to implement.
(d) Time efficiency.
(e) Handle big data, incomplete data, or missing values.
(f) Insensitive to irrelevant features and noise.

Decision tree (DT) classifiers are useful in solving regression and problems involving the classification of categorical and numerical variables [41]. Using DT, data assortment is divided into small sets while each connected to the DT. The DT has decision nodes (DN) and root nodes (RN). The DN may have two or three segments. The leaf node is a judgment, whereas the highest DN that corresponds to the RN becomes the highest predictor. The decision tree has the following merits:

(a) Easy understand, interpret, visualize, and require less training period.
(b) Handles nonlinear parameters efficiently.
(c) Robust to outliers and can handle them automatically.

Some of the limitations of this algorithm includes overfitting, affected by noise and not suitable for very complex and robust datasets. In order to overcome these limitations, random forest model was adopted too since it does not depend on a single tree, as it creates a frets of trees and takes the decision based on the vote count.

2.5 Control of Storage Processes Using Classification Models

Bayesian models play critical role in the groups of algorithms for classification of high dimensional dataset. They are generally fast and could be used to implement intelligent decision making seamlessly. Naïve Bayes classification is a simple form of supervised learning directed toward identifying all data points with a label. Data points with similar labels are considered members of same class. The labels can be greater than or equal to two (2). Postharvest datasets on environmental and biological conditions could be measured using sensor-based IoT-enabled devices deployed in a warehouse, store, or transportation van where farm products are kept after harvest. The data may include temperature(°C), relative humidity (%), oxygen level (%), CO_2 level (%), water loss rate (mg/m^2 h), nitrogen oxide (NO), etc.

Bayesian learning and decision trees may be integrated into a postharvest storage system to provide intelligent interventions for the control and monitoring of fluctuations in environmental parameters around the storage facility. They could provide the following functionalities:

(a) Optimization of environmental conditions such as temperature, relative humidity, O_2/CO_2 levels.
(b) Detection of onset of microbial activity using feedback from sensors.

3 Methodology for the Design of an Intelligent IoT-Driven Postharvest Fruit Storage Facility

The methodology follows the evolutionary prototyping approach and depicted in the conceptual diagrams in Figs. 1, 2 and 3, respectively. The phases are divided into two: pre-prototyping and the actual prototyping phases. The pre-prototyping includes the prestorage activities and the building of prediction models. A very

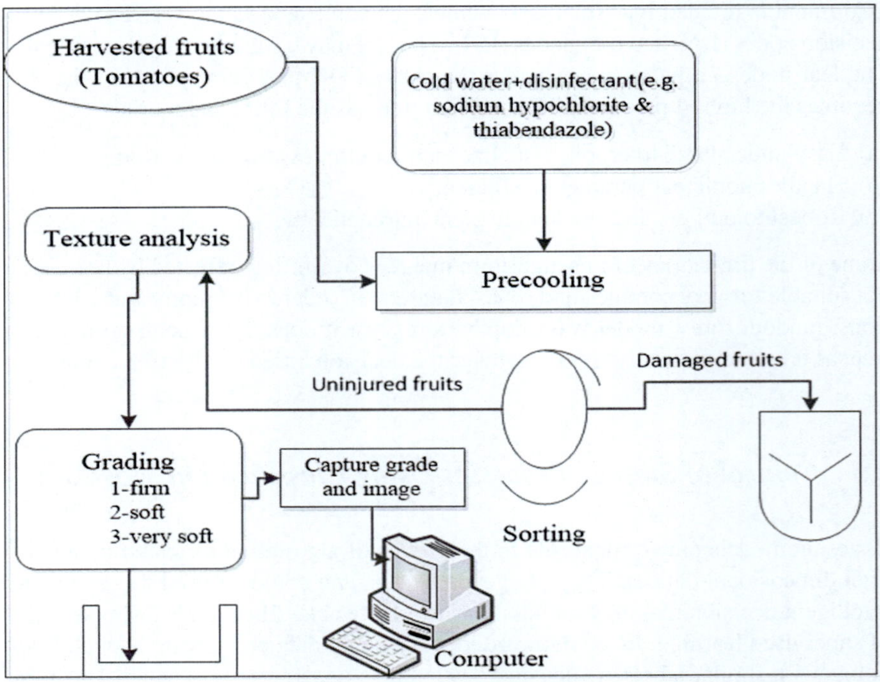

Fig. 1 Prestorage actions on freshly harvested tomato fruits

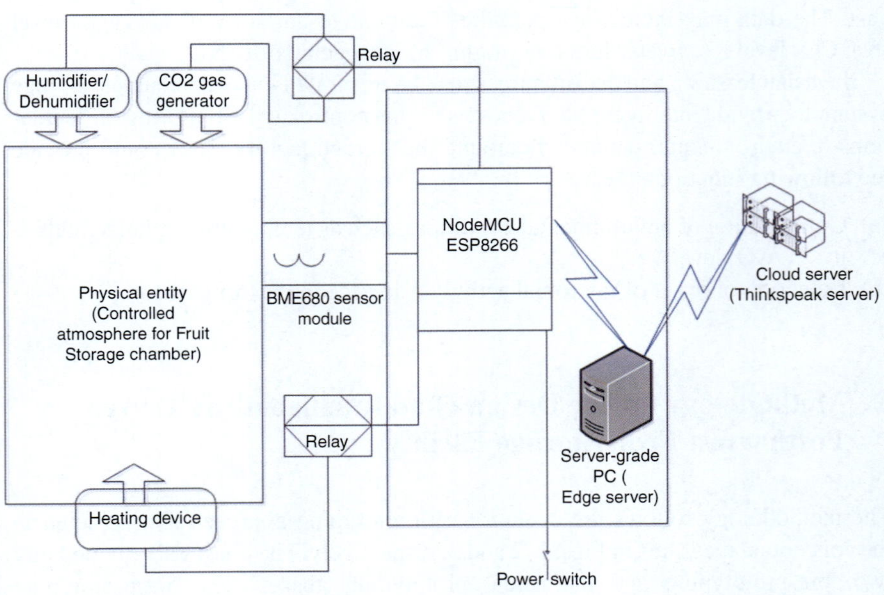

Fig. 2 Conceptual design of the IoT platform

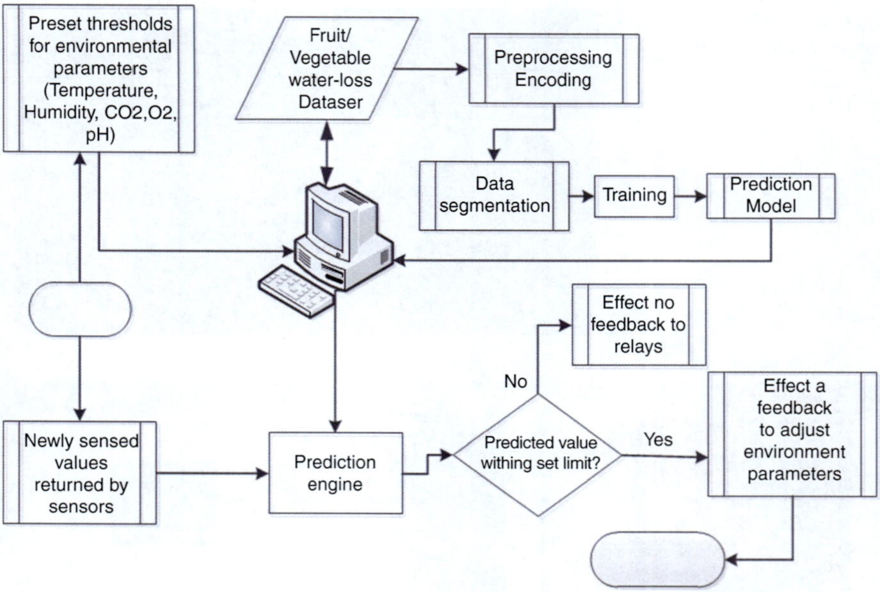

Fig. 3 Conceptual design of the analytics component of the intelligent IoT-enabled storage facility

important part of the model building process is the evolution of the water-loss dataset. The dataset is derived from previous studies involving the measurement of transpiration rates of freshly harvested tomatoes under various temperature, humidity, CO_2, and O_2 levels. The models would be built to predict the water-loss rate, a major factor in the preservation of postharvest fruits and vegetables. The dataset comprises 2500 instances with the following variables: temperature (T), relative humidity (H), CO_2(C) level and O_2 (O) level, and water loss rate (WLR). The WLR is the dependent variables, whereas all others are independent. For each of the independent variables, the values are categorized as low or high and encoded as 1 or 0.

The materials used for prototyping include: breadboard and jumper wires, BME680 sensor pack (4 in 1 module for temperature, humidity, pressure, and gases) or DHT22 AM2302 Module (for temperature and humidity sensor combo), ESP8266-12 integrated board module (see Fig. 4), Arduino IDE, Server-grade PC (HP elite 2.8Ghz core i7 PC @16GB RAM), and USB cables.

Figure 1 shows the prestorage activities on freshly harvested cherry tomatoes from locally cultivated crops from South–South Nigeria. The prestorage activities are fundamental to both automated and nonautomated tomato fruit storage facilities. In most scenarios, sorting and grading are done manually but may be automated using machine learning model (MLM) wherein models are created to enable image analysis and classification. Prior to the use of MLM, the grading of uninjured fruits

Fig. 4 A section of uninjured tomato fruits

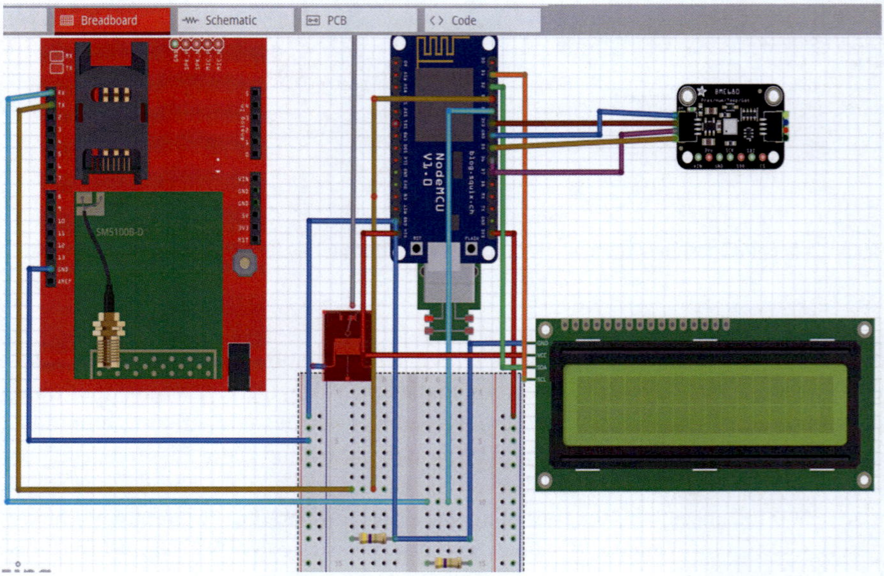

Fig. 5 Schematic of the IoT-enabled client device

(see Fig. 4) would be done using the texture analyzer and in this study the samples (analyzed and classified into grades 1, 2, and 3, respectively) comprises 2000 fruits. The images of the fruits are also captured and stored as "GIF" formats. The grades 1, 2, and 3 represent the degree of firmness and "1" is the highest grade. The IoT prototyping system is built around the NodeMCU ESP8266 microcontroller (with built-in WIFI), a very low-cost prototyping chip compatible with the Arduino IDE. Notable is the BME680 sensor suite (see Fig. 5) that provides a digital multi-purpose sensing function for humidity, pressure, temperature, and gases. However, the BME680 sensor by reason of its applicability is not tied to any particular gas such as CO_2 hence must be calibrated with a specific gas and concentration where such gas may include CO_2, O_2, etc. The communication between the sensor and the microcontroller may be established using a serial peripheral interface (SPI) or Inter-Integrated Circuit (I2C) protocols.

Figure 2 shows the conceptual IoT platform design. In the diagram, a hypothetical physical entity represents the enclosed storage chamber, which may adopt different technologies and sizes. The chamber is connected to through air vents to humidifier/dehumidifier, CO_2 and O_2 generators, then to the environmental sensors, which are attached to microcontroller. Two processing endpoints are used: an edge computer, which performs immediate processing of the sense data, and the remote server located on the Internet. In this prototyping, the *ThingSpeak* server was used as the endpoint from the IoT-enabled client device. It provides an easy IoT prototyping with various development platforms like ESP8266 and Arduino. The essence of the edge computer is to foster high speed processing of data transmitted to by the sensors and to enable faster response times to critically changing environmental parameters.

Figure 3 is a logical diagram of the analytics subsystem. The function of the analytics subsystem are twofold: provision of access to other components of the IoT-based system, data gathering from sensors (through the ESP8266 microcontroller), analysis, prediction, relay of data to the remote server, and assisted control of the storage facilities. The ESP8266 specification and interfacing are presented in Tables 1 and 2.

3.1 Microcontroller Interfacing and Specification

The important characteristics of the ESP8266 are presented in Table 1. Figure 5 shows the schematic of the client IoT-enabled device. Table 2 provides the various interfacing requirements.

Table 1 ESP8266-12 microcontroller features

Processor	L106 32-bit ESP-12E @ 80-160 Hz
Firmware	NodeMCU
Data pins	16 GPIO for interfacing with sensors, switches, LEDs, etc.
ADC channel	1(10-bit) accessible through A0
Communication	UART, SPI, I2C, SPI
RAM	4 MB
SPI pins	4 (SCK, CS, MISO, MOSI) for SPI communication
I2C pins	Available
UART pins	2.
Power supply	3.3 V (max)

Table 2 Component interfacing with ESP8266-12

ESP8266-12	16 × 2 I2C LCD	**GSM Sim900A**	**BME680**
3.3 V		Uses separate power input 4.7–5 V (5 V adapter is needed)	VCC
D1(GPIO5)	SCL		
D2(GPIO4)	SDA		
GND	GND	GND	GND
VIN	VCC		
D3		TX	
D4		RX	
D5			SCL
D7			SDA

Table 3 Dataset description

S/N	Variables	Sub-label
1	Temperature(°C)	Low(0), high(1)
2	Relative humidity (%)	Low(0), high(1)
3	O2 level(%)	Low(0), high(1)
4	CO2 level(%)	Low(0), high(1)
5	Water loss rate(mg/m^2 s)	Low(1), high(1)

3.2 *Data Specification and Model Development*

Table 3 shows a sample specification of the dataset used for building the prediction models. The water loss rate is the dependent variable, hence the value to be predicted in both models. Table 4 shows a subset of the sample dataset comprising 2500 occurrences.

Prior to integration into the IoT system, the performance of the MLMs was evaluated using the following computational parameters: prediction accuracy, false positive rate, false negative rate, true negative rate, true positive rate, precision/recall, and ROC curve.

The Bayes probability theorem describes the probability of a feature as a function of prior knowledge of states connected to that feature. It is an equation that describes relationship of conditional probabilities of statistical quantities. For instance, if the probability of high rate of water loss in a postharvest store is connected to oxygen level in the store, then Bayes probability theorem can be handy to water loss rate accurately using the oxygen level data. Bayes probability theorem is the basis for several strands of Bayes algorithms in supervised machine learning algorithms. The basic assumption of the Naive Bayes algorithms is that no correlation exists between features in the dataset used in model training. Naïve Bayes has an advantage of performing efficiently well on small number of training dataset compared with competing classifier algorithms like decision trees, support vector

Table 4 Sample dataset from ThingSpeak cloud server

Date	Temperature (°C)	Relative humidity (%)	O_2 level (%)	CO_2 level (%)	Water loss rate (mg/m^2 h)
6/25/2019	6	90	4	20	0.4
6/29/2019	10	97	3	19	0.25
7/4/2019	7	96	5	13	0.5
7/5/2019	11	96	2	15	0.2
8/25/2019	9	95	5	14	0.4
8/26/2019	10	94	4	16	0.4
9/5/2019	10	96	2	17	0.3
9/8/2019	10	96	2	18	0.15
9/15/2019	11	97	2	14	0.35
9/21/2019	11	93	5	18	0.5
10/7/2019	11	96	2	16	0.2
10/11/2019	12	93	6	15	0.4
10/26/2019	12	97	7	19	0.5
10/28/2019	12	96	3	22	0.2
11/3/2019	13	96	5	14	0.4
11/16/2019	17	94	6	20	0.6
12/19/2019	17	96	3	15	0.5
1/21/2020	20	95	2	14	0.7
1/25/2020	20	96	5	16	0.8

machine, random forest, logistic regression, etc. The word "naïve" suggests that every pair of features in the dataset is independent of each other. The continuous values associated with each class are distributed according to a normal distribution, which assumption by Gaussian. The interest is the determination of the probability of a label based on measured features from IoT-enabled sensors using Bayesian classification algorithm. This may be written as $P(L \mid \text{features})$.

The theorem is useful in projecting this in terms of quantities that could be computed directly:

$$P(L \mid features) = \frac{P(features \mid L)P(L)}{P(features)} \tag{1}$$

The features under consideration are temperature (°C), relative humidity (%), O_2 level (%), and CO_2 level (%), while the label is the rate of water loss within the postharvest storage environment. The water loss label can be LOW or HIGH at every instance. We consider the two labels as WL_{HIGH} and WL_{LOW} for HIGH and LOW water losses instances, respectively. In taking the decision, we compute the compute the ratio of the posterior probabilities for each label:

$$\frac{P(WL_{HIGH} \mid features)}{P(WL_{LOW} \mid features)} = \frac{P(features \mid WL_{HIGH})P(WL_{HIGH})}{P(features \mid WL_{LOW})P(WL_{LOW})} \tag{2}$$

Then, there is need for model that computes $P(features \mid WL_i)$ for each label where $i = HIGH, LOW$). The model is generative in that it specifies the hypothetical random method that produces the data. Training of the Bayesian classifier deals with setting up each label for the generative model. Though the general version of such a training step is a herculean task, however several assumptions are used to ameliorate this. The Naïve Bayes adequately provides for such assumptions.

4 Results

4.1 DT and NBC Modeling Results

Figure 6 shows the boxplot grouped by water loss label with temperature and relative humidity, respectively. The boxplot is a useful graphical tool for viewing the continuous unimodal dataset distribution. It displays the details on the location, spread, and skewness of the data [25]. In Fig. 6, temperature and relative humidity are biased toward high water loss.

Figure 7 shows the 2-D scatter visual exploration plot of measure data for temperature, relative humidity, O_2, and CO_2 level from the sensor attached to the post-harvest store. Scatter plots show the variation of data points based on different features relatively to water loss labels HIGH and LOW. The correlation plot of the

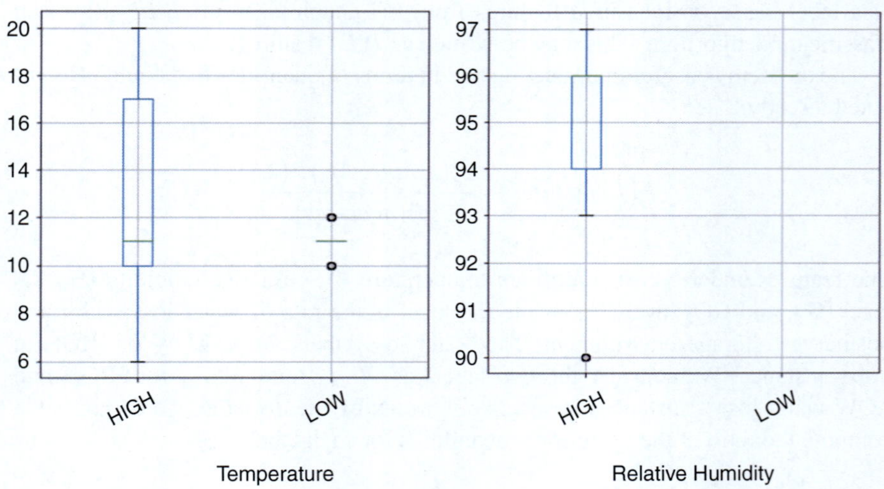

Fig. 6 Boxplot Grouped by Water Loss Label with Temperature and Relative Humidity respectively

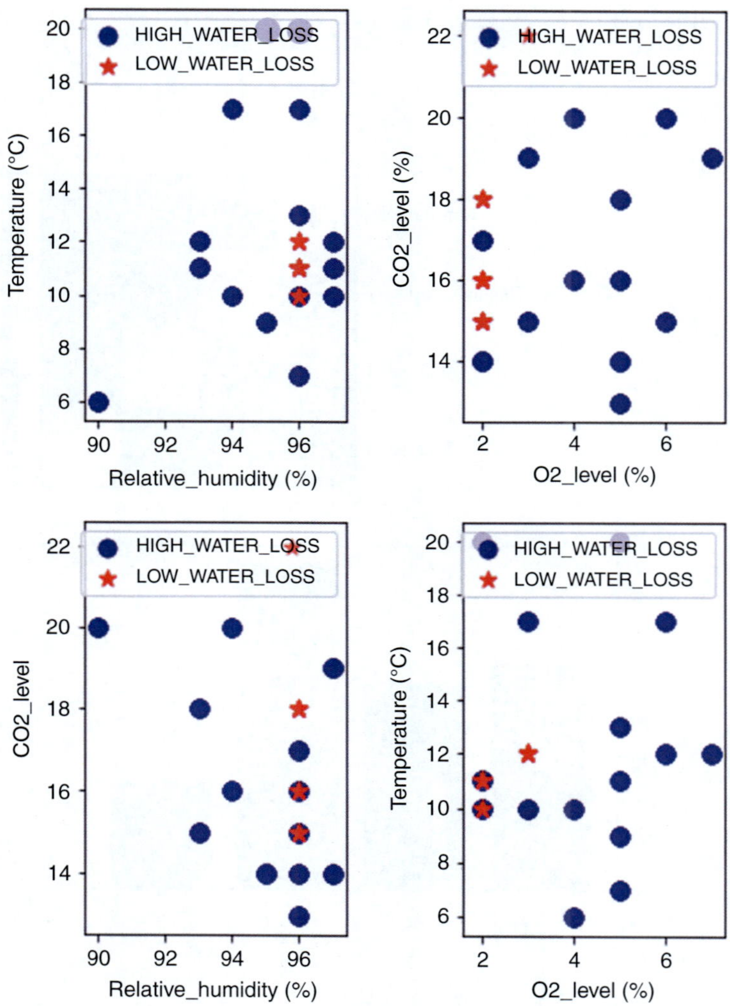

Fig. 7 Scatter Plot of Measured Features (Temperature, Relative Humidity, Oxygen and CO_2 level)

measured features as shown in Fig. 8 indicates that the relative humidity is highly correlated with oxygen and CO_2 levels, respectively. The figure is supportive in feature selection during model creation and prevention of multicollinearity in linear models. Figure 9 is the decision tree classification model for measured features. Importance of variables as computed is given as 0.378639 for O_2 level, 0.260233 for temperature, 0.238828 for relative humidity, and 0.122300 for CO_2 level. From Fig. 9, it obvious that O_2 level is the preferred feature. The feature plot indicates that O_2 level is the most important feature for splitting the data followed by temperature, relative humidity, and CO_2 level, respectively.

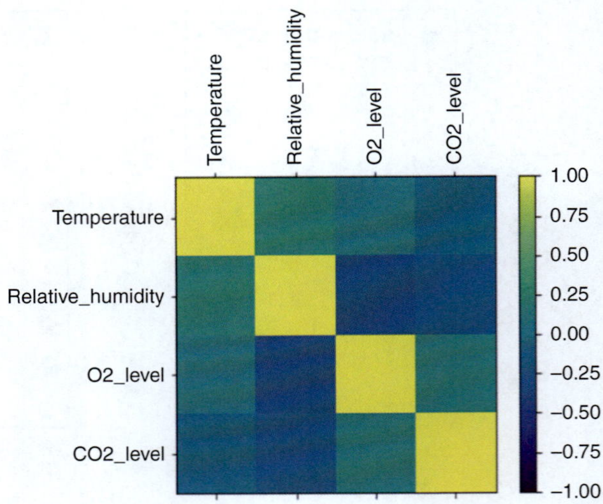

Fig. 8 Correlation plot of the measure features (temperature, relative humidity, oxygen, and CO_2 level)

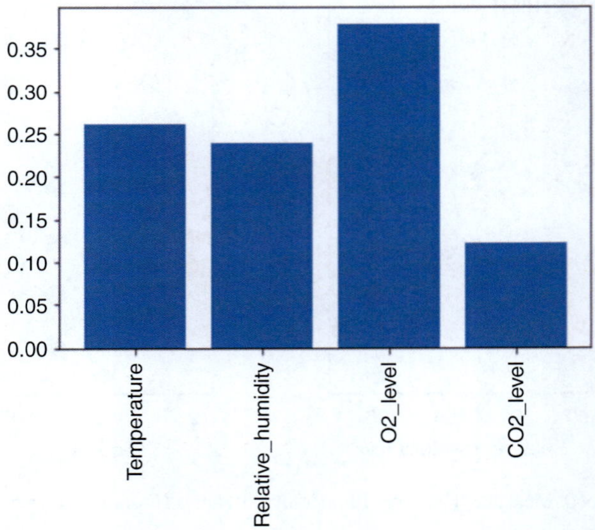

Fig. 9 Decision tree classification model on important features

Decision tree (DT) classification model base on the important features is shown in Fig. 9 while Fig. 10 shows the Decision Tree (DT). Table 5 shows the summary of the model for the decision tree and Naive Bayes (NB). NB performs better on the dataset with 78.95% of correctly classified instances when compared to the DT with 63.16% correctly classified instances. The model may be improved for better accuracy as the number of dataset instances increases.

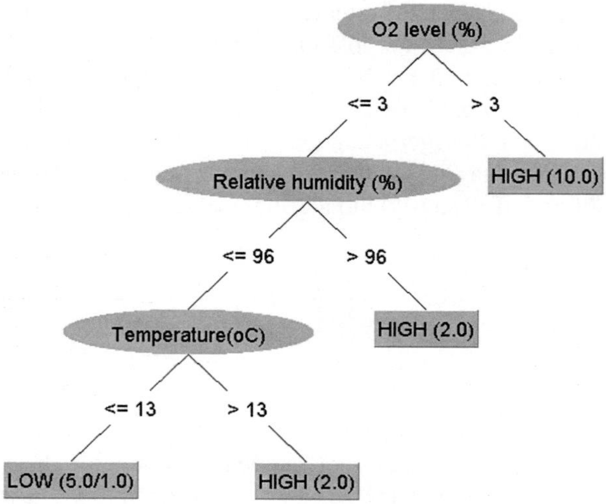

Fig. 10 Decision tree with J48 algorithm

Table 5 Summary of models' test performance

		Classifiers	
		Naïve Bayes	Decision tree
Correctly classified instances (%)		78.9474	63.1579
Relative absolute error (%)		58.8174	96.88%
Root mean squared error		0.4487	0.5416
ROC area	WL_{HIGH}	0.803	0.745
	WL_{LOW}	0.773	0.3
	Weighted average	0.797	0.652

4.2 *Proposed Implementation*

The prototype developed in this chapter is designed to communicate with ThingSpeak cloud platform, however, storage facilities may require a private cloud server or a leased platform for required functions such as data analysis and predictive analytics as well as remote messaging capabilities for cellular interconnectivity. The proposed server infrastructure would have the following components: database server (for storing data generated within the system and data received from the environmental sensor through the web), web server functions (for easy access via the web), and analytics functions (to analyze inputs from remote storage facility.

Going forward, the sensor deployed is considered effective for small storage site deployments. Large infrastructure installations would require more efficient sensors. From the site, temperature, oxygen, CO2, and relative humidity values would be captured and transmitted to remote sites without human intervention. These values would be used to predict the water loss rate, which is a major determinant in postharvest fruit damage. When the remote system (server) receives the data, it uses it to make a prediction on the water loss rate using the Bayesian model. If the predicted value is beyond a certain threshold, it sends a message to the ESP8266 microcontroller. The message would contain the action to be performed, for instance, a high-water loss rate could trigger an action to lower the temperature and increase the humidity through a relay switch that turns on the humidifier. The control would involve a cycle of periodic (hourly) monitoring, messaging, analysis, and feedback control.

5 Conclusion

This chapter presents a study on the application of Bayesian learning and decision trees toward enhancing the control of environmental variables such as temperature, relative humidity, CO_2, and O_2, which have been described as very fundamental to the preservation campaigns on freshly harvested tomatoes in the tropics. Several researches and investigations have revealed that significant proportions of harvested produce including vegetables and fruits perish soon after harvest owing to lack of or poor storage facilities. Postharvest fruits are very sensitive to environmental conditions, and their preservation have been found to be dependent on these factors. A multidisciplinary approach is undertaken in this study to examine the implications of infusing machine learning technology into existing storage fruit storage infrastructure. Requirements analysis on hardware and software for prototyping provided a direction toward understanding the construction of workable prototype. With the prototype, the study concludes that IoT could be extensively be applied to critical postharvest storage infrastructure. More so, the application of Bayesian learning to provide intelligent control during the instrumentation and operation of a postharvest storage site is likely to offer more reliability and accuracy compared with the decision trees.

References

1. Adetunji, C. O., Arowora, K. A., Fawole, O. B., Adetunji, J. B., & Olagbaju, A. R. (2013a). Effect of edible coatings of carboxy methyl cellulose and corn starch on cucumber stored at ambient temperature. *Asian Journal of Agriculture and Biology, 1*(3), 133–140.
2. Adetunji, C. O., Fawole, O. B., Arowora, K. A., Adetunji, J. B., Agbaje, A. B., & Ogundare, M. O. (2013b). Effects of hydrophilic plasticizers added to chitosan coating for extending the storage life of Citrus sinensis. *South Asian Journal of Experimental Biology, 3*(3), 131–136.
3. Adetunji, C. O., Arowora, K. A., Fawole, O. B., & Adetunji, J. B. (2013c). Effects of coatings on storability of carrot under evaporative coolant system. *Albanian Journal of Agricultural Sciences, 12*(3), 485–493.
4. Adetunji, C. O., Omojowo, F. S., & Ajayi, E. S. (2014a). Effects of Opuntia cactus mucilage extract and storage under evaporative coolant system on the shelf life of Carica papaya fruits. *Journal Of Agrobiotechnology, 5*(2014), 49–66.

5. Adetunji, C. O., Ogundare, M. O., Ogunkunle, A. T. J., Kolawole, O. M., & Adetunji, J. B. (2014b). Effects of edible coatings from xanthum gum produced from Xanthomonas campestris pammel on the shelf life of Carica papaya Linn fruits. *Asian Journal of Agriculture and Biology, 2*(1), 8–13.
6. Adetunji, C. O., Williams, J. O., Olayemi, F. F., & Omojowo, F. S. (2014c). Microbiological evaluation of an edible antimicrobial coatings on mangoes fruit stored under evaporative coolant system (ECS). *Asian Journal of Agriculture and Biology, 2*(1), 20–27.
7. Adetunji, C. O., Adejumo, I. S., Afolabi, I. S., Adetunji, J. B., & Ajisejiri, E. S. (2018). Prolonging the shelf-life of 'Agege Sweet' Orange with chitosan-rhamnolipid coating. *Horticulture, Environment, and Biotechnology, 59*(5), 687–697. https://doi.org/10.1007/s13580-018-0083-2
8. Adetunji, C. O., Ojediran, J. O., Adetunji, J. B., & Owa, S. O. (2019). Influence of chitosan edible coating on postharvest qualities of Capsicum annum L. during storage in evaporative cooling system. *Croatian Journal of Food Science and Technology, 11*(1), 1–8. https://doi.org/10.17508/CJFST.2019.11.1.09
9. Aghdam, M. S., Kakavand, F., Rabiei, V., Zaare-Nahandi, F., & Razavi, F. (2019). γ-Aminobutyric acid and nitric oxide treatments preserve sensory and nutritional quality of cornelian cherry fruits during postharvest cold storage by delaying softening and enhancing phenols accumulation. *Scientia Horticulturae, 246*, 812–817. https://doi.org/10.1016/j.scienta.2018.11.064
10. Aji, P. W., Ahmad, C. K., Della, M. P. M., Risky, P. A., Sandika, M. P., Sulton, A. K., & Youngga, R. N. (2019). Naïve Bayes classifier for journal quartile classification. *International Journal of Recent Contributions from Engineering, Science & IT, 7*(2). https://doi.org/10.3991/ijes.v7i2.10659
11. Ali, S., Nawaz, A., Ejaz, S., Haider, S. T.-A., Alam, M. W., & Javed, H. U. (2019). Effects of hydrogen sulfide on postharvest physiology of fruits and vegetables: An overview. *Scientia Horticulturae, 243*, 290–299. https://doi.org/10.1016/j.scienta.2018.08.037
12. Ali, S., Khan, A. S., Anjum, M. A., Nawaz, A., Naz, S., Ejaz, S., & Hussain, S. (2020). Effect of postharvest oxalic acid application on enzymatic browning and quality of lotus (Nelumbo nuciferaGaertn.) root slices. *Food Chemistry, 312*, 126051. https://doi.org/10.1016/j.foodchem.2019.126051
13. Arah, I. K., Ahorbo, G. K., Anku, E. K., Kumah, E. K., & Amaglo, H. (2016). Postharvest handling practices and treatment methods for tomato handlers in developing countries: A mini review. *Advances in Agriculture, 2016*, 6436945. https://doi.org/10.1155/2016/6436945
14. Beaudry, R. M. (1999). Effect of O2 and CO2 partial pressure on selected phenomena affecting fruit and vegetable quality. *Postharvest Biology and Technology, 15*(3), 293–303. https://doi.org/10.1016/S0925-5214(98)00092-1
15. Bhargava, N., Mor, R. S., Kumar, K., & Sharanagat, V. S. (2021). Advances in application of ultrasound in food processing: A review. *Ultrasonics Sonochemistry, 70*, 105293. https://doi.org/10.1016/j.ultsonch.2020.105293
16. Burhan, M., Rehman, R. A., Khan, B., & Kim, B. S. (2018). IoT elements, layered architectures and security issues: A comprehensive survey. *Sensors, 18*(9), 2796.
17. Díaz-Pérez, J. C. (2019). Chapter 8: Transpiration (E. M. B. T.-P. P. and B. of F. and V. Yahia (ed.); pp. 157–173). Woodhead Publishing. https://doi.org/10.1016/B978-0-12-813278-4.00008-7.
18. Duan, Y., Wang, G.-B., Fawole, O. A., Verboven, P., Zhang, X.-R., Wu, D., Opara, U. L., Nicolai, B., & Chen, K. (2020). Postharvest precooling of fruit and vegetables: A review. *Trends in Food Science & Technology, 100*, 278–291. https://doi.org/10.1016/j.tifs.2020.04.027
19. Dumont, M.-J., Orsat, V., & Raghavan, V. (2016). 7 - reducing postharvest losses. In C. B. T.-E. T. for P. F. S. Madramootoo (Ed.), Woodhead publishing series in food science, technology and nutrition (pp. 135–156). Woodhead Publishing. https://doi.org/10.1016/B978-1-78242-335-5.00007-X.
20. FAO. (1989). *Prevention of post-harvest food losses fruits, vegetables and root crops a training manual*. Food and Agriculture Organization of the United Nations.

21. Francois-Lavet, V., Henderson, P., Islam, R., Bellemare, M. G., & Pineau, J. (2018). An introduction to deep reinforcement learning. *Foundations and Trends® in Machine Learning, 11*(3–4), 219–354.
22. Gupta, C., & Prakash, D. (2019). Chapter 10: Safety of fresh fruits and vegetables (R. L. Singh & S. B. T.-F. S. and H. H. Mondal (eds.); pp. 249–283). Academic. https://doi.org/10.1016/B978-0-12-816333-7.00010-2.
23. Hailu, G., & Derbew, B. (2015). Extent, causes and reduction strategies of postharvest losses of fresh fruits and vegetables – A review. *Journal of Biology, Agriculture and Healthcare, 5*(5), 49–63.
24. Huang, D., Tian, W., Feng, J., & Zhu, S. (2020). Interaction between nitric oxide and storage temperature on sphingolipid metabolism of postharvest peach fruit. *Plant Physiology and Biochemistry, 151*, 60–68. https://doi.org/10.1016/j.plaphy.2020.03.012
25. Hubert, M. & Vandervieren, E. (2007) An adjusted boxplot for skewed distributions. Computational Statistics and Data Analysis (2008) doi:https://doi.org/10.1016/j.csda.. 11 .008.
26. ISO/IEC. (2018). *Internet of things (IoT)—reference architecture* (ISO/IEC 30141:2018). National Standards of America.
27. Jiang, Q., Zhang, M., & Xu, B. (2020). Application of ultrasonic technology in postharvested fruits and vegetables storage: A review. *Ultrasonics Sonochemistry, 69*, 105261. https://doi.org/10.1016/j.ultsonch.2020.105261
28. Kong, L., Zhang, Y., Ye, Z. Q., Liu, X. Q., Zhao, S. Q., Wei, L., & Gao, G. (2007). CPC: Assess the protein-coding potential of transcripts using sequence features and support vector machine. *Nucleic Acids Research, 35*, 345–349.
29. Li, J., Sun, Q., Sun, Y., Chen, B., Wu, X., & Le, T. (2019). Improvement of banana postharvest quality using a novel soybean protein isolate/cinnamaldehyde/zinc oxide bionanocomposite coating strategy. *Scientia Horticulturae, 258*, 108786. https://doi.org/10.1016/j.scienta.2019.108786
30. Lin, J., Yu, W., Zhang, N., Yang, X., Zhang, H., & Zhao, W. (2017). A survey on internet of things: Architecture, enabling technologies, security and privacy, and applications. *IEEE Internet of Things Journal, 4*(5), 1125–1142. https://doi.org/10.1109/JIOT.2017.2683200
31. Lin, M., Fang, S., Zhao, X., Liang, X., & Wu, D. (2020). Natamycin-loaded zein nanoparticles stabilized by carboxymethyl chitosan: Evaluation of colloidal/chemical performance and application in postharvest treatments. *Food Hydrocolloids, 106*, 105871. https://doi.org/10.1016/j.foodhyd.2020.105871
32. Liu, W.-Y. (2014). Effect of different temperatures and parameters analysis of the storage life of fresh cucumber and tomato using controlled atmosphere technology. *American Journal of Food Technology, 9*, 117–126.
33. Liu, D.-K., Xu, C.-C., Guo, C.-X., & Zhang, X.-X. (2020). Sub-zero temperature preservation of fruits and vegetables: A review. *Journal of Food Engineering, 275*, 109881. https://doi.org/10.1016/j.jfoodeng.2019.109881
34. Lufu, R., Ambaw, A., & Opara, U. L. (2020). Water loss of fresh fruit: Influencing pre-harvest, harvest and postharvest factors. *Scientia Horticulturae, 272*, 109519. https://doi.org/10.1016/j.scienta.2020.109519
35. Mackowiak, S. D., Zauber, H., Bielow, C., Thiel, D., Kutz, K., Calviello, L., Mastrobuoni, G., Rajewsky, N., Kempa, S., Selbach, M., et al. (2015). Extensive identification and analysis of conserved small ORFs in animals. *Genome Biology, 16*, 179.
36. Mahajan, P. V, Pathak, N., Bovi, G. G., Ntsoane, M. L., Jalali, A., Keshri, N., Rux, G., Praeger, U., & Geyer, M. (2021). 3.01 - Recent advances on packaging and storage Technologies for the Preservation of fresh produce (K. Knoerzer & K. B. T.-I. F. P. T. Muthukumarappan (eds.); pp. 1–21). Elsevier. https://doi.org/10.1016/B978-0-08-100596-5.23040-0.
37. Manyozo, F. N., Ambuko, J., Hutchinson, M. J., & Kamanula, J. F. (2018). Effectiveness of evaporative cooling technologies to preserve the postharvest quality of tomato. *International Journal of Agronomy and Agricultural Research, 13*(2), 114–127.
38. Mostafidi, M., Sanjabi, M. R., Shirkhan, F., & Zahedi, M. T. (2020). A review of recent trends in the development of the microbial safety of fruits and vegetables. *Trends in Food Science & Technology, 103*, 321–332. https://doi.org/10.1016/j.tifs.2020.07.009

39. Nwankwo, W., & Olayinka, A. S. (2019). Implementing a risk management and X-ray cargo scanning document management prototype. *International Journal of Scientific and Technology Research, 8*(9), 93–105.
40. Nwankwo, W., & Ukhurebor, K. E. (2020a). Data Centres: A prescriptive model for green and eco-friendly environment in the cement industry in Nigeria. *International Journal of Scientific and Technology Research, 9*(5), 239–244.
41. Nwankwo, W., & Ukhurebor, K. E. (2020b). Web forum and social media: A model for automatic removal of fake media using multilayered neural networks. *International Journal of Scientific and Technology Research, 9*(1).
42. Nwankwo, W., Olayinka, A. S., & Ukhurebor, K. E. (2019). The urban traffic congestion problem in Benin City and the search for an ICT-improved solution. *International Journal of Science and Technology, 8*(12), 65–72.
43. Rabiei, V., Kakavand, F., Zaare-Nahandi, F., Razavi, F., & Aghdam, M. S. (2019). Nitric oxide and γ-aminobutyric acid treatments delay senescence of cornelian cherry fruits during postharvest cold storage by enhancing antioxidant system activity. *Scientia Horticulturae, 243*, 268–273. https://doi.org/10.1016/j.scienta.2018.08.034
44. Sagar, N. A., & Pareek, S. (2020). Chapter 19: Safe storage and preservation techniques in commercialized agriculture. In C. Egbuna & B. B. T.-N. R. for P. Sawicka (Eds.), *Disease and weed control* (pp. 221–234). Academic. https://doi.org/10.1016/B978-0-12-819304-4.00019-1
45. Sethi, P., & Sarangi, S. R. (2017). Internet of things: Architectures, protocols, and applications. *Journal of Electrical and Computer Engineering, 2017*, 9324035. https://doi.org/10.1155/2017/9324035
46. Shi, Y., Ding, Y., & Yang, S. (2018). Molecular regulation of CBF signaling in cold acclimation. *Trends in Plant Science, 23*(7), 623–637. https://doi.org/10.1016/j.tplants.2018.04.002
47. Sommano, S. R., Chanasut, U., & Kumpoun, W. (2020). 3 - Enzymatic browning and its amelioration in fresh-cut tropical fruits (M. W. B. T.-F.-C. F. and V. Siddiqui (ed.); pp. 51–76). Academic. https://doi.org/10.1016/B978-0-12-816184-5.00003-3.
48. Sucheta, Singla, G., Chaturvedi, K., & Sandhu, P. P. (2020). 2 – Status and recent trends in fresh-cut fruits and vegetables (M. W. B. T.-F.-C. F. and V. Siddiqui (ed.); pp. 17–49). Academic. https://doi.org/10.1016/B978-0-12-816184-5.00002-1.
49. Sun, X., Baldwin, E., & Bai, J. (2019). Applications of gaseous chlorine dioxide on postharvest handling and storage of fruits and vegetables – A review. *Food Control, 95*, 18–26. https://doi.org/10.1016/j.foodcont.2018.07.044
50. UN. World Urbanization Prospects: The 2018 Revision. World Urbanization Prospects: The 2018 Revision (2019). https://doi.org/10.18356/b9e995fe-en.
51. Verploegen, E., Sanogo, O., & Chagomoka, T. (2018). Evaluation of low-cost evaporative cooling Technologies for Improved Vegetable Storage in Mali. https://doi.org/10.1109/GHTC.2018.8601894.
52. Wei, H., Seidi, F., Zhang, T., Jin, Y., & Xiao, H. (2021). Ethylene scavengers for the preservation of fruits and vegetables: A review. *Food Chemistry, 337*, 127750. https://doi.org/10.1016/j.foodchem.2020.127750
53. Yousuf, B., & Qadri, O. S. (2020). 11 – Preservation of fresh-cut fruits and vegetables by edible coatings (M. W. B. T.-F.-C. F. and V. Siddiqui (ed.); pp. 225–242). Academic. https://doi.org/10.1016/B978-0-12-816184-5.00011-2.
54. Zeng, C., Tan, P., & Liu, Z. (2020). Effect of exogenous ARA treatment for improving postharvest quality in cherry tomato (Solanum lycopersicum L.) fruits. *Scientia Horticulturae, 261*, 108959. https://doi.org/10.1016/j.scienta.2019.108959
55. Zhang, W., Cao, J., Fan, X., & Jiang, W. (2020). Applications of nitric oxide and melatonin in improving postharvest fruit quality and the separate and crosstalk biochemical mechanisms. *Trends in Food Science & Technology, 99*, 531–541. https://doi.org/10.1016/j.tifs.2020.03.024
56. Fang, Z., Wang, Y., Peng, L., and Hong, H. (2021). Predicting flood susceptibility using LSTM neural networks. *Journal of Hydrology, 594*, 125734.
57. Arcomano, T., Szunyogh, I., Pathak, J., Wikner, A., Hunt, B.R., and Ott, E. (2020). A Machine Learning-Based Global Atmospheric Forecast Model. *Geophysical Research Letters, 47*(9).

Index

S. Pal et al. (eds.), *Artificial Intelligence-based Internet of Things Systems*, Internet of Things, https://doi.org/10.1007/978-3-030-87059-1

B

E

N

W

Z